Climatic Change and Global Warming of Inland Waters

Climatic Change and Global Warming of Inland Waters

Impacts and Mitigation for Ecosystems and Societies

Edited by

Charles R. Goldman
University of California, USA

Michio Kumagai
Ritsumeikan University, Japan

Richard D. Robarts
World Water and Climate Foundation, Saskatoon, Canada

WILEY-BLACKWELL

A John Wiley & Sons, Ltd., Publication

Library of Congress Cataloging-in-Publication Data

Climatic Change and Global Warming of Inland Waters: Impacts and Mitigation for Ecosystems and Societies / [edited by] Charles R. Goldman, Michio Kumagai, and Richard D. Robarts.
      page cm
   Research solicited from scientists who attended sessions organized by the World Water and Climate Network, WWCN in Nice, France, 2009.
   Includes bibliographical references and index.
   ISBN 978-1-119-96866-5 (hardback)
   1. Climatic changes–Environmental aspects. 2. Climatic changes–Social aspects. 3. Freshwater ecology.   4. Greenhouse gas mitigation. I. Goldman, Charles Remington, 1930- II. Kumagai, Michio. III. Robarts, R. D. (Richard D.)
   QC903.C5484 2013a
   551.48–dc23

                                                                        2012027540

A catalogue record for this book is available from the British Library.

Wiley also publishes its books in a variety of electronic formats. Some content that appears in print may not be available in electronic books.

Set in 10/12 Times by Laserwords Private Limited, Chennai, India
Printed and bound in Singapore by Markono Print Media Pte Ltd

First Impression 2013

Contents

List of Contributors

Adeline Amouroux

School of Architecture, Civil and Environmental Engineering (ENAC), Ecole Polytechnique Fédérale de Lausanne (EPFL), Switzerland

Hans E. Andersen

Department of Bioscience, Aarhus University, Denmark

Heléne Annadotter

Regito Research Center on Water and Health, Vittsjö, Sweden

Rikke Bjerring

Department of Bioscience, Aarhus University, Denmark

Justin D. Brookes

School of Earth and Environmental Science, University of Adelaide, Australia

Robert Coats

Department of Environmental Science and Policy, University of California, Davis, USA

Sudeep Chandra

Aquatic Ecosystems Analysis Laboratory, Department of Natural Resources and Environmental Science, University of Nevada, Reno, USA

Michael T. Coe

The Woods Hole Research Center, Falmouth MA, USA

Mariza Costa-Cabral

Hydrology Futures LLC, Seattle WA, USA

Gertrud Cronberg

Department of Ecology, University of Lund, Sweden

Gabriela da Costa Silva

School of Environment and Sustainability, University of Saskatchewan, Canada

Michael Dettinger

US Geological Survey and Scripps Institute of Oceanography, La Jolla CA, USA

Mary de Winton

National Institute of Water and Atmosphere Ltd (NIWA), New Zealand

Valentina M. Domysheva

Limnological Institute of the Siberian Branch of the Russian Academy of Sciences, Irkutsk, Russia

Balázs M. Fekete

CUNY Environmental CrossRoads Initiative, The City College of New York, City University of New York, New York, USA

Greg J. Fiske

The Woods Hole Research Center, USA

Yoshihiro Fukushima

Department of Environmental Management, Tottori University of Environment Studies, Japan

Marco Gemmer

National Climate Center of the China Meteorological Administration, China

Charles R. Goldman

Department of Environmental Science and Policy, University of California, Davis, USA

Clyde E. Goulden

Institute for Mongolian Biodiversity and Ecological Studies, Academy of Natural Sciences, Philadelphia PA, USA

Munhtuya N. Goulden

Institute for Mongolian Biodiversity and Ecological Studies, Academy of Natural Sciences, Philadelphia PA, USA

Tatiana Gurtovaya

South Russia Centre for Preparation and Implementation of International Projects, Russia

Ora Hadas

Israel Oceanographic and Limnological Research, Yigal Allon Kinneret Limnological Laboratory, Israel

David P. Hamilton

Department of Biological Sciences, University of Waikato, New Zealand

Masayuki Hara

Research Institute for Global Change, Japan Agency for Marine-Earth Science and Technology, Japan

Chris Hendy

Chemistry Department, University of Waikato, New Zealand

Robert M. Holmes

Woods Hole Research Center, USA

Erik Jeppesen

Department of Bioscience, Aarhus University, Denmark; Sino-Danish Centre for Education and Research (SDC), Beijing, China and Greenland Climate Research Centre (GCRC), Greenland Institute of Natural Resources, Nuuk, Greenland

Tong Jiang

National Climate Center of the China Meteorological Administration, China

Liselotte S. Johansson

Department of Bioscience, Aarhus University, Denmark

Torben B. Jørgensen

Department of Bioscience, Aarhus University, Denmark and Limfjordssekretariatet, Nørresundby, Denmark

David Kelly

Cawthron Institute, New Zealand

Fujio Kimura

Research Institute for Global Change, Japan Agency for Marine-Earth Science and Technology, Japan

Daisuke Kitazawa

Underwater Technology Research Center, Institute of Industrial Science, University of Tokyo, Japan

Brian Kronvang

Department of Bioscience, Aarhus University, Denmark

Michio Kumagai

Lake Biwa Sigma Research Center, Ritsumeikan University, Kusatsu, Japan

Søren E. Larsen

Department of Bioscience, Aarhus University, Denmark

Torben L. Lauridsen

Department of Bioscience, Aarhus University, Denmark and Sino-Danish Centre for Education and Research (SDC), Beijing, China

Isabelle Laurion

Center for Northern Studies (CEN), Québec City, Canada and Institut national de la recherche scientifique—Centre Eau Terre Environnement (INRS-ETE), Québec City, Canada

Ulrich Lemmin

School of Architecture, Civil and Environmental Engineering (ENAC), Ecole Polytechnique Fédérale de Lausanne (EPFL), Switzerland

Lone Liboriussen

Department of Bioscience, Aarhus University, Denmark

Xieyao Ma

Research Institute for Global Change, Japan Agency for Marine-Earth Science and Technology, Japan

Sally MacIntyre

Department of Ecology, Evolution and Marine Biology, University of California Santa Barbara, USA

Chris McBride

Department of Biological Sciences, University of Waikato, New Zealand

James W. McClelland

Department of Marine Science, The University of Texas at Austin, USA

John M. Melack

Bren School of Environmental Science and Management, University of California, Santa Barbara CA, USA

Ka Lai Ngai

Aquatic Ecosystems Analysis Laboratory, Department of Natural Resources and Environmental Science, University of Nevada, Reno, USA

Ami Nishri

Israel Oceanographic and Limnological Research, Yigal Allon Kinneret Limnological Laboratory, Israel

Kirsten Olrik

Laboratory of Environmental Biology Aps, Hellerup, Denmark

Ilia Ostrovsky

Israel Oceanographic and Limnological Research, Yigal Allon Kinneret Limnological Laboratory, Israel

Deniz Özkundakci

Department of Biological Sciences, University of Waikato, New Zealand

Reinhard Pienitz

Center for Northern Studies (CEN), Québec City; Takuvik Joint International Laboratory, Université Laval Canada-Centre national de la recherche scientifique (CNRS) (France) Canada and Département de géographie, Université Laval, Québec City, Canada

Antonio Quesada

Department Biología, Universidad Autónoma de Madrid, Spain

John Reuter

Department of Environmental Science and Policy, University of California, Davis, USA

Alon Rimmer

Israel Oceanographic and Limnological Research, Yigal Allon Kinneret Limnological Laboratory, Israel

John Riverson

Tetra Tech, Inc., Fairfax VA, USA

Richard D. Robarts

World Water and Climate Network, Saskatoon, Canada

Goloka Sahoo

Department of Civil and Environmental Engineering, University of California, Davis, USA

Mitsuru Sakamoto

School of Environmental Science, The University of Shiga Prefecture, Japan

Yoshinobu Sato

Water Resources Research Center, Disaster Prevention Research Institute, Kyoto University, Japan

Marc Schallenberg

Department of Zoology, University of Otago, New Zealand

Geoffrey Schladow

Department of Civil and Environmental Engineering, University of California, Davis, USA

Martin Schmid

Eawag, Surface Waters—Research and Management, Kastanienbaum, Switzerland

Alexander I. Shiklomanov

Water Systems Analysis Group, Institute for the Study of Earth, Oceans and Space, University of New Hampshire, USA

Michael N. Shimaraev

Limnological Institute of the Siberian Branch of the Russian Academy of Sciences, Irkutsk, Russia

Dominic Skinner

School of Earth and Environmental Science, University of Adelaide, Australia

Martin Søndergaard

Department of Bioscience, Aarhus University, Denmark

Robert G. M. Spencer

The Woods Hole Research Center, USA

Eugene Stakhiv

IJC Upper Lakes Study and UNESCO-ICIWaRM, Institute for Water Resources, Alexandria, VA, USA

Buda Su

National Climate Center of the China Meteorological Administration, China

Assaf Sukenik

Israel Oceanographic and Limnological Research, Yigal Allon Kinneret Limnological Laboratory, Israel

Hiroyasu Takenaka

Faculty of Engineering, Kinki University, Higashi-Hiroshima, Japan

Suzanne E. Tank

Department of Geography, York University, Canada

Dennis Trolle
Department of Bioscience, Aarhus University, Denmark

Narumi K. Tsugeki
Senior Research Fellow Center, Ehime University, Japan

Jotaro Urabe
Graduate School of Life Sciences, Tohoku University, Japan

David Velázquez
Dept. Biología, Universidad Autónoma de Madrid, Spain

Piet Verburg
National Institute of Water and Atmosphere Ltd (NIWA), New Zealand

Warwick F. Vincent
Center for Northern Studies (CEN), Québec City, Canada; Takuvik Joint International Laboratory, Université Laval, Canada—Centre national de la recherche scientifique (CNRS) (France) and Département de biologie, Université Laval, Québec City, Canada

Marley J. Waiser
formerly of the Global Institute for Water Security, University of Saskatchewan, Canada

Katey M. Walter Anthony
Water and Environmental Research Center, University of Alaska, Fairbanks, USA

Marion E. Wittmann
Department of Biological Sciences, University of Notre Dame, USA

Brent Wolfe
Northwest Hydraulic Consultants, Sacramento CA, USA

Alfred Wüest
Eawag, Surface Waters—Research and Management, Kastanienbaum, Switzerland

Yosef Z. Yacobi
Israel Oceanographic and Limnological Research, Yigal Allon Kinneret Limnological Laboratory, Israel

Wei Ye
Department of Biological Sciences, University of Waikato, New Zealand

Takao Yoshikane
Research Institute for Global Change, Japan Agency for Marine-Earth Science and Technology, Japan

Kohei Yoshiyama
River Basin Research Center, Gifu University, Japan

Alexander V. Zhulidov
South Russia Centre for Preparation and Implementation of International Projects, Russia

Tamar Zohary
Israel Oceanographic and Limnological Research, Yigal Allon Kinneret Limnological Laboratory, Israel

Preface

The World Water and Climate Network (WWCN) was created during the Third World Water Forum in Kyoto Japan in 2003. Its objective is to gather and exchange science-based information on the current and future conditions of our limited surface freshwaters contained in the lakes and rivers of the world. Since then the impact of global climatic change and associated warming has resulted in WWCN sessions organized in conjunction with three international conferences in France, the United States, and China. A fourth session was held in conjunction with the meeting of the American Society of Limnology and Oceanography at Otsu, Japan, in 2012. An important feature of the WWCN activities has been to attract and provide advanced training and scientific exchange for promising young graduate students competitively selected from about ten different countries. Bringing them together for several days to attend student meetings and then to present their research at an international scientific conference provides a stimulating intellectual platform for them to exchange information and make contact with old and new generations of aquatic ecologists, limnologists, hydrologists modelers, and environmental engineers. The meetings take place at a very important and formative time in the students' career development. Following the WWCN in Nice, France in 2009 the editors recruited experienced scientists from the meetings to assemble information on the impact of climatic change on the world's inland waters culminating in a book that Wiley-Blackwell agreed to publish. The following chapters are the result of this initiative.

Little doubt remains in the scientific community that the planet's climate is changing. Notably, the earth continued to gain heat during 2005–2010, a period marked by the strongest solar minimum recorded since accurate monitoring began. In other words, we now have irrefutable evidence that the sun is not the only important nor even perhaps the dominant factor forcing this climate change and that carbon dioxide, the classic greenhouse gas, is being joined by increasing amounts of methane. This gas is over 20 times as potent a greenhouse gas as carbon dioxide and will continue to increase as permafrost melts. Furthermore, analysis of ice cores indicates that the current, rapid rate of warming has not occurred in the last 800 000 years. Some natural and healthy scientific debate will doubtless continue on the exact role of human activity, especially regarding the large but practically unmeasured role of human-made atmospheric aerosols and the feedback loop of water vapor rising from the warming oceans. Major volcanic eruptions could temporarily reduce the global warming trend, as may have occurred following the Novarupta eruption on the Alaska Peninsula on

June 6, 1912, and as was recorded following the Mount Pinatubo eruption in 1991 from the ash blown into the atmosphere. Regardless of periodic volcanic events and the need for refinements of existing models, the evidence for anthropogenic warming is overwhelming. We need to accept the fact that, in the coming years, we face a dangerously warming world accompanied by rising sea levels and the damaging weather extremes that are already occurring. The evidence indicates that the globe is now warming at a rate sufficient to greatly alter the quantity and quality of fresh and marine waters and thus the lives of plants, animals and humans in the current century and beyond. In the chapters that follow we concentrate on the changes that have occurred and are likely to occur to our vital surface inland waters, both fresh and saline, as warming proceeds. We recognized that these changes vary considerably in different locations and have therefore selected authors from around the globe who are experienced with a wide range of different ecosystems. Where possible, recommendations for reducing the anticipated negative impacts of global warming on aquatic ecosystems are included.

The authors of the following chapters are limnologists, hydrologists, modelers and environmental engineers. We explore, through the contributions of these talented scientists from many countries, the impacts that climate change and the associated warming effect have had and are likely to have on lakes, rivers, wetlands, and their watersheds. The chapters, with input from 80 authors, are organized from northern latitudes to the more southern regions since the most extreme conditions of climatic change have already been well documented in the Arctic. In contrast the least impact of climate change for inland waters appears to be in the southern hemisphere where the temperature of New Zealand lakes has not measurably changed. Here the ratio of surrounding ocean to land mass is particularly great. Together with the many lake studies included in this book, three major river systems and their watersheds are included. We have, through selection of the contributors, been able to explore the subject in Siberia and other far northern region bordering the Arctic Ocean, Europe, Asia, Africa, the Middle East, and North and South America, New Zealand and Antarctica. Following the first 20 chapters on lakes and rivers are two unique chapters, 21 and 22, on the impact of global warming on society with the final three chapters dealing with possible mitigation of negative impacts. Chapter 25 presents a new technology that may prove helpful where oxygen depletion in lakes occurs as eutrophication is stimulated by increasing water temperatures.

Chapters 1 and 2 of the book deal with the physical-chemical and biological impacts of climate change on Arctic rivers and lakes. The aquatic resources of this region are vitally important to the indigenous people of the north who have depended on them for millennia and who now face the impacts of extremely rapid climatic change. The rivers of the far north have a profound influence on the Polar Sea, which receives them, since their volumetric contribution is relatively great when compared with the rivers discharging into the much larger world oceans. Chapter 2 is a comprehensive coverage of existent and future changes in the enormous number of large and small fresh water bodies to be found in the permafrost areas of the high arctic. Here conditions have already brought about extensive changes. Chapter 3 covers Lake Baikal, the oldest deepest lake in the world located in Russian Siberia. Its enormous volume of 24 000 cubic kilometers of largely unpolluted water is equivalent to all the water in the United States Great Lakes and may well turn out be Russia's most valuable natural resource. In this lake the annual period and extent of ice cover and thickness is already being reduced with a variety of impacts on local conditions including the use of the lake's ice cover for north and south automotive and truck transport. Chapters 4 and 5 cover two major Chinese rivers, the Yellow and Yangtze, as well as their enormous watersheds. With the huge and growing population in the region dependent upon the water from these two great rivers it is not surprising that the Chinese government is attempting to predict their future water yield on the basis of existing studies. These efforts have underscored the necessity of collecting accurate longer term data to improve prediction on floods and droughts. This monitoring data is urgently needed for improving global water-resource management. This theme reappears in a number of chapters at a time when institutions and governments have been reducing funding for important research-driven data collection so essential for improving the predictive capacity of models used to plan for the future.

Chapters 6 through Chapter 9 all involve conditions in Lake Biwa, the largest lake in Japan, which provides domestic water for over 14 million people and has undergone considerable eutrophication from nutrient input from its large urban and agricultural watershed. Chapter 6 deals with the human impacts on Lake Biwa and the additional stress of warming. Chapter 7 examines the changes in the plankton population as eutrophication has proceeded. Anoxic dead zones now occur in the near bottom waters as the biological oxygen demand of decaying plant and animal material exceeds the replenishment of oxygen from photosynthesis and seasonal mixing. Chapter 8 provides a numerical simulation of deep Japanese lakes and their future mixing as warming increases their water column stability and their resistance to complete mixing. Chapter 9 is a modeling paper with the potential to predict future lake conditions.

Chapters 10 and 11 take the reader to the Scandinavian lakes of Denmark, with Chapter 11 extending the coverage to a wide range of other lake types from different climate zones of the world. All of these studies underscore the increasing importance of eutrophication control. In the shallow phosphorus-rich Danish lakes, the warming effect has promoted dominance of the cyanobacteria, while in the deeper stratified lakes the dinoflagellates are dominant. These important changes in plankton dynamics should serve to revitalize the ongoing struggle by lake managers to slow or reverse the progress of eutrophication which, for years, has degraded the water quality of so many of the world's lakes. Chapter 12 concerns the mid-latitude lakes of Europe with the emphasis on the warming of Lake Geneva, where Professor Forel first gave birth to and named the field of limnology. Warming in this region has tended to promote more radical weather conditions as the atmospheric boundary layer has warmed. The author reports on the importance of ice breakup in Lake Constance as further evidence of progressive lake warming.

Moving to Canada, in Chapter 13 the author examines the wetlands of the prairie pot hole region. Warming is progressing there with definite increases in violent weather conditions and the threat that some of these small wetlands may dry up if rainfall is significantly reduced in the area. Chapter 14 takes the reader to the western United States and the half-century of data collected on the intensively studied Lake Tahoe located between California and Nevada, near the crest of the Sierra Nevada. In the earliest stages of eutrophication Tahoe's entire water volume has warmed a degree in 30 years with warming of the surface water occurring at an astonishing ten times as fast as the whole lake. Although still one of the world's clearest large lakes, it is likely to be subjected to more frequent floods, increased water shed erosion, sediment transport, further eutrophication, and more frequent lowering of lake level below its natural rim.

Chapter 15 notes that, as the world's lakes warm, their biota and associated food webs are changed. The half-century of data at Lake Tahoe shows that, as it warms, it has become increasing vulnerable to invasion of non-native species and has, unfortunately, experienced the intentional and accidental introduction of a variety of warm-water fish and most recently the Eurasian clam. The phenomenon of invasive species is occurring across major landscapes as warm water plants and animals are gradually able to extend their northward range often displacing the endemic northern flora and fauna in their movement.

Chapter 16 concerns Lake Kinneret in Israel, an arid region where water demand for irrigation and municipal water supplies is intense and the threat of increased salinity is always present. Recent reductions in rainfall there may already mirror the expected impacts of climatic change. The authors report on major shifts in the dominant phytoplankton where, in recent years, serious cyanobacterial blooms have already altered the food web. This occurred in conjunction with inputs of organic carbon and nitrogen from degraded wetland peat.

Chapter 17 deals with the Amazon river, boasting a highly productive fishery, which, to a large degree, is based on Varzia lakes along its flood plain. These shallow lakes, which are important as fish nursery areas, are likely to be impacted by climate change as the dynamics of flooding may be altered. Chapter 18 concerns the African Great Lakes, which span the famous Rift Valley. Paleo-limnological proxies indicate that Lake Victoria is heating faster now than at any time in the last 2000 years. These are of enormous economic importance to Africa and have been intensively studied over the years. The author shows how mixing dynamic in the African lakes are testimony to their complexity and are greatly influenced by wind and ocean mixing conditions. Carbon dioxide accumulation in some volcanic African lakes has resulted in a serious threat to those humans and animals living downwind or at a lower elevation. The eruption of gas from volcanic Lake Nyos, in the Cameroon, for example, was responsible for the death of many local people and their cattle, and the lake is currently being degassed to prevent another eruption.

Chapter 19 considers New Zealand and the southern hemisphere, where temperatures are greatly moderated by the enormous water surface of the southern oceans. Studies of North Island Lakes Taupo and Rotorua indicate that, to date, there has not yet been detectable warming. Lake Taupo, like Lake Tahoe, may experience less frequent mixing should warming gradually overcome the influence of the surrounding ocean. Chapter 20 involves the extreme climate of Antarctica, a barren frozen land, which, like the northern polar region, is expected to undergo major changes with loss of sea ice and penguin habitat. Warming on the Alaska Peninsula has been 2.5 °C since 1945, which is reported to be five times faster than the global average. The highly sensitive coastal systems have low biological diversity and, as sentinel habitats, they are likely to show early response to even slight warming and may begin to be invaded by temperate zone species.

Chapters 21 and 22 consider some of the important societal aspects of water and climatic change. In the pasture lands of Mongolia, where Genghis Khan once raised an army of horseman that conquered much of the known world, the descendant herders of the already affected seasonal pastures conditions are forced to change their herding management or give up their ancestral life style. Climate change has altered the availability of grazing and water supplies and severe sudden rainstorms are now more frequent. Chapter 22 deals with the growing problems associated with the many megacities that are emerging throughout the world as populations increase and more people are migrating to the cities. Only a few cities have taken the essential steps to meet some of the challenge of managing water, sewage and solid waste for the swelling numbers of inhabitants, which may double again by midcentury. The cities that have planned ahead are valuable examples of what can and must be done if a reasonable quality of life in large cities is to be maintained.

Chapters 23 and 24 provide recommendations for mitigating at least some of the global challenges facing our warming planet. Chapter 25 reports on an interesting technology for decomposing water into hydrogen and oxygen by electrolysis as a possible means of restoring oxygen to the depths of lakes impacted by eutrophication and climatic warming. The hydrogen produced can serve as a valuable by product of the process.

As noted above, Japan's Lake Biwa, an extremely important water source for Japanese citizens, is already threatened by a changing plankton community and an increase in the production of anoxic near bottom conditions fatal to some fish species. It would appear that our longstanding battle to reduce the greening of lakes by reducing nutrient loading may turn out to be one of the most cost-effective means of mitigating or at least reducing the negative impacts of lake warming. Strong evidence exists for the displacement of endemic species as movement of warm water species north increases because colder lakes and streams warm sufficiently to form more attractive habitat for invasion.

The editors hope that the following chapters will present ideas that will stimulate new approaches to what can and should be done to mitigate the more harmful aspects of climate change as we face the relentless warming of the planet and its effect on our vital and limited surface fresh water and related drinking water and food supplies. The impacts differ greatly in different regions of the world. Global conditions are inseparably linked to global atmospheric greenhouse gases, which can only be controlled by a rapid worldwide acceptance of alternative energy sources together with the emergence of a new carbon-based environmental ethic that drastically reduces greenhouse gas emission from controllable anthropogenic fossil fuel sources. This must be done before we pass a tipping point, or point of no return. Water, like air, is more essential than oil or any other commodity to life and is frequently contaminated or in very limited supply in many regions of the world. Climate change is altering the balance of rain, floods, and droughts, and our water-dependent food supplies essential to feed a dramatically and, in fact, frighteningly increasing world population. Current and anticipated shortages of unpolluted water are certain to be among the most serious challenges ever faced on earth by plants, man and animals alike.

The likelihood of future conflicts over increasingly limited sources of water now and in the future could cause immense human suffering through major and minor wars, famine, pestilence, and death. These were once portrayed as the "Four Horsemen of the Apocalypse." Unfortunately, these four symbolic riders of the world's skies are now joined by a fifth, representing global warming and pollution of water supplies from the combination of industrial, agricultural, and domestic sources. The long conflict over Kashmir has much of its basis in the immense importance of its water-yielding mountains to both India and Pakistan. There is much to be done in a timeframe that now appears to be becoming ever shorter through the biologically mitigated feedback loops that will further accelerate the rate of global warming. Examples from the Arctic include the loss of albido as the ice cap melts and the melting of permafrost with carbon dioxide and methane release from activation of bacterial decomposition. The release of methane, which as we have seen is a far more potent greenhouse gas than carbon dioxide, will accelerate further melting and the associated release of gas to the atmosphere. Atmospheric deposition is influencing the nutrient and pollution loading of our surface waters and their watershed and should provide the motivation to bring world governments together in combating it. During a recent summer, half the dust falling on Lake Tahoe, according to the research of Drs Snyder and Cahill (personal communication), had travelled across the Pacific from China. It is the hope of the authors and editors alike that the chapters contained in this volume will bring further attention to the seriousness of climatic change to the future of our civilization. In offering some suggestions for mitigating some of the negative impacts of climatic change we hope to inspire some new and promising research, management, and political action before irreparable damage is done to our inland waters. These after all form our global life support system and to a great extent their condition and availability will determine the quality of life for this and future generations.

The editors are indebted to Ms. Ayaka Kawai Tawada for her expeditious, skillful and always cheerful coordination and tracking of the various stages of the manuscripts and correspondence with the authors—our work was made easier and more efficient with her help.

Charles R. Goldman
Michio Kumagai
Richard D. Robarts

Part I

Impacts on Physical, Chemical, and Biological Processes

1

Climate Change Impacts on the Hydrology and Biogeochemistry of Arctic Rivers

Robert M. Holmes[1], Michael T. Coe[1], Greg J. Fiske[1], Tatiana Gurtovaya[2], James W. McClelland[3], Alexander I. Shiklomanov[4], Robert G. M. Spencer[1], Suzanne E. Tank[5], and Alexander V. Zhulidov[2]

[1]*Woods Hole Research Center, USA*
[2]*South Russia Centre for Preparation and Implementation of International Projects, Russia*
[3]*Department of Marine Science, The University of Texas at Austin, USA*
[4]*Water Systems Analysis Group, Institute for the Study of Earth, Oceans and Space, University of New Hampshire, USA*
[5]*Department of Geography, York University, Canada*

1.1 Introduction

Rivers integrate. Moreover, they integrate over a fixed and definable area (the watershed), so their discharge and chemistry at any given point is a function of upstream processes in both terrestrial and aquatic environments. As a consequence, changes in river discharge and chemistry can be powerful indicators of climate change impacts at the scale of whole watersheds.

The Arctic is central to considerations of climate change impacts. Warming is greatest in the Arctic and the Arctic is particularly sensitive to warming, so climate change has already dramatically impacted terrestrial, freshwater, and marine components of the Arctic system (Hinzman *et al.* 2005; Serreze, Holland and Stroeve 2007; White *et al.* 2007; see also Chapter 2). Arctic soils contain vast quantities of ancient organic matter that may be released as permafrost thaws, fueling a positive feedback loop and exacerbating warming (Zimov, Schuur and Chapin 2006; Schuur *et al.* 2008). The

Climatic Change and Global Warming of Inland Waters: Impacts and Mitigation for Ecosystems and Societies, First Edition. Edited by Charles R. Goldman, Michio Kumagai and Richard D. Robarts.
© 2013 John Wiley & Sons, Ltd. Published 2013 by John Wiley & Sons, Ltd.

Arctic also contains several of Earth's largest rivers. These rivers exert a disproportionate influence on the ocean because they transport more than 10% of global river discharge into the Arctic Ocean, which contains only ~1% of global ocean volume (McClelland *et al.* 2012). Arctic rivers drain vast watersheds, and trends in their discharge and chemistry are already providing strong insights into changes occurring over the Arctic landmass (Peterson *et al.* 2002; Holmes *et al.* 2012).

Here we describe the pan-Arctic watershed and highlight important characteristics of its principal rivers, and then provide an overview of available data related to the discharge and chemistry of Arctic rivers. Recognizing that comprehensive observational data are fundamental to understanding and responding to Arctic change (SEARCH 2005; Jeffries *et al.* 2007), we focus on multiyear time series, as these datasets are essential for detection of trends that may be related to climate change. We then consider projections of future changes in Arctic river discharge and chemistry. Our analysis is pan-Arctic in scale, with an emphasis on the largest Arctic rivers in Russia, Canada, and Alaska.

1.2   The pan-Arctic watershed

There are many different ways to define the spatial extent of the Arctic. Depending on one's definition, the area of the Arctic landmass can range from under $10 \times 10^6 \, \text{km}^2$ to over $20 \times 10^6 \, \text{km}^2$. For example, a rather restrictive definition was chosen by the Circumpolar Arctic Vegetation Map project, which considered the Arctic to be the region north of the tree line. Based on this definition, the Arctic covers an area of only $7.1 \times 10^6 \, \text{km}^2$, including Greenland and the Canadian Archipelago (Walker *et al.* 2005). At the other extreme, some studies have bounded the Arctic based on the drainage area of the Arctic Ocean, Hudson Bay, and part of the Bering Sea, which encompasses an area of about $20.5 \times 10^6 \, \text{km}^2$ (Holmes *et al.* 2001). When Greenland and the Canadian Archipelago are included, the Arctic grows to about $24 \times 10^6 \, \text{km}^2$ (McGuire *et al.* 2009). Depending on one's interests, any of these definitions may be appropriate.

For this chapter, we follow the definition now used by the Arctic Great Rivers Observatory (www.arcticgreatrivers.org), which defines the Arctic based on hydrology and considers the pan-Arctic watershed as the region draining into the Arctic Ocean plus the watersheds of the Yukon River and rivers entering the Bering Sea north of the Yukon River (Figure 1.1). Based on this definition, the pan-Arctic watershed covers $16.8 \times 10^6 \, \text{km}^2$ (Table 1.1). This region is smaller than the most expansive definitions of the Arctic because it does not include Greenland, the Hudson Bay drainage, or the Canadian Archipelago, but it does include the principal rivers entering the Arctic Ocean—which in several cases have watersheds extending well below 60° N (Figure 1.1).

There are 14 rivers in the pan-Arctic watershed that have mean annual discharges exceeding $25 \, \text{km}^3 \, \text{y}^{-1}$ (Table 1.1). Remarkably, 12 of these rivers are in Russia. The Yenisey, Lena, and Ob' are each among Earth's largest rivers, having mean annual discharges exceeding $400 \, \text{km}^3$ (Table 1.1). Six rivers in the pan-Arctic watershed have basin areas exceeding $500\,000 \, \text{km}^2$ (the Ob', Yenisey, Lena, Mackenzie, Yukon, and Kolyma). Combined, the watersheds of these "*Big 6*" Arctic rivers cover $11.2 \times 10^6 \, \text{km}^2$, or 67% of the pan-Arctic watershed (Table 1.1, Figure 1.1). The next

Table 1.1 Characteristics of the Arctic's largest watersheds. The rivers designated as the "Big Six" have been part of the multinational PARTNERS and Arctic Great Rivers Observatory (Arctic-GRO) projects since 2003 (www.arcticgreatrivers.org), and the "Middle 8" are the next eight largest watersheds.

Watershed	Watershed area[a] (10⁶ km²)	Discharge[b] (km³ y⁻¹)	Runoff (mm y⁻¹)	Sediment yield[c] (t km⁻² y⁻¹)	Tundra[d] (%)	Permafrost[e] (%)	Continuous permafrost (%)	Discontinuous permafrost (%)	Sporadic permafrost (%)	Isolated permafrost (%)	Population[f] (no. people)	Population density (people/km²)
"Big Six"												
Ob′	2.95	427	145	6	1	26	2	4	9	11	28 063 236	9.51
Yenisey	2.56	673	263	2	1	88	33	11	19	25	8 056 579	3.14
Lena	2.40	588	245	8	1	99	79	11	6	3	1 077 226	0.45
Mackenzie	1.75	316	181	74	1	82	16	29	27	10	465 338	0.27
Yukon	0.83	208	251	72	6	99	23	66	10	0	146 411	0.18
Kolyma	0.65	136	209	19	4	100	100	0	0	0	77 764	0.12
"Middle 8"												
Indigirka	0.34	54	159	36	7	100	100	0	0	0	29 332	0.09
Pechora	0.31	164	529	38	23	39	13	4	11	11	556 183	1.78
Severnaya Dvina	0.29	104	369	12	0	0	0	0	0	0	1 286 552	4.46
Khatanga	0.29	108	372		8	100	100	0	0	0	14 930	0.05
Yana	0.23	39	170	18	3	100	100	0	0	0	23 435	0.10
Olenek	0.22	48	218	6	12	100	100	0	0	0	4 368	0.02
Taz	0.15	43	287	7	0	100	6	64	30	0	14 793	0.10
Pur	0.10	32	320	7	0	100	4	76	20	0	259 955	2.73
Pan-Arctic WS	16.8	~3700	~220		15	76	45	13	10	8	42 241 712	2.52
"Big 6"	11.2	2348	210	21	1.4	75	35	16	13	11	37 886 554	3.40
"Middle 8"	1.9	592	312	19	8	76	60	9	5	2	2 189 538	1.13
Remainder	3.7	~760	~205		61	83	73	8	2	0	2 165 620	0.58

[a]Watershed areas are given for the entire watershed, not just the area above the downstream-most gauging stations.

[b]Mean annual discharge is based on observational records for the downstream-most gauging station, extrapolated to the entire watershed assuming that the unmonitored portion of the watershed has the same runoff as the monitored region of the watershed. Discharge values for the "Big Six" rivers are from Holmes *et al.* 2012 and are averages for the 1999–2008 period. Discharge from the Pechora, Severnaya Dvina, Yana, and Olenek are also averages from 1999–2008. Recent observational discharge records are not available for the other rivers, so the values gives represent averages from the following periods: Indigirka 1937–1998, Khatanga 1965–1991, Taz 1962–1996, Pur 1939–2003.

[c]Sediment yield values are from Holmes *et al.* 2002.

[d]The tundra classification comes from the Circumpolar Arctic Vegetation Map (Walker *et al.* 2005) and includes all categories of tundra.

[e]Permafrost extent and classification comes from International Permafrost Association's Circum-Arctic Map of Permafrost and Ground Ice Conditions.

[f]Population statistics come from the LandScan 2009 Global Population database (www.ornl.gov/landscan).

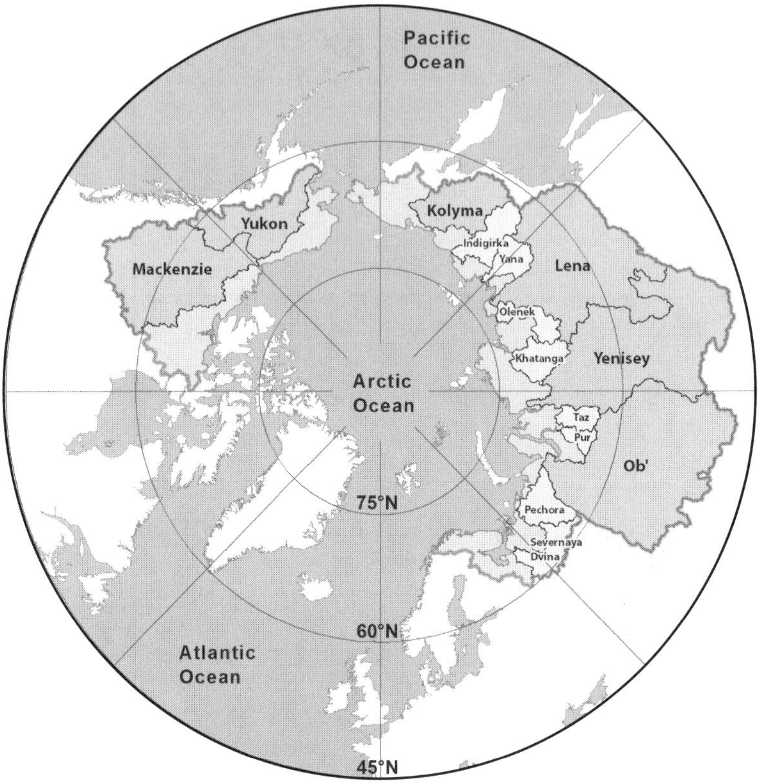

Figure 1.1 Map of the 16.8×10^6 km^2 pan-Arctic watershed, showing its major rivers. Together the basins of the "Big 6" rivers (shown in dark gray) cover 67% of the pan-Arctic watershed, whereas the watersheds of the "Middle 8" rivers (shown in light gray) cover 11% of the pan-Arctic watershed (see Table 1.1). The dark gray line indicates the boundary of the pan-Arctic watershed.

eight largest Arctic watersheds (the "*Middle 8*") together only cover an additional 1.9×10^6 km^2, much less than the basin area of the Ob', Yenisey, or Lena rivers alone. This demonstrates the importance of inclusion of the largest rivers in order to achieve pan-Arctic synthesis, but also highlights the challenge of scaling-up further because each additional river beyond the *Big 6* achieves only incremental gains.

Extrapolation to the remaining third of the pan-Arctic watershed not encompassed by the *Big 6* is tenuous, given that the regions are fundamentally different in many ways. For example, whereas on average the watersheds of the *Big 6* rivers are only 1.4% tundra, the *Middle 8* rivers have 8% tundra and the remainder of the pan-Arctic watershed is 61% tundra (Figure 1.2, Table 1.1). Similarly, continuous permafrost increases from 35% in the *Big 6* to 60% in the *Middle 8* to 73% in the remainder, and human population density decreases moving from the *Big 6* to *Middle 8* and then the remainder of the pan-Arctic watershed (Figure 1.2, Table 1.1). Thus, strategic sampling of smaller rivers combined with modeling will be required to better constrain estimates of contemporary and future biogeochemical fluxes from the pan-Arctic watershed to the Arctic Ocean, or understand processes occurring throughout the pan-Arctic watershed. Regardless of scale, capturing seasonality is critical for estimating

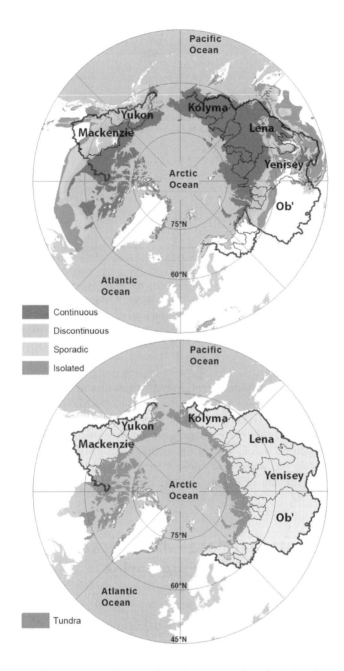

Figure 1.2 Distribution of Arctic permafrost (upper panel) and tundra (lower panel). Continuous permafrost indicates that >90% of the land surface is underlain by permafrost, discontinuous indicates 50–90%, sporadic indicates 10–50%, and isolate indicates <10% of the region contains permafrost. (See insert for color representation.)

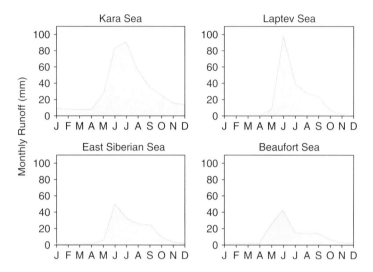

Figure 1.3 River runoff climatologies (mm/month) for four regions around the pan-Arctic domain. Data are average monthly runoff for 1990–1999. Regional runoff climatologies were calculated using the following rivers: Kara Sea (Norilka, Ob, Pur, and Yenisey), Laptev Sea (Anabar, Lena, Olenek, and Yana). East Siberian Sea (Alazeya, Indigirka, Kolyma), Beaufort Sea (Anderson, Kuparuk, Mackenzie). Adapted from McClelland *et al.* (2012).

biogeochemical fluxes from Arctic rivers. Runoff is low around the pan-Arctic domain during the winter and increases to peak values during the May/June timeframe as a consequence of snowmelt (Figure 1.3). The importance of summer runoff (the proportion of annual runoff that occurs during the summer months) varies from region to region, but in general values during the summer months are intermediate relative to those observed in the winter and spring. These seasonal variations in runoff are accompanied by significant variations in water chemistry (discussed in detail below).

1.3 Observational data—historical to contemporary time series

1.3.1 Discharge

Major efforts to monitor river water discharge began in the 1930s on the Eurasian side of the Arctic and the 1960s on the North American side. The records resulting from these monitoring efforts have provided a rich source of data for analyses of variability over a wide range of spatial and temporal scales (e.g., Peterson *et al.* 2002; Yang *et al.* 2002; Yang *et al.* 2003; Déry *et al.* 2005; Déry and Wood 2005; McClelland *et al.* 2006; Yang *et al.* 2007; Déry *et al.* 2009; Shiklomanov and Lammers 2009; Overeem and Syvitski 2010; Rawlins *et al.* 2010). River discharge is currently increasing around much of the pan-Arctic domain, with a strong upward trend in annual values beginning in the 1960s for the Eurasian rivers (Peterson *et al.* 2002) and the late 1980s for the North American rivers (Déry *et al.* 2009). However, quantitative estimates of changes in Arctic river discharge are strongly dependent on the specific rivers and timeframes considered. For example, McClelland *et al.* (2006) analyzed data from 16 Eurasian rivers and 14 North American rivers draining into the Arctic Ocean from 1964 through 2000 and found that annual river discharge increased at a rate of 5.6 km^3 per year. Scaled to include ungauged areas within the Arctic Ocean watershed, the increase in annual river discharge was estimated to be 7.4 km^3 per year. Over this timeframe, the net change in river input to the Arctic Ocean reflected a relatively large increase from Eurasia ($\sim$10%) moderated by a smaller decrease from North America ($\sim$6%). McClelland *et al.* (2006) also examined discharge from 42 rivers in the Hudson Bay region and identified a decrease in annual river discharge of 2.4 km^3 per year, amounting to a $\sim$12% change over the 1964–2000 time period. In contrast, Overeem and Syvitski (2010) analyzed data from nine Eurasian rivers and ten North American rivers within the pan-Arctic domain from 1977 through 2007 and found that river discharge increased in both regions (net change of $\sim$10%) over this more recent timeframe. Given the substantial variability in discharge that Arctic rivers exhibit from year to year, as well as potential linkages between river discharge and longer-term modes of variability in the Arctic such as the North Atlantic Oscillation (Peterson *et al.* 2006), analyses that cover broader time periods are most conducive to examining potential climate change effects. One effort that may be particularly useful in this regard is the annually updated Arctic Report Card (http://www.arctic.noaa.gov/reportcard/). Discharge data for the six largest Eurasian Arctic rivers (Ob', Yenisey, Lena, Severnaya Dvina, and Pechora) as well as four North American Arctic rivers (Yukon, Mackenzie, Peel, and Back) is being tracked as part of this effort (Shiklomanov, http://www.arctic.noaa.gov/reportcard/rivers.html). The time series for the Eurasian rivers begins in 1936 and the time series for the North American rivers begins in 1970. At present the records extend through 2008 and both show overall positive trends in river discharge.

The observed increases in river discharge are broadly consistent with increased net precipitation over the pan-Arctic watershed that is predicted by climate models as a consequence of anthropogenic greenhouse-gas emissions (Wu, Wood and Stott 2005; Holland, Finnis and Serreze 2006; Nohara *et al.* 2006). However, major uncertainties with respect to historical Arctic climate data have hindered empirical assessments of precipitation-discharge linkages in the region (Rawlins *et al.* 2006). A recent synthesis of observed and modeled climate data across the pan-Arctic pointed toward intensification of the freshwater cycle (Rawlins *et al.* 2010), but the relative importance of changes in precipitation and evaporation with respect to trends in river discharge remained unclear. Thawing permafrost does not have great potential to influence river discharge through direct release of frozen water stocks (McClelland *et al.* 2004), but changes in evaporation associated with permafrost thaw may be an important consideration (Hinzman *et al.* 2005). For example, deeper flow paths through the landscape as permafrost thaws will, to some degree, oppose the effect of warmer temperatures on evaporation. Changes in the seasonality of precipitation (i.e., relative amounts of snow versus rainfall) may also have a strong influence on the annual water balance (Rawlins *et al.* 2006).

1.3.2 Biogeochemistry

In contrast with the discharge records discussed above, time-series data for water chemistry of Arctic rivers are relatively scarce. While the discussion below highlights key examples of observed changes in water chemistry, we are just beginning to develop suitable data sets for tracking changes in river-borne constituents at the pan-Arctic scale. For the largest Arctic rivers, we now have a robust baseline with respect to water chemistry that is essential for evaluating future changes, but (in most cases) we do not yet have long enough data sets to identify ongoing changes that may be linked to climate change.

Eurasia   The vast majority of river water entering the Arctic Ocean comes from Eurasia, most of it from Russia. During the Soviet period, Russia maintained an extensive water-quality monitoring network as part of the Unified Federal Service for Observation and Control of Environmental Pollution (OGSNK). The resulting data were not widely available during the Soviet period, largely for political reasons (Zhulidov *et al.* 2000). In the post-Soviet era, however, the data have become more readily available, and sample collection has continued, albeit at a much reduced level. Unfortunately, few of the publications based on these data present time series. Instead, constituent concentrations or fluxes are generally presented as a single value (usually mean annual) with incomplete information about how the value was derived (for example, Telang *et al.* 1991; Gordeev *et al.* 1996; Kimstach, Meybeck and Baroudy 1998; Rachold *et al.* 2004). Although these publications provide clues about spatial differences among rivers, they are of relatively little use for detecting change over time, especially when details of sampling dates or sampling and analytical protocols were not given.

In the hope of establishing a baseline against which to evaluate future changes, Holmes *et al.* (2000) investigated concentrations and fluxes of nitrate, ammonium, and phosphate for 15 Russian Arctic rivers. Using the OGSNK database, they presented monthly averages, generally based on 10–20 years of data. Striking patterns in the data, in particular remarkably high ammonium concentrations, led to uncertainty about the reliability of the OGNSK data (Holmes *et al.* 2000). An expedition to the Ob' and Yenisey rivers in June 2000 was organized to collect new samples in order to evaluate the quality of the long-term data set (Holmes *et al.* 2001). Four different groups, including the laboratories that produced the long-term data set, independently analyzed the Ob' and Yenisey river samples that were collected. Unfortunately, whereas three of the groups obtained similar results, the laboratories that produced the long-term data set found very different results, at least with respect to ammonium (Holmes *et al.* 2001). The unavoidable conclusion was that there were significant data quality issues with the OGSNK dataset. This was one of the motivations for the PARTNERS Project (described in more detail below), which began a standardized program to collect and analyze samples for the six largest Arctic rivers in Russia, Canada, and Alaska in 2003.

Time-series data have been published for suspended sediment flux from the eight largest Arctic rivers, six of which are in Russia (Holmes *et al.* 2002). The database was most complete for the Ob' River, with sampling dating back to the 1930s, whereas other rivers had shorter or less complete datasets. Dramatic differences are apparent when comparing sediment yield for the different rivers. For example, the two largest North American rivers (the Mackenzie and Yukon) together transport ~180 MT of suspended sediment per year, whereas the Arctic's largest river (the Yenisey), with a greater annual discharge than the Mackenzie and Yukon combined, transports only about 5 MT of suspended sediment per year (Holmes *et al.* 2002). Long-term trends were not apparent in the data, except for the Yenisey River where there was a pronounced decrease in sediment flux resulting from the construction of a dam at Krasnoyarsk in 1967 (Holmes *et al.* 2002).

Several studies have obtained information about the biogeochemical composition of Russian Arctic rivers by sampling their plumes or entering their mouths during oceanographic cruises. For example, Lobbes, Fitznar, and Kattner (2000) obtained organic matter and nutrient data from 12 rivers that were sampled during a series of oceanographic cruises during the mid-1990s. Similarly, the plumes of the Ob' and Yenisey rivers were sampled during oceanographic cruises carried out by the SIRRO project (Kohler *et al.* 2003). Semiletov and colleagues (Semiletov *et al.* 2011) sampled the Lena River during oceanographic cruises, and they also conducted sampling trips along the length of the Lena River. Although these and other studies on Russian Arctic rivers have provided important snapshots in time, few have collected the seasonal and multiannual data that are needed to investigate the impacts of climate change on river biogeochemistry.

Alaska   One of the few long-term Arctic datasets with respect to geochemical measurements exists for the upper Kuparuk River on the North Slope (Alaska), which has been studied since 1978 (McClelland *et al.* 2007). Research on the upper Kuparuk River in the late 1970s and early 1980s examined concentrations of nutrients and organic matter and the export of these constituents (Peterson, Hobbie and Corliss 1986; Peterson *et al.* 1992). A recent study using the long-term dataset from the upper Kuparuk River found that annual nitrate export increased by approximately fivefold and annual dissolved organic carbon (DOC) export decreased by about 50% from 1991 to 2001 (McClelland *et al.* 2007). The reported decrease in DOC export was centered on the freshet in May and was principally credited to a decline in river discharge. Conversely, increased nitrate export occurred throughout the spring-summer from May to September and was mainly as a result of increasing concentrations (McClelland *et al.* 2007). The underlying mechanism responsible for the increase in nitrate concentrations in the upper Kuparuk remains unclear, but it is hypothesized to relate to changes in soils and vegetation associated with warming in the region.

The Yukon River has been the focus of a number of biogeochemistry orientated studies in recent years and was also studied by the U.S. Geological Survey from 1978–1980. Striegl *et al.* (2005) found a decrease in discharge-normalized DOC export by the Yukon during the growing season when comparing data from 1978–1980 versus 2001–2003. Within parts of the Yukon River basin the proportion of annual discharge derived from groundwater was increasing. For example, at Eagle Village groundwater increased from ~15% of annual flow in the early 1950s to >20% today (Walvoord and Striegl 2007). In the Striegl *et al.* (2005) study, groundwater near the mouth of the Yukon had an average DOC concentration of ~2.5 mg L^{-1} and DIC concentration of 50 mg L^{-1} versus mean annual river water DOC concentrations of ~8.0 mg L^{-1} and DIC concentrations of 24 mg L^{-1}. Therefore, increased groundwater contribution to river discharge resulted in decreased DOC export and increased HCO$_3^-$ export when normalized to discharge as described by Striegl *et al.* (2005).

In recent years, carbon has been a particular focus of research on the Yukon River. Loads and yields of dissolved and particulate organic and inorganic carbon have been derived for the mouth of the river and several tributaries (Striegl *et al.* 2007; Spencer *et al.* 2009; Guo *et al.* 2012). Carbon dynamics within the Yukon River watershed have also been examined via stable and radiocarbon isotopes (Guo and Macdonald 2006; Guo, Ping and Macdonald 2007; Raymond *et al.* 2007; Striegl *et al.* 2007), and through studies examining organic matter composition (Gueguen *et al.* 2006; Cai, Guo, and Douglas 2008; Elmquist *et al.* 2008; Spencer *et al.* 2008). Nutrient loads and yields have also been derived for the mouth of the Yukon as well as for a number of upstream mainstem sites and tributaries (Guo *et al.* 2004; Dornblaser and Striegl 2007; Cai *et al.* 2008; Guo *et al.* 2012). Toxic substances such as mercury (Hg) have also recently been investigated in the Yukon River, providing a benchmark for the future as thawing of permafrost may accelerate the mobilization of Hg increasing export and potential Hg methylation in Arctic regions (Schuster *et al.* 2011). Therefore, recent years have seen a number of biogeochemical studies within the Yukon River, providing a boon of information for studies in the future to examine how this system responds to climate change.

Canada The Mackenzie River is by far the most studied of any of the Canadian Arctic rivers. Despite this, there are few long-term time-series records of its biogeochemistry. The PARTNERS and Arctic-GRO projects, which are discussed in more detail below, have collected biogeochemical measurements on the Mackenzie since 2003, sampling where the Mackenzie River empties into the Mackenzie River Delta before its ultimate delivery to the Arctic Ocean (McClelland *et al.* 2006). Similarly, the Water Survey of Canada maintains a long-term database for total suspended solids (TSS) at this same location (www.wsc.eg.gc.ca). In contrast to the rarity of long-term time series, however, several authors have compiled detailed within-year datasets for a variety of biogeochemical constituents within the Mackenzie basin. Although their use is limited for discerning long-term trends, these within-year records do provide a clear understanding of the overall biogeochemistry of this system, and how the concentrations of various constituents change across the hydrograph. Weekly measurements during the open water season from the East Channel of the Mackenzie River Delta exist for a suite of constituents, including TSS, DOC, major ions, pH, and dissolved and particulate nutrients (Anema *et al.* 1990; Lesack *et al.* 1998; Tank, Lesack and McQueen 2009; Gareis, Lesack and Bothwell 2010; Tank *et al.* 2011). Emmerton (2006) and Emmerton, Lesack and Vincent (2008) have also collected weekly within-year measurements for a broad suite of constituents on the Mackenzie River, in addition to the Arctic Red and Peel Rivers (the two other major tributaries to the Mackenzie River Delta).

In addition to detailed measurements over time, several authors have undertaken extensive synoptic surveys of biogeochemistry across the Mackenzie catchment, by collecting single time-point measurements on multiple tributaries within the larger river basin. For example, Yi *et al.* (2010) undertook extensive synoptic and time-series surveys of $\delta^{18}O$ and δ^2H within the Mackenzie catchment to explore basin-wide hydrology and quantify the importance of evaporation and snowmelt in the broader catchment. Similarly, Reeder, Hitchon, and Levinson (1972) and Millot *et al.* (2002, 2003) used major ion data to examine weathering processes within the larger Mackenzie basin. These latter authors both found a wide variation in major ionic composition, with variations in overall composition and elemental ratios that reflected the underlying lithological variation throughout the Mackenzie River basin (Millot *et al.* 2003).

Outside of the Mackenzie system, the majority of time-series biogeochemical measurements for Canadian Arctic rivers have been collected by government agencies, and published in government documents. Apart from the government literature, biogeochemical surveys do exist for selected locations, although the measurement record typically ranges from intensive within-year surveys to records extending a few years in length. For example, a series of studies have been conducted in two adjacent streams on Melville Island, where measurements of DOC, dissolved organic nitrogen, dissolved inorganic nitrogen, and suspended sediments were used to explore between-catchment processes and elucidate the relative importance of stormflow, baseflow, and snowmelt to overall constituent fluxes (Lafreniere and Lamoureux 2008; Lewis *et al.* 2011). Similarly, on Nunavut's Boothia Peninsula, time series measurements of suspended sediments were used to quantify controls on sediment yields and its relationship with discharge in three proximate streams (Forbes and Lamoureux 2005). In the southern Yukon Territory, within-year time series measurements were used to explore controls on DOC export in a catchment underlain by discontinuous permafrost (Carey 2003). Finally, a multiyear time series was used to explore controls on mercury flux from the Nelson and Churchill Rivers, and included a characterization of DOC flux (Kirk and Louis 2009).

Data for Canadian Arctic riverine biogeochemistry presented in government publications ranges from series of individual measurements to averages of constituent concentrations measured over a given period of record, with a wide range in the number of observations and years of measurement available depending on the river and constituent of interest (e.g., Environment Canada 1976, 1978, 1982). Some of this constituent concentration data has also been summarized in the GEMS/GLORI database as mean constituent concentrations per river (www.gemswater.org; Meybeck and Ragu 1995), along with data for other world rivers. These data compilations established the underlying biogeochemistry of a suite of Canadian Arctic rivers, and revealed a large degree of variation in Canadian Arctic riverine biogeochemistry. For the purposes of analyzing trends over time, however, data for Canadian Arctic rivers is often difficult to access and assess. For example, much of the collected data has not been published to date, and measurement frequency and length of data series often differs between rivers and constituents. Where government data has been compiled in government documents, long-term means rather than individual data points are often presented, and analytical methods may not be reported or differ between studies. Increasingly, however, the raw data are available at GEMStat (www.gemstat.org), which also includes information on methods.

1.3.3 Pan-Arctic assessments

The only cohesive, multinational assessment of Arctic riverine biogeochemistry to date was initiated in 2003 as the PARTNERS Project (Pan-Arctic River Transport of Nutrients, Organic Matter, and Suspended Sediments), which in 2009 evolved into the Arctic Great Rivers Observatory (Arctic-GRO; www.arcticgreatrivers.org). These projects collect biogeochemical data for the six largest rivers draining to the Arctic Ocean: the Ob', Yenisey, Lena and Kolyma rivers in Russia, and the Yukon and Mackenzie rivers in North America (Figure 1.1, Table 1.1). Despite a number of previous and ongoing studies that have examined Arctic riverine biogeochemistry, an understanding of processes at the pan-Arctic scale has been hampered by differences

Table 1.2 Flow weighted concentrations and minimum and maximum measured concentrations (in parentheses) for selected constituents measured as part of the PARTNERS Project (2003–2007). Flow weighted concentrations are calculated from modeled yearly constituent flux, and are derived from the data presented in Holmes et al. (2012) and Tank et al. (in press). Minimum and maximum measured concentrations are taken directly from the PARTNERS dataset (www.arcticgreatrivers.org).

	Ob'	Yenisey	Lena	Kolyma	Yukon	Mackenzie
DIN ($\mu g\,L^{-1}$)	200.7 (0.2–858.0)	79.8 (3.2–224.6)	57.4 (5.4–242.0)	60.2 (0.8–141.4)	123.7 (59.0–387.8)	89.3 (37.1–174.0)
DON ($\mu g\,L^{-1}$)	256.6 (99.1–421.0)	173.8 (62.0–306.6)	232.5 (121.3–379.5)	155.6 (43.7–259.3)	226.7 (75.0–460.2)	103.3 (49.6–144.6)
TDP ($\mu g\,L^{-1}$)	40.4 (15.2–62.6)	15.2 (7.5–22.7)	10.4 (4.1–18.0)	10.1 (4.5–14.7)	9.8 (2.0–18.0)	10.0 (7.8–11.2)
Si ($mg\,L^{-1}$)	3.4 (1.8–8.9)	2.7 (1.9–3.8)	2.3 (1.6–3.8)	2.5 (1.8–3.8)	3.3 (2.2–6.5)	1.9 (1.5–2.2)
DOC ($mg\,C\,L^{-1}$)	9.6 (5.5–12.0)	7.3 (2.9–13.0)	9.8 (5.2–14.8)	7.4 (2.9–10.3)	7.1 (2.2–15.1)	4.6 (3.1–5.7)
Ca ($mg\,L^{-1}$)	15.8 (10.9–37.1)	17.9 (7.6–30.3)	15.5 (11.8–28.4)	11.2 (8.7–14.8)	31.0 (22.4–48.7)	35.2 (28.4–38.5)
Na ($mg\,L^{-1}$)	6.3 (4.1–13.3)	6.4 (2.6–9.6)	9.0 (3.2–41.5)	1.5 (1.1–2.2)	2.6 (1.7–4.0)	7.7 (4.8–12.7)
Mg ($mg\,L^{-1}$)	4.2 (2.9–9.1)	3.8 (1.9–5.6)	4.5 (3.3–9.0)	2.4 (1.8–3.4)	7.4 (4.2–11.7)	9.6 (7.9–10.5)
Cl ($mg\,L^{-1}$)	5.5 (3.1–12.3)	9.7 (2.6–16.1)	17.5 (4.6–81.4)	0.3 (0.1–0.6)	0.9 (0.5–1.3)	10.1 (5.2–22.1)
Sr ($\mu g\,L^{-1}$)	100.5 (63.9–218.8)	132.0 (55.6–224.2)	112.8 (83.3–286.5)	56.5 (41.1–70.6)	125.2 (80.2–208.0)	190.1 (113.3–271.6)
SO_4 ($mg\,S\,L^{-1}$)	2.5 (1.3–5.0)	3.1 (1.1–5.7)	4.0 (1.7–11.8)	3.4 (2.0–6.3)	10.3 (4.3–18.7)	15.7 (11.1–18.7)

in methodology between studies and a tendency to collect samples only during the summer months (Holmes *et al.* 2000; McClelland *et al.* 2008). Similarly, the short timespan of many previous studies necessarily limits their usefulness for assessing trends over time. Because the PARTNERS and Arctic-GRO projects have a consistent sampling and analytical scheme across all rivers, and collect samples regularly across the spring freshet, summertime period of higher biological activity, and wintertime (under-ice) period, their datasets have been useful for establishing a rigorous baseline for biogeochemistry on large Arctic rivers.

Sampling for PARTNERS occurred between 2003 and 2007. The Arctic-GRO project has been in operation since 2009 and is currently funded through 2016. Although the time series of these projects is still relatively short, it is detailed enough to quantify seasonal and spatial trends in biogeochemical flux from these large Arctic rivers and to assess interannual variability. These datasets have now been used to determine the between-river variation in numerous constituents (see Table 1.2). For example, constituents such as alkalinity, barium, and calcium show elevated mean concentrations in North American rivers compared to their Russian counterparts, while many constituents show clear between-river variation at the continental scale (Andersen *et al.* 2007; Raymond *et al.* 2007; Cooper *et al.* 2008; McClelland *et al.* 2008; Zimmerman *et al.* 2009; Holmes *et al.* 2012; Tank *et al.* in press). Conversely, seasonal within-constituent variation in concentration are strikingly coherent across these six large rivers, with organic constituents increasing at elevated discharge, and inorganic constituents diluting at high flow (Figure 1.3; Raymond *et al.* 2007; Holmes *et al.* 2012; Mann *et al.* 2012).

By focusing measurements across the full seasonal cycle, these databases have significantly refined our estimates of constituent flux from large Arctic rivers. For instance, because organic constituents increase with increasing flow (Figure 1.4), quantifying the concentration of organic constituents during the high-flow spring freshet is critical for determining their yearly flux. Recent estimates using the PARTNERS dataset have calculated that 46% of DON flux, and 49% of DOC flux from these six rivers occurs during the two months surrounding peak flow (Holmes *et al.* 2012), and PARTNERS-based estimates of pan-Arctic riverine DOC flux are substantially higher than previous estimates based largely on summertime measurements alone (Raymond *et al.* 2007; Holmes *et al.* 2012). Even for constituents that do not become more

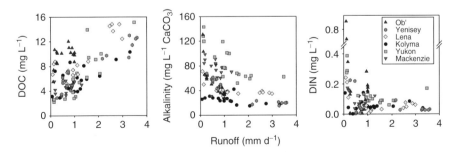

Figure 1.4   Measured concentrations of DOC, alkalinity, and DIN for each of the six PARTNERS rivers, plotted against runoff for the day of the measurement. In these large Arctic rivers, inorganic carbon is always greater than 90% of total alkalinity (Tank *et al.* in press).

concentrated with increasing flow, a lack of spring and wintertime measurements created considerable uncertainty in estimates of their flux, and calculations of flux using seasonally-representative data have clarified many flux estimates (McGuire *et al.* 2009; Tank *et al.* in press). PARTNERS-derived estimates of seasonal and annual constituent flux have also been used to examine the importance of riverine biogeochemistry to processes at the pan-Arctic scale, such as DOC degradation in the Arctic Ocean (Manizza *et al.* 2009), the relative importance of inorganic and organic riverine nitrogen for Arctic Ocean productivity (Tank *et al.* 2012), and the importance of factors such as permafrost and runoff for regulating dissolved inorganic C fluxes from the Arctic basin (Tank *et al.* in press).

1.4 Projections of future fluxes

1.4.1 Discharge

Continental runoff, driven by changes in net precipitation, is expected to increase at high latitudes with global warming. Different studies focusing on the Arctic have predicted a wide range of responses depending on the specific models and greenhouse gas emissions scenarios under consideration, but there is broad agreement among the studies that river discharge will increase (Miller and Russell 2000; Holland *et al.* 2006; Holland *et al.* 2007; Rawlins *et al.* 2010). Although precipitation and evaporation both increase with warming, the projected increases in precipitation outpace the increases in evaporation. Based on results from ten models that participated in the Intergovernmental Panel on Climate Change Fourth Assessment Report, Holland *et al.* (2007) estimated a net precipitation increase of 16% over the pan-Arctic watershed from 1950 through 2050. This included a 4% change from 1950 to 2000 that agreed well with the overall change in pan-Arctic river discharge that was observed during the latter half of the twentieth century (McClelland *et al.* 2006). While the suite of model results all indicated an increase in net precipitation, broad differences in the magnitudes of change among models pointed to a large degree of uncertainty in the precise trajectory of future increases (Holland *et al.* 2007).

1.4.2 Carbon

Of particular relevance with respect to the Arctic and climatic change is organic carbon (OC) cycling and storage as current estimates state that northern permafrost soils contain approximately one-half of global soil carbon (Gorham 1991; Tarnocai *et al.* 2009). A recent study estimated that northern soils contain almost 1700 Pg of OC, 88% of which is in permafrost (Tarnocai *et al.* 2009). Permafrost covers most of the pan-Arctic watershed (Table 1.1, Figure 1.2) and although it can be quite variable in depth (from tens to hundreds of meters), there is always a seasonally thawed active layer at the surface that varies in thickness from centimeters to meters (Brown *et al.* 1998; Frey and McClelland 2009). Increasing air temperatures result in permafrost thaw and degradation, which includes a deepening of the active layer, talik formation, thermokarst development, expansion and creation of thaw lakes, lateral permafrost thawing and a northward migration of the southern permafrost boundary (Zhang *et al.* 2005; Frey and McClelland 2009). Such degradation of permafrost has a number of impacts on hydrology, ecosystem dynamics and biogeochemical cycling in the Arctic and is leading to a significant impact on carbon biogeochemistry across the pan-Arctic including the mobilization of previously stored OC. As the OC that has been locked away in permafrost thaws into the contemporary carbon cycle, much of it may be metabolized by microorganisms in soils, exported into rivers and the ocean where it is also metabolized by microorganisms, ultimately resulting in the transfer of a significant portion of this large carbon reservoir to the atmosphere leading to a positive feedback on climate change (Striegl *et al.* 2005; Holmes *et al.* 2008; Osburn *et al.* 2009).

The response to climate change of OC export from Arctic rivers remains unclear with both an increase in export (Frey and Smith 2005; Guo *et al.* 2007) and a decrease in export predicted (Striegl *et al.* 2005; McClelland *et al.* 2007; Walvoord and Striegl 2007). In any case, because permafrost soils store a large amount of ancient carbon (Gorham 1991; Zimov *et al.* 2006; Zimov, Schuur, and Chapin 2006; Tarnocai *et al.* 2009), as they thaw the age of OC in rivers and streams may be useful as an integrator of the degree of permafrost degradation within the watershed. Currently, Arctic rivers have been shown to export predominantly modern DOC from recently fixed plant material and organic-rich surface soils (Amon and Meon 2004; Benner *et al.* 2004; Neff *et al.* 2006; Raymond *et al.* 2007). Seasonal trends in the radiocarbon age of DOC have been observed in the six largest Arctic rivers as well as tributaries, with an enrichment in Δ^{14}C-DOC (more modern DOC) during the spring flush period in comparison to late-summer DOC (Guo and Macdonald 2006; Neff *et al.* 2006; Raymond *et al.* 2007). Such a trend supports the idea of an increase in age of DOC exported through the summer months as the active layer thaw depth increases and a greater proportion of ancient carbon is mobilized from permafrost soils. Guo *et al.* (2007) reported data from three Arctic watersheds in North America (Yukon, Sagavanirktok and Mackenzie Rivers) and showed that particulate organic carbon (POC) may increase in age with permafrost degradation and resulting river-bank erosion, and hypothesized that future variability in DOC age could be due to changes in plant ecology (a shift from tundra to leaf-bearing plants). Although there is no general consensus it seems that the majority of evidence is pointing toward the fact that as the Arctic warms and permafrost degrades, concurrent increases in active layer depth will result in a greater mobilization of ancient OC into Arctic rivers. To date there is little evidence for an increase in aged DOC in Arctic rivers and streams, yet numerous studies have highlighted long-term permafrost degradation and a deepening in active layer (Frauenfeld *et al.* 2004; Oelke *et al.* 2004; Payette *et al.* 2004; Zhang *et al.* 2005; Osterkamp 2007).

While the release of DOC from soils to aquatic flow paths largely represents a net source of CO_2 to the atmosphere, much of the DIC (the sum of $CO_{2(aq)}/H_2CO_3$, HCO_3^- and CO_3^{2-}) found in freshwater represents a net CO_2 sink. Of the DIC species, bicarbonate (HCO_3^-) and carbonate (CO_3^{2-}; which is minimal at pH < 9) are predominantly derived from chemical weathering, which is one of the primary sinks for CO_2 on land. Chemical weathering causes CO_2 dissolved in water to be transformed to bicarbonate, with all the bicarbonate produced during silicate rock weathering being derived from CO_2 fixation, and half of the bicarbonate produced during carbonate rock weathering being derived from CO_2 and the remainder from the dissolution of carbonate rock. Thus, the production of bicarbonate along the aquatic continuum can serve as a counterbalance to the CO_2 loss that occurs when DOC is mobilized from soils in the organic component of the aquatic C cycle.

Both increases in runoff and decreases in permafrost extent are expected to increase bicarbonate flux from northern catchments. Although bicarbonate concentrations within catchments dilute with increasing runoff (Figure 1.4), the dilution in large Arctic rivers and elsewhere is not great enough to offset overall increases in water yield; within rivers, yearly bicarbonate fluxes increase with increasing yearly runoff (Raymond and Oh 2007; Tank *et al.* in press). Because of these seasonal concentration differences (dilution during the spring freshet, and high concentrations during wintertime low flows) any seasonal variation in the expected increase in Arctic river discharge will have an important effect on the magnitude by which bicarbonate flux increases. Increased discharge that occurs across the hydrograph as a result of an overall intensification of the hydrologic cycle (Rawlins *et al.* 2010), for example, should have relatively less of an impact per unit volume than increases that are concentrated during the winter months, which capture the highest bicarbonate concentrations (Figure 1.3; Walvoord and Striegl 2007; St. Jacques and Sauchyn 2009). Such recent increases in wintertime base flow have been documented for multiple catchments in North America, and appear to result from decreased permafrost, which acts as an effective barrier to recharge (Walvoord and Striegl 2007; St. Jacques and Sauchyn 2009; Ge *et al.* 2011). Because permafrost also restricts surface water to rock interaction (e.g., Frey and McClelland 2009), its degradation may increase bicarbonate flux via two related mechanisms: first, by increasing overall runoff through an increased contribution of (bicarbonate-rich) groundwater to surface flux; and second, by allowing increased weathering as a result of increased water percolation to deeper mineral soils.

Relatively few studies have directly assessed the effect of climate-related change on bicarbonate flux. In Alaska's Yukon River basin, an increase in summertime bicarbonate flux between the 1970s and 2000s (Striegl *et al.* 2005, 2007) occurred alongside increasing base flows (presumably resulting from decreases in permafrost) during this same period (Walvoord and Striegl 2007). In Alaska's North Slope region, a steady increase in alkalinity (largely composed of bicarbonate) has been documented in Toolik Lake, and attributed to a deepening active layer, and thus greater weathering of mineral soils (Hinzman *et al.* 2005). Using the PARTNERS dataset, Tank *et al.* (in press) showed that bicarbonate flux from these large catchments increased with increasing runoff and decreasing permafrost. Studies that have focused on other weathering constituents (Ca, Na, Mg), or total inorganic solids, have shown these fluxes to be higher in low permafrost catchments, indicating that bicarbonate should follow these same trends (MacLean *et al.* 1999; Petrone *et al.* 2006; Frey, Siegel and Smith 2007). Further work to quantify the magnitude of the bicarbonate response to changing climate, and in particular the potential effect of changes in weathering rate on C sequestration, will greatly improve our ability to predict future changes in the C cycle at the landscape scale.

1.4.3 Major ions

Generally, predictions for the flux of major ions from Arctic rivers with changing climate are similar to the predictions for bicarbonate, above, with increased fluxes likely to result from both increases in runoff and decreases in permafrost extent. For runoff, for example, modeled yearly fluxes of both Ca and Na increased within the PARTNERS rivers with increasing yearly discharge. The effect of permafrost on the flux of major ions has been previously reviewed by Frey and McClelland (2009). As mentioned above, several authors have examined trends in major ion concentration and flux across gradients of permafrost extent, and found total inorganic solid concentrations (the sum of eight anion and cation species; Frey, Siegel and Smith 2007), and major cation fluxes (Na, K, Ca, Mg; MacLean et al. 1999; Petrone et al. 2006) increased with decreasing permafrost extent, which likely reflected a decreased contribution of groundwater to surface flow and decreased water-rock interactions with increasing permafrost extent (MacLean et al. 1999; Frey and McClelland 2009). Increases in active layer depth have also been linked to increased fluxes of major ions, and changes in the relative contribution of various ions to overall flux. For example, Keller, Blum, and Kling (2010) found increasing fluxes of Ca relative to Na and Ba in river water in northern Alaska between 1994 and 2004, which indicated increasing thaw depths and the exposure of more readily weatherable carbonate rocks, which are elevated in Ca relative to Na and Ba. Other authors have found that permafrost was enriched in total soluble cations (Kokelj and Burn 2005) and selected cations (Ca and K; Keller, Blum and Kling 2007) compared to the active layer, which indicated that weathering of these constituents will increase with increasing active layer depths.

1.5 Conclusions

Rivers integrate processes occurring throughout their watersheds, so trends in river chemistry and discharge have great capacity for tracking widespread terrestrial change. In addition to diagnosing impacts of climate change and other disturbances on land, altered land-to-ocean hydrologic and biogeochemical fluxes also have profound impacts on the chemistry, biology, and physics of the ocean, particularly the Arctic Ocean, which receives a disproportionate supply of river water in comparison to other ocean basins.

Despite the recent strides in our understanding of the biogeochemistry of Arctic rivers, there are still numerous gaps in our knowledge. Most obviously, there are few datasets that are yet long enough to detect trends over time, which on Arctic rivers have typically required decadal or longer time series (Peterson et al. 2002; Striegl et al. 2005; McClelland et al. 2006; Déry et al. 2009). Second, although the PARTNERS and Arctic-GRO projects' focus on large rivers has allowed for a dataset that captures greater than 50% of the runoff from the Arctic basin (Table 1.1), it misses much of the Arctic's most northerly regions (Table 1.1, Figure 1.2). Thus, using this dataset to understand riverine biogeochemistry across the pan-Arctic, and its potential for future change, results in estimates that are relatively uncertain. In particular, processes occurring in catchments largely underlain by tundra, continuous permafrost (Table 1.1), or where rapid transit time through relatively small catchments limits within-catchment biogeochemical cycling, are not well captured by these data. To date, there have been no systematic efforts to compare the few time series measurements that do exist for these watershed types and regions (McClelland et al. 2007; Keller, Blum and Kling 2010; Lewis, Lafreniere, and Lamoureux 2011) to patterns in concentration and flux at the mouths of large rivers.

To fully understand how Arctic riverine biogeochemistry will change with changing climate, a continued focus on the collection of time-series data, coupled with increased efforts to establish time series of biogeochemistry for rivers across the continuum of

Arctic landscapes, is clearly needed. Each river water sample can tell an important story, but the stories can only be read if the samples are collected. There should also be a concerted effort to archive time-series samples from numerous Arctic rivers, large and small. The archived samples would then be available for analysis in the future as new questions are asked or new analytical methods are developed.

Finally, improved understanding of the current rates of change and projections of future change will require the application of a new generation of integrated earth system models. Given the multitude of changes already underway in the Arctic and the tight coupling of atmospheric, terrestrial, and aquatic components of the Arctic, application of models that explicitly link above and below ground cycles of water, energy, carbon, and nutrients with global and regional atmospheric circulation models is urgently needed. Only coupled land-surface atmosphere models will be able to provide mechanistic understanding of the complex ways in which the Arctic will respond and contribute to global change in the coming decades.

Ironically, although the Arctic is a large producer of fossil fuels, there is little that can be done in the Arctic to mitigate the impacts of anthropogenic climate change. Instead, societal actions outside of the Arctic, mainly reducing fossil fuel combustion and halting tropical deforestation, will be required to dampen the impacts of climate change on Arctic ecosystems.

1.6 Acknowledgments

This work was funded by the US National Science Foundation as part of the Arctic Great Rivers Observatory (NSF-0732522 and NSF-1107774) and the Global Rivers Project (NSF-0851101).

References

Amon, R.M.W. and Meon, B. (2004) The biogeochemistry of dissolved organic matter and nutrients in two large Arctic estuaries and potential implications for our understanding of the Arctic Ocean system. *Mar. Chem.*, **92**, 311–30.

Andersen, M.B., Stirling, C.H., Porcelli, D., *et al.* (2007) The tracing of riverine U in Arctic seawater with very precise U-234/U-238 measurements. *Earth Planet. Sci. Lett.*, **259**, 171–85.

Anema, C., Hecky, R.E., Fee, E.J., *et al.* (1990) *Water chemistry of some lakes and channels in the Mackenzie Delta and on the Tuktoyaktuk Peninsula, NWT*, 1986, Minister of Supply and Services Canada, Ottawa, Canada.

Benner, R., Benitez-Nelson, B., Kaiser, K., and Amon, R.M.W. (2004) Export of young terrigenous dissolved organic carbon from rivers to the Arctic Ocean. *Geophys. Res. Lett.*, **31**, L05305, doi: 05310.01029/02003GL019251.

Brown, J., Ferrians, O.J., Heginbottom, J.A., and Melnikov, E.S. (1998) Circum-arctic map of permafrost and ground ice conditions. National Snow and Ice Data Center/World Data Center for Glaciology, Boulder, CO.

Cai, Y.H., Guo, L.D., and Douglas, T.A. (2008) Temporal variations in organic carbon species and fluxes from the Chena River, *Alaska. Limnol. Oceanogr*, **53**, 1408–19.

Cai, Y.H., Guo, L.D., Douglas, T.A., and Whitledge, T.E. (2008) Seasonal variations in nutrient concentrations and speciation in the Chena River, Alaska. *J. Geophys. Res.-Biogeosci.*, **113**, G03035, 03010.01029/02008JG000733.

Carey, S.K. (2003) Dissolved organic carbon fluxes in a discontinuous permafrost subarctic alpine catchment. *Permafrost and Periglacial Processes*, **14**, 161–71.

Cooper, L.W., McClelland, J.W., Holmes, R.M., *et al.* (2008) Flow-weighted values of runoff tracers (delta 18O, DOC, Ba, alkalinity) from the six largest Arctic rivers. *Geophys. Res. Lett.*, **35**, L18606, doi: 18610.11029/12008GL035007.

Déry, S.J., Hernandez-Henriquez, M.A., Burford, J.E., and Wood, E.F. (2009) Observational evidence of an intensifying hydrological cycle in northern Canada. *Geophys. Res. Lett.*, **36**, L13402, doi: 13410.11029/12009GL038852.

Déry, S.J., Stieglitz, M., McKenna, E.C., and Wood. E.F. (2005) Characteristics and trends of river discharge into Hudson, James, and Ungava bays, 1964–2000. *J. Climate*, **18**, 2540–57.

Déry, S.J. and Wood, E.F. (2005) Decreasing river discharge in northern Canada. *Geophys. Res. Lett.*, **32**, L10401, doi: 10410.11029/12005GL022845.

Dornblaser, M.M. and Striegl, R.G. (2007) Nutrient (N, P) loads and yields at multiple scales and subbasin types in the Yukon River basin, Alaska. *J. Geophys. Res.—Biogeosci.*, **112**, G04S57, doi: 10.1029/2006JG000366.

Elmquist, M., Semiletov, I., Guo, L.D., and Gustafsson, O. (2008) Pan-Arctic patterns in black carbon sources and fluvial discharges deduced from radiocarbon and PAH source apportionment markers in estuarine surface sediments. *Global Biogeochem. Cycles*, **22**, GB2018, 2010.1029/2007GB002994.

Emmerton, C.A. (2006) Downstream nutrient changes through the Mackenzie River Delta and Estuary, Western Canadian Arctic. Simon Fraser University, Burnaby, BC, Canada.

Emmerton, C.A., Lesack, L.F.W., and Vincent, W.F. (2008) Mackenzie River nutrient delivery to the Arctic Ocean and effects of the Mackenzie Delta during open water conditions. *Global Biogeochem. Cycles*, **22**, GB1024, doi: 1010.1029/2006GB002856.

Environment Canada (1976) *Water Quality Data: Northwest Territories 1960–1973*, Minister of Supply and Services Canada, Ottawa, Ontario.

Environment Canada (1978) *Water Quality Data: Manitoba 1961–1976*, Minister of Supply and Services Canada, Ottawa, Canada.

Environment Canada (1982) *Detailed Surface Water Quality Data: Northwest Territories 1977–1979*, Minister of Supply and Services Canada, Ottawa, Ontario.

Forbes, A.C., and Lamoureux, S.F. (2005) Climatic controls on streamflow and suspended sediment transport in three large middle arctic catchments, Boothia Peninsula, Nunavut, Canada. *Arct. Ant. Alp. Res.*, **37**, 304–315.

Frauenfeld, O.W., Zhang, T.J., Barry, R.G., and Gilichinsky, D. (2004) Interdecadal changes in seasonal freeze and thaw depths in Russia. *J. Geophys. Res.—Atmos.*, **109**, D05101, 05110.01029/02003JD004245.

Frey, K.E. and McClelland. J.W. (2009) Impacts of permafrost degradation on arctic river biogeochemistry. *Hydrol. Proc.*, **23**, 169–82.

Frey, K.E. Siegel, D.I., and Smith, L.C. (2007) Geochemistry of west Siberian streams and their potential response to permafrost degradation. *Water Resour. Res.*, **43**, W03406, doi: 03410.01029/02006WR004902.

Frey, K.E., and Smith, L.C. (2005) Amplified carbon release from vast West Siberian peatlands by 2100. *Geophys. Res. Lett.*, **32**, L09401, doi: 09410.01029/02004GL022025.

Gareis, J.A.L., Lesack, L.F.W., and Bothwell, M.L. (2010) Attenuation of in situ UV radiation in Mackenzie Delta lakes with varying dissolved organic matter compositions. *Water Resour. Res.*, **46**, W09516, doi: 09510.01029/02009WR008747

Ge, S.M., McKenzie, J., Voss, C., and Wu, Q.B. (2011) Exchange of groundwater and surface-water mediated by permafrost response to seasonal and long term air temperature variation. *Geophys. Res. Lett.* **38**, L14402, doi: 14410.11029/12011GL047911.

Gordeev, V.V., Martin, J.M., Sidorov, I.S., and Sidorova, M.V. (1996) A reassessment of the Eurasian river input of water, sediment, major elements, and nutrients to the Arctic Ocean. *Amer. J. Sci.*, **296**, 664–91.

Gorham, E. (1991) Northern Peatlands—role in the carbon-cycle and probable responses to climatic warming. *Ecol. Appl.*, **1**, 182–95.

Gueguen, C., Guo, L.D., Wang, D., *et al.* (2006) Chemical characteristics and origin of dissolved organic matter in the Yukon River. *Biogeochemistry*, **77**, 139–55.

Guo, L., Cai, Y., Belzile, C., and Macdonald, R.W. (2012) Sources and export fluxes of inorganic and organic carbon and nutrient species from the seasonally ice-covered Yukon River. *Biogeochemistry*, **107**, 187–206, doi: 110.1007/s10533-10010-19545-z.

Guo, L.D., and Macdonald, R.W. (2006) Source and transport of terrigenous organic matter in the upper Yukon River: Evidence from isotope (delta C-13, Delta C-14, and delta N-15) composition of dissolved, colloidal, and particulate phases. *Global Biogeochem. Cycles*, **20**, GB2011, doi: 2010.1029/2005GB002593.

Guo, L.D., Ping, C.L., and Macdonald, R.W. (2007) Mobilization pathways of organic carbon from permafrost to arctic rivers in a changing climate. *Geophys. Res. Lett.*, **34**, L13603, doi: 13610.11029/12007GL030689

Guo, L.D., Zhang, J.Z., and Gueguen, C. (2004) Speciation and fluxes of nutrients (N, P, Si) from the upper Yukon River. *Global Biogeochem. Cycles*, **18**, GB1038, doi: 1010.1029/2003GB002152.

Hinzman, L.D., Bettez, N.D., Bolton, W.R., *et al.* (2005) Evidence and implications of recent climate change in northern Alaska and other arctic regions. *Clim. Change*, **72**, 251–98.

Holland, M.M., Finnis, J., Barrett, A.P., and Serreze, M.C. (2007) Projected changes in arctic ocean freshwater budgets. *J. Geophys. Res.—Biogeosci.*, **112**, G04S55, doi: 10.1029/2006JG000354.

Holland, M.M., Finnis, J., and Serreze, M.C. (2006) Simulated Arctic Ocean freshwater budgets in the twentieth and twenty-first centuries. *J. Climate*, **19**, 6221–42.

Holmes, R.M., McClelland, J.W., Peterson, B.J., *et al.* (2002) A circumpolar perspective on fluvial sediment flux to the Arctic Ocean. *Global Biogeochem. Cycles*, **16**, 1098, doi: 1010.1029/2001GB001849.

Holmes, R.M., McClelland, J.W., Peterson, B.J., *et al.* (2012) Seasonal and annual fluxes of nutrients and organic matter from large rivers to the Arctic Ocean and surrounding seas. *Estuar. Coasts*, **35**, 369–382, DOI 10.10007/s12237-011-9386-6.

Holmes, R.M., McClelland, J.W., Raymond, P.A., *et al.* (2008) Lability of DOC transported by Alaskan rivers to the arctic ocean. *Geophys. Res. Lett.*, **35**, L03402, doi: 03410.01029/02007gl032837.

Holmes, R.M., Peterson, B.J., Gordeev, V.V., *et al.* (2000) Flux of nutrients from Russian rivers to the Arctic Ocean: Can we establish a baseline against which to judge future changes? *Wat. Resour. Res.*, **36**, 2309–20.

Holmes, R.M., Peterson, B.J., Zhulidov, A.V., *et al.* (2001) Nutrient chemistry of the Ob' and Yenisey Rivers, Siberia: results from June 2000 expedition and evaluation of long-term data sets. *Mar. Chem.*, **75**, 219–27.

Jeffries, M.O., Korsmo, F., Calder, J., and Crane, K. (2007) Arctic observing network: toward a US contribution to pan-Arctic observing. *Arctic Res. of the United States*, **21**, 1–94, http://www.nsf.gov/pubs/2008/nsf0842/index.jsp.

Keller, K., Blum, J.D., and Kling, G.W. (2007) Geochemistry of soils and streams on surfaces of varying ages in arctic *Alaska*. *Arct. Ant. Alp. Res.*, **39**, 84–98.

Keller, K., Blum, J.D., and Kling, G.W. (2010) Stream geochemistry as an indicator of increasing permafrost thaw depth in an arctic watershed. *Chem. Geol.*, **273**, 76–81.

Kimstach, V., Meybeck, M., and Baroudy, E. (eds.) (1998) *A Water Quality Assessment of the Former Soviet Union*, Routledge, London.

Kirk, J. L. and Louis, V.L.S. (2009) Multiyear Total and Methyl Mercury Exports from Two Major Sub-Arctic Rivers Draining into Hudson Bay, *Canada. Environ. Sci. Tech.*, **43**, 2254–61.

Kohler, H., Meon, B., Gordeev, V., *et al.* (2003) Dissolved organic matter (DOM) in the estuaries of the Ob and Yenisei and the adjacent Kara Sea, Russia, in *Siberian River*

Run-off in the Kara Sea (eds. R. Stein, K. Fahl, D. K. Fütterer, *et al.*), Elsevier, Amsterdam, pp. 281–308.

Kokelj, S.V. and Burn, C.R. (2005) Geochemistry of the active layer and near-surface permafrost, Mackenzie delta region, *Northwest Territories. Canada. Can. J. Earth Sci.*, **42**, 37–48.

Lafreniere, M. and Lamoureux, S. (2008) Seasonal dynamics of dissolved nitrogen exports from two High Arctic watersheds, Melville Island, *Canada. Hydrol. Res.*, **39**, 323–35.

Lesack, L.F.W., Marsh, P., and Hecky, R.E. (1998) Spatial and temporal dynamics of major solute chemistry among Mackenzie Delta lakes. *Limnol. Oceanogr.*, **43**, 1530–43.

Lewis, T., Lafreniere, M.J., and Lamoureux, S.F. (2011) Hydrochemical and sedimentary responses of paired High Arctic watersheds to unusual climate and permafrost disturbance, Cape Bounty, Melville Island, *Canada. Hydrol. Proc.*, doi: 10.1002/hyp.8335.

Lobbes, J.M., Fitznar, H.P., and Kattner, G. (2000) Biogeochemical characteristics of dissolved and particulate organic matter in Russian rivers entering the Arctic Ocean. *Geochim. Cosmochim. Ac.*, **64**, 2973–83.

MacLean, R., Oswood, M.W., Irons, J.G., and McDowell, W.H. (1999) The effect of permafrost on stream biogeochemistry: a case study of two streams in the Alaskan (USA) taiga. *Biogeochemistry*, **47**, 239–67.

Manizza, M., Follows, M.J., Dutkiewicz, S., *et al.* (2009) Modeling transport and fate of riverine dissolved organic carbon in the Arctic Ocean. *Global Biogeochem. Cycles*, **23**, doi: 10.1029/2008gb003396.

Mann, P.J., Davidova, A. Zimov, N., *et al.* (2012) Controls on the composition and lability of dissolved organic matter in Siberia's Kolyma River basin. *J. Geophys. Res.*, **117**, G01028, doi: 10.1029/2011JG001798.

McClelland, J.W., Dery, S.J., Peterson, B.J., *et al.* (2006) A pan-arctic evaluation of changes in river discharge during the latter half of the 20th century. *Geophys. Res. Lett.*, **33**, L06715, doi: 06710.01029/02006GL025753.

McClelland, J.W., Holmes, R.M., Dunton, K.H., and Macdonald, R. (2012) *The Arctic Ocean estuary. Estuar. Coasts* **35**, 353–68, doi: 10.1007/s12237-12010-19357-12233.

McClelland, J.W., Holmes, R.M., Peterson, B.J., *et al.* (2008) Development of a pan-arctic database for river chemistry. *EOS*, **89**, 217–18.

McClelland, J.W., Holmes, R.M., Peterson, B.J., and Stieglitz, M. (2004) Increasing river discharge in the Eurasian Arctic: Consideration of dams, permafrost thaw, and fires as potential agents of change. *J. Geophys. Res.—Atmos.*, **109**, D18102, doi: 18110.11029/12004 JD004583.

McClelland, J.W., Stieglitz, M., Pan, F., *et al.* (2007) Recent changes in nitrate and dissolved organic carbon export from the upper Kuparuk River, North Slope, Alaska. *J. Geophys. Res.*, **112**, G04S60, doi: 10.1029/2006JG000371.

McGuire, A.D., Anderson, L.G., Christensen, T.R., *et al.* (2009) Sensitivity of the carbon cycle in the Arctic to climate change. *Ecol. Monogr.*, **79**, 523–55.

Meybeck, M. and Ragu, A. (1995) *River Discharges to the Oceans: An Assessment of Suspended Solids, Major Ions and Nutrients*. UNEP Report, UNEP, Nairobi.

Miller, J.R. and Russell, G.L. (2000) Projected impact of climate change on the freshwater and salt budgets of the Arctic Ocean by a global climate model. *Geophys. Res. Lett.*, **27**, 1183–6.

Millot, R., Gaillardet, J., Dupre, B., and Allegre, C.J. (2002) The global control of silicate weathering rates and the coupling with physical erosion: new insights from rivers of the Canadian Shield. *Earth Planet. Sci. Lett.*, **196**, 83–98.

Millot, R., Gaillardet, J., Dupre, B., and Allegre, C.J. (2003) Northern latitude chemical weathering rates: Clues from the Mackenzie River Basin, Canada. *Geochim. Cosmochim. Ac.*, **67**, 1305–29.

Neff, J.C., Finlay, J.C., Zimov, S.A., *et al.* (2006) Seasonal changes in the age and structure of dissolved organic carbon in Siberian rivers and streams. *Geophys. Res. Lett.*, **33**, L23401, doi: 23410.21029/22006GL028222.

Nohara, D., Kitoh, A., Hosaka, M., and Oki, T. (2006) Impact of climate change on river discharge projected by multimodel ensemble. *J. Hydromet.*, **7**, 1076–89.

Oelke, C., Zhang, T.J., and Serreze, M.C. (2004) Modeling evidence for recent warming of the Arctic soil thermal regime. *Geophys. Res. Lett.*, **31**, L07208, doi: 07210.01029/02003 GL019300.

Osburn, C.L., Retamal, L., and Vincent, W.F. (2009) Photoreactivity of chromophoric dissolved organic matter transported by the Mackenzie River to the Beaufort Sea. *Mar. Chem.* **115**, 10–20.

Osterkamp, T.E. (2007) Characteristics of the recent warming of permafrost in Alaska. *J. Geophys. Res.—Earth Surface*, **112**, F02S02, doi: 10.1029/2006JF000578.

Overeem, I. and Syvitski, J.P.M. (2010) Shifting Discharge Peaks in Arctic Rivers, 1977–2007. *Geografiska Annaler Series A—Phys. Geog.*, **92A**, 285–96.

Payette, S., Delwaide, A., Caccianiga, M., and Beauchemin, M. (2004) Accelerated thawing of subarctic peatland permafrost over the last 50 years. *Geophys. Res. Lett.*, **31**, L18208, doi: 18210.11029/12004GL020358.

Peterson, B.J., Corliss, T., Kriet, K., and Hobbie, J.E. (1992) Nitrogen and phosphorus concentrations and export for the upper Kuparuk River on the North Slope of Alaska. *Hydrobiologia*, **240**, 61–9.

Peterson, B.J., Hobbie, J.E., and Corliss, T.L. (1986) Carbon flow in a tundra stream ecosystem. *Can. J. Fish. Aquat. Sci.*, **43**, 1259–70.

Peterson, B.J., Holmes, R.M., McClelland, J.W., *et al.* (2002) Increasing river discharge to the Arctic Ocean. *Science*, **298**, 2171–3.

Peterson, B.J., McClelland, J., Curry, R., *et al.* (2006) Trajectory shifts in the Arctic and subarctic freshwater cycle. *Science*, **313**, 1061–6.

Petrone, K.C., Jones, J.B., Hinzman, L.D., and Boone, R.D. (2006) Seasonal export of carbon, nitrogen, and major solutes from Alaskan catchments with discontinuous permafrost. *J. Geophys. Res.*, **111**, G02020, doi: 01029/02005JG000055.

Rachold, V., Eicken, H., Gordeev, V.V., *et al.* (2004) Modern terrigenous organic carbon input to the Arctic Ocean, in *The Organic Carbon Cycle in the Arctic Ocean* (eds R. Stein and R.W. Macdonald), Springer-Verlag, Berlin.

Rawlins, M.A., Frolking, S., Lammers, R.B., and Vorosmarty, C.J. (2006) Effects of uncertainty in climate inputs on simulated evapotranspiration and runoff in the Western Arctic. *Earth Interact*, **10**, 1–18. doi: http://dx.doi.org/10.1175/EI1182.1171.

Rawlins, M.A., Steele, M., Holland, M.M., *et al.* (2010) Analysis of the Arctic system for freshwater cycle intensification: observations and expectations. *J. Clim.*, **23**, 5715–37.

Raymond, P.A., McClelland, J.W., Holmes, R.M., *et al.* (2007) Flux and age of dissolved organic carbon exported to the Arctic Ocean: A carbon isotopic study of the five largest arctic rivers. *Global Biogeochem. Cycles*, **21**, GB4011, doi: 4010.1029/2007GB002934.

Raymond, P.A., and Oh, N.H. (2007) An empirical study of climatic controls on riverine C export from three major U.S. watersheds. *Global Biogeochem. Cycles*, **21**, GB2022, doi: 2010.1029/2006GB002783.

Reeder, S.W., Hitchon, B., and Levinson, A.A. (1972) Hydrogeochemistry of surface waters of Mackenzie River Drainage Basin, Canada .1. Factors controlling inorganic composition. *Geochim. Cosmochim. Ac.*, **36**, 825–65.

Schuster, P.F., Striegl, R.G., Aiken, G.R., *et al.* (2011) Mercury export from the Yukon River Basin and potential response to a changing climate. *Environ. Sci. Tech.*, **45**, 9262–7.

Schuur, E.A.G., Bockheim, J., Canadell, J.G., *et al.* (2008) Vulnerability of permafrost carbon to climate change: Implications for the global carbon cycle. *Bioscience*, **58**, 701–14.

SEARCH (2005) *Study of Environmental Arctic Change: Plans for Implementation During the International Polar Year and Beyond*, Arctic Research Consortium of the United States (ARCUS), Fairbanks, Alaska.

Semiletov, I.P., Pipko, I.I, Shakhova, N.E., *et al.* (2011) Carbon transport by the Lena River from its headwaters to the Arctic Ocean, with emphasis on fluvial input of terrestrial particulate organic carbon vs. carbon transport by coastal erosion. *Biogeosciences*, **8**, 2407–26.

Serreze, M.C., Holland, M.M., and Stroeve, J. (2007) Perspectives on the Arctic's shrinking sea-ice cover. *Science*, **315**, 1533–6.

Shiklomanov, A.I., and Lammers, R.B. (2009) Record Russian river discharge in 2007 and the limits of analysis. *Environ. Res. Lett.*, **4**, 045015, doi: 045010.041088/041748-049326.

Spencer, R.G.M., Aiken, G.R., Butler, K.D., *et al.* (2009) Utilizing chromophoric dissolved organic matter measurements to derive export and reactivity of dissolved organic carbon exported to the Arctic Ocean: A case study of the Yukon River, Alaska. *Geophys. Res. Lett.*, **36**, L06401, 06410.01029/02008gl036831.

Spencer, R.G.M., Aiken, G.R., Wickland, K.P., *et al.* (2008) Seasonal and spatial variability in dissolved organic matter quantity and composition from the Yukon River basin, Alaska. *Global Biogeochem. Cycles*, **22**, GB4002, doi: 4010.1029/2008GB003231.

St. Jacques, J.M. and Sauchyn, D.J. (2009) Increasing winter baseflow and mean annual streamflow from possible permafrost thawing in the Northwest Territories, Canada. *Geophys. Res. Lett.* **36**, L01401, doi: 01410.01029/02008GL035822.

Striegl, R.G., Aiken, G.R., Dornblaser, M.M., *et al.* (2005) A decrease in discharge-normalized DOC export by the Yukon River during summer through autumn. *Geophys. Res. Lett.*, **32**, L21413, doi: 21410.21029/22005gl024413.

Striegl, R.G., Dornblaser, M.M., Aiken, G.R., *et al.* (2007) Carbon export and cycling by the Yukon, Tanana, and Porcupine rivers, Alaska, 2001–2005. *Water Resour. Res.*, **43**, W02411, doi: 02410.01029/02006WR005201.

Tank, S.E., Lesack, L.F.W., Gareis, J.A.L., *et al.* (2011) Multiple tracers demonstrate distinct sources of dissolved organic matter to lakes of the Mackenzie Delta, western Canadian Arctic. *Limnol. Oceanogr.*, **56**, 1297–309.

Tank, S.E., Lesack, L.F.W., and McQueen, D.J. (2009) Elevated pH regulates bacterial carbon cycling in lakes with high photosynthetic activity. *Ecology*, **90**, 1910–22.

Tank, S.E., Manizza, M., Holmes, R.M., *et al.* (2012) The processing and impact of dissolved riverine nitrogen in the Arctic Ocean. *Estuar. Coasts*, **35**, 401–15, doi: 10.1007/s12237-12011-19417-12233.

Tank, S.E., Raymond, P.A., Striegl, R.G., *et al.* (in press) A land-to-ocean perspective on the magnitude, source and implication of DIC flux from major arctic rivers to the Arctic Ocean. *Global Biogeochem. Cycles*.

Tarnocai, C., Canadell, J.G., Schuur, E.A.G., *et al.* (2009) Soil organic carbon pools in the northern circumpolar permafrost region. *Global Biogeochem. Cycles*, **23**, GB2023, doi: 2010.1029/2008GB003327.

Telang, S.A., Pocklington, R., Naidu, A.S., *et al.* (1991) Carbon and mineral transport in major North American, Russian arctic, and Siberian rivers: the St. Lawrence, the Mackenzie, the Yukon, the arctic Alaskan rivers, the arctic basin rivers in the Soviet Union, and the Yenisei, in *Biogeochemistry of Major World Rivers* (eds. E. T. Degens, S. Kempe, and J. E. Richey), John Wiley & Sons, Inc., New York, pp. 75–104.

Walker, D.A., Raynolds, M.K., Daniels, F.J.A., *et al.* (2005) The Circumpolar Arctic vegetation map. *J. Veget. Sci.*, **16**, 267–82.

Walvoord, M.A. and Striegl, R.G. (2007) Increased groundwater to stream discharge from permafrost thawing in the Yukon River basin: Potential impacts on lateral export of carbon and nitrogen. *Geophys. Res. Lett.*, **34**, L12402, 12410.11029/12007gl030216.

White, D., Hinzman, L., Alessa, L. *et al.* (2007) The arctic freshwater system: changes and impacts. *J. Geophys. Res.*, **112**, G04S54, doi: 10.1029/2006JG000353.

Wu, P., Wood, R., and Stott, P. (2005) Human influence on increasing Arctic river discharge. *Geophys. Res. Lett.*, **32**, L02703, doi: 02710.01029/02004GL021570.

Yang, D., Kane, D.L., Hinzman, L.D., *et al.* (2002) Siberian Lena River hydrologic regime and recent change. *J. Geophys. Res.*, **107**, D23, 4694, doi: 4610.10292002JD10002542.

Yang, D., Robinson, D., Zhao, Y., *et al.* (2003) Streamflow response to seasonal snow cover extent changes in large Siberian watersheds. *J. Geophys. Res.*, **108**, D18, 4578, doi: 1029/2002JD003149.

Yang, D.Q., Zhao, Y.Y., Armstrong, R., *et al.* (2007) Streamflow response to seasonal snow cover mass changes over large Siberian watersheds. *J. Geophys. Res.—Earth Surface*, **112**, F02S22, doi: 10.1029/2006JF000518.

Yi, Y., Gibson, J.J., Helie, J.F., and Dick, T.A. (2010) Synoptic and time-series stable isotope surveys of the Mackenzie River from Great Slave Lake to the Arctic Ocean, 2003 to 2006. *J. Hydrol.*, **383**, 223–32.

Zhang, T., Frauenfeld, O.W., Serreze, M.C., *et al.* (2005) Spatial and temporal variability in active layer thickness over the Russian Arctic drainage basin. *J. Geophys. Res.*, **110**, D16101, doi: 11029/12004JD005642.

Zhulidov, A.V., Khlobystov, V.V., Robarts, R.D., and Pavlov, D.F. (2000) Critical analysis of water quality monitoring in the Russian Federation and former Soviet Union. *Can. J. Fish. Aquat. Sci.*, **57**, 1932–9.

Zimmermann, B., Porcelli, D., Frank, M., *et al.* (2009) Hafnium isotopes in Arctic Ocean water. *Geochimic. Cosmochim. Ac.*, **73**, 3218–33.

Zimov, S.A., Davydov, S.P., Zimova, G.M., *et al.* (2006a) Permafrost carbon: Stock and decomposability of a globally significant carbon pool. *Geophys. Res. Lett.*, **33**, L20502, doi: 20510.21029/22006GL027484.

Zimov, S.A., Schuur, E.A.G., and Chapin, F.S. (2006b) Permafrost and the global carbon budget. *Science* **312**, 1612–13.

2

Climate Impacts on Arctic Lake Ecosystems

Warwick F. Vincent[1,2,3], Isabelle Laurion[1,4], Reinhard Pienitz[1,2,5], and Katey M. Walter Anthony[6]

[1]Center for Northern Studies (CEN), Québec City, Canada
[2]Takuvik Joint International Laboratory, Université Laval Canada—Centre national de la recherche scientifique (CNRS) (France)
[3]Département de biologie, Université Laval, Québec City, Canada
[4]Institut national de la recherche scientifique—Centre Eau Terre Environnement (INRS-ETE), Québec City, Canada
[5]Département de géographie, Université Laval, Québec City, Canada
[6]Water and Environmental Research Center, University of Alaska, Fairbanks, USA

2.1 Introduction

Lakes and ponds are major features of the Arctic landscape, and span a diverse range of environmental conditions, from dilute, glacier-fed meltwaters to nutrient-rich tundra ponds and perennially ice-capped, stratified lakes with anoxic bottom waters. According to the global lakes and wetlands data base, the majority of the world's lakes with surface areas in the range 0.1 to 50 km^2 occur above latitude 45.5 °N (Lehner and Döll 2004), and 73% of these lie within the permafrost zone (Smith, Sheng, and MacDonald 2007). The most abundant freshwater ecosystems in the north are small and shallow; however, their total area and volume is substantial (Rautio *et al.* 2011 and references therein), and collectively they may influence biogeochemical dynamics at a global scale (Walter *et al.*, 2006). In parts of the Arctic, these numerous shallow waters can account for up to 90% of the total land surface area (Pienitz, Doran, and Lamoureux 2008).

Thermokarst processes (permafrost thawing and erosion) play an important role in many of these lake systems throughout the Arctic, and over a wide range of soil and climate regimes (e.g., Jørgenson and Osterkamp 2005; Jones *et al.* 2011). Grosse *et al.* (2011) estimate that more than 61 000 lakes >0.1 km^2 with a total lake area of

Climatic Change and Global Warming of Inland Waters: Impacts and Mitigation for Ecosystems and Societies, First Edition. Edited by Charles R. Goldman, Michio Kumagai and Richard D. Robarts.
© 2013 John Wiley & Sons, Ltd. Published 2013 by John Wiley & Sons, Ltd.

more than $200\,000\,km^2$ occur in the circumarctic region in permafrost with high to moderate ground ice content, and are likely to be thermokarst lakes. Using correction factors to account for smaller lakes too, the total thermokarst lake area is likely to be in the range $250\,000–380\,000\,km^2$. Beringian thermokarst lakes, defined here as located within the largely unglaciated region from the Mackenzie River, Canada, west to the Lena River, Russia, constitute about 30% ($75\,000–114\,000\,km^2$) of the total pan-Arctic thermokarst lake area.

Large lakes $>500\,km^2$ are also found throughout the circumpolar Arctic. One of the largest above the Arctic Circle is Lake Taymyr (lat. 74.1 °N, $4560\,km^2$, average depth of 2.5 m; Robarts *et al.* 1999) in northern Russia. Large, deep lakes in the Canadian North include Great Bear Lake (lat. 65–67 °N, area of $114\,717\,km^2$, maximum depth of 446 m), Nettilling Lake (66.5 °N, $5066\,km^2$, 132 m; Oliver 1964), Amudjuak Lake (64 °N, $3115\,km^2$, maximum depth unknown), Lake Hazen (81.8 °N, $542\,km^2$, 267 m; Köck *et al.* 2012); Lac à l'Eau Claire (Clearwater Lake; 56.2 °N, $1239\,km^2$, 178 m; Milot-Roy and Vincent 1994), and Pingualuit Crater Lake (61.5 °N, $9\,km^2$, 267 m). Like the latter two lakes, El'gygytgyn Lake (67.5°N, $110\,km^2$, 174 m), in Siberia, also lies in a meteoritic impact crater, and has attracted considerable paleolimnological interest (Melles *et al.* 2007).

Despite this great limnological diversity, northern lakes also have a number of features in common (Vincent, Hobbie, and Laybourn-Parry 2008). Firstly, as a result of their high latitude location, these ecosystems experience extreme seasonal variations in incident solar radiation. Above the Arctic Circle, this translates into three months of continuous winter darkness and three months of continuous light in summer, which in turn give rise to high-amplitude fluctuations in primary production and all related food-web processes. Secondly, these seasonal effects are compounded by snow and ice, which cover these lakes for at least six months each year. For a small and decreasing number of extreme Arctic lakes, thick perennial ice persists throughout the year. The solid ice cap over the lakes influences all aspects of their limnology, including the availability of light for photosynthesis, rates of gas exchange with the atmosphere, interactions with the surrounding watershed, and their stratification and mixing regimes. Thirdly, persistent low temperatures exert a strong control on all physiological and ecological processes within high latitude lakes, and also in their surrounding catchments. This effect on chemical and biochemical reaction rates contributes to a fourth feature—the slow rates of soil-weathering processes in Arctic catchments. These lakes thereby receive only sparse inputs of nutrients, which maintain their water columns in an oligotrophic status of low algal biomass. As a result of all of these constraints, and exacerbated by their remoteness from temperate latitudes, an additional feature of northern ecosystems is their low biodiversity of aquatic plants and animals, many of which are specialized towards extreme cold, low energy supply, and oligotrophy.

There is now compelling evidence of rising atmospheric temperatures at a planetary scale, and the greatest amplitude of change has been recorded at high northern latitudes (IPCC 2007). While global average annual air temperatures have increased by around 0.4 °C since the early 1990s, the North American Arctic over the same period has warmed by 2.1 °C (ACIA 2005). These observations are consistent with results from global circulation models, which predict that the most severe ongoing warming will be in the Arctic, to temperatures up to 8 °C above present values by the end of the twenty-first century. This magnitude of change will have drastic effects on Arctic

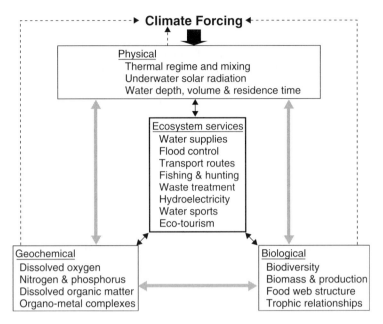

Figure 2.1 Impacts of climate change on northern lakes and their ecosystem services. Dotted lines indicate positive feedback effects. Modified from Vincent (2009).

freshwater ecosystems, given that many of their limnological features are dependent on prolonged sub-zero air temperatures each year.

The aim of the present review is to summarize the primary mechanisms of climate-induced change in northern high latitude lakes (Figure 2.1). We first examine the direct physical impacts of climate warming, ranging from loss of ice cover to complete loss of freshwater ecosystems by drainage or evaporation. We then review the biogeochemical mechanisms of change including shifts in organic matter loading and de-oxygenation, the formation and release of microbial methane from thermokarst lakes associated with permafrost thaw, the biological response mechanisms including loss of cold-adapted taxa and arrival of invasive species, and the local anthropogenic effects accompanying climate change. We conclude this review with a brief analysis of strategies to reduce the impacts of climate warming on northern lakes and their vital ecosystem services.

2.2 Physical impacts of climate change

The circumpolar Arctic has experienced large spatial variations in climate in the past, and similarly large variations in the physical impact of climate on lakes are to be expected among different sectors of the North. There is paleolimnological evidence of such variability, with shifts in diatom community structure in many lakes over the last century consistent with climate change, but to differing extents among sites and sectors (Smol *et al*. 2005). The climate and lake characteristics of northern Québec (Nunavik)-Labrador (Nunatsiavut) region in the low Arctic appear to have been especially stable for hundreds of years, perhaps longer (Pienitz *et al*. 2004), but from the early 1990s onwards have been undergoing rapid change (Bhiry *et al*. 2011).

The most drastic limnological impact of climate change is the complete loss of certain aquatic ecosystems. This may occur through geomorphological effects such as the breaching of an ice dam, or erosion of permafrost soils. For example, many epishelf lakes (freshwater lakes underlain by seawater and dammed by ice shelves) occurred along the northern coast of Ellesmere Island in the Canadian High Arctic in the early twentieth century (Veillette *et al*. 2008), but the warming and break-up of the northern ice shelves has resulted in their drainage and loss (e.g., Mueller *et al*. 2003), and now only one such ecosystem is known to occur in the Arctic (Veillette *et al*. 2011a). Thermokarst lakes on permafrost soils are expanding in size and number in certain parts of the Arctic (e.g., Payette *et al*. 2004), but at other locations they have been observed to suddenly drain as a result of thawing and erosion (Smith *et al*. 2005; Jones *et al*. 2011). Thaw lakes appear to have natural cycles of expansion, erosion, drainage, and reformation (Kessler, Plug, and Walter Anthony 2012; van Huissteden *et al*. 2011), which may accelerate under warmer climate conditions.

Shifts in water balance will also give rise to major changes in lake extent and persistence. Over the past 50 years there has been an overall trend of increasing snow precipitation in the Arctic, but with large spatial variability in current and projected future trends (Brown and Mote 2009). Global climate models predict that there will continue to be large differences in precipitation trends among regions, with decreases in snow water equivalent over Scandinavia and Alaska, no change over the boreal forest region, and increased precipitation over northern Siberia (by 15–30%) and the Canadian Arctic Archipelago (Brown and Mote 2009). However, these trends are offset by warmer temperatures in summer and decreased duration of ice cover, both of which favor water loss by evaporation. Lakes at several locations appear to have shifted to a negative net precipitation-evaporation balance, and for some pond waters, this has led to complete drying up, perhaps for the first time in millennia (Smol and Douglas 2007). Many high Arctic wetlands are dependent on perennial snow banks and glaciers, and are vulnerable to the rapid warming of the cryosphere. The potential changes in northern wetlands and lake extent as a result of increased evaporation and potential drainage are a major source of uncertainty for models of methane release from Arctic permafrost (Koven *et al*. 2011).

The loss of ice cover, or increased duration of ice-free conditions has wide-ranging effects on the limnology of northern waters (Vincent, Hobbie, and Laybourn-Parry 2008; Mueller *et al*. 2009; Prowse *et al*. 2011; see also Chapter 3). The increased availability of light for photosynthesis may enhance the annual rate of primary production, and may also result in a vertical extension of the active photosynthetic communities to deeper parts of the water column, thereby allowing a greater proportion of the total volume of the lake to be available for net primary production. An example of this latter effect is given in Antoniades *et al*. (2009) who concluded that the deep layer of photosynthetic sulfur bacteria in high Arctic meromictic Lake A appeared to be more extensive and active under conditions of decreased snow and ice cover. The absence of ice also allows wind-induced mixing that may entrain nutrients from deeper in the water column. For example, during 2008, an unusually warm year, Lake A lost its perennial ice cover (Vincent *et al*. 2009), and the halocline slightly deepened, bringing up nutrients that stimulated phytoplankton production at the base of the mixolimnion (Veillette *et al*. 2011b). Loss of ice may also result in changes in algal community structure, for example an increase in the ratio of planktonic to benthic diatom taxa (Smol 1988).

For phytoplankton communities in the surface waters of Arctic lakes, loss of snow and ice cover may also result in damage by bright ultraviolet radiation (Gareis, Lesack, and Bothwell 2010). Exposure to UV radiation has multiple impacts on phytoplankton cells and communities, although these may be offset by a variety of photoprotection and repair strategies (Vincent and Roy 1993). Model calculations indicate that loss of ice and its overlying snow can have a greater effect on increasing underwater UV exposure than stratospheric ozone depletion (Vincent, Rautio, and Pienitz 2007). In a perennially ice-covered lake in the High Arctic, for example, experimental removal of snow resulted in a thirteenfold increase in the photosynthetically active radiation (PAR) beneath the ice, but also a sixteenfold increase in biological UV exposure (Belzile *et al.* 2001).

Earlier breakup dates for ice cover and increased duration of exposure of the water column to incident sunlight will also result in radiative heating, and the water columns will thereby have more time to heat up. This is particularly important for high latitude lakes where temperatures are often at or below the 3.98 °C critical threshold of maximum density of water. Even modest warming can shift a lake that stratifies during winter but not summer (cold monomictic) to one that stratifies during both seasons, with periods of mixing in fall and spring (dimictic). We have observed, for example, that Char Lake, the classic cold monomictic lake studied during the International Biological Program in the 1970s (Schindler *et al.* 1974) has recently warmed above the maximum density threshold, and is now thermally stratified in summer. For deep windswept lakes that are only weakly stratified and readily mixed by storms during summer (for instance, the western basin of Clearwater Lake, Nunavik, Canada; Milot-Roy and Vincent 1994), the nonlinear decrease in water density with warming may result in less frequent episodes of summer mixing, or even a shift from polymixis to dimixis. Shorter periods of mixing in spring and autumn can be expected in thermokarst ponds, which already have stratified waters for a large portion of the year (Laurion *et al.* 2010).

Increased stratification has a variety of effects, including positive feedbacks via warming of the surface layer, possible increases in phytoplankton and zooplankton growth rates, and an increased propensity to deep water oxygen depletion. Such changes in stratification and mixing have been inferred from fossil diatom records in Lapland, which suggest an increase in biological productivity and a shift in the zooplankton community toward cladocerans (Sorvari, Korhola, and Thompson 2002). Such shifts can also potentially result in the increased retention of contaminants within Arctic food webs (Chételat and Amyot 2009). Additionally, warming can impair coldwater fish oxythermal habitats, especially when combined with eutrophication (Jacobson, Stefan, and Pereira 2010).

2.3 Biogeochemical impacts of climate change

Climate warming affects the biogeochemistry of lake ecosystems, directly by increasing the reaction rates of all chemical and biochemical processes, and indirectly by a variety of effects on water column and catchment processes. As noted above, one such example of the latter is the depletion of hypolimnetic oxygen under conditions of increased stratification, and less exchange of gases between the bottom waters of the lake and the atmosphere during stratified periods, with peak gas emissions at spring melt and autumnal overturn (Striegl *et al.* 2001; Kortelainen *et al.* 2006). Under extreme conditions, and exacerbated by increased nutrient and organic carbon inputs (see below), these bottom waters may be driven to anoxia. This in turn may result in the liberation of inorganic phosphorus, iron and manganese from the sediments, thereby stimulating additional biological production (Wetzel 2001).

The warming of Arctic catchments will have a variety of effects. Firstly, increased thawing of the permafrost results in increased erosion and transport of tundra soils and organic carbon to thermokarst lakes, and thereby increases the microbial production of carbon dioxide and methane (Walter *et al.* 2006; Mazéas, von Fischer, and Rhew 2009; Laurion *et al.* 2010). This source of organic matter results in the formation and release of methane with radiocarbon ages ranging from modern to several thousand years old, reflecting the age of the Holocene soils decomposing in near surface lake sediments (Walter *et al.* 2008). Of greater concern is the mobilization of hundreds of teragrams of permafrost organic carbon when permafrost thaws underneath lakes (see also Chapter 1). The zone of thaw beneath a lake, called a talik or thaw bulb, is an anaerobic environment in which microbes readily decompose organic matter that was locked up in permafrost for tens of thousands of years (Figure 2.2). This thawing results in the rapid production and emission of methane and carbon dioxide predominately in the form of ebullition (bubbling) and with radiocarbon ages of 30 000–43 000 years (Walter *et al.* 2006, 2008). Given the tremendous size of the permafrost carbon pool (~1700 petagrams; Tarnocai *et al.* 2009), which is more than twice the size of the atmospheric carbon pool, permafrost thaw associated with thermokarst lake cycles could result in the release of more than 50 000 teragrams of ^{14}C-depleted methane in the future (Walter, Smith, and Chapin 2007; Walter Anthony 2009). This is more than ten times the amount of methane in the current atmosphere. The release of this potent greenhouse gas from thermokarst lakes sets up a positive feedback cycle in which methane causes global climate warming, which in turn causes permafrost to thaw, and more methane to be formed and released.

Secondly, thawing of the tundra may also release and mobilize inorganic nutrients (see also Chapter 1), previously immobilized in deep permafrost soil horizons, which then stimulate biogeochemical processes in the receiving waters. Thirdly, there is increasing evidence that Arctic warming is leading to increased plant growth across the tundra, with the northward expansion of shrubs and trees (Tape, Sturm, and Racine 2006; Hudson and Henry 2009; Grant *et al.* 2011). Snow over these erect plants has a much lower albedo than snow on tundra, and their expansion over northern landscapes is likely to cause a positive feedback of increased heating, as well as increased soil weathering during higher temperatures, the expansion of root biomass, and increased microbial activity in the rhizosphere (Vincent, Hobbie, and Laybourn-Parry 2008). The appearance and densification of shrubs and trees in the landscape can also result in a substantial increase in terrestrial organic carbon production and its export to lakes as dissolved and particulate organic matter. Paleolimnological studies suggest that these large changes may have occurred in the past as a result of climate-induced tree line migration, resulting in changes in the colored dissolved organic matter (CDOM) content of lakes, and associated shifts in underwater spectral UV and PAR irradiance (Pienitz and Vincent 2000; Saulnier-Talbot, Pienitz, and

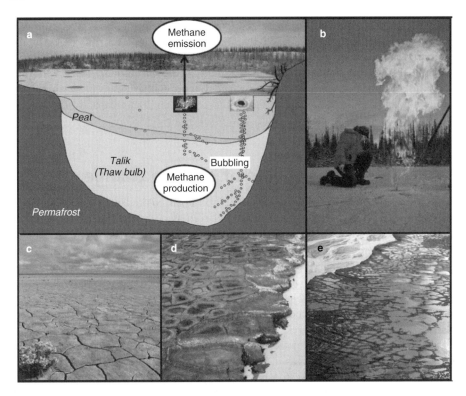

Figure 2.2 Arctic thaw lakes in the changing Arctic. (a) These waterbodies are biogeochemical hotspots on the tundra in which soil and lake organic matter is broken down by microbial activity in the thaw zone beneath the lake, resulting in the liberation of methane and carbon dioxide. Large quantities of these gases are released to the atmosphere via bubbling, which can produce and maintain holes in the ice. Modified from Walter *et al.* (2007). (b) The methane can accumulate as gas pockets beneath the ice, such as here in an Alaska lake where the gas has been vented through a hole made in the ice and then ignited. Photocredit: Todd Paris, November 2009; from Walter Anthony *et al.* (2010). Reproduced with permission. (c) In parts of the Arctic, thaw lakes are expanding in number and size, while in other areas, such as here in the Nettilling Lake region of Baffin Island, landscape erosion has resulted in complete drainage of some waterbodies. Photocredit: Reinhard Pienitz, August 2010. (d) Long-term as well as interannual variations in climate strongly affect the water balance and persistence of lakes on the permafrost. Many of these polygon ponds on Bylot Island, Canada evaporated to dryness in a warm, low precipitation year. Photocredit: Isabelle Laurion, July 2007. (e) The Bylot Island polygon ponds were numerous and extensive during a preceding cool, wet year. Photocredit: Isabelle Laurion, July 2005. (See insert for color representation.)

Vincent 2003). Changes in DOC may also influence the heat budget of lakes and the extent of stratification (Caplanne and Laurion 2008).

The processes described here may lead to a decreased rate of photosynthesis as a result of increased shading by CDOM and terrigenous particles (as in Watanabe *et al.* 2011), although this may be partially offset by increased nutrients and decreased exposure of the phytoplankton to damaging UV radiation. The increased allochthonous input of dissolved and particulate organic carbon by erosion and terrestrial plant growth, in combination with the increased inorganic nutrient supply, is likely to favor increased rates of heterotrophic activity and microbial food web processes in the receiving waters (Sobek *et al.* 2003; Sobek, Tranvik, and Cole, 2005). In combination, this implies that the photosynthesis to respiration ratio in northern lakes will shift downwards in the future, and that these ecosystems will become increasingly net emitters of carbon dioxide. Depending on the acid-neutralizing capacity (alkalinity) of the lake waters, such changes could also result in decreased pH.

2.4   Biological impacts of climate change

The warming of northern ecosystems is likely to impair cold-adapted specialists, such as psychrophilic microbes (Vincent 2010) and cold stenothermal fauna, for example Arctic char (see below). In the marine environment, changes across specific thresholds have led to a complete regime shift in food-web structure (Grebmeier *et al.* 2006). Similar discontinuities may be expected in the future in Arctic freshwater systems, especially with the arrival of invasive generalist species from the South. The local biodiversity of northern lakes may increase in terms of species richness, but at the expense of Arctic endemic species that may be driven to extinction by competition, parasitism or direct thermal stress (Vincent *et al.* 2011).

Paleolimnological analyses of sediment cores from lakes throughout the circumpolar Arctic have shown large changes in the composition of diatom communities over the last century, likely in response to climate change. These floristic changes varied in magnitude and exact timing (Smol *et al.* 2005; Rühland, Paterson, and Smol 2008), and ongoing changes in community structure at all trophic levels are similarly likely to vary greatly among different sectors of the Arctic.

The displacement of native fish in northern lakes is of particular concern to Inuit and First Nations communities, and will be an increasing priority of research and monitoring. For example, a modeling study of the range distribution of smallmouth bass (*Micropterus dolomieu*) showed that it could potentially invade some additional 25 000 northern lakes and, because of its strongly negative effects on other fish, cause the extirpation of four native cyprinid species from these lakes (Sharma *et al.* 2007).

2.5   Human impacts of climate change

Northern lakes provide a variety of key ecosystem services including transport routes, drinking-water supplies, habitats for aquatic wildlife of traditional value to northern communities, and water for industries including hydroelectricity, recreational fishing, eco-tourism and mining. The influence of climate change on water supply and quality is increasingly viewed with concern by Inuit and other indigenous communities (Moquin 2005). Additionally, the warmer temperatures may allow invading species to survive and complete their life cycles, causing the extinction of native biota and serious impairment of traditional hunting and fishing practices (Vincent *et al.* 2011).

Lake and river ice in the north provide winter transport routes that are important to northern indigenous people for access to their traditional hunting and fishing areas, as well as for the heavy transport of goods to remote communities and industries such as mining centers. For example, the Tibbitt to Contwoyto Winter Road in northern Canada, passes over 495 km of frozen tundra, lakes, and rivers, and is estimated to have an economic contribution of more than one billion US$ annually (Prowse *et al.* 2011). These ice roads and traditional routes are now subject to earlier break-up and unseasonal warming, which is reducing their economic value and creating dangerous conditions for freight haulers and northern communities (Ford *et al.* 2008).

Several climate-related effects may influence the future quality of drinking water in the North. Ongoing permafrost degradation may cause a rise in turbidity and dissolved organic matter levels in the raw source waters, which will require vigilance to ensure that chlorination and other disinfection treatments are adjusted appropriately, and that local residents that use the raw water are adequately advised. In a survey of Nunavik Inuit households in 2004, 29% of the consumed water was raw water taken directly from creeks, rivers and lakes (Martin *et al.* 2009). The arrival of new species may also bring with them disease-related problems for drinking water; for example the northward migration and increased population densities of beavers and associated protozoan parasites in eastern Canada (Jarema *et al.* 2009). High-latitude freshwaters are mostly free of toxic bloom-forming cyanobacteria that create a broad spectrum of water quality problems in the temperate zone, but ongoing warming and stratification of Arctic and subarctic lakes combined with climate-related increases in nutrient loading may encourage the development of these taxa (Vincent 2009).

Most northern communities are now equipped with water treatment plants but many do not have water reticulation systems because of permafrost soils and the difficulty of maintaining pipes and flowing water in their harsh winter climates. Throughout many parts of the north, water is delivered daily by truck to houses where it is kept in large tanks. The microbiological cleanliness of storage systems for this water will require increased attention as temperatures warm in the future (Martin *et al.* 2009).

Arctic char (*Salvelinus alpinus*) is a key element of the traditional diet of Inuit and other northern indigenous people, and changes in this and other fish species associated with climate change may have impacts on northern culture and health. Arctic char appears to be especially sensitive to high temperatures relative to other salmonids, and is also the most tolerant of low temperatures (Baroudy and Elliott 1994). With ongoing climate change, Arctic char will be unlikely to survive in the warmer surface and littoral waters of many northern lakes. The arrival of southern species such as Atlantic salmon (*Salmo salar*) and brook trout (*Salvelinus fontinalis*) could reduce or replace Arctic char, and the latter will likely be displaced from some habitats. There may also be climate-induced changes in the migratory behavior of Arctic char, which would potentially result in changes in their productivity and population size distribution (Reist *et al.* 2006; Power, Reist, and Dempson 2008).

Hydroelectricity plays an important role in some northern economies and will require careful attention to climate-related shifts in water supply, specifically the current and future magnitude of changes in precipitation gains and evaporative losses (likely to increase with warmer water temperatures, and longer ice-free conditions), as well as changes in water plant species and density that may influence storage volume and operating protocols. Future changes in reservoir ice conditions may also affect hydroelectric operations (Prowse *et al.* 2011).

The warming climate will also be accompanied by improved access to resources in the Arctic, and thereby increased human activities. There are many examples of severe local pollution effects of human development on northern lakes in the past (e.g., Laperrière *et al.* 2008; Antoniades *et al.* 2011b), and the current expansion of economic and resource extraction activities in the Arctic will require increasing vigilance and appropriate water management strategies to avoid and minimize such impacts in the future.

2.6   Conclusions

As a result of climate change, the northern landscape has now entered a state of rapid transition, and Arctic freshwater ecosystems are beginning to show shifts in their physical, biogeochemical and biological properties. These ongoing changes will affect the ecosystem services that they are able to provide to northern residents, industries and society at large. Many of these impacts are interconnected (Figure 2.1). For example, changes in climate affect water temperature and thereby stratification, which may lead to anoxia, a contraction of habitat for biota requiring highly oxygenated waters, and a deterioration in fishing yield for northern communities. Conversely, ongoing fishing pressure at some locations where food webs are already under thermal stress may hasten the demise of fish stocks and the more rapid establishment of invasive species from the south. Several of these effects may also result in positive feedbacks, at a variety of scales; for example, the local effects of stratification enhancing additional warming of the surface waters of lakes, and the global effects of increasing greenhouse gas emissions from northern waters.

Given that climate is such a powerful agent of change for aquatic ecosystems, it follows that the only effective strategy to minimize harm will be to reduce the rate and endpoint of global warming. This is especially urgent for Arctic ecosystems given that instrumental data records and global circulation models converge on the prediction that it is the highest northern latitudes that will continue to experience the fastest and most extreme increases in temperature. There is a compelling body of evidence that northern environments are already passing across major thresholds of change, and that this is due to the rise in greenhouse gases in the atmosphere caused by human activities. Arctic ice shelves, for example, appear to have broken up in the past (Antoniades *et al.* 2011a), but their current episode of collapse is co-occurring with the collapse of Antarctic ice shelves, implying an unprecedented, synchronized phase of polar deglaciation that would be consistent with human-induced, global climate change (Hodgson 2011). Northern glaciers are currently experiencing attrition at sharply accelerated rates (Gardner *et al.* 2011). Similarly, analyses based on sea ice, climate and ocean proxies imply that the current loss of Arctic sea ice (Perovich and Richter-Menge 2009) is without precedent for 1450 years, and is the result of increased advection of warm Atlantic water into the Arctic basin, also consistent with anthropogenically induced warming (Kinnard *et al.* 2011). The Arctic Ocean is predicted to be seasonally ice-free within decades (Wang and Overland 2009), and this could be a tipping point that triggers widespread degradation of permafrost, with implications for lake water quality, mobilization of permafrost organic carbon, and accelerated methane release.

Although climate mitigation, defined as the reduction of greenhouse gas emissions to the atmosphere, is an urgent priority to avoid dangerous excursions in climate, the unabated year-by-year increases in emission rates imply that adequate control at a planetary level seems unattainable in the short term. Current modeling analyses predict that if emissions continue to increase, a temperature threshold of 2 °C would probably be exceeded over large parts of Eurasia, north Africa and Canada by 2040, and possibly as early as 2030 (Joshi *et al.* 2011). At several locations in the circumpolar north, it appears that these thresholds have already been exceeded (for instance, Hudson Bay, Canada; Bhiry *et al.* 2011). In tandem with global efforts to slow the rates of emission, additional efforts are required at regional and local scales to minimize and manage impacts on natural waters and their surrounding ecosystems.

The most effective approach at the regional scale is that of conservation. High latitude ecosystems have less biodiversity, and therefore less functional redundancy, than those at temperate latitudes and are therefore inherently more sensitive to perturbation (Post *et al.* 2009). The creation of high latitude parks and other conservation zones provides a strategy to reduce the effects of multiple stressors that are superimposed on northern biota in a warming climate and to provide refugia for vulnerable species. These conservation practices may range from local protection from human activities, for example around municipal water supplies, to the creation of large-scale wilderness zones to preserve biological communities and entire ecosystems that are at risk. These northern parks and other protected areas are likely to come under increasing economic pressure as the drive to extract oil, gas and mineral resources from the Arctic continues to accelerate.

At the local scale, the ongoing effects of climate change must be addressed by adaptation strategies. The first requirement is an adequate surveillance system to monitor, communicate, and respond to changes. Strategies to address the increasing safety issues for ice roads include reductions in maximum allowable loads to be transported, modifications to the methods used for ice road construction, and rescheduling to concentrate transport during the coldest part of winter (Prowse *et al.* 2009). Satellite remote sensing (RADARSAT) has been implemented in northern Canada to provide timely warnings of unsafe ice conditions (Gauthier *et al.* 2010), and offers considerable potential throughout the circumpolar north in the future. For drinking water supplies, adequate surveillance and advisories are also critical to ensure water quality and safety. These essential resources require the development of integrated freshwater management plans, which include consideration of alternate water sources as traditional supplies change in quantity or quality. The likely shifting of thermal conditions in northern lakes and reservoirs to those conducive to growth by noxious cyanobacteria will also require ongoing attention, with emphasis on catchment control of phosphorus and other nutrient sources. For hydroelectric reservoirs, shifting ice conditions will have both positive and negative effects, and may require adaptive changes in operating procedures, with attention to minimize deleterious impacts associated with ice jams and ice breakup downstream of the spillway (Prowse *et al.* 2011). Fisheries management plans will also need to be adapted to the changes in migration and productivity of northern fish populations with ongoing climate change. The potential arrival of invasive species will create particularly challenging problems for ecosystem management, including fisheries, and will require increased surveillance and prevention measures as road access to the north continues to develop, along with increased industrial, eco-tourism and recreational boating activities.

In summary, northern aquatic ecosystems are a rich resource of enormous cultural, economic and ecological value. These waters are now undergoing rapid changes in their physical, biogeochemical and biological properties, and their abilities to provide ecosystem services are beginning to be compromised. These changes are likely to continue and to be amplified in the foreseeable future. Local adaptation strategies need to be put in place, and regional conservation and management plans are an important priority to reduce the effects of multiple stressors. At the planetary scale, ongoing efforts are required to reduce greenhouse gas emissions and slow global warming, allowing more time for adaptation. In all of these respects, northern ecosystems are an early warning system of major change that will occur throughout the global environment, and they are a natural laboratory in which to develop appropriate strategies to manage the world's precious, and increasingly vulnerable, freshwater resources.

2.7   Acknowledgments

We thank the editors for the opportunity to contribute to this volume and for their assistance and encouragement. We also acknowledge the funding and logistics agencies that support our northern research, including the Natural Sciences and Engineering Research Council, the National Science Foundation, the Fonds de recherche du Québec—Nature et technologies, the Network of Centres of Excellence program ArcticNet, the Polar Continental Shelf Project of Natural Resources Canada, Aboriginal Affairs and Northern Development Canada, and the Canada Research Chairs program.

References

ACIA (2005) *Arctic Climate Impact Assessment*, Cambridge University Press, Cambridge.

Antoniades, D., Francus, P., Pienitz, R., *et al.* (2011a) Holocene dynamics of the Arctic's largest ice shelf. *Proc. Natl. Acad. Sci. USA*, **108**, 18899–904.

Antoniades, D., Michelutti, N., Quinlan, R., *et al.* (2011b) Cultural eutrophication, anoxia, and ecosystem recovery in Meretta Lake, High Arctic Canada. *Limnol. Oceanogr.*, **56**, 639–50.

Antoniades, D., Veillette, J., Martineau, M.-J., *et al.* (2009) Bacterial dominance of phototrophic communities in a High Arctic lake and its implications for paleoclimate analysis. *Polar Sci.*, **3**, 147–61.

Baroudy, E. and Elliott, J.M. (1994) The critical thermal limits for juvenile Arctic charr, *Salvelinus alpinus*. *J. Fish Biol.*, **45**, 1041–53, doi: 10.1111/j.1095-8649.1994.tb01071.x.

Belzile, C., Vincent, W.F., Gibson, J.A.E., and Van Hove, P. (2001) Bio-optical characteristics of the snow, ice, and water column of a perennially ice-covered lake in the High Arctic. *Can. J. Fish. Aquat. Sci.*, **58**, 2405–2418, doi: 10.1139/cjfas-58-12-2405.

Bhiry, N., Delwaide, A., Allard, M., *et al.* (2011) Environmental change in the Great Whale River region, Hudson Bay: Five decades of multidisciplinary research by Centre d'études nordiques (CEN). *Ecoscience*, **18**, 182–203, doi: 10.2980/18-3-3469.

Brown, R.D. and Mote, P.W. (2009) The response of Northern Hemisphere snow cover to a changing climate. *J. Clim.*, **22**, 2124–45.

Caplanne, S. and Laurion, I. (2008) Effect of chromophoric dissolved organic matter on epilimnetic stratification in lakes. *Aquat. Sci.*, **70**, 123–33, doi: 10.1007/s00027-007-7006-0.

Chételat, J. and Amyot, M. (2009) Elevated methylmercury in High Arctic *Daphnia* and the role of productivity in controlling their distribution. *Global Change Biol.*, **15**, 706–18, doi: 10.1111/j.1365-2486.2008.01729.x.

Ford, J.D., Pearce, T., Gilligan, J., *et al.* (2008) Climate change and hazards associated with ice use in northern Canada. *Arct. Ant. Alp. Res.*, **40**, 647–59.

Gardner, A.S., Moholdt, G., Wouters, B., *et al.* (2011) Sharply increased mass loss from glaciers and ice caps in the Canadian Arctic Archipelago. *Nature*, **473**, 357–60.

Gareis, J.A.L., Lesack, L.F.W., and Bothwell, M.L. (2010) Attenuation of in situ UV radiation in Mackenzie Delta lakes with varying dissolved organic matter compositions. *Water Resour. Res.*, **46**, W09516, doi: 10.1029/2009WR008747.

Gauthier, Y., Tremblay, M., Bernier, M., and Furgal, C. (2010) Adaptation of a radar-based river ice mapping technology to the Nunavik context. *Can. J. Remote Sensing*, **36**, S168–S185.

Grant, R.F., Humphreys, E.R., Lafleur, P.M., and Dimitrov, D.D. (2011) Ecological controls on net ecosystem productivity of a mesic arctic tundra under current and future climates. *J. Geophys. Res.*, **116**, G01031, doi: 10.1029/2010JG001555.

Grebmeier, J.M., Overland, J.E., Moore, S.E., *et al.* (2006) A major ecosystem shift in the northern Bering Sea. *Science*, **311**, 1461–1464.

Grosse, G., Romanovsky, V.E., Jorgenson, T., *et al.* (2011) Vulnerability and feedbacks of permafrost to climate change. *Eos Trans. AGU*, **92**, 73–4.

Hodgson, D.A. (2011) First synchronous retreat of ice shelves marks a new phase of polar deglaciation. *Proc. Natl. Acad. Sci. USA*, **108**, 18859–60.

Hudson, J.M.G. and Henry, G.H.R. (2009) Increased plant biomass in a High Arctic heath community from 1981 to 2008. *Ecology*, **90**, 2657–663.

IPCC (2007) *Intergovernmental Panel on Climate Change, Fourth Assessment Report (AR4) of the United Nations*, IPCC, Geneva.

Jacobson, P.C., Stefan, H.G., and Pereira, D.L. (2010) Coldwater fish oxythermal habitat in Minnesota lakes: influence of total phosphorus, July air temperature, and relative depth. *Can. J. Fish. Aquat. Sci.*, **67**, 2002–13.

Jarema, S.I., Samson, J., McGill, B.J., and Humphries, M.M. (2009) Variation in abundance across a species' range predicts climate change responses in the range interior will exceed those at the edge: a case study with North American beaver. *Global Change Biol.*, **15**, 508–22.

Jones, B., Grosse, G., Arp, C.D., *et al.* (2011) Modern thermokarst lake dynamics in the continuous permafrost zone, northern Seward Peninsula, Alaska. *J. Geophys Res.*, **116**, G00M03, doi:10.1029/2011JG001666.

Jorgenson, M.T. and Osterkamp, T.E. (2005) Response of boreal ecosystems to varying modes of permafrost degradation. *Can. J. Forest Res.*, **35**, 2100–11.

Joshi, M., Hawkins, E., Sutton, R., *et al.* (2011) Projections of when temperature change will exceed 2 °C above pre-industrial levels. *Nat. Clim. Change*, **1**, 407–12.

Kessler, M.A., Plug, L.J., and Walter Anthony, K.M. (2012) Simulating the decadal to millennial scale dynamics of morphology and carbon mobilization of a thermokarst lake in N.W. Alaska. *J. Geophys. Res.*, **117**, G00M06, doi: 10.1029/2011JG001796.

Kinnard, C., Zdanowicz, C.M., Fisher, D.A., *et al.* (2011) Reconstructed changes in Arctic sea ice over the past 1450 years. *Nature*, **479**, 509–12.

Köck, G., Muir, D., Yang, F., *et al.* (2012) Bathymetry and sediment geochemistry of Lake Hazen (Quttinirpaaq National Park, Ellesmere Island, Nunavut). *Arctic*, **65**, 56–66.

Kortelainen, P., Rantakari, M., Huttunen, J.T., *et al.* (2006) Sediment respiration and lake trophic state are important predictors of large CO_2 evasion from small boreal lakes. *Global Change Biol.*, **12**, 1554–67.

Koven, C.D., Ringeval, B., Friedlingstein, P., *et al.* (2011) Permafrost carbon-climate feedbacks accelerate global warming. *Proc. Natl. Acad. Sci. USA*, **108**, 14769–74.

Laperrière, L., Fallu, M.-A., Hausmann, S., *et al.* (2008) Paleolimnological evidence of mining and demographic impacts on Lac Dauriat, Schefferville (subarctic Québec, Canada). *J. Paleolimnol.*, **40**, 309–24.

Laurion, I., Vincent, W.F., Retamal, L., *et al.* (2010) Variability in greenhouse gas emissions from permafrost thaw ponds. *Limnol. Oceanogr.*, **55**, 115–33.

Lehner B, and Döll, P. (2004) Development and validation of a global database of lakes, reservoirs and wetlands. *J. Hydrol.*, **296**, 1–22.

Martin D., Belanger, D., Gosselin, P., *et al.* (2009) Drinking water and potential threats to human health in Nunavik: adaptation strategies under climate change conditions. *Arctic*, **60**, 195–202.

Mazéas, O., von Fischer, J.C., and Rhew, R.C. (2009) Impact of terrestrial carbon input on methane emissions from an Alaskan Arctic lake. *Geophys. Res. Lett.*, **36**, L18501, doi: 10.1029/2009GL039861.

Melles, M., Brigham-Grette, J., Glushkova, O.Y., *et al.* (2007) Sedimentary geochemistry of a pilot core from Lake El'gygytgyn—a sensitive record of climate variability in the East Siberian Arctic during the past three climate cycles. *J. Paleolimnol.*, **37**, 89–104.

Milot-Roy, V. and Vincent, W.F. (1994) Ultraviolet radiation effects on photosynthesis: the importance of near-surface thermoclines in a subarctic lake. *Arch. Hydrobiol.*, **43**, 171–84.

Moquin, H. (2005) Freshwater in Inuit communities. *ITK Environment Bulletin*, **3**, 4–8.

Mueller, D.R., Van Hove, P., Antoniades, D., *et al.* (2009) High Arctic lakes as sentinel ecosystems: cascading regime shifts in climate, ice-cover and mixing. *Limnol. Oceanogr.*, **54**, 2371–85.

Mueller, D. R., Vincent, W.F., and Jeffries, M.O. (2003) Break-up of the largest Arctic ice shelf and associated loss of an epishelf lake. *Geophys. Res. Lett.*, **30**, 2031. doi: 10.1029/2003GL017931.

Oliver, D.R. (1964) A limnological investigation of a large Arctic lake: Nettilling Lake, Baffin Island. *Arctic*, **17**, 65–144.

Payette, S., Delwaide, A., Caccianiga, M., and Beauchemin, M. (2004) Accelerated thawing of subarctic peatland permafrost over the last 50 years. *Geophys. Res. Lett.*, **31**, L18208.

Perovich, D.K. and Richter-Menge, J.A. (2009) Loss of sea ice in the Arctic. *Ann. Rev. Mar. Sci.*, **1**, 417–41.

Pienitz R., Doran, P.T., and Lamoureux, S.F. (2008) Origin and geomorphology of lakes in the polar regions, in *Polar Lakes and Rivers—Limnology of Arctic and Antarctic Aquatic Ecosystems* (eds. W.F. Vincent and J. Laybourn-Parry), Oxford University Press, Oxford, pp. 25–41.

Pienitz, R., Saulnier-Talbot, É., Fallu, M.-A., *et al.* (2004) Long-term climate stability in the Québec-Labrador (Canada) region: Evidence from paleolimnological studies. Proceedings of the Arctic Climate Impact Assessment (ACIA), Arctic Monitoring and Assessment Programme, (AMAP), Reykjavik, Iceland, 9–12 November.

Pienitz, R. and Vincent, W.F. (2000) Effect of climate change relative to ozone depletion on UV exposure in subarctic lakes. *Nature*, **404**, 484–7.

Post, E., Forchhammer, M.C., Bret-Harte, M.S., *et al.* (2009) Ecological dynamics across the arctic associated with recent climate change. *Science*, **325**; 1355–8.

Power, M., Reist, J.D., and Dempson, J.B. (2008) Fish in high-latitude lakes, in *Polar Lakes and Rivers—Limnology of Arctic and Antarctic Aquatic Ecosystems* (eds. W.F. Vincent and J. Laybourn-Parry), Oxford University Press, Oxford, pp. 249–69.

Prowse, T., Alfredsen, K., Beltaos, S., *et al.* (2011) Effects of changes in arctic lake and river ice. *Ambio*, **40**, 63–74.

Prowse, T., Furgal, C., Chouinard, R., *et al.* (2009) Implications of climate change for economic development in northern Canada: Energy, resource, and transportation sectors. *Ambio*, **38**, 272–81.

Rautio, M., Dufresne, F., Laurion, I., *et al.* (2011) Shallow freshwater ecosystems of the circumpolar Arctic. *Écoscience*, **18**, 204–22.

Reist, J.D., Wrona, F.J., Prowse, T.D., *et al.* (2006) An overview of the effects of climate change on selected Arctic freshwater and anadromous fishes. *Ambio*, **35**, 381–7.

Robarts, R.D., Zhulidov, A.V., Zhulidova, O.V., *et al.* (1999) Biogeography and limnology of the Lake Taymyr-wetland system, Russian Arctic: an ecological synthesis. *Monograph. Stud.*, **121**, 159–200.

Rühland, K., Paterson, A.M., and Smol, J.P. (2008) Hemispheric-scale patterns of climate-related shifts in planktonic diatoms from North American and European lakes. *Global Change Biol.*, **14**, 2740–54.

Saulnier-Talbot, É., Pienitz, R., and Vincent, W.F. (2003) Holocene lake succession and palaeo-optics of a subarctic lake, northern Québec, Canada. *The Holocene*, **13**, 517–26.

Schindler, D.W., Welch, H.E., Kalff, J., *et al.* (1974) Physical and chemical limnology of Char Lake, Cornwallis Island (75 °N lat). *J. Fish. Res. Bd. Can.*, **31**, 585–607.

Sharma, S., Jackson, D.A., Minns, C.K., and Shuter, B.J. (2007) Will northern fish populations be in hot water because of climate change? *Global Change Biol.*, **13**, 2052–64.

Smith, L.C., Sheng, Y., and MacDonald, G.M. (2007) A first pan-arctic assessment of the influence of glaciation, permafrost, topography and peatlands on Northern Hemisphere lake distribution. *Perm. Periglac. Process.*, **18**, 201–8.

Smith, L.C., Sheng, Y., MacDonald, G.M., and Hinzman, L.D. (2005) Disappearing Arctic lakes. *Science*, **308**, 1429.

Smol, J.P. (1988) Paleoclimate proxy data from freshwater arctic diatoms. *Verh. Internat. Verein. Limnol.*, **23**, 837–44.

Smol, J.P., and Douglas, M.S.V. (2007) Crossing the final ecological threshold in high Arctic ponds. *Proc. Natl. Acad. Sci. USA*, **104**, 12395–7.

Smol, J.P., Wolfe, A.P., Birks, H.J.B., *et al.* (2005) Climate-driven regime shifts in the biological communities of arctic lakes. *Proc. Natl. Acad. Sci. USA*, **102**, 4397–402.

Sobek, S., Algesten, G., Bergström, A.-K., *et al.* (2003) The catchment and climate regulation of pCO_2 in boreal lakes. *Global Change Biol.*, **9**, 630–41.

Sobek, S., Tranvik, L.J., and Cole, J.J. (2005) Temperature independence of carbon dioxide supersaturation in global lakes. *Glob. Biogeochem. Cycles*, **19**, GB2003, doi: 10.1029/2004GB002264.

Sorvari, S., Korhola, A., and Thompson, R. (2002) Lake diatom response to recent Arctic warming in Finnish Lapland. *Glob. Change Biol.*, **8**, 153–63.

Striegl, R.G., Kortelainen, P., Chanton, J.P., *et al.* (2001) Carbon dioxide partial pressure and ^{13}C content of north temperate and boreal lakes at spring ice melt. *Limnol. Oceanogr.*, **46**, 941–5.

Tape, K., Sturm, M., and Racine, C. (2006) The evidence for shrub expansion in Northern Alaska and the Pan-Arctic. *Global Change Biol.*, **12**, 686–702.

Tarnocai, C., Canadell, J.G., Schuur, E.A.G., *et al.* (2009) Soil organic carbon pools in circumpolar permafrost region. *Global Biogeochem. Cycles*, **23**, GB2023, doi: 10.1029/2008GB003327.

van Huissteden, J., Berrittella, C., Parmentier, F.J.W., *et al.* (2011) Methane emissions from permafrost thaw lakes limited by lake drainage. *Nature Climate Change*, **1**, 1–5.

Veillette, J., Lovejoy, C., Potvin, M., *et al.* (2011a) Milne Fiord epishelf lake: a coastal Arctic ecosystem vulnerable to climate change. *Ecoscience*, **18**, 304–16.

Veillette, J., Martineau, M.-J., Antoniades, D. *et al.* (2011b) Effects of loss of perennial lake ice on mixing and phytoplankton dynamics: Insights from High Arctic Canada. *Annals Glaciol.*, **51**, 56–70.

Veillette, J., Mueller, D.R., Antoniades, D., and Vincent, W.F. (2008) Arctic epishelf lakes as sentinel ecosystems: past, present and future. *J. Geophys. Res.—Biogeosc.*, **113**, G04014, doi: 10.1029/2008JG000730.

Vincent, W.F. (2009) Effects of climate change on lakes, in *Encyclopedia of Inland Waters* (ed. G.E. Likens), Elsevier, Oxford, vol. 3, pp. 55–60.

Vincent, W.F. (2010) Microbial ecosystem responses to rapid climate change in the Arctic. *ISME J.*, **4**, 1089–91.

Vincent, W.F., Callaghan, T.V., Dahl-Jensen, D., *et al.* (2011) Ecological implications of changes in the Arctic cryosphere. *Ambio*, **40**, 87–99.

Vincent, W.F., Hobbie, J.E., and Laybourn-Parry, J. (2008) Introduction to the limnology of high latitude lake and river ecosystems, in *Polar Lakes and Rivers—Limnology of Arctic and Antarctic Aquatic Ecosystems* (eds. W.F. Vincent and J. Laybourn-Parry), Oxford University Press, Oxford, pp. 1–23.

Vincent, W.F., Rautio, M., and Pienitz, R. (2007) Climate control of underwater UV exposure in polar and alpine aquatic ecosystems, in *Arctic Alpine Ecosystems and People in a*

Changing Environment (eds. J.B. Orbaek, R. Kallenborn, I. Tombre, E., *et al.*), Springer, Berlin, pp. 227–49.

Vincent, W.F. and Roy, S. (1993) Solar UV-B effects on aquatic primary production: damage, repair and recovery. *Environ. Rev.*, **1**, 1–12.

Vincent, W.F., Whyte, L.G., Lovejoy, C., *et al.* (2009) Arctic microbial ecosystems and impacts of extreme warming during the International Polar Year. *Polar Sci.*, **3**, 171–80.

Walter, K.M., Chanton, J.P., Chapin III,, F.S., *et al.* (2008) Methane production and bubble emissions from arctic lakes: isotopic implications for source pathways and ages. *J. Geophys. Res.*, **113**, doi: 10.1029/2007JG000569.

Walter, K.M., Smith, L.C., and Chapin III., F.S. (2007) Methane bubbling from northern lakes: present and future contribution to the global CH_4 budget. *Phil. Trans. R. Soc. A.*, **365**, 1657–76.

Walter, K.M., Zimov, S.A., Chanton, J.P., *et al.* (2006) Methane bubbling from Siberian thaw lakes as a positive feedback to climate warming. *Nature*, **443**, 71–5.

Walter Anthony, K.M. (2009) Methane: a menace surfaces. *Scient. Amer.*, **301**, 68–75.

Walter Anthony, K.M., Vas, D.A., Brosius, L., *et al.* (2010) Estimating methane emissions from northern lakes using ice-bubble surveys. *Limnol. Oceanogr. Meth.*, **8**, 592–609.

Wang, M., and Overland, J.E. (2009) A sea ice free summer Arctic within 30 years? *Geophys. Res. Lett.*, **36**, L07502; doi: 10.1029/2009GL037820.

Watanabe, S., Laurion, I., Pienitz, R., *et al.* (2011) Optical diversity of thaw lakes in discontinuous permafrost: a model system for water color analysis. *J. Geophys. Res.—Biogeosci.*, **116**, G02003, doi:10.1029/2010JG001380.

Wetzel, R.G. (2001) *Limnology, Lake and River Ecosystems*, 3rd edn. Elsevier Science, San Diego CA.

3

Trends in Hydrological and Hydrochemical Processes in Lake Baikal under Conditions of Modern Climate Change

Michael N. Shimaraev and Valentina M. Domysheva
Limnological Institute of the Siberian Branch of the Russian Academy of Sciences, Irkutsk, Russia

3.1 Introduction

Lake Baikal, situated at 456 m above sea level in East Siberia, is the deepest (1642 m) and largest (23 615 km³) (Table 3.1) oligotrophic freshwater lake on Earth. The lake extends for about 636 km from southwest to northeast, with an average width of 49 km. It is divided by underwater ridges into three main basins: the southern basin (maximum depth 1461 m), the central basin (1642 m) and the northern basin (904 m) (Table 3.1). The climate of the surrounding territory is sharply continental with severe winters and moderately warm summers. The huge thermal capacity of the lake affects local climate within the lake depression. In 1996, Lake Baikal was declared a United Nations World Heritage Site.

Global climatic warming in the twentieth century (IPCC: Climate Change 1996) induced changes in hydrological conditions and, as a result, impacted biological processes and water quality in lake ecosystems in Europe, Asia, North America and Africa (Magnuson *et al.* 1997; Filatov 1998; Scheffer *et al.* 2001; Gronskaya *et al.* 2002; Kumagai *et al.* 2003; O'Reilly *et al.* 2003; Anneville *et al.* 2007).

Earlier, significant changes in climate and some hydrological processes occurring in Lake Baikal, which is situated almost in the center of Eurasia, were recorded over different time intervals (Verbolov *et al.* 1965; Afanasyev 1967; Shimaraev

Climatic Change and Global Warming of Inland Waters: Impacts and Mitigation for Ecosystems and Societies,
First Edition. Edited by Charles R. Goldman, Michio Kumagai and Richard D. Robarts.
© 2013 John Wiley & Sons, Ltd. Published 2013 by John Wiley & Sons, Ltd.

Table 3.1   Morphometric data of Lake Baikal (Data from Sherstyankin *et al.* 2003).

	Southern basin	Central basin	Northern basin	Total
Maximal depth (m)	1461	1642	904	–
Mean depth (m)	853.4	856.7	598.4	–
Surface area (km^2)	7432	10 600	13 690	31 722
Volume (km^3)	6342.7	9080.6	8192.1	23 615.4

et al. 1994). Data from instrumental observations of air temperature (since 1896) and ice phenomena in southern Baikal (since 1868), as well as information on the annual water inflow into Lake Baikal (since 1901), allowed us to follow climatic influences on hydrological processes for the past 100 years (Shimaraev *et al.* 1994; Shimaraev and Domysheva 2002; Shimaraev *et al.* 2002). Observations of surface water temperature at shore sites (since 1940) and temperature of the entire water column of the lake (since 1971) were less continuous.

3.2   Air temperature

The general trend of air-temperature changes in the Baikal area (Figure 3.1) matched those of global temperature variation: an increase beginning in the late 1910s to the middle of the century, a decrease by the early 1970s, and a new (more significant)

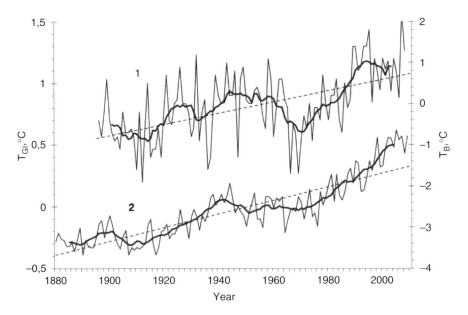

Figure 3.1   The current and 11-year running average annual air temperature air at Babushkin Station, southern Baikal (T_B, 1), global temperature (T_G, 2) and trends.

increase toward the end of the century. The positive trend of annual temperatures in the lake area ($1.2\,°C\ ka^{-1}$) was twice as high as the average trend for the Earth ($0.6\,°C\ ka^{-1}$). Statistical analysis has revealed short-term (2–7 years) and long-term intrasecular (about 20 years) cycles with pronounced increasing and decreasing phases. In the twentieth century, two complete cycles (1912–1936 and 1937–1969) were observed with odd phases of two incomplete cycles (a decreasing phase from 1896 to 1911 and an increasing phase beginning from 1970). The increasing temperature phase at the end of the century was characterized by an anomalously significant duration (nearly 25 years) and an air-temperature rise (by $2.1\,°C$). A tendency for annual air temperature to decrease, despite high values, was observed after 1995.

A significant linear trend of increasing air temperature was characteristic of all seasons from 1896 to 2008 (Figure 3.2): 1.9, 1.5, 1.1, and $0.66\,°C$ per 100 years in winter, spring, summer, and autumn, respectively. The maximum temperature rise over the century (by $2.1–2.2\,°C$) was recorded in December and January, whereas the minimum one ($0.1–0.5\,°C$) was in August, September and October. An intrasecular cyclicity was specific depending on the season. Temperatures in autumn and winter had a short decreasing phase from the 1950s to the early 1970s and a continuous rising phase to the mid 1990s. Thereafter temperature decreased and rose again by 2007–2008. In spring and summer the temperature decrease was observed from the 1940–1950s to the early 1980s, and the phase of its next rise ended only at the beginning of the twenty-first century (Figure 3.2). Considering intrasecular fluctuations as a natural cyclicity, one can conclude that significant warming from 1970–1995 resulted from the positive phase in the current intrasecular climate cycle compared to the "secular" trend caused by natural and anthropogenic factors.

3.3   Surface water temperature (T_W)

The response to global warming in the twentieth century was a temperature rise in the surface layers of Lake Baikal that occurred simultaneously with the air temperature rise (Shimaraev and Domysheva 2002; Troitskaya, Shimaraev, and Tsekhanovsky 2003). In southern Baikal (Listvyanka Settlement), average surface water temperature for May–September decreased insignificantly in the 1950s–1970s, and then rose abruptly by the mid-1990s (Figure 3.3). The temperature in other areas of the lake changed in a similar way. It rose faster in central and northern Baikal than in southern Baikal (by $0.54–0.60\,°C$ and $0.25–0.35\,°C$ per 10 years, respectively). The temperature of the last warmest decade in the twentieth century exceeded that of the coldest decade (1964–1975) by $0.9–1.5\,°C$ in the southern basin and by $1.8–2\,°C$ in the central and northern basins of the lake. The periods with water temperatures above 1, 4 and $10\,°C$ became longer by 0.4, 0.3 and 0.17 days per 10 years, respectively (Figure 3.4). Spring homothermy came earlier and autumn came later, therefore, the period with direct temperature stratification was longer. In past years there were some cases when the surface temperature reached $18–20\,°C$ even in deep areas of the lake.

Data from systematic thermal surveys of the entire lake were used for the analysis of temperature changes of the water column using a constant sampling grid since 1972. From 1972–1992 these surveys were carried out in June–December. In March

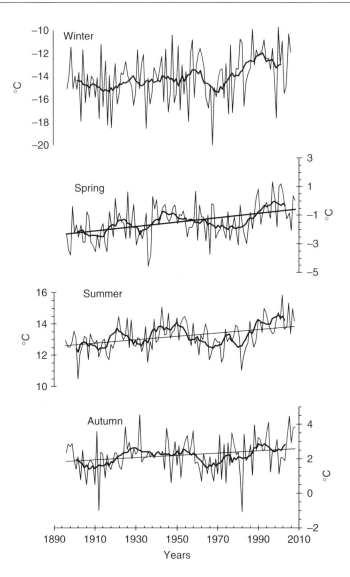

Figure 3.2 The current and 11-year seasonal running average values and trends of air temperature at Babushkin Station.

they were performed every year in southern Baikal, from 1972–1983 in northern Baikal, and in central Baikal only in 1972. Temperature was measured with a mercury thermometer to $\pm 2 \times 10^{-2}\,°C$ every 5–50 m down to 300 m and then every 100 m to the bottom. Between 1995 and 2009 in every year except 1996, one survey was carried out in June–September using a SBE-25 thermistor probe (to $\pm 2.0 \times 10^{-3}\,°C$, resolution $1.0 \times 10^{-4}\,°C$) and at depth intervals of (± 30–60 cm). The data obtained from 1972–1992 and 1995–2007 were considered as two separate sequences due to discrepancies in precision of the measurements (Shimaraev *et al.* 2009).

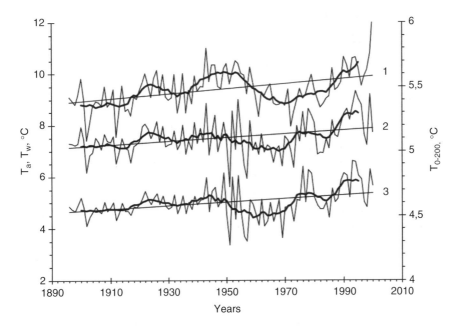

Figure 3.3 Average temperature of air (1), surface water (2) and calculated average temperature of the 0–200 m layer (3) in May–October (Southern Baikal).

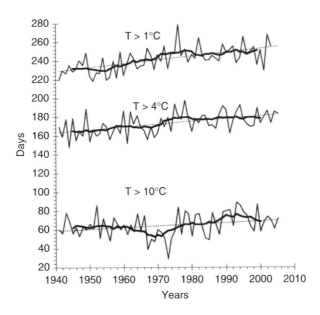

Figure 3.4 Duration of periods with surface water temperatures above 1, 4 and 10 °C.

3.4   Water temperature of the Lake Baikal upper zone

The response of waters of the upper (to 300 m) and deep zones of Lake Baikal to climatic changes was different due to different mechanisms of heat transfer (Shimaraev 1996). In spring and autumn, free temperature convection in the upper zone (Vereshchagin 1936; Weis, Carmack, and Koropalov, 1991; Shimaraev and Granin, 1991) together with dynamic processes provided relatively fast heat exchange between the water and atmosphere. Therefore, long-term changes of average water temperature in the 0–300 m layer during a warm period (May–September) and surface water temperature have to be similar. The relationship between average temperature of the 0–300 m layer zone and T_w is presented by Verbolov's empirical model. The application of this model, as well as the dependence between T_w and air temperature makes it possible to assess thermal characteristics of southern Baikal for those years where there was no monitoring of surface water temperature (prior to 1941). The results of the calculation of surface water temperature (to 1941) and average temperature of the 0–300 m layer demonstrated a positive long-term trend in their changes during a warm period (Figure 3.3).

The results of observations for 1972–1992 (a period of the most significant warming) were used for reconstructing real temperature changes in the pelagic zone of Lake Baikal. Warming caused a noticeable rise of temperature in the 0–100 m and 100–200 m layers (in southern Baikal also in the 200–300 m layer). Average temperature in the 0–300 m layer increased by 0.05–0.10 °C in June–September. Thus, warming caused a temperature rise in the upper layers of the lake, the lower boundary of which corresponded to the depth of active seasonal mixing under the influence of wind and free temperature convection (Shimaraev and Granin 1991; Weiss *et al.* 1991).

3.5   Deep water zone temperature

Temperature of deep waters (below 300–500 m) was affected by two processes: turbulent heat transfer from upper layers and deep temperature convection in spring and early winter, which caused cooling of water by intrusions from the upper lake layers. The activity of these differently directed heat fluxes had little to do with heat exchange on the surface and depended mainly on the dynamic influence of the atmosphere on deeper waters.

Between 1972 and 1991, temperature fluctuations were low in the deep zone (<0.01–0.001 °C per 10 years) and statistically unreliable (Table 3.2). Inter-annual temperature fluctuations (6–10 years) in deep layers with an amplitude of about 0.05–0.07 °C were more evident (Figure 3.5). They were caused mainly by the activity of deep convection depending on the wind velocity in months before ice formation (December–January) and after ice breakup (May) (Shimaraev *et al.* 2009). In 1995–2009, small but significant temperature changes were recorded mainly in the deepest layers of some basins (Table 3.2). In southern Baikal, temperature decreased in the 100 near-bottom layer, whereas it increased in central Baikal deeper than 600 m and in northern Baikal near the bottom (Figure 3.5). These changes were not attributable to climate change but to changes in the ratio between turbulent heat

Table 3.2 Water temperature changes (upper number—$°$C/ten years) and coefficients of determination for linear regressions of these changes over time in deep water layers of Lake Baikal in 1972–1992 and 1995–2009.

	Southern Baikal			Central Baikal			Northern Baikal	
Layers (m)	**1972–1992**	**1995–2007**	**Layers (m)**	**1972–1992**	**1995–2007**	**Layers (m)**	**1972–1992**	**1995–2007**
200–400	0.002 $R^2 = 0.004$	−0.016 $R^2 = 0.098$	200–400	−0.01 $R^2 = 0.042$	0.006 $R^2 = 0.008$	200–400	−0.001 $R^2 = 0,001$	−0.006 $R^2 = 0.037$
400–600	−0.007 $R^2 = 0.064$	−0.9 $R^2 = 0.143$	400–600	−0.008 $R^2 = 0.047$	−0.007 $R^2 = 0.051$	400–600	0.001 $R^2 = 0.002$	−0.006 $R^2 = 0.06$
600–1200	$8 \cdot 10^{-4}$ $R^2 = 0.002$	**−0.012** $\mathbf{R^2 = 0.347}$	600–1400	$-4 \cdot 10^{-3}$ $R^2 = 0.017$	**0.01** $\mathbf{R^2 = 0.4}$	500–700	$1 \cdot 10^{-4}$ $R^2 = 1 \cdot 10^{-5}$	−0.001 $R^2 = 0.003$
1200–1400	0.002 $R^2 = 0.014.$	**−0.02** $\mathbf{R^2 = 0.370}$	1400–1600	0.007 $R^2 = 0.045$	**0.028** $\mathbf{R^2 = 0.467}$	700–900	0.003 $R^2 = 0.010$	0.006 $R^2 = 0.153$
200–100 from bottom		**−0.015** $\mathbf{R^2 = 0.482}$			**0.026** $\mathbf{R^2 = 0.755}$			0.002 $R^2 = 0.019$
100 above bottom		**−0.026** $\mathbf{R^2 = 0.318}$			0.028 $R^2 = 0.259$			0.008 $R^2 = 0.15$
5 m above bottom		**−0.035** $\mathbf{R^2 = 0.293}$			0.03 $R^2 = 0.079$			0.014 $R^2 = 0.073$

Bold font indicates $<0.05 - <0.01$.

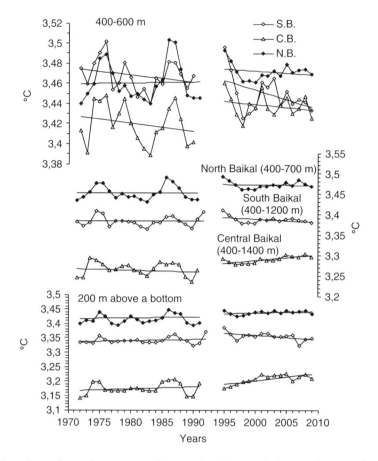

Figure 3.5   June–September average of temperature in separate layers of deep basins in Lake Baikal in 1972–1992 and temperature of these layers based on single measurements during the period June–September in 1995–2009. The data for separate basins were derived by averaging measurements at 5–7 constant deep-water stations.

flow and intrusive flows of cold waters. Calculations of the quantitative effect of cold near-bottom intrusions in some basins showed that in 1993–2009 this process occurred more often and more actively in southern Baikal than in other parts of the lake. This is the reason for a tendency of temperature differences between basins (Shimaraev *et al.* 2011). For 1972–2009, no significant tendencies were observed in long-term temperature changes in deep layers of Lake Baikal that were characteristic of waters of the upper zone. Thus, differences in mechanisms of heat transfer affected the response of waters of the upper (up to 300 m) and deep water zones of Lake Baikal to climatic changes (Shimaraev 1996). Warming for the last three decades of the twentieth century that caused the temperature increases of the surface and waters of the upper zone of Lake Baikal did not affect temperature of the deep zone. It means that water warming due to the heat transfer from upper layers and heat exchange with the bottom was compensated by intrusions of cold waters during reverse temperature stratification. In the present-day period, the regularity of cold intrusions renewed deep waters and retained oxygenated conditions in these waters.

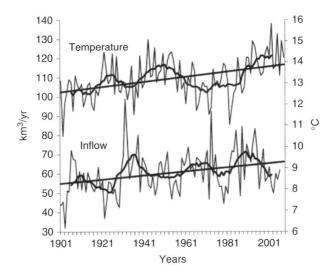

Figure 3.6 Eleven-year current running average values and trends of average annual air temperature (Babushkin Station) and total annual water inflow into Lake Baikal.

3.6 River inflows to Lake Baikal

The main characteristic feature of inflow variations during the century was its increase with climate warming having a positive trend of 10.5 km³ year⁻¹ per 100 year (Figure 3.6). The inflow rise calculated per 0.1 °C of annual air temperature increase was about 1.5% of its long-term value. It almost corresponded to the analogous coefficient (2%) for the total summer and autumn atmospheric precipitation in the Transbaikal region adjacent to the lake (Groisman 1981). Intrasecular cycles with a duration of 20–28 years were distinguished in air temperature and inflow variations: 1904–1929, 1930–1958, 1959–1979, with the latest cycle beginning in 1980 and probably ending in 2003. The comparison between these cycles and cycles of temperature revealed a clearly pronounced counterphase character. The relationship between anomalies of inflow and air temperature in summer compared to their trend was negative with a correlation coefficient $r = -0.3$ ($p < 0.05$). This phenomenon was likely caused by intrasecular cyclicity of processes related to zonal westward air-mass transfer in warm seasons when atmospheric precipitation provided about 80% of the annual total water inflow into the lake. The intensification of these processes increased the amount of precipitation and decreased the air temperature, whereas weakening caused the rise of temperature and evaporation from the basin surface and decrease of atmospheric precipitation and river inflow into the lake.

3.7 Lake Baikal ice regime

Warming caused "softening" of ice conditions at Lake Baikal (Figure 3.7) beginning from the mid-nineteenth century (Magnuson *et al.* 2000). Freezing started later and ice breakup earlier. During the period of observations from 1860 to 2010 in southern Baikal (Settlement Listvyanka), the duration of freezing changed by 10 days in

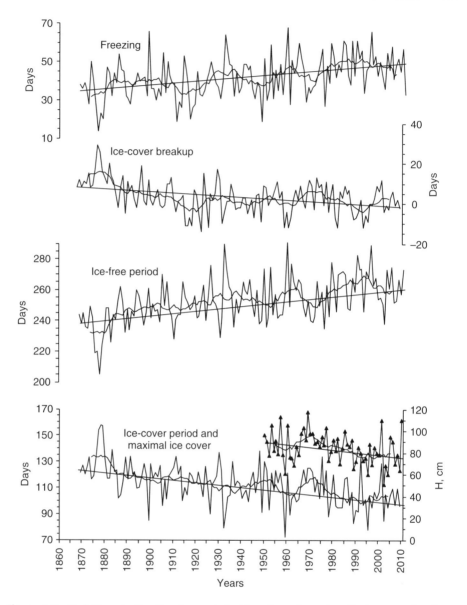

Figure 3.7 Variations in the freezing periods (days from December 1), ice-cover break-up periods (days from May 1), ice-free period duration, ice-cover period duration and maximal ice cover thickness, H (Listvyanka Settlement, Southern Baikal) based on current values, eleven-year smoothed values, and trends.

100 years and ice breakup by seven days. The duration of the ice-free period was extended and, accordingly, the ice-cover period was reduced by 17 days. The maximal ice thickness decreased in 1950–2000 on average by 2.4 cm every 10 years. In the period of significant warming in 1970–1995, the rate of change of ice processes abruptly increased reaching 43 days per 100 years for ice formation and 61 days per 100 days for ice break-up. Freezing was 10 days later and ice breakup 15 days earlier.

The duration of ice cover was 25 days shorter. Ice thickness decreased on average by 8.8 cm per 10 years. However, beginning from the mid-1990s to mid-2010, observations from onshore sites and from satellites showed tendencies to earlier ice formation, later ice break-up and an increase of the ice cover period (Kouraev *et al.* 2007). These changes are attributed to cyclic climatic fluctuations of the past years.

The major meteorological factor causing fluctuations in terms of freezing (D_{fr}) was air temperature (T_a) in November-December by affecting the intensity of heat losses from the water surface. The relationship between these characteristics is described by the following equation for Lake Baikal (1896–2010): $D_{fr} = 4.26T_a + 75$ ($R^2 = 0.57$, $p < 0.001$), where D_{fr} is the number of days from December 1 to the date of freezing. Temperature conditions in spring also affected the process of ice breakup. However, the correlation between the terms of ice breakup and air temperature was not high (Livingston 1999). This was attributed to the effect of both thermal and dynamic factors (wind) on ice breakup (Kouraev *et al.* 2007; Shimaraev 2008), as well as ice thickness, which depends on air temperature in winter months.

3.8   Effect of atmospheric circulation on climatic and hydrological processes at Lake Baikal during the present period

The most significant changes of climatic characteristics and hydrological processes at Lake Baikal were characterized by variability of macrocirculation processes in the northern hemisphere. Zonal circulation and the Siberian High (SH) are known to lead to large-scale anomalies in winter air temperatures and atmospheric precipitation over all of Northern Eurasia (Hurrell 1995). Such processes significantly influenced the long-term dynamics of ice phenomena (Livingstone 1999; Todd and Mackay 2003; Shimaraev 2008) and temperature of the upper water layers in spring and summer. Additional studies were performed on the effect of these and other circulation mechanisms on climatic characteristics and hydrological processes at Lake Baikal during anomalous warming in the second half of the twentieth and beginning of the twenty-first centuries. Among preferred modes of variability at large scales in the northern hemisphere we used indices of North-Atlantic oscillations (NAO) and Arctic Oscillations (AO) for 1950–2008 (Shimaraev and Starygina 2010). A SH index was also used. Data on air temperature, wind velocity, atmospheric precipitation, and river inflow into Lake Baikal (for 1950–2006) averaged from the observations at ten Baikal meteorological stations in 1968–2008 were used in the analysis (Figure 3.8).

Over this period, warming was caused by intensification of zonal transfer that significantly affected the air temperature, T_a, in winter, spring and autumn and on annual temperature. The relationship of T_a with the AO index ($r = +0.5-0.6$, $p < 0.05-0.01$) was closer in these seasons, whereas that with NAO appeared to be statistically significant only for annual and winter values ($r = +0.4-0.5$, $p < 0.05-0.01$). In summer, correlation coefficients of T_a and AO and NAO indices were negative and less significant. The effect of a Siberian high-pressure maximum on temperature was revealed in winter and spring ($r = -0.47$ to -0.66, $p < 0.01$). Index fluctuations of zonal circulation, especially AO, determined major change tendencies of annual values of atmospheric precipitation ($r = 0.4$, $p < 0.01$), river inflow ($r = 0.29$, $p < 0.01$), and wind

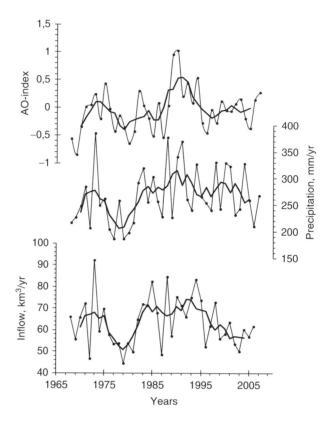

Figure 3.8   Arctic Oscillation (AO) Index (average annual values), the annual sum of an atmospheric precipitation and total river inflow to Baikal. The current and five-year smoothed values are shown.

velocity (r = 0.43, p<0.01). Intensification of zonal circulation from the beginning of 1968 to the mid-1990s occurred together with abrupt weakening of the Siberian high-pressure maximum (Shimaraev and Starygina 2010). This created conditions for deeper penetration of air masses from the Atlantic to the Baikal region during the cold season. These conditions resulted in abnormally high increases of air tempera-ture (annually by 1.9 °C and in winter almost by 8 °C) and surface water temperature in summer, and increases in total atmospheric precipitation and riverine inflow into Lake Baikal by 20%. After 1995, the reduction of zonal circulation activity caused changes in the trends of air temperature, precipitation and inflow (Figure 3.8).

3.9   Problems of possible hydrological condition changes at Lake Baikal in the twenty-first century are attributed to global change

Data on ice regimes in some lakes and rivers in the northern hemisphere (Magnuson *et al.* 2000) indirectly prove that warming has occurred for the past 150 years (or even for 450 years according to observations at Lake Suwa in Japan). This trend was related to the influence of natural factors (volcanic and solar activity, interaction between ocean and atmosphere, and so forth) and was likely to also persist in the twenty-first century. The effect of natural factors and progressive increases in concentrations of anthropogenic greenhouse gases in the atmosphere have been taken into consideration in estimation of recent and forthcoming climatic change trends. Digital modeling of global climatic evolution in the twenty-first century showed that the rise of global air temperature in the near-surface layer, as compared to that of the pre-industrial period, could reach 2 °C by 2025 and 4 °C by 2100 (Kondratyev 1992; Kostina 1997). However, it has been suggested that the factual temperature rise could reach only half of these values due to the thickness of clouds, which increases the reflection of solar radiation (Kostina 1997).

If we accept these rough estimates, the relationship between regional (Lake Baikal) and global temperature increase (2:1) and variations of freezing duration versus temperature (15 days/1 °C) in the twentieth century may serve as the basis for a forecast. In this case, annual air temperature on Lake Baikal was expected to rise by 2 °C from 1900 to 2025 and by 4 °C by 2100, whereas the ice cover period would decrease by 1 and 2 months, respectively, being 56–60 days in the southern and central basins of Lake Baikal and 76 days in the northern basin by the end of the century. Since interannual variations in the freezing period exceed 40–50 days in 10% of cases, one can predict winters with short freezing or unstable periods at the end of the twenty-first century. The same estimates were obtained using a climate change scenario averaged on values of air temperature increment from 2000 to 2100. According to the paleoclimatic data and calculations of the model of general atmospheric circulation, a twofold increase in the atmospheric CO_2 concentration could cause an increase of atmospheric precipitation by 5–10% at the latitude of Lake Baikal in the northern hemisphere (Budyko 1986; Kondratyev 1992). The riverine inflow increment value would most likely to remain within the same interval of 5–10%.

Possible changes of the lake's temperature regime were studied with quantitative ratios between annual air temperature and temperature of Lake Baikal's surface water and that of the 0–300 m layer during the 1970–1995 warming. With an increase of air temperature of 2 and 4 °C, the temperature of the lake surface was expected to rise by 1.3 and 2.7 °C, correspondingly, in the 0–100 m layer, by 0.8 and 1.7 °C, and in the 100–200 m layer by 0.2 and 0.5 °C. Average temperature of the 0–300 m layer in June–December may increase insignificantly (by 0.05–0.1 °C). The period with direct temperature stratification may be prolonged by 9–19 days due to earlier (in spring) and later (in autumn) transition of surface temperature via 4 °C (temperature of maximal density). The terms of development of free temperature convection in spring and autumn may also be shifted. However, these changes cannot affect renewal processes of deep waters which will provide annual nutrient inflow into the trophogenic layer and oxygen aeration of deep layers of Lake Baikal.

3.10   Nutrient changes in Lake Baikal—silica

Temperature rise of the upper water layers, extension of the period with direct temperature stratification and shortening of the ice-cover period (Figure 3.7) have affected biological processes in the trophogenic layer of Lake Baikal. Significant increases of chlorophyll *a* concentration and primary production were recorded in southern Baikal near Bolshiye Koty in the last quarter of the twentieth century. At the same time, the structure of the zooplankton community changed with a decreased abundance of copepods and rotifers and an increased abundance of cladoceran (Hampton *et al.* 2008). Similar changes occurred in the zooplankton composition in different basins of Lake Baikal in 1961–1995 (Afanasyeva and Shimaraev 2006). All these changes may be very important for food-chain dynamics and the nutrient cycles and may affect the content of nutrients in the water column of Lake Baikal.

Of particular interest was the behavior of dissolved silica. The role of Si in the Baikal ecosystem is governed by its participation in the vital activity of producing diatom frustules and formation of bottom sediments containing remnants of dead diatomaceous plankton. Biogenic silica in bottom sediments is considered one of the main proxies of climatic changes in Inner Asia for the past hundreds of thousands and million years (Grachev *et al.* 1997; Baikal Drilling Project Group 1998).

Let us consider variations in the dissolved Si content during anomalous warming in the Baikal area at the end of the twentieth century (Domysheva *et al.* 1998; Domysheva 2001; Shimaraev and Domysheva 2002, 2004). An increase of phytoplankton production was observed in the lake during that period.

According to long-term observations (Votintsev 1961; Domysheva *et al.* 1998), the amount of Si brought annually by river inflow into Lake Baikal is $8.8 \, \text{g} \, \text{Si} \, \text{m}^{-2}$, including $1.4 \, \text{g} \, \text{Si} \, \text{m}^{-2}$ supplied by the Angara River. The value of annual silica accumulation is $7.4 \, \text{g} \, \text{Si} \, \text{m}^{-2}$. The input of silica into bottom sediments may be considered as an average for many years, as silica concentration in the water of the major tributaries has remained almost permanent for the past 40–50 years (Sorokovikova *et al.* 2001). Estimations of silica concentration in the water column suggest that its annual consumption for the production of diatoms exceeded the supply of this element from rivers—about $48 \, \text{g} \, \text{Si} \, \text{m}^{-2}$ (Domysheva *et al.* 1998; Domysheva 2001). This is because of partial regeneration of silica from dead diatoms with its consequent return to the upper layers due to vertical water mixing. These data are annual average values. However, in the years of intensive diatom growth, the amount of silica involved in the turnover and its content in the water column, as well as its supply to bottom sediments, were significantly different from mean values. Under these conditions, deep layers are the major source of silica supply into the upper trophogenic layer. Elevated concentrations of silica are due to mineralization of its compounds with the decay of sinking diatom frustules and input of riverine waters enriched with silica (Votintsev 1961; Domysheva *et al.* 1998).

Because of the lack of representative data, little attention was paid to the problem of annual silica variations in Lake Baikal. Measurements carried out in the 1950s and 1960s near Listvyanka in southern Baikal revealed annual silica variations, which were much higher than its possible consumption by diatoms (Votintsev and Glazunov 1963; Votintsev 1965). Later studies (Shimaraev *et al.* 1999), however, showed that those local changes were caused by vertical and horizontal redistribution of silica under the effect of dynamic processes. More attention was paid to seasonal Si variations associated with seasonal fluctuations in diatom production (Votintsev and Glazunov 1963; Votintsev 1965). Diatom production reaches a higher peak under the ice in early spring with another peak immediately after ice out. As a result, Si concentrations in the upper $300 \, \text{m}$ layer are lowest in April–May and highest in early winter. New data allowed us to investigate annual variations of silica concentration within the entire water column of Baikal.

Detailed data on Si content in the Baikal water column were obtained during annual detailed surveys between 1993 and 2001. Observations were performed at 21 sites along a longitudinal profile, including 6–8 sites in each basin of the lake (Figure 3.9). Water samples were collected at depths of 0, 25, 50, and 100 m with 100–200 m intervals down to the bottom. Mean values of Si concentrations for each basin are presented in Table 3.3. Morphometric data were taken into account for calculating average values of Si concentration in different basins of the lake (Shimaraev *et al.* 1994). Observations were carried out from May to October when the main peak of the spring diatom

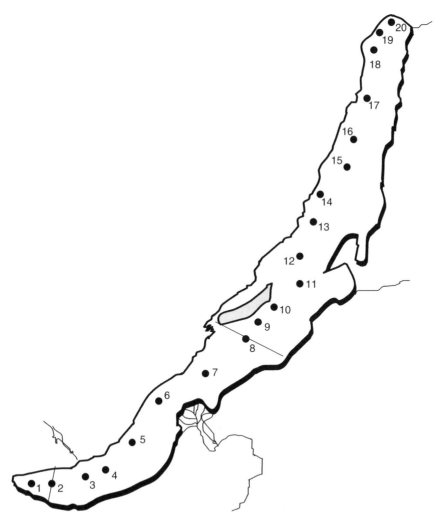

Figure 3.9 Water sampling stations for silica analysis between 1993 and 2009 (1–20). Solid lines transect Marituy-Solzan (2) and Boldakovo-Olkhon Gate (8) at which more frequent measurements were made.

Table 3.3 Average silica concentrations in three sub-basins of Lake Baikal in 1993–2001 (mg Si L^{-1}).

Southern Baikal

Depth (m)	1994 October	1995 May	1996 September	1997 July	1998 September	1999 September	2000 September	2001 September
0	0.68	0.74	0.70	0.41	0.60	0.31	0.50	0.41
25	0.67	0.77	0.67	0.39	0.58	0.28	0.29	0.39
50	0.56	0.76	0.61	0.38	0.49	0.30	0.31	0.38
100	0.60	0.75	0.64	0.40	0.50	0.40	0.37	0.40
200	0.74	0.74	0.67	0.64	0.51	0.43	0.43	0.45
300	0.76	0.80	0.72	0.70	0.56	0.53	0.52	0.55
400	0.82	0.81	0.77	0.75	0.61	0.58	0.52	0.52
600	0.83	0.87	0.80	0.84	0.72	0.73	0.62	0.64
800	0.88	0.99	0.87	0.93	0.84	0.78	0.85	0.74
1000	0.97	1.13	0.99	1.00	0.91	0.92	0.88	0.92
1200	–	1.17	1.20	1.05	1.00	1.00	0.82	0.82
1300	1.12	1.32	1.25	1.02	0.95	0.99	1.08	0.94
1400	1.14	1.22	–	–	–	–	–	1.06
Average[*]	0.84	0.91	0.82	0.78	0.70	0.65	0.63	0.63

Central Baikal

Depth (m)	1993 May	1994 October	1995 May	1997 July	1998 September	1999 September	2000 September	2001 September
0	0.91	0.83	0.88	0.77	0.91	0.39	0.51	0.53
25	0.91	0.84	0.91	0.76	0.79	0.33	0.46	0.52
50	0.92	0.83	0.90	0.78	0.79	0.35	0.42	0.54
100	0.92	0.86	0.92	0.78	0.79	0.38	0.44	0.53
200	0.92	0.87	0.92	0.79	0.78	0.55	0.49	0.56
300	0.92	0.89	0.95	0.89	0.79	0.70	0.61	0.59
400	0.94	0.90	0.95	0.94	0.82	0.77	0.69	0.65
600	1.05	0.98	1.03	0.97	0.90	0.84	0.77	0.79
700	1.11	1.03	0.91	1.00	0.91	0.86	0.78	0.78
800	1.17	1.08	1.04	1.02	0.92	0.88	0.79	0.78
900	1.24	1.10	1.09	1.05	1.00	0.88	0.82	0.89
1000	1.30	1.13	1.14	1.08	1.08	0.87	0.85	1.00
1200	1.50	1.37	1.40	1.26	1.24	0.97	0.88	1.14
1300	1.56	1.25	1.42	1.35	1.34	1.00	0.79	1.18
1400	1.62	1.50	1.63	1.45	1.52	1.03	0.76	1.23
1500	1.69	1.53	1.38	1.50	1.58	1.13	0.66	1.40
1600	1.75	1.39	1.70	1.57	–	1.45	0.74	1.46
Average[*]	1.11	1.02	1.05	0.99	0.95	0.77	0.69	0.78

Table 3.3   (*continued*)

Northern Baikal

Depth (m)	1993 May	1994 October	1995 May	1996 September	1997 July	1998 September	1999 September	2000 September	2001 September
0	0.94	0.86	0.95	0.88	0.77	0.74	0.51	0.67	0.55
25	0.96	0.86	0.95	0.88	0.78	0.69	0.51	0.64	0.53
50	0.96	0.86	0.96	0.87	0.78	0.69	0.53	0.66	0.51
100	0.95	0.89	0.95	0.87	0.78	0.72	0.63	0.66	0.55
200	0.95	0.90	0.95	0.88	0.81	0.75	0.74	0.66	0.62
300	0.96	0.93	0.94	0.92	0.91	0.80	0.80	0.74	0.68
400	0.98	0.91	1.00	0.96	0.98	0.82	0.87	0.81	0.74
600	1.04	1.01	1.05	1.05	1.04	0.94	0.94	0.91	0.89
700	1.10	1.09	1.15	1.09	1.07	1.01	1.06	1.04	0.93
800	1.12	1.10	1.16	1.18	1.12	1.03	1.06	0.96	0.97
900	1.32	1.09	1.32	1.35	1.17	1.04	1.05	0.86	0.92
Average[*]	1.00	0.94	1.01	0.96	0.92	0.83	0.81	0.79	0.72

Silica concentrations were measured spectrophotometrically to a precision of $0.01-0.02\,\text{mg Si L}^{-1}$ with a relative error of 2%.
Dashes = no data.
[*]Weighted volumetric mean for each water layer.

bloom was finishing or was over. Therefore, the data obtained allowed us to define the trend of annual variations in the Si content during the observation period.

The entire lake was characterized by a decrease in Si content during 1995–2001 (Figure 3.10). This was also confirmed by results of independent seasonal observations carried out at the Listvyanka–Tankhoy transect (southern Baikal) at the same time (Domysheva 2001). The trend was for Si content to decrease within the entire water column, becoming slightly attenuated in deeper zones (Figure 3.11). At depths of

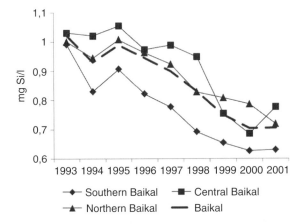

Figure 3.10   Variation of average weighted concentrations of silica in water masses of Baikal from 1993 to 2001.

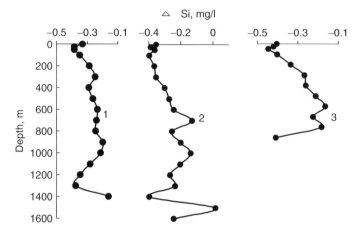

Figure 3.11 Increase of silica content (Δ mg Si L^{-1}) in water masses of Southern (1), Central (2) and Northern Baikal (3) from 1995 to 2001.

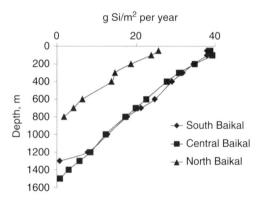

Figure 3.12 Vertical transfer of silica (g Si m^{-2} per year) in three basins of Baikal in 1995–2001.

100–200 m above the bottom, negative incremental concentration values could again increase in the southern and northern basins owing to the supply of Si-poor waters from upper layers of the lake into the near-bottom zone as a result of deep convection in early spring and late autumn (Weiss, Carmack, and Koropalov, 1991; Domysheva *et al.* 1998). Using data on Si reserve variations at different levels, we calculated average values of Si supply from lower layers to the upper layers (Figure 3.12). They decreased with depth due to the fall of the water volume area ratio below a particular isobath (from 0.73 at the surface to 0.07 at a depth of 1500 m) and to a lower intensity of vertical water mixing.

Average values of vertical silica flows from deep layers to upper zones for 1995–2001 were unequal at different sites of the lake. In the northern basin, they appeared to be almost 40% lower than those in the southern and central areas

(Figure 3.12). Coefficients of vertical mixing calculated from data on the element flow and its distribution (ca. $20\,cm^2\,s^{-1}$ in the southern and central basins and $11\,cm^2\,s^{-1}$ in the northern part) at the boundary of upper and deep water lake zones showed decreased activity of mixing processes in the northern part of the lake. Coefficients of vertical water mixing, K_z, were calculated from the data of long-term water temperature observations using the method of "thermal flow". At the interface of upper and deep zones (depth is $250\,m$), K_z (17–$18\,cm^2\,s^{-1}$) in the southern and central basins was two times higher than that in the northern basin. These results corresponded to K_z values obtained with the method of "silica flow" (Shimaraev, Troitskaya, and Domysheva 2003).

Between 1995 and 2001, Si reserves in the lake decreased by $200 \pm 18\,g\,m^{-2}$, which is 28% with regard to initial reserves ($\sim730\,g\,m^{-2}$). Such changes could only be caused by intensification of productivity of large diatoms and increased release of Si from the water column into bottom sediments. The value of Si consumed was equal to the sum of Si loss in the water column and its input with riverine waters (minus the loss with the Angara River flow). It was $41\,g\,m^{-2}$, which was 5.6 times higher than the annual average value ($7.4\,g\,m^{-2}$). The assumption of increased diatom productivity was supported by observations for phytoplankton in southern Baikal. For instance, maximal algal biomass near Bolshiye Koty was 26–30 times higher in the summer and autumn of 1990–1997 relative to 1964–1974. This was accompanied by its growth in spring when the development of diatoms was maximal: 1.7 times higher in 1990–1997 and 2.8 times in 1995–1997 when reserves began to decrease. High diatom productivity was more frequently observed in the 1990s in comparison with the previous periods: 1990, 1994, 1996, and 1997 (Bondarenko 1997, 1999), as well as 1999 and 2000.

Since the end of the 1940s, the influx of nitrogen and phosphorus compounds with river waters has gradually increased (Sorokovikova *et al.* 2001). However, no increase of their concentrations has been recorded in the lake because of active vertical water mixing. Therefore, the increase in compounds mentioned above can hardly be responsible for the changes in algal biomass. The algal bloom was probably caused by changes in abiotic conditions of the lake. Long-term dynamics of phytoplankton development (about 7 to 10 years) are attributed to climatic changes (Shimaraev 1971; Shimaraev *et al.* 1994). This is evident from the correlation of variations in phytoplankton abundance and biomass with ice cover period, solar radiation penetrating the water column, and water temperature. Calculations based on the empirical biomass-surface water temperature relationship (Shimaraev *et al.* 1994) show that annual phytoplankton biomass, which is mostly produced by large diatoms, should be approximately five times higher during 1990–1997 relative to 1964–1974. This was consistent with the increase in Si consumption in the second half of the 1990s.

We assumed that changes in Si reserves in Baikal usually reflected long-term dynamics of climatic variations in this area. In this connection, let us consider the dynamics of climatic changes (Figures 3.2 and 3.3) and data on diatoms in southern Baikal for the second half of the twentieth century. In the 1950s and 1960s, corresponding to the maximum of the intrasecular warming cycle and partially its recession curve (Figure 3.2), intervals between years with the highest diatom productivity were 3–4 years (Antipova 1969, Popovskaya 1987, 1991). In 1950, 1953, 1957 and 1960, the high productivity of diatom algae was mainly provided by the development of large-celled diatom species, *Aulacoseria baikalensis* and *A. islandica*. In this case, the consumption of Si should have increased, whereas its content in the water column should have decreased. During the period of low air temperatures after 1968, high productivity of *A. baikalensis* was registered only in 1974, 1982, and 1990 (every 6 to 8 years), with a simultaneous decline in its development (Popovskaya 1991). The abundance of small celled diatom species, *Nitzschia acicularis* (Bondarenko 1999), whose frustules usually dissolved after death, increased during this period. The inverse correlation of biomass with the abundance of diatoms was also typical of the whole phytoplankton group between 1950 and 1980 (Kuzevanova 1986). Such conditions caused the reduction in the consumption of Si and growth of its reserves in the lake. In the early 1990s, dynamics of *Aulacoseria* development were close to that observed in the 1950–1960s due to warming that was more prominent than in the 1950s (Bondarenko 1997). Other diatom species developed and Si reserves of the lake declined significantly. Thus, Si reserves should be restored at the beginning of the current century due to the cyclicity of changes in water and temperature. However, if the common trend of global warming were preserved, changes in the lake ecosystem that were typical of the 1990s could be even more prominent during the periods of higher peaks of the subsequent intrasecular climatic cycles.

The results obtained assume that climatic changes affect the content of dissolved silica within the entire water column of Lake Baikal (Shimaraev and Domysheva 2004). In the years close to the maximum intrasecular temperature cycles characterized by a very high level of diatom productivity (as in the 1990s), the amount of Si involved in the ecosystem cycle substantially exceeded its inflow by rivers. At that time, deep waters of the lake were the main source of Si sustaining the growth of plankton. Consequently, Si reserves decreased within the entire water column. Observations show that after 2001 its reserves have been restored. Lost Si was probably replenished by river waters and its production from decomposing diatoms descending to the bottom or those decomposing in the water column.

The cyclicity of dissolved silica content may be attributed to the response of diatoms to modern intrasecular climatic changes. Data on dynamics of dissolved silica suggested significant changes in the rate of its accumulation in bottom sediments of Lake Baikal, even within historically short time intervals that are comparable with the climatic rhythms of decades.

3.11   Concentration dynamics of phosphate, nitrate and dissolved oxygen in Lake Baikal

Dynamics of the concentrations of other nutrients and oxygen was studied from the data on single one-month surveys from June to September. Weighted average concentrations of these elements was calculated in the same way as for silica (5–7 sites in each part of the lake) from their average concentrations at some depths taking into account the layer volume of the basins.

Weighted average concentrations of phosphate were close to those of silica dynamics with a decrease from 1994 to 1997–2002 and an increase to the level of the early 1990s in the next years by 2009 (Figure 3.13). The range of change was about $0.015\,\text{mg}\,\text{L}^{-1}$ (40% of the average concentration). Changes in nitrate concentrations (its minimum was recorded in 1994 and 1999–2001, Figure 3.13)

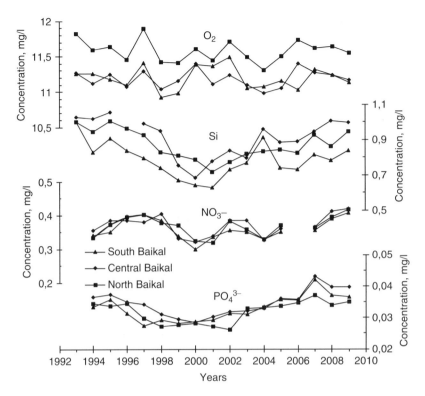

Figure 3.13 Change of average concentration of silica, nitrate and phosphate and dissolved oxygen in three basins of Lake Baikal from 1993 to 2009.

were more complicated. Fluctuation range was $0.1\,mg\,L^{-1}$ or 27% of the average nitrate concentration. The minimum nutrient concentrations in 1998–2001 may be attributed to the fact that these data were obtained in September when the development of the summer algal complex was accompanied by a significant drop (sometimes to analytical zero) of nitrate and phosphate content in the surface water layer. However, this decrease did not exceed 510% of the maximum.

Since the 1990s, weighted average concentrations of dissolved oxygen in the water in some basins have remained almost at the same level changing in some years only by $0.5\,mg\,L^{-1}$, on average. Its constant value in the deep water zone is supported by regularity of water renewal due to deep convection. One of the reasons for the registered fluctuations of these concentrations may have been changes of hydrological and biological conditions in certain years, which affected gas exchange between water and the atmosphere, intensity of algal photosynthesis, and solubility of oxygen depending on temperature of the upper layers.

As for the major ions, it was very difficult to detect possible recent changes of their concentrations in Lake Baikal (Grachev 2002), which are caused by processes occurring in its basin (Votintsev 1973). This is accounted for by the volume of Baikal water, which is much larger (about 400 times) than annual river flow, and by relatively fast vertical mixing of water masses (10–30 years) (Weiss, Carmack, and Koropalov 1991; Hohmann *et al.* 1998; Kodenev *et al.* 1998).

3.12   Acknowledgements

The study was supported by grant 09-05-00222 of the Russian Foundation for Basic Research. N.A. Bondarenko is acknowledged for providing data on the phytoplankton.

References

Afanasyev, A.N. (1967) *Fluctuations of Hydrometeorological Regime on the Territory of the USSR*, Nauka, Moscow, p. 230.

Afanasyeva, E.L. and Shimaraev, M.N. (2006) Long-term zooplankton variations in the pelagial of Lake Baikal under global warming, in *Aquatic Ecology at the Dawn of the XXI Century* (eds. A.F. Alimov and V.V. Bul'on), KMK Scientific Press Ltd., Moscow, pp. 253–65.

Anneville, O., Ju. C., Molinero, S., *et al.* (2007) Long-term changes in the copepod community of Lake Geneva. *J. Plank. Res.*, **29**: (Suppl 1): i49–i59, doi: 10.1093/plankt/fbl066.

Antipova, N.I. (1969) *Seasonal and Annual Dynamics of Phytoplankton. Abstract, candidate thesis*, Irkutsk University, Irkutsk, p. 33.

Baikal Drilling Project Group (1998) A continuous record of climate changes for the last five million years from the bottom sediments of Lake Baikal. *Geologiya i Geofizika (Russian Geology and Geophysics)*, **39**, 139–56.

Bondarenko, N.A. (1997) *Structure and Production of Phytoplankton in Lake Baikal [in Russian]. Abstract, candidate thesis*, Institute of Biology of Inner Waters, Borok, p. 22.

Bondarenko, N.A. (1999) Floral shift in the phytoplankton of Lake Baikal, *Siberia: Recent dominance of Nitzschia acicularis. Plankton Biol. Ecol.*, **46**, 18–23.

Budyko, M.I. (1986) Anthropogenic climatic changes. *Priroda (Nature)*, **8**, 14–21.

Domysheva, V.M. (2001) *Distribution and Dynamics of Oxygen and Biogenic Elements in Pelagic Baikal [in Russian]. Abstract, candidate thesis*. Institute of Geography, Irkutsk, p. 22.

Domysheva, V.M., Shimaraev, M.N., Gorbunova, L.A., *et al.* (1998) Silica in Lake Baikal. *Geografia i Prirodnye Resursy*, **4**, 73–81.

Filatov, N.N. (1998) Level fluctuations of large European lakes and climate change. *Dokl. RAN*, **359**, 255–7.

Grachev, M.A., Likhoshway, Y.V., Vorobyeva, S.S. *et al.* (1997) Signals of paleoclimates of the Upper Pleistocene in the sediments of Lake Baikal. *Geologiya i Geofizika (Russian Geology and Geophysics)*, **38**, 957–80 (994–1016).

Groisman, P.Y. (1981) Empiric estimates of relationship between processes of glacials and interglacials with moistening regime of the USSR territory. *Izv. Akad. Nauk SSSR, Ser. Geogr.* **5**, 86–95.

Gronskaya, T.P., Lemeshko, N.A., Arvola, L., and Jarvinen, M. (2002) Lakes of European Russia and Finland as indicators of climate change. *World Resource Review*, **14**, 189–203.

Hampton, S.E., Izmest'eva, L.R., Moore, M.V., *et al.* (2008) Sixty years of environmental change in the world's largest fresh water lake–Lake Baikal, Siberia. *Global Change Biol.*, **14**, 1–12, doi: 10.1111/j.1365-2486.2008.01616.x.

Hohmann, R., Hofer, M., Kipfer, R., *et al.* (1998) Distribution of helium and tritium in Lake Baikal. *J. Geophys. Res.*, **103**,12823–38.

Hurrell, J.W. (1995) Decadal trends in the North Atlantic Oscillation: regional temperatures and precipitation. *Science*, **269**, 676–9.

IPCC: Climate Change (1996) *The Science of Climate Change*, Cambridge, Cambridge University Press.

Kodenev, G.G., Shimaraev, M.N., and Shishmarev, A.T. (1998) Determination of terms of renewal of Baikal deep waters with chemical tracers. *Geologiya i Geofizika (Russian Geology and Geophysics)*, **39**, 842–50.

Kondratyev, K.Y. (1992) *Global Climate*, Nauka, St. Petersburg, p. 359.

Kostina, E.E. (1997) *Global Climate Change and its Possible Consequences. Review*, Dalnauka, Vladivostok, p. 102.

Kouraev, A.V., Semovski, S.V., Shimaraev, M.N., *et al.* (2007) The ice regime of Lake Baikal from historical and satellite data: Relation to air temperature, dynamical, and other factors. *Limnol. Oceanogr.*, **52**, 1268–86.

Kumagai, M., Vincent, W.F., Ishikawa, K., and Aota, Y. (2003) Lessons from Lake Biwa and other Asian lakes: Global and local perspectives, in *Freshwater Management* (eds. M. Kumagai, W.F. Vincent), Springer, Tokyo, pp. 1–23.

Kuzevanova, E.N. (1986) *Long-term dynamics of phytoplankton and zooplankton in South Baikal [in Russian]. Abstract, candidate thesis*, Irkutsk University, Irkutsk, p. 23.

Livingstone, D.M. (1999) Ice break in Southern Baikal and its relationship with local and regional air temperatures in Siberia and the North Atlantic Oscillation. *Limnol. Oceanogr.*, **44**, 1486–97.

Magnuson, J.J., Robertson, D.M., Benson, B.J., *et al.* (2000) Historical trends in lake and river ice cover in the Northern Hemisphere. *Science*, **289**, 1743–6.

Magnuson, J.J., Webster, K.E., Assel, R.A., *et al.* (1997) Potential effects of climate changes on aquatic systems: Laurentian Great Lakes and Precambrian Shield Region. *Hydrol. Proc.*, **11**, 825–71.

O'Reilly, C.M., Alin, S.R., Plisner, P.D., *et al.* (2003) Climate change decreases aquatic ecosystem productivity of Lake Tanganyika, Africa. *Nature*, **24**, 766–8.

Panagiotopoulos, F., Shahgedanova, M., Hannachi, A., and Stephenson D.B. (2005) Observed trends and teleconnections of the Siberian High: A recently declining center of action. *J. Climate*, **18**, 1411–22.

Popovskaya, G.I. (1987) Phytoplankton in the world's deepest lake, in *Marine and Freshwater Plankton* (ed. B.L. Gutelmakher), Zoological Institute, Leningrad, pp. 107–16. (In Russian.)

Popovskaya, G.I. (1991) *Phytoplankton of Lake Baikal and its Long-Term Dynamics (1958–1990). Doctoral thesis*, United Institute of Geology, Geophysics and Mineralogy SB RAS, Novosibirsk, p. 32.

Scheffer, M., Carpenter, S., Folke, J.A., and Walker, B. (2001) Catastrophic shifts in ecosystems. *Nature*, **413**, 591–6.

Sherstyankin, P.P., Khlystov, O.M., Alekseev, S.P., *et al.* (2003) Electronic bathymetric map of Lake Baikal. VIII International Conference, Recent Methods and Means of Oceanological Investigations. *Materials of the Conference*. Part I. Moscow, pp. 265–73.

Shimaraev, M.N. (1971) Hydrometeorological factors and population dynamics of Baikal plankton, in *Limnology of Deltaic Regions of Lake Baikal* [in Russian] (eds. N.A. Florensov and B.F. Lut), Nauka. Leningrad, pp. 259–67.

Shimaraev, M.N. (1996) Heat and Mass Exchange in Lake Baikal. Abstract, doctoral thesis, *Institute of Geography*, Irkutsk [in Russian], p. 48.

Shimaraev, M.N. (2008) On effect of North-Atlantic Oscillations (NAO) on ice-thermal processes at Lake Baikal. *Doklady Akademii Nauk*, **423**, 397–400.

Shimaraev, M.N. and Domysheva, V.M. (2002) Dynamics of dissolved silica in Lake Baikal. *Doklady Earth Sciences*, **387**, 1075–8.

Shimaraev, M.N. and Domysheva, V.M. (2004) Climate and long-term silica dynamics in Lake Baikal. *Geologiya i Geofizika (Russian Geology and Geophysics)*, **45**, 310–16.

Shimaraev, M.N., Domysheva, V.M., Verbolov, V.I., *et al.* (1999) Hydrophysical processes and distribution of dissolved silica in Lake Baikal. *Geologiya i Geofizika (Russian Geology and Geophyslcs)*, **40**, 1502–5.

Shimaraev, M.N. and Granin, N.G. (1991) Temperature stratification and the mechanisms of convection in Lake Baikal. *Doklady Akademii Nauk SSSR*, **321**, 831–5.

Shimaraev, M.N., Kuimova, L.N., Sinyukovich, V.N., and Tsekhanovsky, V.B. (2002) Manifestation of global climatic changes in Lake Baikal during the twentieth century. *Doklady Earth Sciences*, **383**, 288–91.

Shimaraev, M.N. and Starygina, L.N. (2010) Zonal circulation of atmosphere and climate and hydrological processes at Lake Baikal (1968–2007). *Geog. Nat. Res.*, **3**, 62–8.

Shimaraev, M.N., Troitskaya, E.S., and Domysheva, V.M. (2003) Intensity of vertical changes in each basin of Lake Baikal. *Geog. Nat. Res.*, **3**, 68–73.

Shimaraev, M.N., Troitskaya, Y.S., and Gnatovsky, R.Y. (2009) Modern climate changes and deep water temperature of Lake Baikal. *Doklady Earth Sciences*, **427**, 804–8. Published in Russian in *Doklady Akademii Nauk*, **426**, 685–9.

Shimaraev, M.N., Verbolov, V.I., Granin, N.G., and Sherstyankin, P.P. (1994) *Physical Limnology of Lake Baikal: A Review*, BICER, Irkutsk-Okayama, p. 81.

Shimaraev, M.N., Zhdanov, A.A., Gnatovsky, R.Y., *et al.* (2011) Specific characteristics of near-bottom intrusions at Lake Baikal for 1993–2009. *Water Res.*, **38**, 163–8.

Sorokovikova, L.M., Sinyukovich, V.N., Khodhzer, T.V., *et al.* (2001) Input of biogenic elements and organic matter into Lake Baikal. *Meteorologiya i Gidrologiya*, **4**, 78–86.

Thompson, D.W.J. and Wallace, J.M. (1998) The Arctic Oscillation signature in the wintertime—geopotential height and temperature fields. *Geophys. Res. Lett.*, **25**, 1297–300.

Todd, M.C. and Mackay, A.W. (2003) Large-scale climate controls on Lake Baikal ice cover. *J. Clim.*, **16**, 3186–99.

Troitskaya, E.S., Shimaraev, M.N., and Tsekhanovsky, V.V. (2003) Long-term changes of surface water temperature in Lake Baikal. *Geog. Nat. Res.*, **2**, 47–50.

Troitskaya, Y.S. and Shimaraev, M.N. (2005) Transparency by Secci disk and water temperature in Southern Baikal. *Atmos. Oceanic Optics*, **18**, 120–3.

Vereshchagin, G.Y (1936) Fundamental features of vertical distribution of water mass dynamics in Lake Baikal, in Akademiku Vernadskomu (ed. N. Nalivkin), Academy of Sciences Publishers, Moscow, pp. 120730.

Votintsev, K.K. (1961) *Gidrokhimiya Ozera Baikal (Hydrochemistry of Lake Baikal)*, Akad. Nauk SSSR, Moscow, p. 311.

Votintsev, K.K. (1965) Hydrochemical conditions in pelagic Baikal, in Limnological studies of Lake Baikal and some Lakes in Mongolia *[in Russian]*, (eds. G.I. Galazy and M.Y. Bekman), Nauka, Moscow, pp. 71–114.

Votintsev, K.K. 1973. *Prognosis of water quality of Lake Baikal till 2000*. Abstract Book V Conference of Geographers from Siberia and Far East, Irkutsk, pp. 39–40.

Votintsev, K.K. and Glazunov, I.V. (1963) Hydrochemical regime of Lake Baikal in the region of Listvyanka Village, in Hydrochemical studies of Lake Baikal *[in Russian]*. (ed. G. I. Galazy), Izd. AN SSSR, Moscow, pp. 3–56.

Weiss, R.F., Carmack, E.C., and Koropalov V.M. (1991) Deep-water renewal and biological production in Lake Baikal. *Nature*, **349**, 665–9.

4

Hydrological Analysis of the Yellow River Basin, China

Xieyao Ma[1], Yoshinobu Sato[2], Takao Yoshikane[1], Masayuki Hara[1], Fujio Kimura[1], and Yoshihiro Fukushima[3]

[1]*Research Institute for Global Change, Japan Agency for Marine-Earth Science and Technology, Japan*
[2]*Water Resources Research Center, Disaster Prevention Research Institute, Kyoto University, Japan*
[3]*Department of Environmental Management, Tottori University of Environment Studies, Japan*

4.1 Introduction

The Yellow River plays a critical role in socioeconomic development and ecological maintenance over a wide area of northwest China (Figure 4.1). Located between 32–42 °N and 96–119 °E, the river drains an area of approximately 0.752 million km^2. Most of the basin is arid or semi-arid, with a population of 130 million, most of whom are farmers and rural residents. Statistical records show that the current state of water diversion and usage of the available river runoff is 39.5 billion m^3 and water consumption is 30.7 billion m^3. Agricultural irrigation is the main water use, with an annual diverted water volume of 36.2 billion m^3 and consumption of 28.4 billion m^3. This accounts for 92% of the total water consumption (Yellow River Conservancy Commission, http://www.yellowriver.gov.cn/).

The hydrological environment of the Yellow River has changed markedly with changes in climate (Liu and Zheng 2004; Yang *et al*. 2004; Sato *et al*. 2008) and increasing anthropogenic water use (Sato *et al*. 2009; Ma *et al*. 2010). Discharge has been drastically decreased and zero-flow events have appeared at outlets every year since the 1970s. Ma *et al*. (2010) reported that water use for irrigation has not changed, although precipitation decreased in the 1990s. Sato *et al*. (2009) noted that a soil and water conservation project in the Loess Plateau decreased the available water resources by approximately 10–50%.

Climatic Change and Global Warming of Inland Waters: Impacts and Mitigation for Ecosystems and Societies, First Edition. Edited by Charles R. Goldman, Michio Kumagai and Richard D. Robarts.
© 2013 John Wiley & Sons, Ltd. Published 2013 by John Wiley & Sons, Ltd.

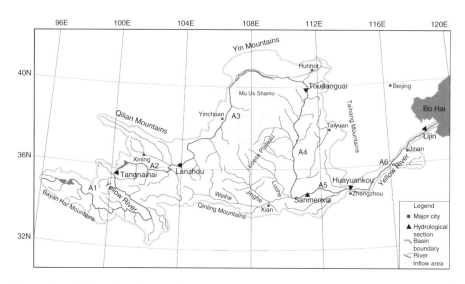

Figure 4.1 Yellow River basin and its sub-basins. A1, Tangnaihai watershed, 122 000 km²; A2, interwatershed area between the Tangnaihai and Lanzhou, 100 551 km²; A3, interwatershed area between the Lanzhou and Toudaoguai, 163 415 km²; A4, interwatershed area between the Toudaoguai and Sanmenxia, 302 455 km²; A5, interwatershed area between the Sanmenxia and Huayuankou, 41 616 km²; A6, interwatershed area between the Huayuankou and Lijin, 21 833 km².

Since 1998 to avoid regional drying of the river, a minimum discharge has been required at several checkpoints by executive order. However, until now the true nature of the problem of water shortages was not clear. Precipitation is the only input item in the river basin water budget. An accurate grasp of precipitation changes over the basin could help us understand changes in the water cycle and allow taking prompt measures to deal with problems. Therefore, numerical modeling of precipitation is necessary to predict future climate change and water resources in the basin. Here we discuss changes in precipitation over the Yellow River basin based on observed data. Dynamical downscaling using a regional climate model was performed. We describe the results of numerical experiments for a two-decade period (from 1980 to 1997).

The river runoff observed at Huayuankou and Lijin in the 1990s had decreased to almost half and one-third of that in the 1960s, respectively. Hence, drying in the lower reaches of the Yellow River basin became more serious in the 1990s. The main factors inducing water shortages in the lower Yellow River basin were recognized as an increase in water consumption within the lower reaches and a decrease in river water supplied from upstream. However, the contributions of these two factors have not been clarified quantitatively in previous studies. We attempted to clarify the mechanisms of drying in the Yellow River basin using a long-term (1960–2000) hydrological simulation.

4.2   Climate and land use in the Yellow River basin

The Yellow River basin has a continental climate. In winter, the basin is dominated by dry, cold high-pressure systems from Mongolia. In summer, subtropical high-pressure of the western Pacific affects the basin. The temperature in the basin is higher in the southeast than in the northwest, and lower in the hills than on the plain. The annual mean temperature is 1–8 °C in the upper reaches, 8–14 °C in the middle reaches, and 12–14 °C in the lower reaches. According to the analysis of satellite data obtained in 2000, the main land cover is grassland and open shrubland (Ma *et al.* 2010). Irrigated areas are widely distributed over the whole basin. Farming is single-cropping in the area above Lanzhou and inter- or multiple-cropping elsewhere.

Precipitation is mainly from rainfall, with only a small proportion of snowfall. The annual rainfall decreases from the southeast to the northwest. The annual average precipitation is 450 mm, with one-third of the basin receiving less than 400 mm. The precipitation distribution is extremely uneven. Annual precipitation data since the 1950s at five stations from the upper to middle reaches are available (Figure 4.2). Dari and Lushi are located in the source areas, on the north slopes of the Bayan Har Mountains and Qinling Mountains, respectively. The annual precipitation there was far greater than at other stations. The area of minimum precipitation was distributed between Lanzhou and Toudaoguai. At Linhe, annual precipitation was <200 mm and the potential evaporation was approximately 2400 mm. Maduo station, located in the upper area, was also dry (annual precipitation <400 mm). In general, the 1960s and 1980s have been recognized as relatively wet decades, and the 1970s

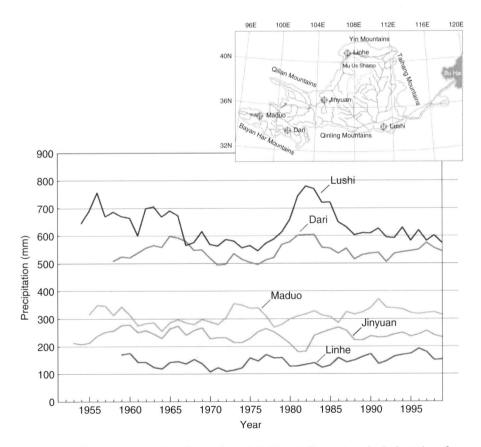

Figure 4.2 Five-year anomalies of annual precipitation at five meteorological stations from 1953 to 1999.

and 1990s as relatively dry. Ma *et al*. (2010) found that the precipitation in the 1990s decreased approximately 35 mm (7%) in the area over Lanzhou and 63 mm (11%) in the middle reaches, compared with the previous three decades (the 1960s to the 1980s).

4.3   Precipitation numerical experiment

A complete compression and nonhydrostatic regional climate model, the Weather Research and Forecasting (WRF) model (version 2.2), was used. The domain center point was set at 35.5 °N and 105 °E (Figure 4.3). The horizontal resolution was 20 km for the long-term run and 10 km for the sensitivity experiment. The National Centers for Environmental Prediction/National Center for Atmospheric Research re-analysis dataset in six-hourly intervals was used as the lateral boundary condition. A precipitation dataset with 0.1° grid resolution (Xie *et al*. 2007) based on gauge observations in the Yellow River basin was used to evaluate the WRF performance there from 1980 to 1997.

To examine spatial differences, we selected four areas: the source area of the Yellow River above Tangnaihai station (A1); the upper reaches between Tangnai-hai and Lanzhou (A2); the arid area between Lanzhou and Toudaoguai (A3); and the semi-arid area between Toudaoguai and Sanmenxia, including the Loess Plateau (A4) (Figure 4.1).

In the WRF model, the combination of microphysics and cumulus schemes may be freely chosen in order to represent the various atmospheric processes that take place in sub-grid scales. For convective parameterization, there are several cumulus options (e.g., the Kain–Fritsch scheme, the Betts–Miller–Janjic scheme, the Grell–Devenyi ensemble scheme, the simplified Arakawa–Schubert scheme, and so forth). We chose the microphysics scheme of Lin *et al*. (1983) and the Kain–Fritsch cumulus scheme for this study through a comparative test. First, a sensitivity experiment with increased horizontal resolution was performed in the relatively wet years of 1989 and 1992 and the dry years of 1986 and 1997.

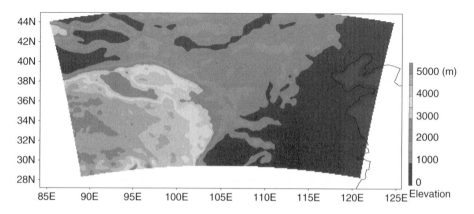

Figure 4.3 Domain of the WRF model for the Yellow River basin. (See insert for color representation.)

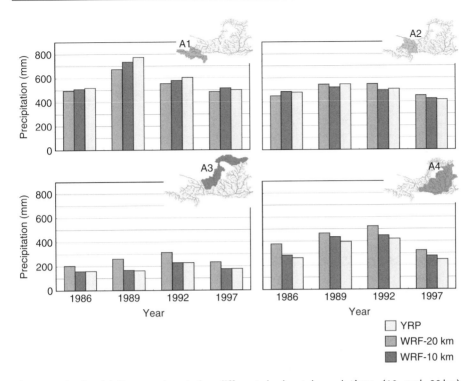

Figure 4.4 Sensitivity experiment for different horizontal resolutions (10 and 20 km), compared to observations (YRP). (See insert for color representation.)

There was a difference in the annual precipitation amount between the dataset and simulations in the wet years (1989 and 1992) and dry years (1986 and 1997) (Figure 4.4). The results showed that precipitation was underestimated in the basin but the tendency of the annual change was roughly the same as in the dataset. It was also clear that there was little difference between the 10 and 20 km resolution calculations. Therefore, we performed the long-term calculation for the period 1980–1997 at a 20 km resolution.

The WRF model underestimated the observed values in A3 and A4 from 1980 to 1997 (Figure 4.5). However, the interannual variations were correctly modeled in almost every watershed.

In addition, to evaluate changes in the spatial distribution of precipitation, we examined the difference between the two periods (1980 to 1989 and 1990 to 1997). The regions with large changes in A1 and A4 were well represented, although there was a difference in the trend in part of the northeast of the basin (Figure 4.6).

Although the estimated precipitation in the Yellow River basin was inaccurate, the calculated precipitation was well distributed spatially, and the model run was stable even over long periods. Thus, an assessment of the climate-change impact would be a worthwhile outcome. In the future, a more detailed investigation including a higher resolution than 10 km should be conducted to improve the accuracy of the precipitation modeling. An integrated water-cycle simulation using a regional climate model and hydrological model will be performed to evaluate the impact of global warming in the Yellow River.

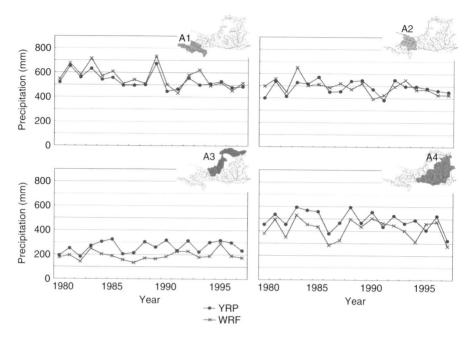

Figure 4.5   Simulated (WRF) and observed (YRP) annual precipitation for four regions from 1980 to 1997 over the Yellow River basin at 20 km horizontal resolution.

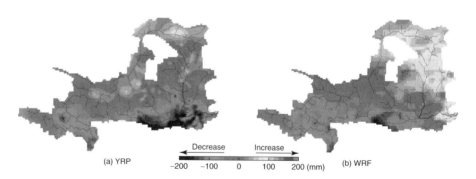

Figure 4.6   Comparison of the decadal annual precipitation difference between the 1990s and 1980s between (a) the YRP observed dataset and (b) the WRF simulation. (See insert for color representation.)

4.4   Numerical experiment of river runoff

For the long-term runoff analysis, we used 41 years (1960–2000) of daily meteorological observation data from 128 stations. We interpolated these data onto a 0.1° grid as input for the semi-distributed hydrological model. Then, to predict the evapotranspiration loss from various land-use types, we applied high-resolution satellite remote sensing data. The remote sensing data included the elevation, a land-surface classification map, and normalized difference vegetation index datasets, which were also converted into the 0.1° grid scale. The hydrological model was based on the soil-vegetation-atmosphere transfer and hydrological cycle (SVAT-HYCY) model, which was composed of the following three submodels: a one-dimensional heat balance model of the land surface, a runoff formation model, and a river-routing network model. However, there were many anthropogenic factors such as irrigation water withdrawals, large reservoir operations, and human-induced land-use changes in the Yellow River basin. We considered these artificial factors in our model by applying simple submodels for irrigation water intake, reservoir operation, and land-use change. The details of the model structure and parameters have been summarized by Sato *et al*. (2007, 2008, 2009).

The performance of the hydrological model simulations indicated the annual runoff from the source area to the lower reaches during the past 40 years (Figure 4.7). In this analysis, the subwatersheds of the Yellow River were extended to six, namely A1 to A4 and A5 to A6 (see Figure 4.1). Although we did not consider the influence of long-term land-use changes in this model simulation, the observed discharges were reasonably captured by the model, except in A4. Therefore, the influence of land-use changes on the long-term water balance of the Yellow River basin should not be severe.

4.5 Hydrological impact due to human-induced land-use changes

Massive land-use changes for soil and water conservation, including land terracing, afforestation (tree and grass planting), and silt-control (check-dam building) projects, have occurred since the 1970s in the middle reaches (A4) of the Yellow River basin. This land surface engineering has increased the amount of rainfall infiltration by reducing the surface overland flow. To clarify the influence of soil and water conservation more precisely, we investigated the influence of the rainfall infiltration (soil permeability) as well as that of vegetation changes.

The potential influences of soil and water conservation on the long-term water balance in the middle reaches of the Yellow River basin were calculated (Figure 4.8). In this analysis, we tried to compare the changes in river runoff, evapotranspiration, and surface (overland) flow due to soil and water conservation by the model simulation. According to these results, we found that the soil and water conservation decreased the river discharge by about 10–50%, increased the evapotranspiration by about 2–13%, and decreased the surface overland flow by about 14–74%. These results suggested that soil and water conservation decreases not only soil erosion by decreasing surface (overland) flow, but also the available water resources in the middle reaches of the Yellow River basin by increasing evapotranspiration loss due to increased vegetation.

4.6 Long-term water balance in the Yellow River basin

There were two drops in the observed annual river runoff in the main stream of the Yellow River (Toudaoguai and Lijin), which probably reflected water withdrawal from the river channel for the large irrigation districts located in these relatively arid regions (Figure 4.9a). Observed values decreased at all hydrological stations

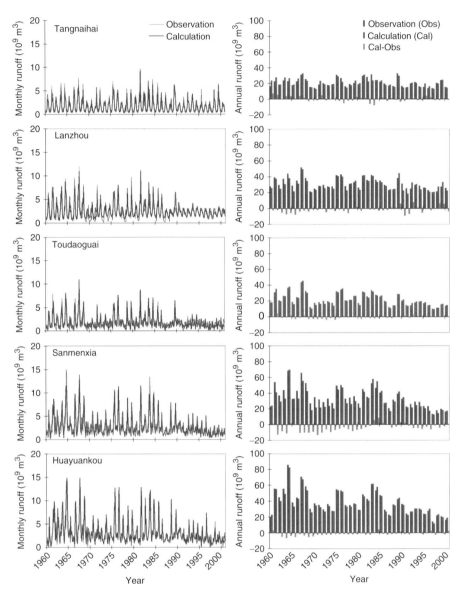

Figure 4.7 Simulated monthly and annual runoffs in the Yellow River basin from 1960 to 2000.

(Figure 4.9b). For example, the observed runoff in the 1990s at Lijin decreased by almost 36 billion m³ compared to that of in the 1960s. The data also indicated that there are hydrological "sources" and "sinks" in the Yellow River basin (Figure 4.9c). From this figure, we can see that most of the river water was supplied from A1 and A4. The amount of water supplied from the middle reaches decreased drastically. However, although the amount of water consumption did not change much at A3, it increased rapidly in the lower reaches (Figure 4.9c). Finally, the water shortages in the lower reaches of the Yellow River basin (e.g., the decrease at Lijin) were induced

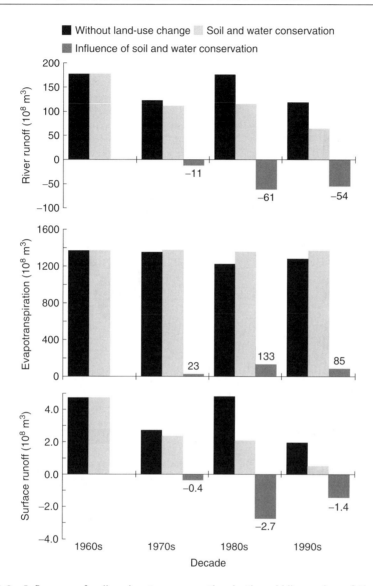

Figure 4.8 Influences of soil and water conservation in the middle reaches of the Yellow River basin.

by the following two factors: (1) an increase in water consumption within the lower reaches (31%) and (2) a decrease in the water supply from upstream of Huayuankou (69%) (Figure 4.9d).

4.7 Water saving effect in the Yellow River basin

Currently, agricultural irrigation is the main water use. Moreover, flood irrigation is still used in most regions (Ma *et al.* 2010). The water use efficiency using flood

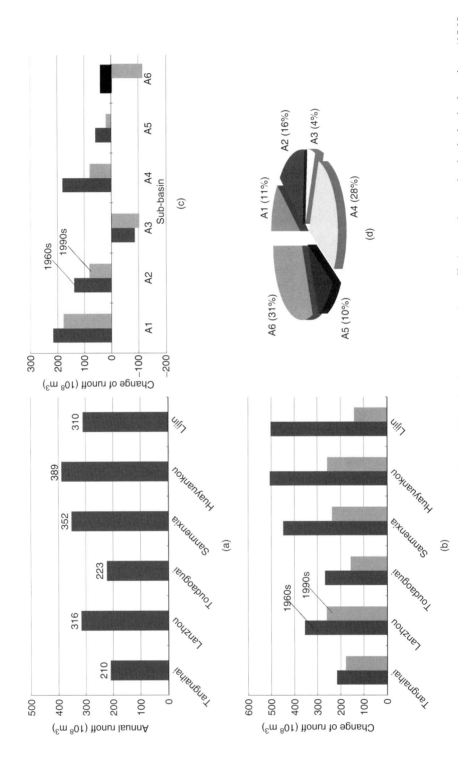

Figure 4.9 Analysis of long-term water balance in the Yellow River basin: (a) annual runoff observed at major hydrological stations (1960 to 2000), (b) runoff change between the 1960s and 1990s, (c) runoff change for each sub-basin, and (d) available water decrease in each

irrigation is <0.5, indicating that approximately 50% of the water taken from the river is wasted. There is a large potential for water saving with flood irrigation.

To evaluate possible effects of water conservation, a numerical experiment was carried out by Ma *et al*. (2010) with an assumption of no changes in irrigated area in the basin for the period 1980–2001. Water saving was calculated by reducing the number of irrigation days because agriculture irrigation takes place using a rotation system for each region. The result indicated that the total water use in the area from A1 to A5 was 20.5 billion m^3 under current conditions. This value of water use would be decreased further by 26.6% (5.5 billion m^3) and 52.1% (10.7 billion m^3) by decreasing the number of irrigation days by 20% and 40%, respectively. Such water saving is feasible by improving maintenance of the canal system and using an efficient irrigation method (e.g., drip irrigation) instead of flood irrigation with a unified management of the dam operation system.

4.8 Conclusions

An obvious cyclic variation in precipitation was detected based on data analysis over the past four decades, from the 1960s to the 1990s, in the source areas of the Yellow River basin. It was confirmed that the basin was relatively wet in the 1960s and 1980s and dry in the 1970s and 1990s. To examine the changes in precipitation over the basin, the regional climate model WRF was run for the period from 1980 to 1997.

A resolution sensitivity experiment was conducted for the annual precipitation amount in four years (1986, 1989, 1992, and 1997). The results showed that the characteristics of each year could be represented over the basin. The simulated precipitation amounts were underestimated compared to the dataset in four regions in the upper to middle reaches, but the difference between the 20 km and 10 km resolutions was limited.

A long-term numerical experiment was run for the period 1980 to 1997 at 20 km resolution. The simulated regional annual precipitation amounts were underestimated in A3 and A4. The interannual variation in precipitation was represented for these regions. The distribution of annual precipitation over the whole basin was almost reproduced. In addition, the estimated regions of decreasing precipitation between the periods 1990–1997 and 1980–1989 agreed closely with the dataset.

To clarify the long-term water balance in the Yellow River basin, a semidistributed hydrological model that could consider both human activity and climate variations was introduced and applied to the whole Yellow River basin. The main findings were as follows: (i) the hydrological impact of land-use changes might not be too severe except in the middle reaches (Loess Plateau), (ii) massive land-use change (soil and water conservation) in the Loess Plateau decreased not only soil erosion but also the available water resources by about 10–50%, and (iii) the relative contributions to water shortages in the Yellow River basin due to the increase in water consumption within the lower reaches and the decrease in the water supply from upstream of Huayuankou were 31% and 69%, respectively.

To mitigate water shortage by climate change, employing water savings in agriculture is a major challenge for the Yellow River basin. The numerical experiment indicated that the effect of water conservation was clear and water use for irrigation would be decreased by 26.6% to 52.1% by improving maintenance of canals and using drip irrigation instead of flood irrigation. Such measures could form the basis for a mitigation strategy to keep the river flow from drying up under future climate change scenarios.

4.9   Acknowledgements

The work on precipitation simulation was supported by the Research Program on Climate Change Adaptation of the Ministry of Education, Culture, Sports, Science and Technology, Japan and by the Environment Research and Technology Development Fund (S-8) of the Ministry of the Environment, Japan.

References

Lin, Y.-L., Farley, R.D., and Orville, H.D. (1983) Bulk parameterization of the snow field in a cloud model. *J. Climate Appl. Meteor.*, **22**, 1065–1092.

Liu, C. and Zheng, H. (2004) Changes in components of the hydrological cycle in the Yellow River basin during the second half of the 20th century. *Hydrol. Process.*, **18**, 2337–45.

Ma, X., Fukushima, Y., Yasunari, T., *et al.* (2010) Examination of the water budget in upstream and midstream regions of the Yellow River. *China Hydrol. Process.*, **24**, 618–30.

Sato, Y., Ma, X., Matsuoka, M., and Fukushima, Y. (2007) Impacts of human activity on long-term water balance in the middle-reach of the Yellow River basin. *IAHS Publication*, **315**, 85–91.

Sato, Y., Ma, X., Xu, J., *et al.* (2008) Analysis of long-term water balance in the source area of the Yellow River basin. *Hydrol. Process.*, **22**, 1618–29.

Sato, Y., Onishi, A., Fukushima, Y., *et al.* (2009) An integrated hydrological model for the long-term water balance analysis of the Yellow River Basin, China, in *From Headwaters to the Ocean* (eds. M. Taniguch, W.C. Burnet, Y. Fukushima, *et al.*), Taylor & Francis Group, London, pp. 209–15.

Xie, P., Yatagai, A., Chen, M., *et al.* (2007) A gauge-based analysis of daily precipitation over East Asia. *J. Hydromet.*, **8**, 607–26.

Yang, D., Li, C., Hu, H., *et al.* (2004) Analysis of water resources variability in the Yellow River of China during the last half century using historical data. *Wat. Resour. Res.*, **40**, W06502, doi:10.1029/2003WR002763.

5

Water Resources under Climate Change in the Yangtze River Basin

Marco Gemmer, Buda Su, and Tong Jiang

National Climate Center of the China Meteorological Administration, China

5.1 Introduction

Both in terms of average discharge and a length of nearly 6400 km, the Yangtze River is the third longest river in the world. It has a basin area of 1.8 million km^2 (see Figure 5.1) that hosts more than 400 million people, who mostly live in the densely populated east of the basin. An example is the Yangtze River Delta Region, which covers 1% of China's land surface, houses 6% of the nation's population, and produces 20% of the nation's GDP. The river flow and the water level of the Yangtze River and its tributaries vary in a broad range. There have been numerous dramatic floods and drought periods in the Yangtze River Basin, from early civilization until recently.

A large part of the Yangtze River Basin has subtropical monsoon climate. The distribution of precipitation over the Yangtze River Basin is very uneven in space and time. The annual mean precipitation in the basin varies from 270–500 mm in the western region to 1600–1900 mm in the southeastern region (Gemmer *et al.* 2008). Rainfall is concentrated during the summer season, and during the flood season from May to October precipitation may make up three-quarters, or more, of the annual total.

An annual climatic feature in the Yangtze River Basin is the onset of the *mei-yu* season (or plum-rain season, as it coincides with fruit maturation season) for 20 days in May (Jiang, Kundzewicz, and Su 2008). This quasi-stationary rain front may become stationary over the middle/lower Yangtze River Basin and can bring abundant precipitation (Zhang, Zhai, and Qian 2005). The onset and length of the *mei-yu* season varies largely with the Southeast Monsoon (from the northwest Pacific) and the Southwest Monsoon (from the Bay of Bengal) (Becker, Gemmer, and Jiang 2006). During the so-called "zero *mei-yu*" years, however, the monsoon precipitation arrives comparatively late and moves quickly northwards to the Huanghe River Basin. These years are

Climatic Change and Global Warming of Inland Waters: Impacts and Mitigation for Ecosystems and Societies,
First Edition. Edited by Charles R. Goldman, Michio Kumagai and Richard D. Robarts.
© 2013 John Wiley & Sons, Ltd. Published 2013 by John Wiley & Sons, Ltd.

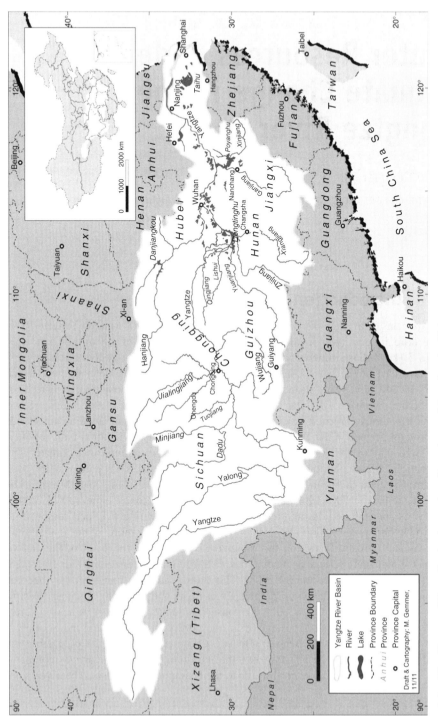

Figure 5.1   The Yangtze River Basin, its main tributaries and location in China.

marked by extensive droughts. The distinct spatio-temporal variation of precipitation often leads to drought/flood hazard in different sub-basins of the Yangtze River at the same time or to a spring drought and late summer flood at the same place (Gemmer *et al*. 2008).

The flood/drought hazard depends on a combination of anthropogenic and natural factors. Deforestation, agricultural land expansion, urbanization, construction of roads, reclamation of wetlands and lakes, and strengthening of embankments of the river has been progressing in the last 50 years due to the growing population pressure and socio-economic development. These human activities reduce the flood storage capacity in the basin, increase the value of the runoff coefficient, and aggravate the flood hazard (Gemmer 2004). In turn, more people rely on the extensive irrigation system and below-average precipitation causes droughts as the reservoirs and rivers do not provide sufficient water for agricultural, domestic, and industrial use.

A significant share of the global economic losses due to floods in the last decades has been recorded in China and, according to the Munich Reinsurance Company (Berz and Kron 2004), the floods on the Yangtze and Songhua rivers in 1998 and the 1996 floods on the Yangtze, Huanghe and Huaihe rivers caused material damage of 30.7 billion USD and 24 billion USD, respectively (nominal 1998 and 1996 price level). Droughts in the north, northeast, and southwest of China in 2009 have caused about 18 billion USD direct economic losses.

Decadal sequences of flood and drought years are historically documented. Floods and droughts can be traced back to the Tang Dynasty (608–906 AD). Between the second and twentieth centuries 178 extreme flood events were recorded in the Yangtze River Basin (Gemmer 2004). Flood events never cover the entire Yangtze River Basin and even local or regional floods can cause tremendous losses even if the main channel is not affected.

Influenced by the increasing summer precipitation and rainstorm trends, the summer runoff and flood discharge of the lower Yangtze region has shown an upward trend in recent decades (Jiang, Kundzewicz, and Su 2008). Both annual and extreme precipitations have shown a positive trend in recent decades in the middle and lower Yangtze reaches. Precipitation extremes are projected to further increase in future, thus increasing the spatio-temporal variability. This might impose further risks for the socio-economic sphere in the Yangtze River Basin as the extensive floods in the east of the basin and drought in the southwest have recently shown.

5.2 Observed changes of precipitation and temperature

5.2.1 Precipitation

China's climate regime has a strong variability in the time scale of decades (Qian *et al*. 2007). Whereas the annual precipitation sums in the Yangtze River Basin do not show any significant long-term trend during the last century, a significant increase of rainfall in months of maximum precipitation between 1950 and 1999 can be detected (Becker, Gemmer, and Jiang 2006). Precipitation in the Yangtze River Basin showed a tendency towards a concentration of summer rainfalls and an increasing number

Table 5.1 Observed precipitation trends in the Yangtze River Basin in the second half of the twentieth century.

Area	Tendency
Upper reaches	Significant increase in July Significant decrease from January to March
Middle reaches	Significant increase in January, June and July Significant decrease in May, September, and October
Lower reaches	Significant increase in January, June, July Significant decrease in April

of heavy rain days within a shorter time period (Gemmer, Becker, and Jiang 2004; Becker, Gemmer, and Jiang 2006; Su, Gemmer, and Jiang 2008).

As highlighted in Table 5.1, precipitation in the middle reaches of the Yangtze River significantly decreased in May and took a significant positive trend in June and July after 1950. The trend of summer precipitation in June and July is critical regarding floods, especially in the middle reaches of the Yangtze River. Changes of the precipitation extremes from May to October in the past 50 years are of direct importance to the occurrences of high and low river flow in the Yangtze River Basin (Gemmer *et al*. 2008). The summer peaks are increasing and the water availability in autumn is decreasing.

Precipitation extremes in the Yangtze River Basin tended to increase during the past 50 years, for example, the average intensity per rain day above $10\,\text{mm}\,\text{d}^{-1}$ increased (Su *et al*. 2005; Su, Jiang, and Jin 2006; Su, Gemmer, and Jiang 2008; Jiang, Kundzewicz, and Su 2008; Chen, Chen, and Ren 2011). In the upper, middle, and lower reaches of the Yangtze River, precipitation trends have shown spatio-temporal disparities. In Wuhan, for example, more than 90% of the rain storms (precipitation $>50\,\text{mm}\,\text{d}^{-1}$) occur during flood season from May–September. Precipitation amounting to 471 mm fell in Wuhan from 21–25 July 1998, and contributed to a discharge increase of the Yangtze River by $11\,000\,\text{m}^3\,\text{s}^{-1}$.

Climate indices that increase winter and late spring droughts (negative monthly precipitation trends, extension of dry spells) shifted to more extremes. Flood hazards (positive monthly precipitation trends in June and July, increase of extreme precipitation) showed significant trends in the Yangtze River Basin in the second half of the twentieth century. This had a direct impact on the availability of surface water resources.

5.2.2 Temperature

We have analyzed daily temperature records of 134 meteorological stations of the China Meteorological Administration in the Yangtze River Basin from 1961–2008. We can conclude that daily minimum temperature increased significantly almost in the entire basin during the reference period. Daily maximum temperature increased mostly in the source region (West) and northern part of the Yangtze River Basin. The annual number of frost days (annual sum of days when daily minimum temperature falls below $0\,^\circ\text{C}$) decreased significantly over the entire basin. The annual occurrence of ice days (annual sum of days when daily maximum temperature falls below $0\,°C$) shows a decreasing trend in the middle and lower reaches of the Yangtze River Basin. The annual occurrence of summer days (annual sum of days when daily maximum temperature rises above $25\,°C$) increased significantly in the Yangtze River Basin except for the source region (west) and the southern parts. The annual occurrence of tropical nights (annual sum of days when daily minimum temperature lies above $20\,°C$) increased in the basin except for the source region.

As daily minimum and maximum temperature has increased since the early 1960s, and the warm spell duration indicator (annual count of days with at least six consecutive days when daily maximum temperature falls above the 90th percentile centered on a five-day window for the base period) increased in the source region and the northern parts of the basin.

The findings on temperature trends are similar to those in other river basins of China—indices for annual, monthly, minimum and maximum temperature—have increased, and the general warming trend can be detected especially in winter.

5.3   Observed changes of streamflow and flood/drought indices

The discharge of the Yangtze River gradually increases downstream following cumulative contribution from numerous tributaries, reaching a maximum at the downstream-most station at Datong. Since the early 1960s, the streamflow of the Yangtze River decreased in the upper reaches and increased in the middle and lower reaches (Zhang *et al.* 2005). For the Yangtze Delta, flood and drought conditions have been reconstructed for more than 1000 years. The periods featured by a transition from a warm period to a cold period indicated a change to dry conditions every 25–28 years, while the transition from cold to warm climate was accompanied by wet conditions (Zhang, Gemmer, and Chen 2008) in the Yangtze Delta. Recently, the period from 1950–1980 showed a higher frequency of drought events.

Wetness conditions in the middle and lower reaches of the Yangtze River and dryness conditions in some tributaries of the upper Yangtze River in Sichuan Province increased (Zhai *et al.* 2010a). The increase of drought was less severe in the upper reaches than in the middle reaches of the Yangtze River (Su, Gemmer, and Jiang 2008). The higher frequency of droughts did not only result from lower precipitation (see Introduction), but also from other natural factors influencing water balance conditions in these basins. This can be reconfirmed by distinct trends in simulated and actual evapotranspiration in the Yangtze River Basin since 1960, which was caused by changes in net radiation, cloud cover, and surface wind speeds (Wang *et al.* 2011). Soil moisture, which is the moisture available for plants in the development stage and defining drought, was related to precipitation as well as to evaporative demand, which in turn was affected by surface air temperature, solar radiation, and other variables. Soil moisture trends were therefore not necessarily consistent with trends in precipitation.

This shows that not only the trends of climate indices are relevant but also the selection of indices describing extremes (e.g., choosing annual or monthly indices) in regional studies. Indeed, dryness/wetness conditions have changed in the sub-basins of the Yangtze River as described above. A fivefold increase of withdrawal

Table 5.2   Streamflow trends of the Yangtze River in the second half of the twentieth century.

Area	Tendency
Upper reaches	Significant increase in February and July Significant decrease in May, October, November
Middle reaches	Significant increase in July Significant decrease in autumn
Lower reaches	Significant increase from January–March, July and August Significant decrease in May

and consumption of water for agricultural, domestic, and industrial use has had relatively small impact on annual and monthly water discharges (Xu *et al.* 2008b). However, the impacts are imminent regionally and on a different timescale.

In the Poyang Lake Basin for instance, the largest freshwater lake in China, which marks the border between the middle and lower reaches of the Yangtze River, streamflow has increased both on annual and seasonal scales since the early 1960s (Zhao *et al.* 2010).

Monthly Yangtze flow records from 1961 to 2000 showed a statistically significant upward trend in February and July, and a downward trend in May, October, and November at Yichang station (Jiang, Kundzewicz, and Su 2008). Upward trends prevailed at the Datong station (January, February, March, July, and August), whereas a downward trend was found only in May. The data also indicated that a significant upward trend at both gauging stations occurred in July—the month with the highest discharge in Yichang and Datong—during which numerous floods have taken place (see Table 5.2).

Flood discharges (accumulated and maximum discharges during the flood season) in the middle and lower reaches of the Yangtze River showed significant upward trends in the last 40 years, which were caused by the increasing intensity and frequency of rainstorms. The annual maximum discharge (during flood season in summer) of the Yangtze River has especially increased significantly (Jiang, Kundzewicz, and Su 2008). The natural runoff increased in summer and decreased greatly in autumn (Xu *et al.* 2008a). This natural change in runoff in the upper Yangtze River basin implies an increasing flood risk in summer and water shortage in autumn.

5.4   Projections

Statistical extrapolations suggest peaks in the frequencies of extreme precipitation events for most of the Yangtze River's sub-basins around the years 2012–2013 and 2017–2018 even though major differences between the sub-basins are expected (Becker *et al.* 2008). Based on the results of General Circulation Models (GCMs) the annual mean precipitation will increase in the Yangtze River Basin until 2040 and the annual cycle shows an extension of the Asian summer monsoon season with increasing rainfall. The coupling between monsoon circulation and monsoon rainfall strengthens (Orlowsky *et al.* 2010).

Heavy precipitation events for single days and pentads are projected to increase in their intensity. A larger fraction of the total annual precipitation is projected to occur during heavy precipitation events (rain days exceeding the 95th percentile) and precipitation is projected to increase in the middle and lower Yangtze River Basin (Piao *et al*. 2010).

High probabilities for the future intensification of interannual variability of precipitation in both winter and summer over the Yangtze River Basin are projected. This means that the interannual standard deviation of precipitation increases by more than by 20% and precipitation extremes increase, in turn aggravating dryness and wetness episodes (Chen, Jiang, and Li 2011).

Future projections of climate extremes derived from an ensemble of coupled general circulation models (CGCMs) diagnose a longer duration of heat waves in the Yangtze River Basin in the twenty-first century than today and more frequent warm nights (Xu *et al*. 2009). The annual average temperature will increase resulting from warming trends during all seasons.

Climate models project wetter conditions over much of China. This also pertains to the principal flood season, June to August. Since the low-lying and broad valley of the lower part of the Yangtze River basin is densely populated, the region is very vulnerable to flooding and inundations cause very high material damage. If the presently observed trend continues into the future, higher flood risk in the lower part of the basin can be expected. Projected dryness/wetness pattern in the future 50 years under the SRES-A2 scenario (high emissions) are similar to the observed one in 1961–2000 (Zhai *et al*. 2009), while an increase of wetness conditions is projected for the Yangtze River Basin (Zhai *et al*. 2010b). Hydrological models driven by future climate simulations predict an overall increase in river runoff in China by 2100, due to the increase in precipitation. From 2020–2040 little change is projected in annual streamflow, but an increase after 2040 becomes more evident after 2060. It is found that runoff responds more evidently to the middle- and high-emission scenarios (A1B and A2), and less evidently to the lower scenario (B1) (Huang, Yang and Chen, 2010). Although no significant trend in annual streamflow is projected for Yichang station (upper/middle reaches) and Datong station (lower reaches), seasonal changes are significant (Zeng *et al*. 2011). Hirabayashi *et al*. (2008) compared the flood risk projections for 2071–2100 with a control period from 1901–2000. What used to be a 100-year flood in China in the control period is likely to occur much more frequently in the future. Under changed climate, these flood events might have a return period of 50 years or less.

The projections of higher surface temperatures in summer and winter (average, maximum and minimum) and the projected increase of precipitation extremes and dryness/wetness patterns in the Yangtze River Basin put water resources management tasks at risk. Seasonal projections derived from Liu *et al*. (2009) show the variation of decadal runoff depth in the Yangtze River Basin as compared to the 2000s under different emission scenarios in the GCM ECHAM5/MPI-OM (Figure 5.2). As can be seen, the projections generally predict drier conditions in autumn and winter and wetter conditions in summer, although the projections vary broadly. The projected runoff trends, also derived from Liu *et al*. (2009) and averaged from 2001–2050 for the Yangtze River Basin under the SRES-A2 scenario (high emissions, the current pathway) show that runoff depth increases significantly in the source area and middle reaches of the basin (Figure 5.3). However, this decreases significantly in the southwest

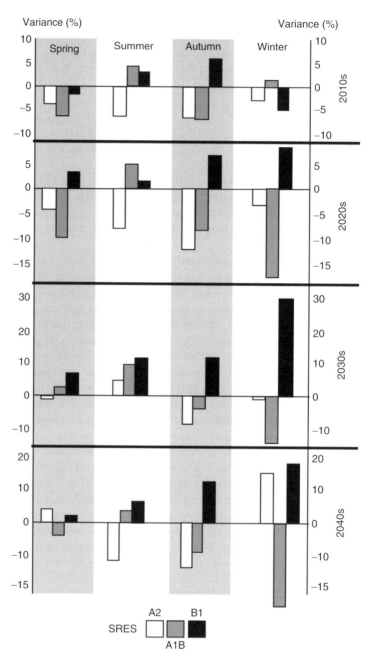

Figure 5.2   Change rates of the ECHAM5-projected seasonal runoff depth of the Yangtze River in decades until 2050 under the three SRES emission scenarios relative to 2000s (2001–2010).

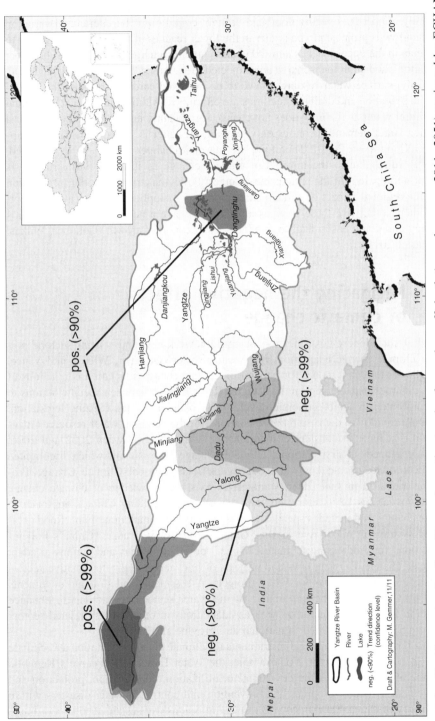

Figure 5.3 Significance levels of positive and negative trends of annual runoff depth (annual average 2011–2051) projected by ECHAM5 for SRES-A2 scenario in the Yangtze River Basin.

of the basin, which is the area recently hit by long-lasting drought episodes. From this figure it can be derived that runoff depth will be strongly reduced in the upper reaches and likely impact the existing flood and drought conditions. The interannual variation of streamflow is projected to be higher in the lower reaches of the Yangtze River (up to 30%) than in the upper reaches until 2050 and winter discharge is projected to be lower while spring and autumn discharge are projected to increase slightly (Zeng *et al*. 2011).

Already, surface water resources have to be managed carefully because the investments in irrigation and drainage facilities are enormous. Although the projected annual agricultural water availability does not change significantly until 2040 and agricultural irrigation water demand increases only slightly, the latter is sensitive to daily precipitation (Xiong *et al*. 2010). Therefore, the seasonal changes have to be investigated. China is vulnerable to the impacts of climate change as its water infrastructure is weak or unprepared for the anticipated changes. Policymakers need better information about the regional impact of climate change on water supplies, and on ways of adapting to them, otherwise these changes are likely to have a negative effect on China's economy and could easily lead to a reduction in economic growth (Gemmer, Wilkes, and Vaucel 2011).

5.5   Mitigating the negative effects of climatic change

In general, the Chinese government is using framework laws and existing sectoral policy for climate change adaptation in the water sector (Gemmer, Wilkes, and Vaucel 2011). So far the main modes for integrating climate change adaptation into actions of water resources management in the Yangtze River Basin have been specific actions to be incorporated in socio-economic development plans under the existing legislation. The future trend is to develop more regulations for the actors in water resources management that have to consider climate change adaptation. Between 2002 and 2006, China developed its first National Climate Change Assessment, which highlighted existing knowledge and key knowledge gaps. The National Climate Change Programme, released in June 2007, summarizes existing knowledge of climate change impacts in various key sectors, outlines the principles guiding China's approach to climate change, and includes specific actions such as the completion of anti-flood engineering systems in major river basins. China's 2008 White Paper, "China's Policies and Actions for Addressing Climate Change," presents policies and actions to adapt to climate change. Existing laws and projects of relevance to climate sensitive sectors are listed, along with areas for future projects. Assessments and strategic documents issued by China stress the importance of the water sector. Key actions to enhance adaptation in the water sector, as well as adaptations in other closely related sectors such as agriculture, are listed in national and provincial plans.

China has many existing legal instruments for climate change adaptation and climate proofing in the water sector. For example, the Water Law of the People's Republic of China of 2002 governs the development, utilization, conservation, protection and management of water resources and prevention and control of water disasters within the territory of China. Special River Basin Commissions were established to coordinate river basin management in seven major river basins, and management plans for each have been drawn up.

In China, many academic studies have been conducted indicating possible changes in flood risks due to climate change, but legally binding climate change adaptation planning has not yet been mainstreamed. Instead, we find in the national and provincial adaptation plans that further investments in water infrastructure are proposed. Standards for these investments that consider future flood return periods have yet to be developed.

The Chinese government has been stepping up the efforts in water resources management by enhancing the water supply capacity for urban and rural areas, flood prevention systems and the construction of farming infrastructure. Based on the twelfth national five-year economic development plan, to 2020, the national investment in water conservancy projects will account for 4 trillion Yuan ($612 billion). In order to overcome the projected variability of surface water resources in the Yangtze River Basin, strict water resource management measures to limit the scale of water exploitation, improve the efficiency of water usage, and curb water pollution have been established. Investment in water conservancy reached 142.7 billion Yuan in 2009, some of which are for the implementation of key water projects and for water resources development, utilization and allocation. China promotes the concept of a water-saving society and pushes for the demand side management of water, market-based water rights, water withdrawal licensing, and institutional mechanisms for accounting. The concepts, however, are not yet settled in overarching legislation.

China's National Defense Planning for Meteorological Hazards (2009–2020) was launched by the China Meteorological Administration (CMA) and approved by the State Council in order to strengthen urban, rural, and coastal areas, major river basins, strategic economic areas, major transportation, and power transmission lines. The plan suggests measures for meteorological disaster prevention and depicts engineering and nonengineering measures in order to improve the ability for meteorological disaster-prevention capacity and also serves climate change adaptation. It can be seen as the outline for an overarching management plan in order to deal with the projected increases in precipitation and temperature extremes.

There is no doubt that flood risk has grown in many places in China and is likely to grow further in the future due to a combination of anthropogenic and climatic factors. Intense precipitation increases in a warming climate. However, reliable and detailed quantification of aggregate flood statistics are very difficult to obtain for the past and the present and nearly impossible to obtain for the future. The country has embarked upon a vigorous and ambitious task to improve flood preparedness by both structural ("hard") and non-structural ("soft") measures. The former refer to such defenses as dikes, dams and flood control reservoirs, diversions, etc. The latter include implementing watershed management (source control), zoning; insurance; flood forecasting/warning system; increasing awareness etc. Structural measures, both dikes and dams of different sizes, have a very long tradition in China and also play a vital role in flood prevention today. The Chinese government is implementing plans to build national weather-radar networks and to use radar data for real-time flood forecasting in order to improve forecasting accuracy. Coupling of radar data with hydrological models helps overcome differences in spatial resolutions of the remotely sensed data and the hydrological models. In 2010, flood hazard mapping guidelines were published as a professional standard by the Ministry of Water Resources. This was based on experiences collected by way of mapping exercises carried out since 2005 in pilot studies. In addition, increasing attention has been put on the practice of drought management.

For example, proposals and plans have been made to improve the drought monitoring and management capacity in the twelfth five-year national economic plan.

In July 2010, specific funds for global climate change research were provided by the Ministry of Sciences and Technology (MOST) for 19 projects, which include impacts of climate change on the cropping system, Tibet plateau research, regional impacts of weather extremes in East Asia and large-scale land use change and its impact on global change. The scientific results can support legislative decisions supporting the mainstreaming of climate change adaptation into existing framework laws for the water resources sector.

Until now, the main focus of climate change adaptation has remained as infrastructure investment. The Chinese government has approved the construction of several large-scale hydropower dams in the upper reaches of the Yangtze River. The benefits are twofold: they are supposed to improve flood and drought management capacity, overcoming the projected seasonal disparities of water availability, and also support China in meeting its Green Energy targets. However, such large-scale hydropower dams are known to have various negative impacts on the functioning of ecosystems and the socio-economy. The Three Gorges Dam (TGD) is a world-famous example that has been discussed among environmental, social, and economic scientists. Although the TGD is located some 3800 km upstream the Yangtze River's mouth from the Yellow Sea, the hydrological impacts and its capacity for flood risk mitigation have been intensively studied since the beginning of the twenty-first century. King, Gemmer, and Wang (2001), King *et al*. (2004) and Gemmer (2004) discussed the potential role of the TGD in the complex flood protection system for the middle and lower reaches of the Yangtze River. It was concluded that the TGD would not have a significant impact on flood peaks in these reaches as the large tributaries connect to the Yangtze River downstream of the TGD. Eight years after the impoundment was gradually started, however, the hydrological situation proved to be more complex and sensitive to seasonal changes in precipitation. Guo *et al*. (2011) used observed streamflow data for assessing the effect of the TGD. The operation of the TGD has affected the stream flow and water level of the Yangtze River with seasonal and spatial variety. The seasonal variation followed the TGD's seasonal impounding and releasing of water, depending on the impounding/releasing rate and the seasonal flow of the river. The most significant effects were confined in the river reach near the TGD, and were as much as five times those of sections downstream. Thus, the effect weakened downstream by inflows to the Yangtze River from downstream tributaries. Changes in the Yangtze River discharge caused by the TGD further altered the interrelationship between the river and Poyang Lake (mentioned before), disturbing the lake basin's hydrological processes and water resources. A major consequence of such changes has been a weakening in the river forcing on the lake, allowing more lake flow to the river from July to March. This situation was exaggerated between autumn 2010 and summer 2011, when the lake area was hit by a large-scale drought that affected nearly all of China causing water scarcity and economic losses in aqua- and agriculture in the Yangtze River Basin.

The focus of previous studies on the impact of the TGD mainly focused on flood control, improved river transportation and hydropower generation but also societal consequences of resettlement and increased erosion and landslides due to resettlement. In 2011, the TGD and other anthropogenic factors were blamed for the severe Yangtze River drought and even mentioned as a driver for regional climate change. The TGD impacted the local climate in a buffer of 10 km only because the surface area of the reservoir is not much larger than the surface of the river area before impoundment (Wu *et al.* 2011). In climatological terms examining the Standardized Precipitation Index, the drought was a natural extreme event in nearly all sub-basins of the Yangtze River, which would have been mitigated but not prevented with a drought-adapted management plan of the TGD. In fact, inadequate management of the TGD was the main reason why little drought alleviation was achieved. The structure of the annual management plan has not been revised for further issues; no designated measures consider explicit drought control or meteorological disaster prevention, except for flooding.

The China Meteorological Administration (CMA) is the specific agency responsible for drought monitoring and recommendations on drought warnings to the Chinese State Council. CMA was not involved in the Three Gorges Constructions Commission to provide guidance to the annual management plan for TGD operations. At the end of the devastating drought, the Chinese State Council mandated the release of additional water from the TGD reservoir, which alleviated the situation to some extent. Pre-disaster measures would have lessened the adverse effects of the drought even further. We recommend that the annual management plan be revised by including additional measures, such as natural hazards control and essential knowledge of related stakeholders. Finally, it should be noted that the TGD is the largest hydropower dam in the world (based on its capacity) but the waterscape in the Yangtze River Basin is dominated by an additional 50 000 reservoirs, which are distributed over all tributaries. Making socio-economic use of the water, the river has always and will always have to be regulated during the diverse seasons. Under the projected impact of climate change and the hydrological impact of additional large-scale hydropower dams, water regulation has to become more advanced, timely, and interactive. It shows that climate change adaptation measures and practices do not only have to consider climate itself, but also other specific factors, such as the economy, technology, as well as social and cultural norms.

References

Becker, S., Gemmer, M., and Jiang, T. (2006) Spatiotemporal analysis of precipitation trends in the Yangtze River catchment. *Stochastic Environ. Res. and Risk Assess.*, **20**, 435–444.

Becker, S., Hartmann, H., Coulibaly, M., *et al.* (2008) Quasi periodicities of extreme precipitation events in the Yangtze River Basin, *China. Theoret. Appl. Climat.*, **94**, 139–52.

Berz, G. and Kron, W. (2004) Überschwemmunskatastrophen und Klimaänderung: Trends und Handlungsoptionen aus (Rück-) Versicherungssicht, in *Warnsignal Klima: Genug Wasser für alle?* (eds. J.L. Lozan, H. Graßl, P. Hupfer, *et al.*) Wissenschaftliche Auswertungen, Hamburg, Germany, pp. 264–9.

Chen, W., Jiang, Z., and Li, L. (2011a) Probabilistic projections of climate change over China under the SRES A1B scenario using 28 AOGCMs. *J. Climate*, **24**, 4741–56.

Chen, Y., Chen, X., and Ren, G. (2011b) Variation of extreme precipitation over large river basins in China. *Adv. Clim. Change Res.*, **2**, 108–14.

Gemmer, M. (2004) *Decision Support for Flood Risk Management at the Yangtze River by GIS/RS-based Flood Damage Estimation*, Shaker, Aachen.

Gemmer, M., Becker, S., and Jiang, T. (2004) Observed monthly precipitation trends in China 1951–2002. *Theoret. Appl. Climat.*, **77**, 39–45.

Gemmer, M., Jiang, T., Su, B., and Kundzewicz, Z.W. (2008) Seasonal precipitation changes in wet season and their influence on flood/drought hazards in the Yangtze River basin, *China. Quatern. Internat.*, **186**, 12–21.

Gemmer, M., Wilkes, A., and Vaucel, L.M. (2011) Governing climate change adaptation in the EU and China: An analysis of formal institutions. *Adv. Climate Change Res.*, **2**, 1–11.

Guo, H., Hu, Q., Zhang, Q., and Feng, S. (2011) Effects of the Three Gorges Dam on Yangtze River flow and river interaction with Poyang Lake, China: 2003–2008. *Journal of Hydrology*, http://dx.doi.org/10.1016/j.jhydrol.2011.11.027 (accessed June 20, 2012).

Hirabayashi, Y., Kanae, S., Emori, S., *et al.* (2008) Global projections of changing risks of floods and droughts in a changing climate. *Hydrol. Sci. J.*, **53**, 754–73.

Huang, Y., Yang, W.F., and Chen, L. (2010) Water resources change in response to climate change in Changjiang River basin. *Hydrolog. Earth Sys. Sci. Discussion*, **7**, 3159–88.

Jiang, T., Kundzewicz, Z.W., and Su, B. (2008) Changes in monthly precipitation and flood hazard in the Yangtze River Basin, China. *Internat. J. Climat.*, **28**, 1471–81.

King, L., Gemmer, M., and Wang, R. (2001) Hochwasserschutz und Landnutzung am Yangtze. *Geogr. Rundschau*, **53**, 28–34.

King, L., Hartmann, H., Gemmer, M., and Becker, S. (2004) Der Drei-Schluchten-Staudamm am Yangtze—Ein Großbauprojekt und seine Bedeutung für den Hochwasserschutz. *Petermanns Geogr. Mitt.*, **148**, 28–35.

Liu, B., Jiang, T., Ren, G., and Fraedrich, K. (2009) Projected surface water resource of the Yangtze River Basin before 2050. *Adv. Climate Change Res.*, **5**, 54–9.

Orlowsky, B., Bothe, O., Fraedrich, K., *et al.* (2010) Future climates from bias-bootstrapped weather analogs: an application to the Yangtze River Basin. *J. Climate*, **23**, 3509–24.

Piao, S., Piais, C., Huang, Y., *et al.* (2010) The impacts of climate change on water resources and agriculture in China. *Nature*, **467**, 43–51.

Qian, W., Lin, X., Zhu, Y., *et al.* (2007) Climatic regime shift and decadal anomalous events in China. *Climatic Change*, **84**, 167–89.

Su, B., Gemmer, M., and Jiang, T. (2008) Spatial and temporal variation of extreme precipitation over the Yangtze River Basin. *Quatern. Internat.*, **186**, 22–31.

Su, B., Jiang, T., and Jin, W. (2006) Recent trends in observed temperature and precipitation extremes in the Yangtze River Basin, *China. Theoret. Appl. Climat.*, **83**, 139–51.

Su, B., Xiao, B., Zhu, D., and Jiang, T. (2005) Trends in frequency of precipitation extremes in the Yangtze River Basin, China: 1960–2003. *Hydrol. Sci. J.*, **50**, 479–92.

Wang, Y., Liu, B., Su, B., *et al.* (2011) Trends of calculated and simulated actual evaporation in the Yangtze River Basin (1961–2007). *J. Climate*, **24**, 4494–507.

Wu, J. Gao, X., Zhang, D., *et al.* (2011) Regional Climate Model Simulation of the Climate Effects of the Three Gorges Reservoir with Specific Application to the Summer 2006 Drought over the Sichuan-Chongqing-Area. *J. Trop. Meteorol.*, **27**, 44–52.

Xiong, W., Holman, I., Lin, E., *et al.* (2010) Climate change, water availability and future cereal production in China. *Agri. Ecosys. Environ.*, **135**, 58–69.

Xu, J., Yang, D., Yi, Y., *et al.* (2008a) Spatial and temporal variation of runoff in the Yangtze River basin during the past 40 years. *Quatern. Internat.*, **186**, 32–42.

Xu, K., Milliman, J.D., Yang, Z., and Xu, H. (2008b) Climatic and anthropogenic impacts on water and sediment discharges from the Yangtze River (Changjiang), 1950–2005, in *Large Rivers: Geomorphology and Management* (ed. A. Gupta), John Wiley & Sons, Ltd, Chichester, pp. 609–26.

Xu, Y., Xu, C., Gao, X., and Luo, Y. (2009) Projected changes in temperature and precipitation extremes over the Yangtze River Basin of China in the 21st century. *Quatern. Internat.*, **208**, 44–52.

Zeng, X., Kundzewicz, Z.W., Zhou, J., and Su, B. (2011) Discharge projection in the Yangtze River basin under different emission scenarios based on the artificial neural networks. *Quatern. Internat.*, http://dx.doi.org/10.1016/j.quaint.2011.06.009.

Zhai, J., Liu, B., Hartmann, H., *et al.* (2010b) Dryness/wetness variations in ten large river basins of China during the first 50 years of the 21st century. *Quatern. Internat.*, **226**, 101–11.

Zhai, J., Su, B., Krysanova, V., *et al.* (2010a) Spatial variation and trends in PDSI and SPI indices and their relation to streamflow in 10 large regions of China. *J. Climate*, **23**, 649–63.

Zhai, J., Zeng, X., Su, B., and Jiang, T. (2009) Patterns of dryness/wetness in China before 2050 projected by the ECHAM5 model [in Chinese]. *Adv. Climate Change Res.*, **5**, 220–5.

Zhang, Q., Gemmer, M., and Chen, J. (2008) Climatic changes and flood/drought risk in the Yangtze Delta, China, during the past millennium. *Quatern. Internat.*, **176–177**, 62–9.

Zhang, Q., Jiang, T., Gemmer, M., and Becker, S. (2005b) Precipitation, temperature and runoff analysis from 1950 to 2002 in the Yangtze basin, *China. Hydrol. Sci.*, **50**, 65–80.

Zhang, Y., Zhai, P., and Qian, Y. (2005a) Variations of Meiyu Indicators in the Yangtze-Huaihe River Basin during 1954–2003. *Acta Meteor. Sinica*, **19**, 479–84.

Zhao, G., Hoermann, G., Fohrer, N., *et al.* (2010) Streamflow trends and climate variability impacts in Poyang Lake Basin, *China. Wat. Resour. Manage.*, **24**, 689–706.

6

Biogeochemical Ecosystem Dynamics in Lake Biwa under Anthropogenic Impacts and Global Warming

Mitsuru Sakamoto

School of Environmental Science, The University of Shiga Prefecture, Japan

6.1 Introduction

Lakes are of supreme importance to our society owing to their ecosystem services but are extremely vulnerable to external impacts. Substances and energy delivered from the atmosphere and watersheds are more-or-less retained in biological, chemical and physical systems in lakes causing gradual or more drastic changes of these systems. Eutrophication and resulting deterioration of lake water quality and biotic communities throughout the world during the last half century exemplify the vulnerability of lake ecosystems.

Recently, increasing concern has arisen about the negative effects of global warming on lakes (Verburg, Hecky, and Kling. 2003; Schindler 2009). Global warming affects lakes by modifying not only hydrological and material dynamics in and from their watersheds but also their thermal regime, such as mixing dynamics, together with their biogeochemical cycles and related biological activities. The dynamics between lakes and their watersheds, which lead to eutrophication, are also affected by global warming and they combine to create many new issues that have emerged to threaten the world's lakes.

To develop sustainable lake-management plans and policies we must carefully and systematically examine reliable scientific information on lake responses to the combined impacts of eutrophication and global warming in different lake types. As a step toward this goal, in this chapter the pelagic biogeochemical dynamics of Lake Biwa are examined with a focus on how the lake has responded to eutrophication and

Climatic Change and Global Warming of Inland Waters: Impacts and Mitigation for Ecosystems and Societies, First Edition. Edited by Charles R. Goldman, Michio Kumagai and Richard D. Robarts.
© 2013 John Wiley & Sons, Ltd. Published 2013 by John Wiley & Sons, Ltd.

global warming during the last half-century. Possible steps for conserving the lake are suggested based on this analysis. Published information from Shiga Prefecture, Shiga Prefecture Fisheries Experiment Station, Lake Biwa Environmental Research Institute and related scientific reports were analyzed to document the responses of Lake Biwa to these external impacts and to explore ways for conserving the lake from further deterioration. The database of the Meteorological Agency of Japan was also used to examine climate change related impacts.

6.2   Environmental features of Lake Biwa

Lake Biwa is the largest freshwater lake located in the center of the main island of Japan. It is 85 m above sea level and is under the influence of the East Asian monsoon climate. The lake, with a surface area of $670\,km^2$, a volume of $27.5\,km^3$ and a maximum depth of 104 m, consists of a large deep North Basin (mean depth, 43 m) and a small shallow South Basin (mean depth, 4 m). The North Basin, occupying approximately 91.5% of the entire lake area, is monomictic with full lake water overturn in winter and thermal stratification from April to September. Before the mid-1900s, the North Basin was oligotrophic. The minimum dissolved oxygen (DO) saturation of 70% at the lake bottom and a vertical distribution of DO without sharp increases or decreases were recorded in the early 1930s (Kobe Marine Observatory 1935). Since then the lake has gradually become eutrophic and the current trophic state of the North Basin and the South Basin are oligo-mesotrophic and mesotrophic, respectively.

The watershed has an area of $3848\,km^2$ of which 62% is hills and mountain regions covered by dense forest and the remainder is low alluvial plains comprised farm lands, and urban and industrial areas. The lake receives an annual water supply of $4.2\,km^3$ via about 460 rivers and discharges the water from one outlet river and two man-made canals giving a lake water residence time of 5.5 years according to recent records (Kinki Regional Bureau, MLLT 2005).

The lake has undergone considerable nutrient enrichment that resulted in a deterioration of lake water quality during the two decades after 1960. Due to reduced loading of waste water by governmental environmental control a significant recovery of water quality occurred after 1985. However, after 1990 new environmental issues such as continued higher air temperature and less precipitation in summer, and deterioration of Lake Biwa's water quality due to the development of cyanobacterial blooms and development of an anoxic hypolimnion, have taken place successively. Global warming was placed under intensive examination as the factor causing these new environmental issues.

6.3   Global warming and its effects on the thermal and hydrological regimes in Lake Biwa

A gradual increase in atmospheric temperature, referred to as global warming, has now progressed around the globe. According to IPCC IV assessment report (IPCC 2007), the global increase of air temperature was 0.56 to 0.92 °C over 100 years (1906 to

2005). Global warming was also associated with a change in the amount and pattern of precipitation. Figure 6.1 shows the 100 year trend of climate at Hikone located on the northeastern shore of Lake Biwa from the database of the Meteorological Agency of Japan (http://data.jma.go.jp/obd/ stats/ etrn/). The graphs indicate that annual air temperature has increased at the mean rate of 0.026 °C year^{-1} and annual precipitation has decreased at a mean rate of 3.23 mm y^{-1} over 50 years from 1960 to 2010. Summer weather after 1960 was characterized by frequent alternating periods of heavier precipitation and less or no precipitation as compared with the situation before 1960.

The first report of the effects of climate change on the thermal regime in Lake Biwa was by Hayami and Fujiwara (1999). They provided a detailed examination of the monitoring data of lake water temperature covering more than 33 years. They found that the interannual water temperature increase rate (0.08–0.12 °C year^{-1}) was greater than that of air temperature and became greater with depth, reaching a maximum increase rate of 0.041 °C year^{-1} in the bottom layer. A warm winter was associated with a larger increase in water temperature in the deeper layer, causing overall warming

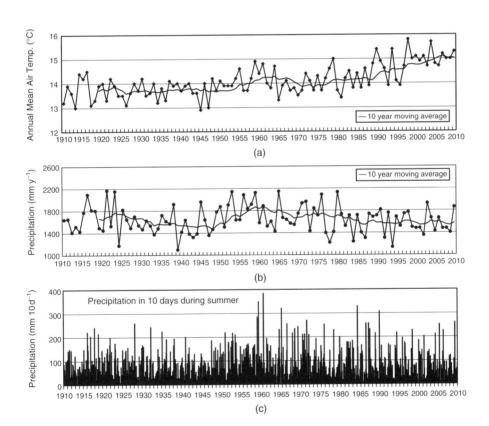

Figure 6.1 Time series changes in annual mean air temperature and annual precipitation during the 100 years from 1910 to 2010 at Hikone located on the northeastern shore of the North Basin. (a) Annual mean air temperature, (b) annual precipitation, and (c) precipitation in the consecutive 10 days during June to September. Adapted from the database of Meteorological Agency, Japan.

of the lake water. They also indicated that the interannual increase in air temperature was larger in winter than in summer, affecting the vertical mixing of the water column resulting in incomplete winter mixing.

Kumagai *et al*. (2006), based on a continuous monitoring survey of limnological parameters at a depth of 95 m in the North Basin, showed that complete winter mixing of the lake occurred in 2001, 2002 and 2003 but was followed by incomplete winter mixing in 2004 and 2005 due to lake water warming, resulting in development of a hypoxic hypolimnion. Global warming also affected the onset time of seasonal lake mixing. Kumagai *et al*. (2006) reported that the start of complete vertical mixing in winter was increasingly delayed each year after 2002. This delay resulted in incomplete winter mixing, accelerating the decrease in DO concentration in the hypolimnion. Through a detailed analysis of water temperature data derived from the monitoring survey by Shiga Prefecture, Tsujimura *et al*. (2010) reported gradual increases in Schmidt's Stability Index and the number of days of lake thermal stratification after 1980, suggesting an influence of increased thermal stability on the decreased DO concentration in the hypolimnion in late fall.

The northern part of the watershed of Lake Biwa has been well known as the site of heavy snow cover every winter. After the 1990s, the amount of snowfall decreased and the inflow rate of snowmelt water into Lake Biwa decreased. Based on a detailed examination of the data of snow cover and DO concentration in the hypolimnion of Lake Biwa, Kumagai and Fushimi (1995) suggested that the decline in the amount of snow cover in the mountain region due to global warming caused decreased inflow of cold snowmelt water rich in oxygen into Lake Biwa and accelerated the decline of DO concentration in the hypolimnion.

In the surface water of the North Basin, a noticeable seasonal change in NO_3 concentration was observed in association with the vertical mixing of the lake water. Using the seasonal change of NO_3 concentration as an indicator of vertical mixing, Sakamoto (2011) indicated an earlier onset of autumnal turnover in the North Basin due to the warming water (Figure 6.2). He indicated that higher NO_3 concentration in the subsurface layer was associated with higher water temperatures after 1990 than observed before 1989 and was responsible for stimulated phytoplankton growth and resulting changes in the dominant species. According to Ichise *et al*. (1996), considerable changes in the seasonal periodicity and flora of phytoplankton were observed in the North Basin after 1990. Hsieh at al. (2010) made a detailed analysis of Lake Biwa monitoring survey data and stated that the observed shift of phytoplankton communities after 1990 was driven by changes in physical mixing of the water column by global warming. Paleolimnological studies on the long-term evolution of the plankton community in Lake Biwa also indicated the community changes were associated with eutrophication and climate change (see also Chapter 7).

Based on the evaluation of long-term changes in the rate of water supply into Lake Biwa, Kinki Regional Bureau, MILT (2005) reported a 30% decrease of the annual water supply rate during the 40 years from 1960 to 2000 in association with decreasing precipitation due to global warming. No information was provided on changes of material supply from the watershed in association with the hydrological changes.

As described above, gradual increases in water temperature in Lake Biwa modified the thermal mixing regime, resulting in the development of a hypoxic hypolimnion and changed the seasonal sequence of phytoplankton dynamics. These changes exerted significant influences on biogeochemical processes in Lake Biwa.

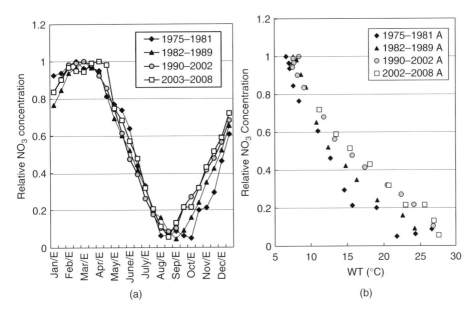

Figure 6.2 (a) Seasonal changes of relative NO$_3$ concentration (relative value of the observed concentration in a given month to the peak concentration in overturn season) and (b) the relationship between relative NO$_3$ concentration and water temperature of the lake water at 0.5 m in the center of the North Basin (from Sakamoto 2011). Aquatic Ecosystem Health & Management by Aquatic Ecosystem Health and Management Society. Reproduced with permission of TAYLOR & FRANCIS INC. in the format republish in a book/textbook via Copyright Clearance Center.

6.4 Biogeochemical evolution in the pelagic system of Lake Biwa

In the biogeochemical dynamics of a pelagic lake ecosystem, phytoplankton primary production, mineralization of organic matter by the activity of heterotrophs, regeneration of mineralized nutrients to phytoplankton, and oxygen consumption associated with heterotrophic metabolism are closely interrelated to create a unique pelagic ecosystem for a given environment. The internal impacts on lakes due to global warming, such as changes in water temperature and lake water mixing, as well as the impact of increasing external nutrient loading, affects these biogeochemical processes and their balance, resulting in a modified pelagic ecosystem. To examine the effects of global warming on the pelagic ecosystem of Lake Biwa, changes in the biogeochemical processes in the North Basin during the last 50 years are traced.

Although a series of long-term determinations of primary production (PP) has not been performed in Lake Biwa, several *ad hoc* studies have provided valuable information on the time series changes of PP. There was a gradual increase in the daily rate of PP during the vegetation season over three decades from 1962 to 1997 (Figure 6.3). To examine the contribution of PP to DO consumption in the hypolimnion, the apparent oxygen utilization rate (AOUR) through the water column was calculated as the mean daily decreased rate of DO in the hypolimnion below

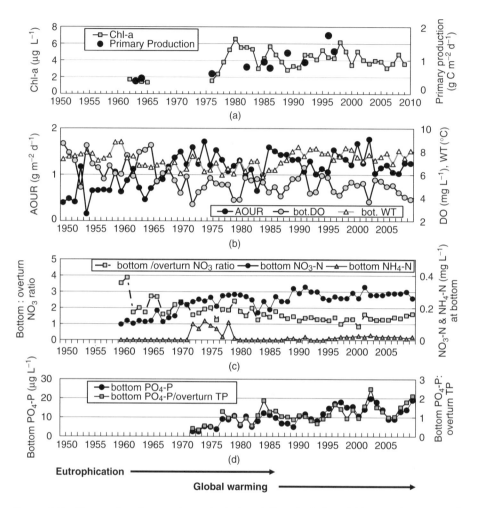

Figure 6.3 Time series changes in ecological and limnological variables at the center of the North Basin. (a) Mean primary production (PP) rate and mean chlorophyll-a (Chl-a) concentration during the vegetation season from April to October; (b) annual mean water temperature (WT) and minimum dissolved oxygen (DO) concentration at the bottom of the hypolimnion and apparent oxygen utilization rate (AOUR) in the hypolimnion below the depth of 25 m during stratification period; (c) the annual maximum concentrations of NO_3-N and NH_4-N at the bottom of the hypolimnion and the bottom:overturn ratio of NO_3 concentration (the ratio of NO_3 concentration at the annual maximum at the bottom to those in the preceding overturn season; (d) the annual maximum concentration of PO_4-P in the bottom of the hypolimnion and the PO_4:TP ratio (the ratio of the annual maximum PO_4-P concentration in the bottom layer to the mean total phosphorus (TP) concentration in the water column during the preceding overturn season). The data are from the references cited in Ishikawa and Kumagai (2007) on primary production, from Saijo *et al.* (1966), and the database of Shiga Prefecture and Shiga Prefecture Fisheries Experiment Station on Chl-a, and Shiga Prefecture Fisheries Experiment Station for other limnological variables.

25 m to the lake bottom at 75 m during the stratification period (from early spring at full overturn to late fall when the DO concentration at the bottom fell to the annual minimum) using the database of Shiga Prefecture Fisheries Experiment Station. The AOUR significantly increased annually over 20 years from 1953 to 1973, followed by relatively higher AOUR rates with a marked year-to-year fluctuation (Figure 6.3). The changes of PP were positively correlated with those of AOUR and negatively with those of annual minimum DO at the bottom (Figure 6.4), suggesting PP increased AOUR and decreased DO at the bottom.

We need to pay attention to the continued appearance of low minimum DO concentration in the hypolimnion of Lake Biwa in late fall after the mid-1990s, irrespective of a recovery of lake water quality. Two processes could be involved: one is increased oxygen consumption due to increased bottom temperature and a second is reduced oxygen supply into lake water from the air due to global warming. The bottom water temperature showed a gradual increase after 1983, followed by continued higher temperature after 1990 (Figure 6.3). Plotting AOUR values against bottom water temperature (Figure 6.5a) indicated that higher AOUR values were associated with bottom water temperature $>7\,°C$. This suggested that increased bottom water temperature due to global warming accelerated hypolimnetic oxygen consumption after 1990. Examination of the relationship between AOUR and low DO concentration in the bottom layer (Figure 6.5b) indicated a significant correlation between AOUR and the annual minimum DO over 40 years from 1950 to 1989 and a slightly different and less significant correlation after 1990. This suggested that other factors were involved in decreasing the annual minimum DO in the hypolimnion after 1990.

As described in the foregoing section, Kumagai *et al.* (2006) reported that full winter lake overturn did not occur in 2004 and 2005 due to warming of the bottom layer of the North Basin. Atmospheric oxygen is a major source of oxygen in lake water, so unless there is a full overturn of a lake in winter, the limited DO reservoir will be consumed quickly by heterotrophic metabolism in the hypolimnion. The database of Shiga Prefecture indicated an increasing annual trend of incomplete winter overturn

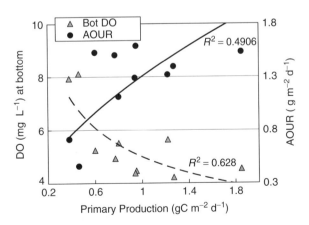

Figure 6.4 Relationship of the mean primary production (PP) rates during the vegetation season to AOUR in the hypolimnion and the annual minimum DO concentration of the bottom layer (1 m above the lake bottom) at the center of the North Basin. The solid and broken lines with R^2 values are regression lines between two variables. From the data sources cited in Figure 6.3.

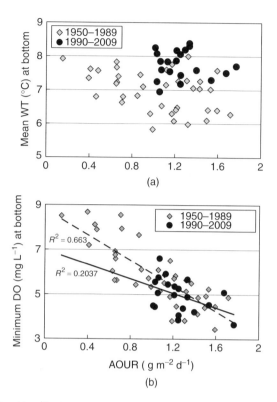

Figure 6.5 Relationship of AOUR to (a) the annual mean bottom water temperature (WT) and (b) the annual minimum dissolved oxygen (DO) concentration in the bottom layer at the center of the North Basin. Adapted from the data base of Shiga Prefecture Fisheries Experiment Station.

in the North Basin after 1980. Prolonged stratification of the lake water under global warming was also responsible for creating a low DO environment. Kumagai *et al*. (2006) reported a belated start to winter full overturn in the North Basin due to warming of the lake after 2000. The database of Shiga Prefecture indicated that a lower minimum DO concentration of $<6\,\mathrm{mg\,L^{-1}}$ was associated with annual stratification periods of >200 days (Figure 6.6).

As noted above, the development of low minimum bottom DO in late fall continued even after 1990 during the recovery of water quality in Lake Biwa. The decline of Chl-a and a decrease in the downward flux of autochthonous organic matter from PP could result in decreased hypolimnetic oxygen consumption. However, the continued development of low minimum hypolimnetic DO suggested another oxygen consumption process was involved after 1990. One possible process was oxygen consumption by bottom sediments. It is well documented that bottom sediments are characterized by relatively high organic matter content and high microbial activity. Through intensive field studies of the sediments in the profundal region of Lake Biwa, Murase and Sakamoto (2000) found that autochthonous organic matter of 8% and 5.5% of annual PP in terms of carbon and nitrogen was supplied to the sediments. Moreover, the sediments also contained allochthonous organic matter at around 38% of total organic matter. In a subsequent study, Murase and Sugimoto (2006) showed

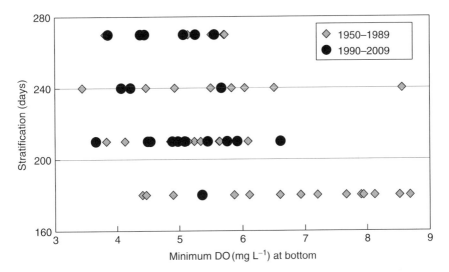

Figure 6.6 Relationships between the annual minimum dissolved oxygen (DO) concentration of the bottom layer and the days of lake water stratification prior to the annual minimum DO concentration in the bottom. From the monthly survey data of Shiga Prefecture Fisheries Experiment Station.

that profundal bottom sediment was responsible for about 20% of DO consumption in the hypolimnion. Urabe *et al.* (2005), who studied the production to respiration ratio of plankton communities in the pelagic region of the North Basin, indicated that the pelagic zone was a system subsidized by an allochthonous organic matter supply characterized by a high respiration: production ratio >1. These reports indicated a strong contribution of heterotrophic metabolism of organic matter not originating from in-lake PP processes to DO consumption in the sediments and hypolimnion. Increased water temperature and prolonged lake water stratification could stimulate these oxygen consumption processes, resulting in the continued development of the low DO environment in late fall in the bottom layer.

6.5 Effects of hypoxic hypolimnion development on biological and geochemical processes in Lake Biwa

In Lake Biwa, long-term monitoring surveys of both water quality and biotic communities have been conducted by prefectural research institutes and universities for about 50 years. These surveys have provided useful information on the status and changes of the ecosystem, especially on the dynamics of biogeochemical and ecological changes in the low oxygen environment of the North Basin's hypolimnion after 1990 under global warming.

Evolutional change of the benthic worm population, the wide distribution of sulfur-oxidizing bacteria, *Thioploca* sp., and the mass occurrence of manganese oxide particles in the oxic-reductive layers of the bottom sediments must be documented as the representative issues closely related to the development of the hypoxic hypolimnion after 1990.

In the North Basin before 1973, benthic animals in the profundal zone deeper than 90 m were dominated by the large tubificid worm, *Branchiuyra sowerbyi*, a species without gills and with resistance to hypoxia. After 1992, this form was completely replaced by the sludge worm, *Tubifex tubifex*, which was small in size and showed greater resistance to hypoxia. These changes pointed to the progress of hypoxia in profundal sediments of the North Basin after 1990 (Nishino *et al.* 2002).

In a monitoring survey in March 1991, Lake Biwa Environmental Research Institute observed a dense mat of *Thioploca* sp. in the profundal region at 60 m depth of the North Basin. Subsequent surveys indicated a wider distribution of this bacterium in higher density on the muddy bottom sediments. *Thioploca* sp. is a gliding filamentous microaerophilic bacterium which can oxidize hydrogen sulfide and occurs in the upper layer of sediments having an inter-phase of oxidizing and reducing horizons at low hydrogen sulfide concentrations (Nishino, Fukui, and Nakajima, 1998). In their intensive survey of the bottom sediments in the North Basin, Maeda and Kawai (1988) detected the occurrence of sulfide and a high sulfate reduction activity. Anoxic sediments covered with oxidizing and reducing layers is an environment suitable to sulfide production and sustain a wide distribution of *Thioploca* sp.

Another important issue related to the development of a hypoxic hypolimnion was the wide distribution of manganese oxide particles on the sediment surface. In the regular monitoring survey by Lake Biwa Environmental Research Institute in early November 2002, an occurrence of small red particles (20 µm in diameter) was found in water >90 m (Ichise *et al.* 2003). On chemical and microscopical analysis of the collected samples, the particles were identified as manganese oxide known as *Metallogeniumu* (Ichise *et al.* 2003). A wide distribution of manganese oxide particles on the sediment surface in deep profundal waters of the North Basin was formed from oxidation of reduced manganese supplied from the anoxic sediments. As well, manganese in concentrations of $20-300 \mu g L^{-1}$ occurred in the lake water just above the sediments before full lake overturn. These observations indicated the occurrence of an interphase of oxic and reductive zones above the sediments after 1990.

According to Miyajima (1994), who studied the dynamics of inorganic nitrogen and manganese at the mud-water interface of the North Basin, availability of oxygen in the lake water greatly controlled oxidation and reduction of inorganic nitrogen and manganese above the bottom sediments. The progress of hypoxia in the bottom layer of Lake Biwa after 1990 should greatly modify oxidative and reductive metabolism at the mud-water interface and result in the wide occurrence of *Thioploca* and *Metallogeniumu*.

Oxidative and reductive dynamics of inorganic nitrogen in the hypolimnion are sensitive to a change in oxygen availability. Therefore the abundance of NO_3 in oxic form and NH_4 in reductive form of inorganic nitrogen could reflect the redox conditions of the bottom layer. Occurrence of NO_3 in high concentrations in association with NH_4 in low concentrations indicated better availability of oxygen in the hypolimnion even after development of *Thioploca* and *Metallogeniumu* just above the surface of North Basin bottom sediments (Figure 6.3).

In hypolimnetic NO_3 dynamics, denitrification is a representative microbial process responsible for consumption of NO_3 in the hypolimnion. On the basis of nitrogen isotope ratio analysis of both bottom sediments and endemic fish specimens

(*Chaenogobius isaza*) from Lake Biwa, Ogawa *et al*. (2001) indicated enhanced denitrification associated with decreasing hypolimnetic DO after 1960. Under the assumption that the ratio of maximum NO_3 concentration in the bottom layer to the initial NO_3 concentration during overturn could reflect the difference between nitrification and denitrification, the time course of the ratio was plotted (Figure 6.3). The ratio showed a significant decline and fell to almost one after 1990. Plotting the ratio against the annual minimum DO concentration and mean water temperature of the bottom indicated the low ratio was associated with low minimum DO concentration and high water temperature (Figure 6.7a,b). This result indicates activated loss of NO_3 by decreased nitrification and increased denitrification in the low DO environment of $<6\,mg\,L^{-1}$ and higher mineralization and nitrification coupled with denitrification in bottom temperatures of $>7\,°C$ with progression of global warming after 1990.

Another important biogeochemical process in hypoxic environments is increased liberation of phosphate phosphorus (PO_4-P) from sediments (Mortimer 1942; Andersen 1982). Although redox-controlled phosphorus release is generally understood, phosphorus release from lake sediments is much more complex than described in pioneering works (Hupfer and Lewandowski 2008). Since microbial processes consume oxygen and liberate phosphorus, it is difficult to distinguish whether oxygen depletion is the result or the cause of phosphorus release. In the North Basin, in association with the development of the hypoxic hypolimnion, a gradual increase in the concentration of bottom layer PO_4 was observed after 1990 (Figure 6.3d). The PO_4-P:TP ratio (the ratio of annual maximum PO_4-P concentration in the bottom layer to the total phosphorus concentration in the water column during the preceding overturn) calculated as a mineralization index also showed an increase and was >1 after 1900–1995, suggesting a change in the mineralization state of phosphorus in the bottom layer of the hypolimnion after 1990–1995. Plotting the PO_4-P:TP ratio against mean water temperature and annual minimum DO concentration of the bottom layer (Figure 6.7c,d) indicated that the higher PO_4:TP ratio after 1990 was associated with higher bottom water temperature $>7\,°C$ and with lower DO concentration $<4\,mg\,L^{-1}$ in the bottom layer. Enhanced mineralization associated with DO consumption due to increased water temperature was likely responsible for the increasing PO_4 concentration.

In the hypolimnion of Lake Biwa, a considerable increase in NO_3 concentration in association with that in PO_4 concentration was observed along with development of lake thermal stratification, reaching a maximum level in late fall (Shiga Prefecture Fisheries Experiment Station 1950–2009). There are many reports on the effect of NO_3 at higher concentrations oxidizing the bottom sediments to stimulate mineralization of organic phosphorus and preventing PO_4 liberation by microbiological and chemical processes (e.g., Andersen 1982; Jensen and Andersen 1992).

Through experimental analysis of phosphorus release from the bottom sediment of Lake Biwa, Sugiyama (2004) indicated that phosphorus liberation from the sediments in the profundal region of the North Basin was mostly due to microbial mineralization of organic matter in the sediment and was less affected by DO depletion in overlying lake water. Using ^{31}P NMR analysis, Takayama, Sato, and Nakajima (2000) calculated that organic phosphorus accounted for 40–84% of phosphorus in the bottom sediment of the profundal region of the North Basin. Together with these published information, the higher ratio of bottom PO_4 to overturn TP found at higher bottom water temperatures after 1990 (Figure 6.3) indicated that increased water temperature due to global warming was largely responsible for the stimulated mineralization of

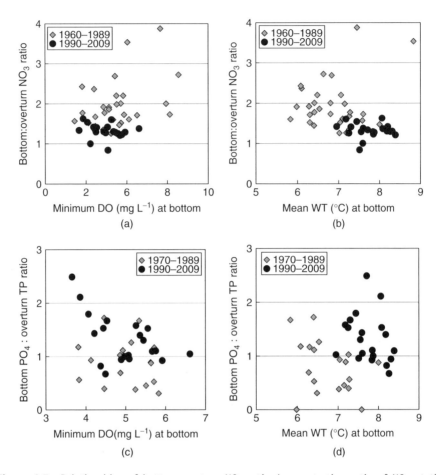

Figure 6.7 Relationships of bottom:overturn NO_3 ratio (concentration ratio of NO_3 at the annual maximum in the bottom of hypolimnion to NO_3 during overturn season) to (a) annual minimum DO concentration and (b) mean water temperature (WT) of the bottom layer, and those of the bottom PO_4-P:overturn TP ratio (concentration ratio of PO_4-P of the bottom layer at annual maximum to TP during overturn season) to (c) the annual minimum DO concentration and (d) mean water temperature of the bottom layer. From the database of Shiga Prefecture Fisheries Experiment Station.

organic matter in the bottom sediments and resulted in increased PO_4 concentrations in the hypolimnion.

6.6 Possible effects of climate change on Lake Biwa and adaptation measures to expected changes

As described, global warming has significantly affected the thermal regime in Lake Biwa and caused several changes in the pelagic ecosystem, especially biogeochemical and ecological processes due to increasing water temperature, changing lake water mixing and developing hypoxic environment after 1990. According to the synthesis report on climate change and its impacts in Japan by MEXT (Ministry of Education, Culture, Sport, Science and Technology) in collaboration with JMA (Japan Meteorological Agency) and MOE (Ministry of the Environment) (2009), the air temperature in Japan is projected to increase at a rate of about $0.2\,^\circ C$ per decade from 2012 to 2032 under all scenarios of global warming. Although there is a variation in temperature increase according to the scenario, temperature is projected to increase by 1.8 to $4.0\,^\circ C$ from the end of the twentieth century to the end of the twenty-first century. It is also expected that annual precipitation will increase by about 5% by the end of the twenty-first century with large interannual variability. Precipitation and days of heavy precipitation in summer are also expected to increase with global warming.

As reviewed above, studies on climate change effects on the thermal regime and biogeochemical processes in Lake Biwa indicated that the impact of global warming on the pelagic ecosystem are considerable although detailed information on the processes is limited. In light of the sensitivity of biogeochemical and biological metabolism to changes in water temperature as observed in the hypolimnion after 1990, progression of climate change could lead to significant adverse effects on the pelagic ecosystem of Lake Biwa.

Two scenarios for the possible effects of climate changes can be described as follows. The first is the accelerated liberation of phosphorus from the bottom sediments due to the progression of global warming. The observation of increased PO_4-P concentration in association with increasing bottom water temperature showed that warming was effective in increasing mineralization activity in the bottom sediments. Continued air temperature increases could cause further increases of bottom water temperature and stimulate phosphorus mineralization in the sediments. Associated increased oxygen consumption could stimulate the creation of a DO-depleted environment. In such an environment, increased PO_4 liberation due to reduction of P-Fe complex in the sediments could be expected if NO_3 is depleted by activated denitrification. An increasing supply of PO_4 from the hypoxic bottom layer could accelerate both phytoplankton growth and DO consumption associated with microbial decomposition of autochthonous organic matter from the epilimnion.

The second scenario is environmental change of the pelagic environment due to increasing loading of TP, TN, and organic matter from the watershed. A considerable increase in days of heavy precipitation has been recorded since 1960 at Hikone irrespective of decreasing annual precipitation (Figure 6.1). The data also suggested that the days of no or less precipitation also increased, resulting in an atypical pattern of alternating days with very heavy precipitation interspersed with days of virtually no precipitation. This altered precipitation pattern could exert significant influences on the terrestrial ecosystem and cause marked increases in nutrient and organic matter loading into the lake from several sources, including forested areas, in the watershed.

On the basis of extensive surveys of nutrient loads from rice paddies, which account for 92% of the farms in the watershed of Lake Biwa, Okubo, and Azuma (2005) indicated larger effluent contributions from rice paddies in the supply of nitrogen and phosphorus during the rainy season, especially after heavy rain. After the mid-1980s, rice-paddy farming had been conducted using a modern irrigation system that separates irrigation water from discharge water. With this system, a considerable discharge of silt and organic matter containing nitrogen and phosphorus into the lake is accelerated by mechanical raking and fertilization during the rainy season in May and June

(Okubo *et al*. 2008). Expected increases in heavy rain associated with global warming should result in an increase in irrigation discharge. Irrespective of a recovery of Lake Biwa from deterioration due to eutrophication, considerable increased discharge of nutrients again could result in a noticeable increase of eutrophication. Compared with the deterioration of the lake ecosystem due to eutrophication from the 1960s to 1980s, higher temperature and stronger thermal stratification of the lake could increase the extent of the hypoxic hypolimnion and result in pronounced modification of the lake ecosystem.

One effective way to adapt to the coming issues and expected worsening effects is to examine both the within lake processes and the watershed processes created by global warming to identify and understand their possible effects on Lake Biwa or lakes generally. This would be followed by exploring ways to remediate the worst consequences of warming. In the case of increasing nutrient and organic matter loading from the watershed, especially from paddy fields by more frequent heavy precipitation, agriculture practices using less fertilizer and less mechanical raking are especially needed. Modification of irrigation systems including accelerated reuse of irrigation water to reduce loading is of paramount importance as an adaptive measure because the water used in growing rice accounts for 70% of the total water use in Shiga Prefecture (Shiga Prefecture 2010).

Possible increases in heavy rain could also cause increased loading of suspended matter in rivers and result in increased silting in the littoral region. Growth of rooted plants in these silted regions could retard the normal flow of the water and modify the littoral ecosystem. Increases in heavy rain can also increase the risk of flooding and lake water level change. The water level of Lake Biwa gradually decreased after 1970 due to both artificial regulations for water supply to downstream societies and by decreased precipitation (Kinki Regional Bureau, MILT 2005). Considerable fluctuation of the water level due to a change in precipitation could have a strong negative impact on the hatching of endemic fishes because they lay eggs on submerged stems and leaves of emergent and submerged plants.

For development of adaptive measures to effectively respond to these expected changes of lake environments, systematic integration of information and possible countermeasures from several partners with close cooperation with local and national governments are required. To accelerate intellectual integration toward the establishment of sustainable countermeasures, close communication between scientists from several disciplines, on scientific processes and technological experiences, is very important. The author hopes this chapter will contribute to toward the establishment of sustainable ways to tackle lake environmental problems under the combined impacts of eutrophication and global warming in countries with the Asian monsoon climate.

References

Andersen, J.M. (1982) Effect of nitrate concentration in lake water on phosphorus release from the sediment. *Water Res.*, **16**, 1119–26.

Hayami, I. and Fujiwara, T. (1999) Recent warming of the deep water in Lake Biwa. [In Japanese with English summary.] *Oceanography in Japan*, **8**, 197–202.

Hsieh, C.H., Ishikawa, K., Sakai, Y., *et al*. (2010) Phytoplankton community reorganization driven by eutrophication and warming in Lake Biwa. *Aquat. Sci.*, **72**, 467–83.

Hupfer, M. and Lewandowski, J. (2008) Oxygen controls the phosphorus release from lake sediments—a long lasting paradigm in limnology. *Internat. Rev. Hydrobiol.*, **93**, 415–32.

Ichise, S., Wakabayashi, T., Matuoka, Y., *et al*. (1996) Sequential change of phytoplankton flora in the North Basin of Lake Biwa 1978–1995 [in Japanese]. *Rep. Shiga Pref. Inst. Pub. Hlth and Environ. Sci.*, **31**, 84–100.

Ichise, S., Wakabayashi, T., Okamoto, T., *et al*. (2003) Manganese oxide particles mediated microbiologically in the profundal region of the North Basin of Lake Biwa [in Japanese]. *Rep. Shiga Pref. Inst. Pub. Hlth & Environ. Sci.*, **38**, 100–5.

IPCC (2007) IPCC Fourth Assessment Report. *Climate Change 2007*, Cambridge University Press, Cambridge.

Ishikawa, T. and Kumagai, M. (2007) Primary production in Lake Biwa [in Japanese], in *Lake Biwa Handbook* (ed. Shiga Prefecture), Shiga Prefecture, Otsu, p. 58–9.

Jensen, H.S. and Andersen, F. (1992) Importance of temperature, nitrate, and pH for phosphate release from aerobic sediments of four shallow, eutrophic lakes. *Limnol. Oceanogr.*, **37**, 577–89.

Kinki Regional Bureau, MILT (2005) Present State and Change of Lake Biwa [in Japanese]. *Kinki Regional Bureau, Ministry of Land, Infrastructure*, Transportation and Tourism, Osaka.

Kobe Marine Observatory (1935) Results of the limnological survey of Lake Biwa-ko in February and March, 1931 [in Japanese]. *J. Oceanogr.*, **7**, 419–59.

Kumagai, M. and Fushimi, H. (1995) Inflow due to snowmelt, in *Physical Processes in a Large Lake: Lake Biwa, Japan* (eds. S. Okuda, J. Imberger and M. Kumagai) American Geophysical Union, Washington, D.C., pp. 129–39.

Kumagai, M., Okubo, K., Jao, C., and Song, X. (2006) Comparative study of Lake Fuxian and Lake Biwa from the view point of geophysical feature and environmental dynamics [in Japanese]. In *Lakes and Drainage Basins in East Asia Monsoon Area* (eds. M. Sakamoto and M. Kumagai), Nagoya University Press, Nagoya, pp. 184–202.

Maeda, H. and A. Kawai, 1988. Hydrogen sulfide production in bottom sediments in the northern and southern Lake Biwa. *Nippon Suisan Gakkaishi*, **54**, 1623–33.

MEXT, JMA, MOE (2009) Climate change and its impacts in Japan. *Synthesis report on observation, projections and impacts assessment of climate change*. Ministry of Education, Culture, Sports and Technology, Japan, Meteorological Agency, Ministry of the Environment, Tokyo.

Miyajima, T. (1994) Mud-water fluxes of inorganic nitrogen and manganese in the pelagic region of Lake Biwa: Seasonal dynamics and impacts on the hypolimnetic metabolism. *Arch. Hydrobiol.*, **130**, 303–24.

Mortimer, C. (1942) The exchange of dissolved substances between mud and water in lakes. *J. Ecol.*, **30**, 147–201.

Murase, J. and Sakamoto, M. (2000) Horizontal distribution of carbon and nitrogen and their isotopic composition in the surface sediment of Lake Biwa. *Limnology*, **1**, 177–84.

Murase, J. and Sugimoto, A. (2006) Gas metabolism in aquatic environments [in Japanese]. *Global Environment and Ecosystem* (eds. H. Takeda and J Urabe), Kyoritushuppan, Tokyo.

Nishino, M., Fukui, M., and Nakajima, T. (1998) Dense mats of *Thioploca*, gliding filamentous sulfur-oxidaizing bacteria in Lake Biwa, Central Japan. *Wat. Res.*, **32**, 953–7.

Nishino, M., Nakajima, T., Tujimura, S., *et al*. (2002) Studies on the ecosystem changes associated with deoxygenation in the profundal region of the North Basin [in Japanese]. *Bull. Lake Biwa Res. Inst.*, **19**, 18–35.

Ogawa, N., Koitabashi, T., Oda, H., *et al*. (2001) Fluctuation of nitrogen isotope ratio of gobiid fish (Isaza) specimens and sediments in Lake Biwa, Japan, during the 20th century. *Limnol. Oceanogr.*, **46**, 1228–36.

Okubo, T. and Azuma, Y. (2005) Pollution loads from the watershed to Lake Biwa and their impacts on the lake water quality [in Japanese]. *Bull. Lake Biwa Research Institute*, **22**, 55–72.

Okubo, T., Tsujimura, S., Kawasaki, E., *et al*. (2008) Effects of loadings from non-point sources on the water quality of Lake Biwa [in Japanese]. *Res. Rep. Lake Biwa Environ. Res. Inst.*, **4**, 50–64.

Saijo, Y., Sakamoto, M., Toyoda, Y., *et al*. (1966) Interim report on the study of lake metabolim [in Japanese], in *Interim Report of BST Survey* (ed. Survey Teams of Biological Resources in Lake Biwa), Kinki Regional Bureau of Ministry of Construction, Osaka, pp. 406–66.

Sakamoto, M. (2011) Limnological responses to changes in the thermal mixing regime in Lake Biwa associated with global warming. *Aquatic Ecosystem Health and Management*, **124**, 214–18

Schindler, D.W. (2009) Lakes as sentinels and integrators for the effects of climate change on watersheds, airsheds, and landscapes. *Limnol. Oceanogr.*, **54**, 2349–53.

Sugiyama, M. (2004) *Nutrient Liberation from Sediments at the Deoxygenated Bottom Layer of Lake Biwa* [In Japanese]. The 2003 Fiscal Year Research Report under the Contract to Lake Biwa Research Institute, Faculty of Integrated Studies, Kyoto University, Kyoto.

Takayama, T., Sato, Y., and Nakajima, S. (2000) Characterization of organic matter, minerals and phosphorus species in the bottom sediments of Lake Biwa using solid-state NMR [In Japanese with English summary]. *Jpn. J. Soc. Soil. Plant Nutrition*, **71**, 194–203.

Tsujimura, S., Aoki, S., Okumua, Y., *et al*. (2010) Analytical survey on the current status on the development of anoxia in Lake Biwa and its effects on the North Basin ecosystem [in Japanese]. *Res. Rep. Lake Biwa Environ. Res. Inst.*, **6**, 70–84.

Urabe, J., Yoshida, T., Bahadur, T.G., *et al*. (2005) The production-to-respiration ratio and its implication in Lake Biwa. *Ecol. Res.*, **20**, 367–75.

Verburg, P., Hecky, R.E., and Kling, H. (2003) Ecological consequence of a century of warming in Lake Tanganika. *Science*, **301**, 505–7.

7

Eutrophication, Warming and Historical Changes of the Plankton Community in Lake Biwa during the Twentieth Century

Narumi K. Tsugeki[1] and Jotaro Urabe[2]

[1]Senior Research Fellow Center, Ehime University, Japan
[2]Graduate School of Life Sciences, Tohoku University, Japan

7.1 Introduction

7.1.1 Background

In a variety of lakes, eutrophication due to increased nutrient loadings from the watershed has induced large changes in plankton communities (Schindler 1978, 2006). In addition, recently evidence has shown that the plankton communities in lakes are likely imposed by putative global environmental changes such as warming (O'Reilly et al. 2003; Jöhnk et al. 2008). Warming in winter may further alter life-history traits of zooplankton through changes in overwintering conditions. Many zooplankton groups such as water fleas (Cladocera) usually survive unfavorable periods of winter or drought as dormant stages (resting eggs), and hatch at the onset of more favorable conditions (e.g., Lampert, Lampert, and Larsson 2010). The keystone herbivore in lakes, Daphnia, are often least abundant in winter when water temperature is low and food supply is limited due to low algal production (e.g., Straile 2002; Winder and Schindler 2004; Yoshida et al. 2001), thus they sometimes survive in the form of resting eggs for overwintering. However, if winter conditions are not unfavorable due to warming for Daphnia as plankters, resting eggs might not be necessary for overwintering. Yet, only a little knowledge has been accumulated on how such global environmental

Climatic Change and Global Warming of Inland Waters: Impacts and Mitigation for Ecosystems and Societies, First Edition. Edited by Charles R. Goldman, Michio Kumagai and Richard D. Robarts.

changes affect community structures and life histories of planktonic organisms in concert with eutrophication, although these provide prime information necessary to lake management under global environmental changes (Magnuson *et al*. 1997).

Lake Biwa is one of the world's few ancient lakes and has a historical record extending back circa 400 000 years (Meyers, Takemura, and Horie 1993). At present it supplies drinking water for 14 million people living in the Kansai region of Japan. As described in Chapter 6, eutrophication has occurred in this lake since the 1960s when regional development plans were enacted as a part of the national development plan (Nakamura 2002; Yamamoto and Nakamura 2004). Furthermore, the mean water temperature has increased over the past several decades, especially with a marked increase in winter (see Chapter 6). Thus, Lake Biwa is an ideal ecosystem for examining effects of eutrophication and warming on lake communities. In this chapter we describe how eutrophication, meteorological conditions and warming have affected the community structure of phytoplankton and life history traits of dominant zooplankton in Lake Biwa using paleolimnological data such as fossil pigments, plankton remains and resting eggs of zooplankton species stored in the lake sediment together with monitoring data.

7.1.2 Paleolimnology

To elucidate how, and to what extent, environmental changes have affected a plankton community in a given lake ecosystem, long-term biological monitoring data are necessary. However, such biological monitoring data are in general absent for most lakes. A paleolimnological approach is one method to overcome a shortage of biological information from the past of a given lake ecosystems because biological and biogeochemical proxies stored in lake sediments can offer clues for reconstructing the past biological communities and food web. For example, algal remains stored in lake sediments can provide historical information on changes in abundance of phytoplankton species, although taxonomic resolution is limited to diatoms and some green algae which possess siliceous frustules or cell walls that can resist degradation for long (Smol *et al*. 2005; Tsugeki *et al*. 2010). In addition, fossil pigments such as chlorophyll derivatives and carotenoids can be used to infer outlines of historical changes in algal communities because algal pigments are often taxon specific (Leavitt *et al*. 1999). Remains of cladocerans, which are one of the major zooplankton groups, are also well preserved in lake sediments (Frey 1986). Body and head parts of hard-shelled cladocerans such as chydorids (Jeppesen *et al*. 2001), are well preserved in lake sediments and some soft-shelled chitinous taxa such as *Daphnia* leave postabdominal claws, which are hard parts of the body. These animals produce resting eggs (ephippial eggs) that are stored in lake sediments for substantial periods (Hairston Jr. 1999; Jankowski and Straile 2003). From sediments accumulated in lake bottoms, we can establish accurate time-depth profiles using isotope geochronology of ^{210}Pb and ^{137}Cs. Thus, if there are data on sediment dating, and abundant biological proxies such as remains, fossil pigments and resting eggs in the sediments, we can estimate annual sedimentation rates or flux for these plankton species or groups and reconstruct changes in plankton communities for the past century. Based on these data, a number of studies have reconstructed past phyto- and zooplankton communities and have successfully identified factors that caused past changes in lake ecosystems (Leavitt, Carpenter and Kitchell 1989; Kerfoot, Robbins, and Weider 1999; Leavitt *et al*. 1999;

Interlandi, Kilham and Theriot 2003). In the following sections, we will show the case of Lake Biwa.

7.2   Factors affecting historical changes in the plankton community

7.2.1   Eutrophication

In Lake Biwa, eutrophication occurred since the 1960s when the first comprehensive national development plans were implemented in Japan (Yamamoto and Nakamura 2004; see also Chapter 6). In accordance with this, the sedimentation rate (annual flux to the lake sediments) of fossil pigments and algal remains increased through the 1960s to the 1970s (Figures 7.1–7.3; Tsugeki *et al.* 2010) although these were highly limited before the 1960s. Specifically, remains of two large green algal species, *Staurastrum* spp. and *Pediastrum biwae*, dramatically increased in the lake sediments since the 1960s (Figure 7.2). These paleolimnological data are consistent with the monitoring report by Negoro (1968) who showed that *Staurastrum dorsidentiferum* var. *ornatum* was rare in 1961, but increased sixfold and became the predominant species in summer by 1966. Algal data collected monthly from 1962 to 1991 by Shiga Prefecture also showed that algal abundance followed the temporal trend of total phosphorus concentration from the 1960s to 1970s (Hsieh *et al.* 2010). During this eutrophication, algal species composition also changed. Some algal species such as *Aphanocapsa* sp. (cyanobacteria) and *Diatoma* sp. (diatom) occurred in the

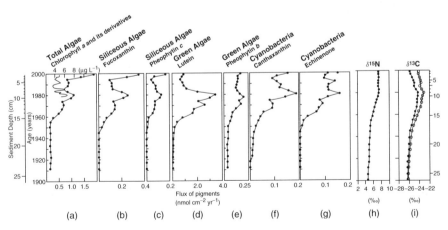

Figure 7.1   Time and sediment depth profiles of annual fluxes (sedimentation rates) for fossil pigments common to all algal taxa, i.e., chlorophyll *a* and its derivatives (a), pigments specific to siliceous algae, i.e., fucoxanthin (b) and pheophytin c (c), pigments specific to green algae, i.e., lutein (d) and pheophytin b (e), and pigments specific to cyanobacteria, i.e., canthaxanthin (f) and echinenone (g), and stable N (h) and C (i) isotope ratios in Lake Biwa. In the upper left panel (a), annual mean of chlorophyll a abundance, estimated directly from the 1979–1999 monitoring data, are inserted by the grey line. The stable C isotope ratios corrected for the Suess effect are indicated by open circles. Data (a)–(g) from Tsugeki *et al.* (2010) and (h)–(i) from Hyodo *et al.* (2008); used with permission.

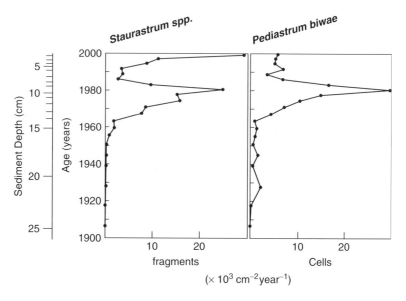

Figure 7.2 Time and sediment depth profiles of annual fluxes for remains of two green algae in Lake Biwa. Data from Tsugeki *et al.* (2010); used with permission.

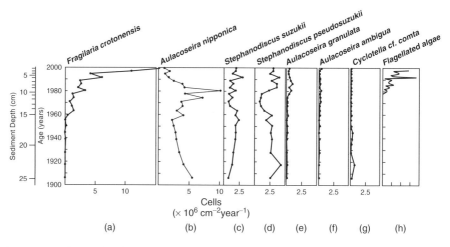

Figure 7.3 Time and sediment depth profiles of annual fluxes for remains of various diatoms (a)–(g) and cell abundance of flagellated algae (*Cryptomonas* spp. and *Rhodomonas* sp.) in winter estimated from the monitoring data (h) in Lake Biwa. Data (a)-(g) from Tsugeki *et al.* (2010) and (h) from Tsugeki *et al.* (2009); used with permission.

1960s but afterwards disappeared (Hsieh *et al.* 2010). Instead, in addition to large green algae such as *S. dorsidentiferum* var. *ornatum* (Tsugeki *et al.* 2010), *Fragilaria capucina* (diatom) and *Oocystis* sp. (green algae) that had not appeared previously became abundant in the 1970s (Hsieh *et al.* 2010). These studies showed that the composition and abundance of the phytoplankton community greatly changed during the eutrophication that had progressed in the 1960s to 1970s.

In parallel with the algal sedimentation rate, those of cladoceran zooplankton remains, *Daphnia* and *Bosmina*, rapidly increased from 1960 to 1980 (Tsugeki, Oda, and Urabe 2003) (Figure 7.4). In Lake Biwa, *Daphnia galeata* was the sole *Daphnia* species until 1999, when *D. pulicaria* appeared (Urabe *et al.* 2003). Thus, *D. galeata* were less abundant before the 1960s. This fact implies that not only primary production but also secondary production increased in Lake Biwa during eutrophication. However, eutrophication did not necessarily increase all organisms in Lake Biwa. According to the sediment records, two rhizopod species, *Difflugia biwae* and *Difflugia brevicolla*, were abundant before 1960, but decreased dramatically thereafter (Tsugeki, Oda, and Urabe 2003). *D. biwae* was first recorded in Lake Biwa (Kawamura 1918) and found to be only distributed in a limited number of lakes in East Asia (Yang and Shen 2005). However, the species was not found in sediments dated after 1980. Since they have not been seen since 1987 (Ichise *et al.* 1996), *D. biwae* might be extinct in Lake Biwa. These rhizopod species usually lived at the lake bottom (Negoro 1954) and appeared as plankton in some seasons (Negoro 1954; Fenchel 1987). In Lake Biwa, the yearly minimum oxygen concentration at the lake bottom at an offshore site decreased from 6 to $3 \, \mathrm{mg} \, \mathrm{L}^{-1}$ in the 1960s (Ishikawa, Narita, and Urabe 2004), due to oxygen consumption through decomposition of sedimented organic matter that increased due to increased primary production (see Chapter 6). Thus, eutrophication has interfered with their life histories by deteriorating environmental conditions at the lake bottom (Tsugeki *et al.* 2003).

Eutrophication during the 1960s–1970s also seems to have affected the relative importance of bottom-up and top-down effects on keystone herbivores in Lake Biwa. In this lake, crustaceans comprised 86.1% of zooplankton biomass on average (Yoshida *et al.* 2001), and among crustaceans *Daphnia galeata* has been one of the dominant species during the last several decades (Okamoto 1984; Miura and Cai 1990; Urabe,

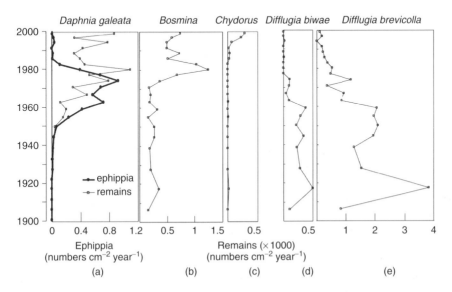

Figure 7.4 Time and sediment depth profiles of annual fluxes of *Daphnia galeata* ephippia (bold line with filled circles) and remains (thin line with open squares) (a), and remains of *Bosmina* (b), *Chydorus* (c), *Difflugia biwae* (d) and *Difflugia brevicolla* (e) in Lake Biwa. Data from Tsugeki *et al.* (2003) and Tsugeki *et al.* (2009); used with permission.

Nakanishi, and Kawabata 1995; Yoshida *et al.* 2001). However, before 1960 *Difflu-gia* and *Bosmina* were relatively abundant but *Daphnia* were minor components of the zooplankton community (Tsugeki, Oda, and Urabe 2003). One may suspect that algal food was not sufficient to support population growth of *Daphnia* due to the olig-otrophic conditions in those days. However, the low abundance of *Daphnia* before the 1960s cannot be explained by food shortage (Tsugeki, Oda, and Urabe 2003) because *Daphnia* are ecologically superior to *Bosmina* in exploitative competition for food (Threlkeld 1976; Goulden, Henry, and Tessier 1982). Rather, a *Daphnia* population might be suppressed due to high loss rates by fish predation (Tsugeki, Oda, and Urabe 2003). It is well known that most planktivorous fish prey selectively on *Daphnia* species compared with *Bosmina* because of their large body size (e.g., Brooks and Dodson 1965; Gliwicz and Pijanowska 1989). A number of previous studies have shown that planktivorous fish in Lake Biwa preferentially prey on *Daphnia* (Sunaga 1964; Nakanishi and Nagoshi 1984; Urabe and Maruyama 1986; Kawabata *et al.* 2002). Note that the annual fishery catch of planktivorous fish in Lake Biwa did not change greatly before and after the 1960s (Yuma, Hosoya, Nagata 1998), suggesting that the predation pressure of planktivorous fish on *Daphnia* was the same level around the 1960s, although food availability for these zooplankton increased due to eutrophication (Tsugeki, Oda, Urabe 2003). Thus, the increase in *Daphnia* abundance after the 1960s in this lake seems to reflect changes in relative strength of top-down and bottom-up forces on the zooplankton community by eutrophication in a way that *Daphnia* could maintain a larger population due to reduction of predation loss relative to the population growth rate.

In Lake Biwa, *Eodiaptomus japonicus* was another dominant zooplankton. Unlike *Daphnia*, however, copepods do not leave any clues about their populations in lake sediments because their bodies and organs are easily decomposed. Therefore, we analyzed zooplankton samples collected monthly by Kyoto University at a south site in the north basin of Lake Biwa from 1966 to 2000 (Figure 7.5). Details of the sampling and enumeration methods are given elsewhere (Tsugeki, Ishida, and Urabe 2009). The time-series data showed that the abundance of *E. japonicas* has gradually increased over the last four decades (Figure 7.5a: r = 0.54, p = 0.003 except for 1967), though such a gradual increasing trend was not apparent in *Daphnia* populations. One may suspect that predation pressure on *E. japonicus* gradually decreased in recent years. In Lake Biwa, *E. japonicus* is known to be preyed upon by fishes (Kawabata *et al.* 2002) and by the cyclopoid copepod *Mesocyclops dissimilis* (Kawabata 1991). However, fish predation on *E. japonicus* seems to have been less intense compared with *Daphnia* because their abundance was relatively high even in summer when predation activities of fish are high (Yoshida *et al.* 2001). The abundance of *M. dissimilis* which preyed on nauplius stages of *E. japonicus* (Kawabata 2006) did not show any increased trends in recent years (Figure 7.5c). Thus, it is difficult to explain the gradual increase of *E. japonicus* abundance for the past 30 years by changes in predation pressure alone.

After the 1980s, total phosphorus (TP) concentration, which limited algal growth (Tezuka 1984; Urabe 1999), was stabilized in Lake Biwa, probably because sewage systems were deployed in the watershed in the late 1970s (Yamada and Nakanishi 1999; see also Chapter 6). Indeed, sedimentation rates of most fossil pigments and green algal remains decreased in the 1980s when Shiga Prefecture enacted the "Ordinance for the Prevention of Eutrophication in Lake Biwa" (Nakamura 2002), although the rates remained higher than those before 1960. Monitoring data also showed that

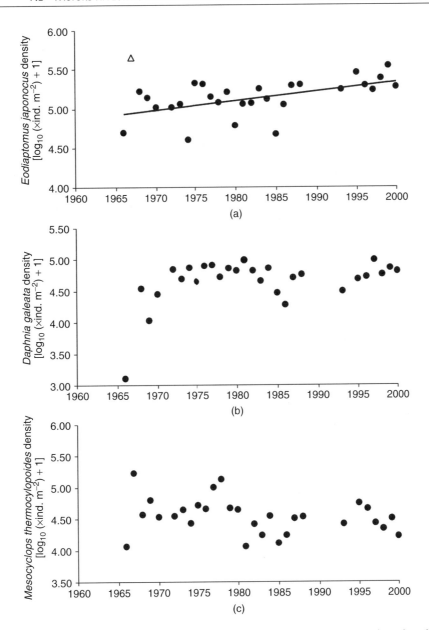

Figure 7.5 Long-term changes in annual mean abundances of *Eodiaptomus japonicus* (a), *Daphnia galeata* (b) and *Mesocyclops dissimilis* (c) estimated from the monthly monitoring data. Blank denotes periods when monthly sampling was not done. Significant regression lines against calendar years are shown.

annual averages of chlorophyll a concentration decreased to a low level in the middle 1980s (Figure 7.1a). Similarly, the d^{15}N value of the lake sediments (Figure 7.1h) and preserved fish samples, which had increased through the 1960s and the 1970s, also stabilized after 1980 because of eutrophication (Hyodo *et al*. 2008; Nakazawa *et al*. 2010). These results indicated that eutrophication had ceased in this lake in the 1980s.

However, nitrogen loading increased somewhat continually, resulting in increased TN relative to TP concentrations. The changes in the relative nutrient concentrations may have indirectly affected *E. japonicas* abundance. Indeed, correlation analyses showed that the abundance of *E. japonicas* correlated positively with TN concentration, but negatively with TP concentration (Table 7.1). Significant correlation between *E. japonicas* abundance and TP concentration in the past 30 years was also reported for time-series zooplankton samples collected by Shiga prefecture at a site in the north basin of Lake Biwa (Hsieh *et al*. 2011). Recent theories on ecological stoichiometry suggest that in general copepods suffer less from P deficiencies in diets compared with *Daphnia* because their body P content relative to N is much smaller than those of *Daphnia* species (Andersen and Hessen 1991; Elser and Urabe 1999; Sterner and Elser 2002). In addition, in contrast to *Daphnia*, copepods can ingest algae in a wide range of cell or colony sizes (Okamoto 1984; Kawabata 1987) and have an ability to discriminate food with different nutrient contents (DeMott 1986, 1988). These results imply that an increase in N loading relative to P may have created feeding conditions that favored *E. japnoics* in Lake Biwa. Alternatively, or additionally, the increase in their abundance may have been related to recent changes in meteorological conditions and temperature regime as discussed below.

7.2.2 Meteorological conditions

Although TP concentration in recent years tended to decrease, it still remains at a high level (Figure 7.6). In addition, concentrations of nitrate and soluble reactive silicate (SRSi) gradually increased even after 1980. Nonetheless, algal abundance decreased after 1980. Thus, although the increase of algal abundance during the period from 1960 to 1980 was mainly due to increased nutrient loading, the decrease in algal abundance occurred in the 1980s could not be attributable to changes in nutrient conditions alone.

In Lake Biwa, annual average wind velocity decreased through the 1960s and 1970s, but then increased during the 1980s (Figure 7.7a). Because wind energy input is one of the critical factors that affect water-mixing regimes, changes in wind regimes likely alter light condition for algal communities (Frenette *et al*. 1996; Petersen *et al*. 1997). It should be noted that since Lake Biwa has a large surface area, the surface mixed layer extends to great depths (10–15 m) even in the productive period from late spring to early fall (Gurung *et al*. 2002). Consequently, light can often be a crucial factor for

Table 7.1 Pearson correlation coefficients for the annual abundance of *Eodiaptomus japonicus* and *Daphnia galeata* in Lake Biwa against limnological and meteorological variables.

Category	Variable	Period	n	E. japonicus	D. galeata
Limnological	Water temperature	1966–2000	35	0.361	0.111
	Total P	1979–2000	22	−0.678*	0.264
	Total N	1979–2000	22	0.557*	0.060
	SRSi	1979–2000	22	0.461	0.140
Meteorogical	Wind velocity	1966–2000	35	−0.037	−0.414
	Precipitation	1966–2000	35	0.190	0.047
	Sunshine duration	1966–2000	35	0.186	0.266

*Significant at the 5% level after FDR correction

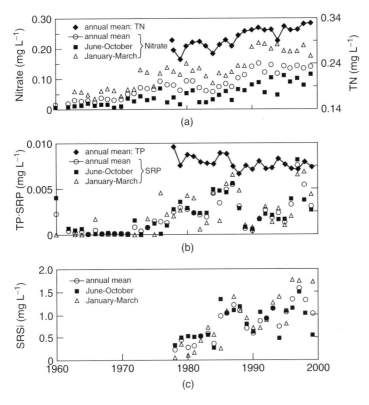

Figure 7.6 Long-term changes in annual mean of TN concentration, and annual and seasonal means of nitrate concentrations (a), annual mean of TP concentration, and annual and seasonal means of SRP concentrations (b), and annual and seasonal means of SRSi concentrations (c) in the surface layer (0–15 m) at a 70 m-deep site in the north basin of Lake Biwa. Data from the Fisheries Experimental Station of Shiga Prefecture (Shiga Prefecture 1963–2000).

limiting algal productivity in Lake Biwa (Urabe *et al.* 1999). In Lake Kinneret, low algal abundance, and primary production rates were observed in years with high wind energy input (Berman and Shteinman 1998), probably because deep mixing exposed algae to low-light environments in deep waters. In accord with the case of Lake Kinneret, multivariate analysis of fossil pigments and algal remains sedimentation rate data by Tsugeki *et al.* (2010) identified wind velocity as the most important explanatory variable for changes in algal assemblages from the 1960s to 1990s in Lake Biwa (Figures 7.8a, c). Similarly, Hsieh *et al.* (2010) showed that temporal changes in phytoplankton cell volume in Lake Biwa were negatively correlated to a wind mixing index, which was calculated from daily maximum wind velocity. Note that wind velocity was relatively low during the period from the mid 1970s to the early 1980s. Thus, together with eutrophication of Lake Biwa the decreased wind velocity may have contributed to the increase in algal abundance in the 1970s (Tsugeki *et al.* 2010). In contrast, because wind energy input increased at a high level after the mid 1980s, it is likely that the increased wind energy caused the decrease in algal abundance after 1980, despite constant phosphorus supply and even increased supplies of nitrogen and silica (Hsieh *et al.* 2010; Tsugeki *et al.* 2010).

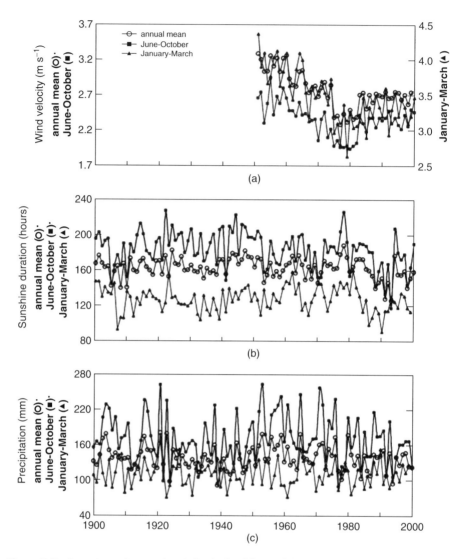

Figure 7.7 Long-term changes in wind velocity (a), sunshine duration (b), and precipitation (c) at Hikone, the site of the meteorological observatory nearest to Lake Biwa. In each panel, the annual mean (open circle), monthly mean in winter (January–March: closed triangle) and the monthly mean in the warm-water season (June–October: closed square) are shown. Data from (Hikone Meteorological Observatory 1993–2000) operated by the Japanese Meteorological Agency (JMA).

Meteorological conditions such as sunshine duration in addition to wind velocity may have affected species composition of the algal community in Lake Biwa. In this lake, large green algae such as *Staurastrum* and *Pediastrum* dominated generally during the productive season (June–October) (Ichise *et al.* 1996). Sedimentation rates of remains and pigments (lutein) of these green algae dramatically increased from the mid-1970s to early 1980s but decreased in the mid 1980s when the sunshine duration

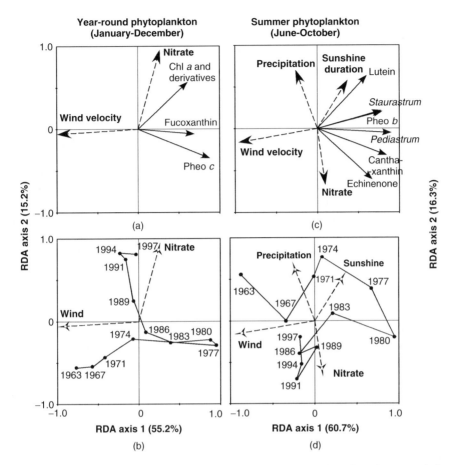

Figure 7.8 Results of redundancy analyses for paleolimnological and environmental data showing year-round phytoplankton assemblage (a, b) and summer phytoplankton assemblage (c, d) with bioplots of fluxes for fossil pigments and remains (solid arrows) and significant explanatory variables (dotted arrows) (a, c), and historical transition of the phytoplankton assemblage scores (b, d). Pigments include chlorophyll a and its derivatives, pheophytin b (Pheo b), pheophytin c (Pheo c), fucoxanthin, lutein, canthaxanthin, and echinenone. Remains include *Staurastrum* spp. (*Staurastrum*) and *Pediastrum biwae* (*Pediastrum*). Data from Tsugeki *et al.* (2010); used with permission.

decreased (Figure 7.7b). However, the sedimentation rate of a pigment (echinenone) characteristic of cyanobacteria remained at a relatively high level even during that period (Figure 7.1). Tsugeki *et al.* (2010) showed that meteorological variables such as sunshine duration also contributed to the variance in the changes in the algal assemblage for the last several decades (Figures 7.8c). Through buoyancy, cyanobacteria could maintain their position within the photic zone. Thus, they would have an advantage over other algae without a buoyancy mechanism under conditions of low light (Scheffer *et al.* 1997, Passarge *et al.* 2006). According to Reynolds (1984), even if an algal species with large cells adapts to low light conditions, they cannot thrive in a water column if they cannot stay in the surface layer by frequent vertical mixing. For this reason, abundance of large green algae such as *Staurastrum* and *Pediastrum* have been limited in years with low sunshine duration and increased wind velocity (Tsugeki *et al.* 2010).

7.2.3   Warming

Global warming is also an important factor affecting algal populations in Lake Biwa. In this lake, the water temperature especially in winter, has been increasing due to warming (Endo *et al.* 1999; Hayami and Fujiwara 1999; Kumagai 2008; for more detail see Chapter 6). *Aulacoseira nipponica*, an endemic diatom species in Lake Biwa (Tsuji and Hoki 2001), appeared predominantly in winter when strong vertical mixing was achieved (Sugawara 1938; Nakanishi 1976; Kawabata 1987). However, the abundance of this species dramatically decreased since the 1980s while flagellated algae, *Cryptomonas* spp. and *Rhodomonas* sp, have increased (Figures 7.3b, h). Because *Aulacoseira* have heavy cells, they generally require turbulent mixing to maintain their position within the photic zone (Reynolds 1984). In contrast to diatoms, flagellated algae can adjust their position in the surface layer with abundant light under static conditions because they can swim and maintain their vertical position (Rhew *et al.* 1999).

An increase in water temperature in winter shortens the period of holomixis and decreases intensity and frequency of vertical water mixing in lakes (Straile *et al.* 2003; Verburg, Hecky, and Kling 2003). Recent winter warming may therefore have created a condition disadvantageous to *A. nipponica* (Tsugeki *et al.* 2010). In support of this inference, analyses of sediments in a wide variety of northern hemisphere lakes showed that the abundance of *Aulacoseira* species decreased in parallel with a warming trend and began to decrease in the 1970s in temperate regions (Rühland, Paterson, and Smol 2008).

Algal composition has also changed even in late winter to spring in Lake Biwa. In recent years, cosmopolitan diatom species *Fragilaria crotonensis*, which appear in the late winter to early spring (Kagami *et al.* 2006) rapidly increased (Figure 7.3a). The sedimentation rate of *F. crotonensis* was positively related with concentrations of nitrate, soluble reactive phosphorus (SRP), and soluble reactive silicate (SRSi) (Tsugeki *et al.* 2010). This species grows well under conditions of high N supply relative to P supply when SRSi is sufficiently supplied (Interlandi and Kilham 1998; Interlandi, Kilham, and Theriot 1999). In Lake Biwa, growth rates of diatom species were often strongly limited by silica from spring to late summer (Kagami *et al.* 2006). According to Goto *et al.* (2007) and Miyajima, Nakano, Nakanishi (1995), silica deficiency in this lake was caused mainly by silica elimination by consumption and sedimentation of *Aulacoseira* when they bloomed during the winter holomixis period. Note that SRSi concentration in Lake Biwa notably increased in the mid 1980s (Figure 7.6c) when *A. nipponica* did not bloom in winter (Ichise *et al.* 1996). This fact suggested that an increased abundance of *F. crotonensis* was induced by not only increased nitrogen loading due to anthropogenic activities, but also by increased surplus of dissolved silicate due to winter warming (Tsugeki *et al.* 2010).

7.3 Winter warming and life history changes in *Daphnia*

7.3.1 Resting egg production

Many zooplankton groups such as Cladocera usually survive unfavorable periods as dormant stages through resting eggs. The typical Cladoceran life cycle is thus cyclical parthenogenesis: females reproduce asexually under favorable conditions and switch to sexual reproduction by producing males under unfavorable conditions. Through sexual reproduction, cladocerans produce resting eggs. The eggs are enveloped by a chitinous cover called ephippium, endure unfavorable periods and can hatch when conditions become favorable again. Ephippia (resting egg pouches) have been stored in lake sediments for several hundred years (see for example Cáceres 1998; Jankowski and Straile 2003; Tsugeki, Ishida, and Urabe 2009). Thus, we can infer historical changes in resting egg production of Cladoceran species by analyzing the number of ephippia stored in lake sediments together with their planktonic remains (Frey 1986; Jeppesen *et al.* 2001). However, there have been few studies which simultaneously examined the ephippia and remains of *Daphnia* except in Lake Biwa (Tsugeki, Ishida, and Urabe 2009).

As mentioned above, *D. galeata* were less abundant before 1960 but increased when eutrophication was progressing in this lake (Chapter 6). Similar to *Daphnia* remains, the sedimentation rate of ephippial abundance increased from 1950 to the mid 1970s (Figure 7.4a). The increase of ephippial abundance from 1950 to the mid-1970s was apparently derived from a substantial increase in population density of *D. galeata* during eutrophication of the lake (Tsugeki *et al.* 2009). A similar temporal trend was also reported in Lake Constance (Jankowski and Straile 2003). However, in Lake Biwa ephippial abundance of *D. galeata* rapidly decreased from the 1980s (Figure 7.4a) even though this species still dominated the zooplankton community afterwards (Okamoto 1984; Urabe 1995; Yoshida *et al.* 2001). Ephippial abundance of *Daphnia* in the lake sediments is often used to infer past population abundance (e.g., Kerfoot, Robbins, and Weider 1999). However, the discrepancy between abundance in *Daphnia* remains and ephippia of Lake Biwa indicated that ephippial abundance was not a useful surrogate for reconstructing past population abundance (Tsugeki *et al.* 2009).

7.3.2 Overwintering

In lake Biwa, as in other lakes (Rellstab and Spaak 2009), winter is the most severe season for *D. galeata* because their population density declines to an annual minimum level (Tsugeki, Ishida, and Urabe 2009). The one sure way to overcome severe winter conditions is by producing resting eggs in advance when environmental conditions are favorable (Lampert, Lampert, and Larsson 2010). However, if environmental conditions are favorable year round, resting egg production may result in decreasing immediate fitness because at one time *Daphnia* produce only a pair of resting eggs (ephippial eggs) that hatch later and sacrifice producing a number of parthenogenetic eggs that hatch without delay (Lynch 1983). The other way for overwintering is to pass the season as planktonic individuals. Although this way has a high extinction risk, these overwintering individuals may gain fitness if they successfully pass the season because they can initiate reproduction as soon as environmental conditions become favorable. These two ways are not mutually exclusive because some individuals of *Daphnia* populations producing resting eggs may overwinter as plankton (Lampert *et al*. 2010). However, if there are no severe seasons in a year, resting-egg production may be decreased because it is no longer essential for overwintering and likely decreases immediate fitness as explained above. In Lake Biwa, resting egg production of *Daphnia* populations estimated from ephippial abundance in the lake's sediments decreased from the 1980s. Interestingly, the number of *D. galeata* individuals collected during winter tended to increase in recent years (Tsugeki, Ishida, and Urabe 2009). Indeed, flux of ephippial eggs to the lake bottom correlated negatively with *Daphnia* abundance in the preceding winter (Figure 7.9). The results suggested that the recent decline of resting egg production by *D. galeata* populations was due to increased success of overwintering as plankton individuals (Tsugeki, Ishida, and Urabe 2009).

If so, why have *D. galeata* become adapt at overwintering successfully as plankton in the water column of Lake Biwa? One possibility is improvement of food conditions for *Daphnia* (Tsugeki, Ishida, and Urabe 2009). As mentioned above, abundance of *Aulacoseira nipponica* decreased with winter warming since the 1980s while flagellated algae such as *Cryptomonas* spp. and *Rhodomonas* sp. have become dominant phytoplankton in winter in Lake Biwa (Figures 7.3b, h). These flagellated algae are edible and highly nutritious to *Daphnia* (Ahlgren *et al*. 1990), whereas *A. nipponica* are unpalatable food because of the large colony size (Urabe *et al*. 1996). This circumstantial evidence suggests that winter warming may have improved food conditions and favored *D. galeata* to overwinter as plankton. As a result, their life history traits have changed to reduced resting egg production (Tsugeki, Ishida, and Urabe 2009).

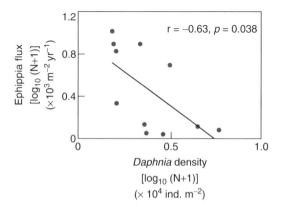

Figure 7.9 The annual flux of *Daphnia* ephippia plotted against mean abundance of *Daphnia galeata* in winter for years corresponding to those for sediment age used in estimation of ephippial flux. Data from Tsugeki *et al*. (2009); used with permission.

7.4 Conclusion

In Lake Biwa, intensive environmental monitoring was not started until 1965 (Mori *et al*. 1967; also reviewed in Chapter 6). Thus, only fragmentary information existed on the plankton community before the 1960s. Paleolimnological analysis of the plankton remains and photosynthetic pigments stored in sediments uncovered changes in the composition and abundance of plankton community over the last 100 years. Before the 1960s, Lake Biwa was oligotrophic and both primary and secondary production were limited. After the 1960s, algal abundance greatly increased indicating that eutrophication was developing at that time. Notable changes that occurred in the 1960s to 1970s were the dramatic increases in abundance of large green algae (*Staurastrum* spp. and *Pediastrum biwae*) and zooplankton *D. galeata*, and decreases in amoeboid protozoans. In the 1980s, changes were still found in the algal community and overwintering of zooplankton species in spite of the cessation of eutrophication. These changes were related to meteorological conditions and winter warming. Note that these historical changes in plankton communities were revealed mainly by analyzing biological and biogeochemical clues stored in the lake sediments. Other than the example of Lake Biwa, many studies using these clues have clarified that climate changes, especially recent warming, caused unique changes in each lake ecosystem with often drastic shifts in dominant plankton (Axford *et al*. 2009; Hobbs *et al*. 2010; Smol and Douglas 2007; Smol *et al*. 2005). According to Hede *et al*. (2010), ephippial egg production of *Daphnia* in a Danish lake remained relatively low during warming periods but increased during cooling events over several thousand years. Together with the case of Lake Biwa, the results imply that life history traits of *Daphnia* are highly vulnerable to climate changes.

Once again, these findings were uncovered by sedimentary records. This demonstrates that paleolimnological methods are a powerful tool to view past plankton communities and even changes in life history traits of planktonic species related to climate changes. However, it has apparently some limitations in uncovering the detailed mechanisms causing these changes. For example, we could not reconstruct historical changes in copepod populations without planktonic samples because these plankton did not leave any clues or proxies in the lake sediments. Thus, long-term monitoring with sample collections is essential to detect changes in lake ecosystems and uncover complex factors causing these changes. Paleolimnological methods serve effectively to complement and supplement long-term monitoring for examining effects of climate changes on lake ecosystems as shown here.

References

Ahlgren, G., Lundstedt, L., Brett, M., and Forsberg, C. (1990) Lipid composition and food quality of some freshwater phytoplankton for cladoceran zooplankters. *J. Plankton Res.*, **12**, 809–18.

Andersen, T. and Hessen, D.O. (1991) Carbon, nitrogen, and phosphorus content of freshwater zooplankton. *Limnol. Oceanogr.*, **36**, 807–14.

Axford, Y., Briner, J.P., Cooke, C.A., *et al*. (2009) Recent changes in a remote Arctic lake are unique within the past 200 000 years. *Proc. Natl. Acad. Sci.*, **106**, 18443–46.

Berman, T. and Shteinman, B. (1998) Phytoplankton development and turbulent mixing in Lake Kinneret (1992–1996). *J. Plankton Res.*, **20**, 709–26.

Brooks, J.L. and Dodson, S.I. (1965) Predation, body size, and composition of plankton. *Science*, **150**, 28–35.

Cáceres, C.E. (1998) Interspecific variation in the abundance, production, and emergence of Daphnia diapausing eggs. *Ecology*, **79**, 1699–710

DeMott, W.R. (1986) The role of taste in food selection by freshwater zooplankton. *Oecologia*, **69**, 334–40.

DeMott, W.R. (1988) Discrimination between algae and artificial particles by freshwater and marine copepods. *Limnol. Oceanogr.*, **33**, 397–408.

Elser, J.J. and Urabe, J. (1999) The stoichiometry of consumer-driven nutrient recycling: theory, observations, and consequences. *Ecology*, **80**, 735–51.

Endo, S., Yamashita, S., Kawakami, M., and Okumura, Y. (1999) Recent warming of Lake Biwa water. *Jpn. J. Limnol.*, **60**, 223–28. (In Japanese with English abstract.)

Fenchel, T. (1987) *Ecology of Protozoa. The Biology of Free-living Phagotrophic Protists*, Springer, Berlin.

Frenette, J.J., Vincent, W., Legendre, L., and Nagata. T. (1996) Size-dependent changes in phytoplankton C and N uptake in the dynamic mixed layer of Lake Biwa. *Freshw. Biol.*, **36**, 221–36.

Frey, D.G. (1986) *Cladocera Analysis*, John Wiley & Sons, Ltd., Chichester.

Gliwicz, Z.M. and Pijanowska, J. (1989) The Role of Predation in Zooplankton Succession. Plankton Ecology: Succession in Plankton Communities, Springer, Berlin, pp. 253–96.

Goto, N., Iwata, T., Akatsuka, T., *et al.* (2007) Environmental factors which influence the sink of silica in the limnetic system of the large monomictic Lake Biwa and its watershed in Japan. *Biogeochemistry*, **84**, 285–95.

Goulden, C.E., Henry, L.L., and Tessier, A.J. (1982) Body size, energy reserves, and competitive ability in three species of Cladocera. *Ecology*, **63**, 1780–9.

Gurung, T.B., Urabe, J., Nozaki, K., *et al.* (2002) Bacterioplankton production in a water column of Lake Biwa. *Lakes Reservoirs: Res.Management*, **7**, 317–23.

Hairston Jr., N.G., Lampert, W., Cáceres, C.E., *et al.* (1999) Rapid evolution revealed by dormant eggs. *Nature*, **401**, 446.

Hayami, Y. and Fujiwara, T. (1999) Recent warming of the deep water in Lake Biwa. *Oceanogr Jpn*, **8**, 197–202. (In Japanese with English abstract.)

Hede, M.U., Rasmussen, P., Noe-Nygaard, N., *et al.* (2010) Multiproxy evidence for terrestrial and aquatic ecosystem responses during the 8.2 ka cold event as recorded at Højby Sø, Denmark. *Quat. Res.*, **73**, 485–96.

Hikone Meteorological Observatory (1993–2000) *Shigaken no kisho*, Hikone Meteorological Observatory, Shiga. (In Japanese.)

Hobbs, W.O., Telford, R.J., Birks, H.J.B., *et al.* (2010) Quantifying recent ecological changes in remote lakes of North America and Greenland using sediment diatom assemblages. *PLoS ONE* **5**, e10026.

Hsieh, C.H., Ishikawa, K., Sakai, Y., *et al.* (2010) Phytoplankton community reorganization driven by eutrophication and warming in Lake Biwa. *Aquat. Sci.—Res. Across Boundaries*, **72**, 467–83.

Hsieh, C.H., Sakai, Y., Ban, S., *et al.* (2011) Eutrophication and warming effects on long-term variation of zooplankton in Lake Biwa. *Biogeosciences*, **8**, 1383–99.

Hyodo, F., Tsugeki, N., Azuma, J., *et al.* (2008) Changes in stable isotopes, lignin-derived phenols, and fossil pigments in sediments of Lake Biwa, Japan: Implications for anthropogenic effects over the last 100 years. *Sci. Total Environ.*, **403**, 139–47.

Ichise, S., Wakabayashi, T., Matsuoka, Y., *et al.* (1996) Succession of phytoplankton in northern basin of Lake Biwa, 1978–1995. *Report of Shiga Prefecture Institute of Pubic Health and Environmental Science*, **31**, 84–100. (In Japanese.)

Interlandi, S.J. and Kilham, S.S. (1998) Assessing the effects of nitrogen deposition on mountain waters: a study of phytoplankton community dynamics. *Wat. Sci. Tech.*, **38**, 139–46.

Interlandi, S.J., Kilham, S.S., and Theriot, E.C. (1999) Responses of phytoplankton to varied resource availability in large lakes of the Greater Yellowstone Ecosystem. *Limnol. Oceanogr.*, **44**, 668–82.

Interlandi, S.J., Kilham, S.S., and Theriot, E.C. (2003) Diatom-chemistry relationships in Yellowstone Lake (Wyoming) sediments: Implications for climatic and aquatic processes research. *Limnol. Oceanogr.*, **48**, 79–92.

Ishikawa, T., Narita, T., and Urabe, J. (2004) Long-term changes in the abundance of *Jesogammarus annandalei* (Tattersall) in Lake Biwa. *Limnol. Oceanogr.*, **49**, 1840–7.

Jankowski, T. and Straile, D. (2003) A comparison of egg-bank and long-term plankton dynamics of two *Daphnia* species, *D. hyalina* and *D. galeata*: Potentials and limits of reconstruction. *Limnol. Oceanogr.*, **48**, 1948–55.

Jeppesen, E., Leavitt, P., De Meester, L., and Jensen, J.P. (2001) Functional ecology and palaeolimnology: Using cladoceran remains to reconstruct anthropogenic impact *Trends Ecol. Evol.*, **16**, 191–98

Jöhnk, K.D., Huisman, J., Sharples, J., *et al.* (2008) Summer heatwaves promote blooms of harmful cyanobacteria. *Global Change Biol.*, **14**, 495–512.

Kagami, M., Gurung, T.B., Yoshida, T., and Urabe, J. (2006) To sink or to be lysed? Contrasting fate of two large phytoplankton species in Lake Biwa. *Limnol. Oceanogr.*, **51**, 2775–86.

Kawabata, K. (1987) Ecology of large phytoplankton in Lake Biwa: population dynamics and food relations with zooplankton. *Bull. Plankton Soc., Jpn.*, **34**, 165–72.

Kawabata, K. (1991) Ontogenetic changes in copepod behaviour: an ambush cyclopoid predator and a calanoid prey. *J. Plankton Res.*, **13**, 27–34.

Kawabata, K. (2006) Clearance rate of the cyclopoid copepod *Mesocyclops dissimilis* on the calanoid copepod *Eodiaptomus japonicus*. *Plankton and Benthos Research*, **1**, 68–70.

Kawabata, K., Narita, T., Nagoshi, M., and Nishino, M. (2002) Stomach contents of the landlocked dwarf ayu in Lake Biwa, Japan. *Limnology*, **3**, 135–42.

Kawamura, T. (1918) *Japanese Freshwater Biology*, *Syoukabou*, Tokyo. (In Japanese.)

Kerfoot, W.C., Robbins, J.A., and Weider, L.J. (1999) A new approach to historical reconstruction: Combining descriptive and experimental paleolimnology. *Limnol. Oceanogr.*, **44**, 1232–47.

Kumagai, M. (2008) Lake Biwa in the context of world lake problem. *Verh. Internat. Verein. Limnol.*, **30**, 1–15.

Lampert, W., Lampert, K.P., and Larsson, P. (2010) Coexisting overwintering strategies in *Daphnia pulex*: A test of genetic differences and growth responses. *Limnol. Oceanogr.*, **55**, 1893–900.

Leavitt, P.R., Carpenter, S.R., and Kitchell, J.F. (1989) Whole-lake experiments: the annual record of fossil pigments and zooplankton. *Limnol. Oceanogr.*, **34**, 700–17.

Leavitt, P.R. and Findlay, D.L. (1994) Comparison of fossil pigments with 20 years of phytoplankton data from eutrophic Lake 227, Experimental Lakes Area, Ontario. *Can. J. Fish. Aquat. Sci.*, **51**, 2286–99.

Leavitt, P.R., Findlay, D.L., Hall, R.I., and Smol, J.P. (1999) Algal responses to dissolved organic carbon loss and pH decline during whole-lake acidification: Evidence from paleolimnology. *Limnol. Oceanogr.*, **44**, 757–73.

Lynch, M. (1983) Ecological genetics of *Daphnia pulex*. *Evolution*, **37**, 358–74.

Magnuson, J., Webster, K.E., Assel, R.A., *et al.* (1997) Potential effects of climate changes on aquatic systems: Laurentian Great Lakes and Precambrian Shield Region. *Hydrol. Processes*, **11**, 825–71.

Meyers, P.A., Takemura, K., and Horie, S. (1993) Reinterpretation of Late Quaternary sediment chronology of Lake Biwa, Japan, from correlation with marine glacial-interglacial cycles. *Quat. Res.*, **39**, 154–62.

Miura, T. and Cai, Q. (1990) *Annual and Seasonal Occurrences of the Zooplankters Observed in the North Basin of Lake Biwa from 1965–1979*, Lake Biwa Research Institute, Otsu.

Miyajima, T., Nakano, S., and Nakanishi, M. (1995) Planktonic diatoms in pelagic silicate cycle in lake Biwa, Japan. *Jpn. J. Limnol.*, **56**, 211–20.

Mori, S. Yamamoto, K., Negoro, K., *et al.* (1967) First report of the regular limnological survey of Lake Biwa (Oct. 1965–Dec. 1966). I. General Remark. *Mem. Fac. Sci. Kyoto Univ. (B)*, **1**, 36–40.

Nakamura, M. (2002) Lake Biwa watershed transformation and the changed water environments. *Verh. Int. Verein. Limnol.*, **28**, 69–83.

Nakanishi, M. (1976) Seasonal variations of chlorophyll a amounts, photosynthesis and production rates of macro-and microphytoplankton in Shiozu Bay, Lake Biwa. *Physiol. Ecol. Japan*, **17**, 535–49.

Nakanishi, N. and Nagoshi, M. (1984) Yearly fluctuation of food habits of the Isaza, *Chaenogobius isaza*. Lake Biwa. *Jpn. J. Limnol.*, **45**, 279–88. (In Japanese with English abstract.)

Nakazawa, T., Sakai, Y., Hsieh, C., *et al.* (2010) Is the relationship between body size and trophic niche position time-invariant in a predatory fish? First stable isotope evidence. *PLoS ONE*, **5**, e9120.

Negoro, K. (1968) Phytoplankton of Lake Biwa, in *Flora ohmiensis* (ed. S. Kitamura), Hoikusya, Osaka. (In Japanese.)

Okamoto, K. (1984) Size-selective feeding of *Daphnia logispina hyalina* and *Eodiaptomus japonicus* on a natural phytoplankton assemblage with the fractionizing method. *Mem. Fac. Sci. Kyoto Univ. (B)*, **9**, 23–40.

O'Reilly, C.M., Alin, S.R., Plisnier, P.D., *et al.* (2003) Climate change decreases aquatic ecosystem productivity of Lake Tanganyika, Africa. *Nature*, **424**, 766–8.

Passarge, J., Hol, S., Escher, M., and Huisman, J. (2006). Competition for nutrients and light: stable coexistence, alternative stable states, or competitive exclusion. *Ecol. Monogr.*, **76**, 57–72.

Petersen, J.E., Chen, C.C., and Kemp, W.M. (1997) Scaling aquatic primary productivity: experiments under nutrient-and light-limited conditions. *Ecology*, **78**, 2326–38.

Rellstab, C. and Spaak, P. (2009) Lake origin determines *Daphnia* population growth under winter conditions. *J. Plankton Res.*, **31**, 261–71.

Reynolds, C.S. (1984) *The Ecology of Freshwater Phytoplankton*, Cambridge University, Cambridge.

Rhew, K., Baca, R.M., Ochs, C.A., and Threlkeld, S.T. (1999) Interaction effects of fish, nutrients, mixing and sediments on. *Freshw. Biol.*, **42**, 99–110.

Rühland, K., Paterson, A.M., and Smol, J.P. (2008) Hemispheric-scale patterns of climate-related shifts in planktonic diatoms from North American and European lakes. *Global Change Biol.*, **14**, 2740–54.

Scheffer, M., Rinaldi, S., Gragnani, A., *et al.* (1997) On the dominance of filamentous cyanobacteria in shallow, turbid lakes. *Ecology*, **78**, 272–82.

Schindler, D.W. (1978) Factors regulating phytoplankton production and standing crop in the world's freshwaters. *Limnol. Oceanogr.*, **23**, 478–86.

Schindler, D.W. (2006) Recent advances in the understanding and management of eutrophication. *Limnol. Oceanogr.*, **51**, 356–63.

Shiga Prefecture (1963–2000) Annual report of the regular observation in Lake Biwa. Fisheries Experimental Station of Shiga Prefecture, Hikone. (In Japanese.)

Smol, J.P. and Douglas, M.S.V. (2007) From controversy to consensus: making the case for recent climate change in the Arctic using lake sediments. *Front. Ecol. Environ.*, **5**, 466–74.

Smol, J.P., Wolfe, A.P., Birks, H.J.B., *et al.* (2005) Climate-driven regime shifts in the biological communities of Arctic lakes. *Proc. Natl. Acad. Sci.*, **102**, 4397.

Sterner, R.W. and Elser, J.J. (2002) *Ecological Stoichiometry: The Biology of Elements from Molecules to the Biosphere*, Princeton University Press, Princeton, NJ.

Straile, D. (2002) North Atlantic Oscillation synchronizes food-web interactions in central European lakes *Proc. Roy. Soc. London—Biol. Sci.*, **269**, 391–5.

Straile, D., Jöhnk, K., and Rossknecht, H. (2003) Complex effects of winter warming on the physicochemical characteristics of a deep lake. *Limnol. Oceanogr.*, **48**, 1432–8.

Sugawara, K. (1938) The seasonal variation of the occurrence of phytoplankton and the circulation of silicon in Lake Biwa. *Jpn. J. Limnol.*, **8**, 434–45. (In Japanese.)

Sunaga, T. (1964) On the seasonal change of food habits of several fishes in Lake Biwa. *Physiol Ecol.*, **12**, 252–8. (In Japanese with English abstract.)

Tezuka, Y. (1984) Seasonal variations of dominant phytoplankton, chlorophyll a and nutrient levels in the pelagic regions of Lake Biwa. *Jpn. J. Limnol.*, **45**, 26–37.

Threlkeld, S.T. (1976). Starvation and the size structure of zooplankton communities. *Freshw. Biol.*, **6**, 489–96.

Tsugeki, N., Oda, H., and Urabe, J. (2003) Fluctuation of the zooplankton community in Lake Biwa during the 20th century: a paleolimnological analysis. *Limnology*, **4**, 101–7.

Tsugeki, N.K., Ishida, S., and Urabe, J. (2009) Sedimentary records of reduction in resting egg production of *Daphnia galeata* in Lake Biwa during the 20th century: A possible effect of winter warming. *J. Paleolimnol.*, **42**, 155–65.

Tsugeki, N.K., Urabe, J., Hayami, Y., *et al.* (2010) Phytoplankton dynamics in Lake Biwa during the 20th century: complex responses to climate variation and changes in nutrient status. *J. Paleolimnol.*, **44**, 69–83.

Tsuji, A. and Hoki, A. (2001) *Centric Diatoms in Lake Biwa*. Lake Biwa Study Monogr., Otsu.

Urabe, J. (1995) Direct and indirect effects of zooplankton on seston stoichiometry. *Ecoscience*, **2**, 286–96.

Urabe, J., Ishida, S., Nishimoto, M., and Weider, L.J. (2003) *Daphnia pulicaria*; a zooplankton species that suddenly appeared in 1999 in the offshore zone of Lake Biwa. *Limnol.*, **4**, 35–41.

Urabe, J., Kawabata, K., Nakanishi, M., and Shimizu, K. (1996) Grazing and food size selection of zooplankton community in Lake Biwa during BITEX'93. *Jpn. J. Limnol.*, **57**, 27–37.

Urabe, J. and Maruyama, T. (1986) Prey selectivity of two cyprinid fishes in Ogochi Reservoir. Bull. *Jpn. Soc. Sci. Fish.*, **52**, 2045–54.

Urabe, J., Nakanishi, M., and Kawabata, K. (1995) Contribution of metazoan plankton to the cycling of nitrogen and phosphorus in Lake Biwa. *Limnol. Oceanogr.*, **40**, 232–41.

Urabe, J., Sekino, T., Nozaki, K., *et al.* (1999) Light, nutrients and primary productivity in Lake Biwa: An evaluation of the current ecosystem situation. *Ecol. Res.*, **14**, 233–42.

Verburg, P., Hecky, R.E., and Kling, H. (2003) Ecological consequences of a century of warming in Lake Tanganyika. *Science*, **301**, 505.

Winder, M. and Schindler, D.E. (2004) Climatic effects on the phenology of lake processes. *Global Change Biol.*, **10**, 1844–56.

Yamada, Y. and Nakanishi, M. (1999) *Regional Development, Urbanization and Water, and Changes in Material Cycling*. Iwanami-Shoten, Tokyo. (In Japanese.)

Yamamoto, K. and Nakamura, M. (2004) An examination of land use controls in the Lake Biwa watershed from the perspective of environmental conservation and management. *Lakes and Reservoirs: Research and Management*, **9**, 217–28.

Yang, J. and Shen, Y. (2005) Morphology, biometry and distribution of *Difflugia biwae* Kawamura, 1918 (Protozoa: Rhizopoda). *Acta Protozool*, **44**, 103–11.

Yoshida, T., Kagami, M., Bahadur Gurung, T., and Urabe, J. (2001) Seasonal succession of zooplankton in the north basin of Lake Biwa. *Aquat. Ecol.*, **35**, 19–29.

Yuma, M., Hosoya, K., and Nagata, Y. (1998) Distribution of the freshwater fishes of Japan: an historical overview. *Environ. Biol. Fishes*, **52**, 97–124.

8

Numerical Simulation of Future Overturn and Ecosystem Impacts for Deep Lakes in Japan

Daisuke Kitazawa
Underwater Technology Research Center, Institute of Industrial Science, University of Tokyo, Japan

8.1 Introduction

The global mean surface temperature has increased by 0.44 °C (0.018 °C year^{-1}) for 25 years from 1981 to 2005 (Intergovernmental Panel on Climate Change 2007, hereafter IPCC). Some Japanese meteorological stations, however, show a higher increasing rate of atmospheric temperature. For example, the meteorological stations around the largest Japanese lake, Lake Biwa, show an increasing rate of the annual mean atmospheric temperature of 0.02–0.07 °C year^{-1}. The atmospheric temperature will almost certainly keep increasing in Japan since IPCC expects the global mean surface temperature to increase by 1.1–6.4 °C until 2100.

Climate change will affect the physical limnology of lakes including water temperature, vertical mixing, stratification, and icing events (see Chapters 2, 3, 9, 12, 14, 16–20). An increase in water temperature due to global warming has already been observed at various depths in many lakes (Schindler *et al*. 1996; Quayle *et al*. 2002; Livingstone 2003; O'Reilly *et al*. 2003; Straile, Johnk, and Rossknecht 2003; Verburg, Hecky, and Kling 2003; Vollmer *et al*. 2005; George 2007). Its rate is different among lakes and between depths in a lake. This difference depends on the location of a lake (latitude and altitude), climate, and topography (water depth and exposure to wind). Generally, lake bottom water temperature does not follow increasing rates of surface water temperature with the result that the water column stratifies earlier in the year and more strongly. In the tropical zone, water overturns less frequently and some not

Climatic Change and Global Warming of Inland Waters: Impacts and Mitigation for Ecosystems and Societies, First Edition. Edited by Charles R. Goldman, Michio Kumagai and Richard D. Robarts.
© 2013 John Wiley & Sons, Ltd. Published 2013 by John Wiley & Sons, Ltd.

at all, with only the upper water becoming well mixed. The depth of the vertical mixed layer in lakes is expected to be reduced due to global warming. In the temperate and southern cold zones, lakes become well mixed once a year from top to bottom at a water temperature of near $>4\,°C$. Surface water warming may lead to a cessation of overturn and consequently eternal stratification will occur. In the temperate regions of the world lake waters overturn two times a year when water temperature passes $4\,°C$. The effect of global warming may limit overturn to once a year as for lakes in the temperate and southern cold zones.

Higher water temperature reduces dissolved oxygen (hereafter DO) solubility in water, while surface water is always saturated by oxygen. It also increases biological processes such as algal growth and decomposition of organic matter that consumes DO in deep water. Prolonged stratification reduces the downward flux of DO. The decrease in DO concentration at the bottom enhances the internal loading of phosphorus from the bottom sediment which if mixed into the upper water column leads to the enhancement of primary production and speeds the process of eutrophication. In contrast, reduced vertical mixing causes a decrease in upward transport of phosphorus and thus primary production. Dissolved oxygen concentrations may be enhanced by lower production and sedimentation of organic matter. Therefore, global warming affects physical limnology and ecosystems in many ways (Livingstone 2008).

Numerical simulation can be a powerful tool to predict changes in physical limnology and ecosystem function as a result of climate change. The effects of global warming for several decades have been examined under scenarios of the increase in carbon dioxide as it influences atmospheric temperature (De Stasio et al. 1996; Peeters et al. 2002; Danis et al. 2004; Matzinger et al. 2007). The predicted results show the enhancement of stratification and changes in the pattern of overturn. These numerical studies have been performed with a one-dimensional vertical model, which assumes horizontal uniformity of the environment. The one-dimensional vertical model is being used to model three-dimensional physical processes, such as wind-driven currents, density-driven currents, and internal waves. One-dimensional models can reproduce well vertical mixing due to water surface cooling and wind stress (Perroud et al. 2009). Differences in these modeling applications affect the accuracy of numerical simulation even if the governing equations and boundary conditions are identical. Three-dimensional numerical model is time consuming but promising with the availability of advanced computers.

This chapter focuses on the effects of climate change in two Japanese deep lakes; Lake Biwa and Lake Ikeda (Figure 8.1). The most distinctive feature of the physical limnology of these lakes is a pronounced annual thermal cycle. The lakes change from a thermally mixed condition in winter to a strongly stratified condition in summer. Stratification decreases DO concentration in the deep waters of these lakes due to bacterial decomposition, reduced downward flux of oxygen, and sediment oxygen demand. These deep waters in Lake Ikeda have been hypoxic or anoxic over the last two decades because overturn did not occur after 1986 (Kumagai et al. 2003). Lake Biwa becomes well mixed in winter every year so that the DO concentration recovers to the saturated level even in deep water. However, the recovery of DO was delayed in the winter during 2006 and 2007 (Yoshimizu et al. 2010; see also Chapter 6). An eventual loss of overturn is now feared in Lake Biwa as has occurred in Lake Ikeda.

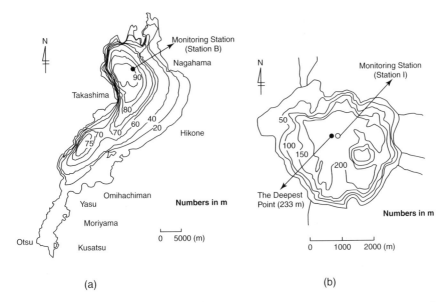

Figure 8.1 Topographies of Lake Biwa and Lake Ikeda and the locations of monitoring stations
(a) Lake Biwa and (b) Lake Ikeda.

Hence, a three-dimensional hydrodynamic and ecosystem coupled model was created
to predict the future change in physical limnology and ecosystem characteristics in
Lake Biwa. The model was calibrated by comparing modeled and observed variables
for the period 1982–1991 both in Lake Biwa and Lake Ikeda. The effects of climate
change were then predicted by the model for Lake Biwa on the basis of each of the
three scenarios (see below) of global warming from IPCC.

8.2   Numerical model

8.2.1   Hydrodynamic model

The hydrodynamic model of Kitazawa, Kumagai, and Hasegawa (2010) is a
nonlinear, three-dimensional, finite-difference model that solves mass, momentum,
heat, chemical, and biological conservation equations of fluid dynamics (Mellor
1996). The model is hydrostatic so that the vertical distribution of pressure is
determined directly from the density field. The density was approximated as a
constant unless it was multiplied by gravity in the buoyancy force (Boussinesq
approximation). The equation of state (Fotonoff and Millard 1983) calculates water
density as a function of water temperature.

   The vertical eddy viscosity and diffusivity coefficients were calculated on the basis
of Richardson's number, which was derived from the fields of current velocity and
water density (Munk and Anderson 1948; Webb 1970). The horizontal eddy viscosity and diffusivity coefficients were assumed to be constants (Akitomo, Kurogi, and
Kumagai 2004).

8.2.2 Ecosystem model

The ecosystem model of Kitazawa *et al*. (2010) is based on the observed features of the lower trophic level food web in Japanese lakes (Ikeda and Adachi 1978; Matsuoka, Goda, and Naito 1986; Taguchi and Nakata 2009). It is a phosphorus and nitrogen-controlled model with nine state variables (Figure 8.2): phytoplankton, cell quotas of phosphorus and nitrogen, zooplankton, particulate and dissolved organic carbon (POC and DOC), dissolved inorganic phosphorus and nitrogen (DIP and DIN), and DO. Material cycles are described by carbon, phosphorus, nitrogen, and oxygen flows, with varying ratios of nutrients and oxygen to carbon for conservation of their mass.

It is assumed in this model that phosphorus and nitrogen are the two limiting nutrients that control primary production. Silicate, trace metals and inorganic carbon were ignored in the model because they are assumed to be in sufficient supply. Cell quotas of phosphorus and nitrogen were taken into account for algal growth under nutrient-limiting conditions (Droop 1974; Lehman, Botkin, and Likens 1975). Phytoplankton and zooplankton are dominated by many species. The phase pattern is associated with the seasonal variation in the environment. Dividing the phytoplankton and zooplankton species into several groups was difficult in preparing many biological parameters so that each plankton species was summarized by one state variable. Bacteria also contribute to the lower trophic level food web mainly through decomposition of organic matter, resulting in mineralization and oxygen consumption. Bacteria are also a food source for various grazers and therefore recycle carbon and nutrients. In addition, under phosphorus-limited conditions they may outcompete phytoplankton for phosphorus (Robarts *et al*. 1998). A part of the effects of bacteria were included

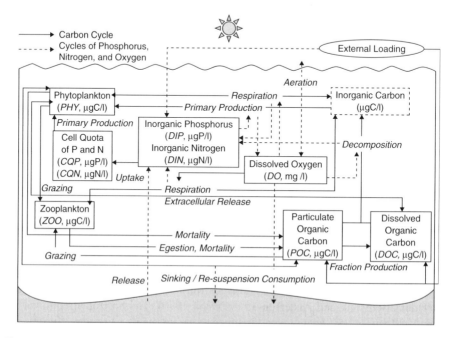

Figure 8.2 Ecosystem model for Japanese lakes. Boxes and arrows represent state variables and fluxes among state variables, respectively.

in the model, while bacteria were not chosen as a state variable explicitly. POC refers to fecal pellets, bacteria, dead phytoplankton and zooplankton. Extracellular carbon released by phytoplankton and a part of POC move into the DOC pool. The sinking, decomposition, mineralization, and oxygen consumption of phytoplankton and POC were taken into account in the food web system. Higher trophic level predators such as large fish, human fishing, and birds were neglected on the assumption that they have only a secondary impact on the food web system. Instead, the effects of these were included implicitly in the mortality rates of phytoplankton and zooplankton.

State variables are connected by fluxes which are formulated by several mathematical equations, including chemical and biological parameters. The time variations in each state variable were calculated from the budget of fluxes each time. This was combined with the time variation in each state variable due to hydrodynamic processes.

8.2.3 Boundary condition

The model used a free-slip lateral boundary condition. Time-dependent wind stress and heat flux were given at the surface. A bulk aerodynamic formation was used to calculate heat and momentum fluxes over the water surface at each grid point. The model did not use heat flux but included friction boundary conditions at the bottom.

8.2.4 Grid system

The hydrodynamic model of Lake Biwa has 37 levels with uniform horizontal and vertical grid sizes of 2 km and 2.5 m, respectively (Figure 8.3a). The bathymetry was derived from the observations at about 250 000 points. The hydrodynamic model of

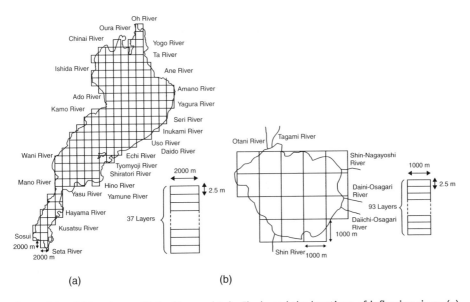

Figure 8.3 Grid systems of Lake Biwa and Lake Ikeda and the locations of inflowing rivers (a) Lake Biwa and (b) Lake Ikeda.

Lake Ikeda has 93 layers with uniform horizontal and vertical grid sizes of 1 km and 2.5 m, respectively (Figure 8.3b). The evaluation points of current velocity, pressure, water temperature, density, and state variables were arranged based on the staggered grid system.

8.2.5   Computational method

The governing equations were discretized by a finite difference scheme. An explicit time integration method of Euler, QUICK (Quadratic Upstream Interpolation for Convective Kinematics), and a second-order central difference method were used for time derivative, advection, eddy viscosity and diffusivity terms, respectively. The time step was 40 seconds for Lake Biwa and 10 seconds for Lake Ikeda.

Each variable was solved sequentially (Figure 8.4). Water-surface elevation was calculated by integrating the continuity equation using the boundary condition at the surface and bottom. The momentum equation in the vertical direction can be integrated from the surface to the bottom under the boundary condition to obtain the pressure. Horizontal current velocities, vertical current velocity, water temperature, and each state variable were solved by using momentum equations, continuity equation, and advection–diffusion equations of water temperature and each state variable, respectively. Water density was calculated by the state equation.

8.2.6   Computational condition

The numerical model was run for the period 1982–1991 for model calibration in Lake Biwa and Lake Ikeda. It starts running in March 1981 to avoid problems with initialization of the field of water temperature and state variables.

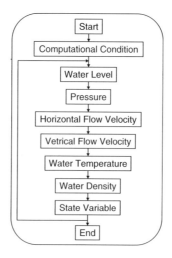

Figure 8.4   Flow chart of the numerical simulation procedure.

The bulk formation requires the meteorological data of atmospheric pressure and temperature, solar radiation, the amount of cloud cover, relative humidity, precipitation, and wind velocity and direction. A part of hourly meteorological data was obtained from Hikone Meteorological Station at Lake Biwa, and from Ibusuki Meteorological Station at Lake Ikeda. Some meteorological data were estimated from hourly data after 1991 at Hikone and Makurazaki Meteorological Stations for Lake Biwa and Lake Ikeda, respectively. Annual budget of water mass was 5 billion m^3 through 27 rivers entering Lake Biwa (Figure 8.3a) while that for Lake Ikeda was 38 million m^3 entering through two rivers (Figure 8.3b). The inflow of water from each river was given based on its catchment area and on the seasonal variation in precipitation. Water temperature of each river was the same as that in the grid adjacent to the river mouth. The total inflow of chemical oxygen demand (COD), total phosphorus (T-P), and total nitrogen (T-N) were assumed to be $16\,790\,t\ year^{-1}$, $400\,t\ year^{-1}$, and $7300\,t\ year^{-1}$, respectively, in Lake Biwa (Shiga Prefecture 2004). They were distributed into three areas; South Basin, and the west and east coasts of the North Basin (Somiya 2000). The total inflow of COD, T-P, and T-N in Lake Ikeda were assumed to be $125\,t\ year^{-1}$, $3.8\,t\ year^{-1}$, and $550\,t\ year^{-1}$, respectively (Kagoshima Prefecture 2001).

The concentration of total organic carbon (TOC) was estimated by multiplying the concentration of COD by 1.1. The concentrations of POC and DOC were distributed by assuming that the ratio of POC and DOC to TOC were 0.4 and 0.6, respectively. The difference between total phosphorus or nitrogen and organic phosphorus or nitrogen was given as the concentration of DIP or DIN. The river water DO concentration was assumed to be saturated throughout a year.

The DO concentration must be sensitive to changes in stratification caused by climate change. The state of the ecosystem was predicted under several scenarios for the period 2011–2110 in Lake Biwa with no warming (Scenario 1), a $0.025\,°C$ year^{-1} increase (Scenario 2), and a $0.05\,°C$ year^{-1} increase in atmospheric temperature (Scenario 3) (Figure 8.5). Hourly meteorological data for the period 1990–2010 were obtained from Hikone Meteorological Station (Lake Biwa). The meteorological data set for two decades was used repeatedly with a linear increase in atmospheric temperature, while the other variables in boundary conditions were not varied.

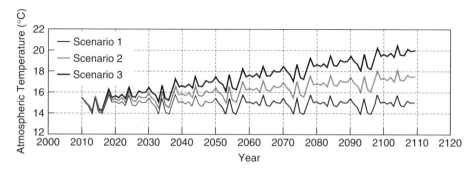

Figure 8.5 Three scenarios of atmospheric temperature increase generated from the prediction of IPCC.

8.3   Comparison with observations

8.3.1   Water temperature

Modeled water temperature at the surface and the bottom of Station B in 1982–1991 was compared with the biweekly observations in Lake Biwa (Figure 8.6a). The most distinctive feature of the physical limnology of Lake Biwa was a pronounced annual thermal cycle. The lake changed from a thermally mixed condition in winter to strongly stratified condition in summer. Lake Biwa becomes well mixed from top to bottom at a water temperature of about 7 °C. In the spring, warming heats the surface water and a thermal front (thermocline) is formed in the water column. A well-developed thermocline persists throughout the summer. In the fall, cooling weakens the stratification until the water column is again mixed from top to bottom. The thermal front disappears entirely in winter.

The model could reproduce the seasonal variation in water temperature, while the modeled water temperature at the bottom was warmer than the observed water temperature in 1984–1985. This discrepancy may be attributed to inaccuracies in the specification of the meteorological data such as wind speed and short-wave solar radiation. This hypothesis should be assessed for more accurate predictions. Overturn occurred every year since the increasing rate of water temperature at the bottom during stratified seasons from April to December was 0.59 °C year^{-1} (observation) or 0.62 °C year^{-1} (simulation), which was larger than that at the surface. Water temperature at

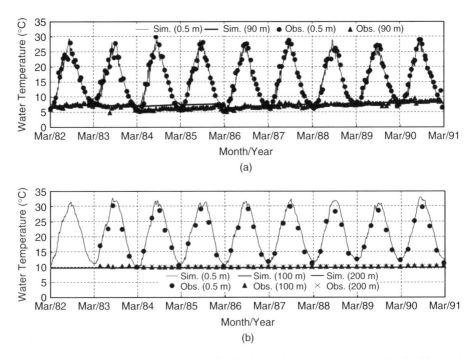

Figure 8.6   Comparison between modeled and observed water temperature of Lake Biwa and Lake Ikeda during the period 1982–1991 (a) Lake Biwa (Station B) and (b) Lake Ikeda (Station I).

the bottom decreased with that at the surface due to surface cooling after Lake Biwa became well mixed.

Modeled water temperature at the surface, at 100 m and 200 m below the surface of Station I in 1982–1991 was compared with the bimonthly observation in Lake Ikeda (Figure 8.6b). Water temperature showed a similar seasonal variation with that in Lake Biwa. However, in Ikeda, lake overturn ceased from 1987 so that stratification remained even in the winter. In contrast to the bottom water temperature in Lake Biwa, water temperature increased little by little at 200 m, and even at 100 m below the surface. Heat was not as well distributed in Lake Ikeda compared to Lake Biwa because of the stronger stratification. This may be attributed to weak development of wind-driven currents, density-driven currents and internal waves resulting from the unique topography of Lake Ikeda.

8.3.2 Dissolved oxygen

The DO concentration is the key variable for the survival of endemic bottom fauna in both lakes. Modeled DO concentration was compared with the observed one in Lake Biwa (Figure 8.7a) and Lake Ikeda (Figure 8.7b). DO concentration at the bottom decreased approximately from 11 mg L^{-1} to 4 mg L^{-1} during stratified seasons in Lake Biwa. It recovered to saturated levels every year due to overturn. The recovery occurred immediately with overturn and which then led to a gradual increase in DO

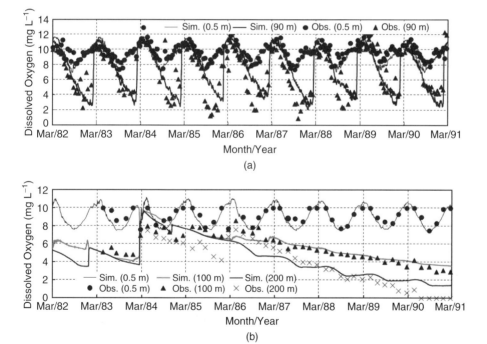

Figure 8.7 Comparison between modeled and observed dissolved oxygen concentrations of Lake Biwa and Lake Ikeda during the period 1982–1991. (a) Lake Biwa (Station B); (b) Lake Ikeda (Station I).

concentration both at the surface and bottom due to decreasing water temperature. The DO concentration was at a maximum when Lake Biwa became well mixed. In contrast, a lack of seasonal overturn prevented the recovery of the DO concentration at 100 m and 200 m below the surface from 1987 in Lake Ikeda. The recovery of DO concentration in 1986 could not be reproduced in the model. This must be attributed to the inaccuracy in reproducing the vertical mixing of the water during the same period. The consumption rate of oxygen was then underestimated at 200 m, while it was reproduced well at 100 m. Hypoxic water enhances the release of phosphorus from sediment. A rising concentration of DIP was observed and was reproduced by the model (not shown here).

8.4   Future prediction

Global warming is expected to reduce vertical mixing and overturn through enhanced stratification in these two studied lakes. Predicted water temperature for the period 2011–2110 showed a similar increase at the surface and bottom of Lake Biwa (Figure 8.8). Under IPCC Scenario 3, the increasing rates of water temperature were 0.045 °C year^{-1} and 0.038 °C year^{-1} at the surface and bottom, respectively, which were smaller than the IPCC projected increasing rate of atmospheric temperature (0.05 °C year^{-1}). The warming of the bottom water was not greatly delayed in respect to that of the surface water. Overturn has occurred every year since the increasing

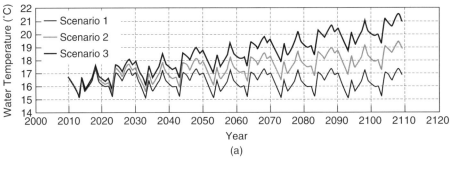

(a)

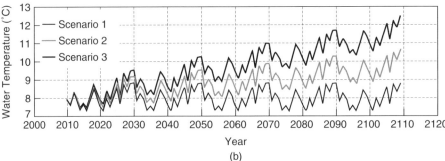

(b)

Figure 8.8   Prediction of water temperature at Station B in Lake Biwa for the period 2010–2110 under three IPCC climate-warming scenarios (a) at the surface and (b) at the bottom.

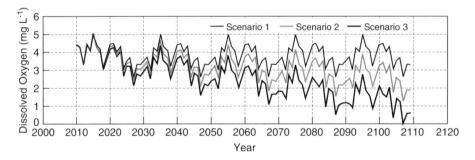

Figure 8.9 Prediction of annually minimal DO concentration at Station B in Lake Biwa for the period 2010–2110 under three IPCC climate-warming scenarios.

rate of the bottom water temperature during stratified seasons was much larger than that of the surface water temperature.

However, the annual minimal DO concentration became lower with global warming and was often completely exhausted after 2100 with Scenario 3 in Lake Biwa (Figure 8.9). Higher water temperature reduced the solubility of oxygen in the water. The saturated concentration of DO was $11.8\,\mathrm{mg\,L^{-1}}$ in 7 °C water and $10.7\,\mathrm{mg\,L^{-1}}$ in 11 °C water. This difference was reflected in the predicted DO concentrations, while the annual minimal DO concentrations decreased at a greater rate than the decreasing rate of saturated DO concentration. This may be attributed to prolonged stratification caused by the difference in the increasing rate of water temperature between the surface and bottom. Overturn was delayed through stronger stratification so that oxygen was consumed at the bottom for a prolonged period. Higher water temperatures also increased biological processes such as algal growth and decomposition of organic matter, which caused increased oxygen consumption at the bottom. Downward flux of DO may have been reduced by stronger stratification. Another possibility of the decrease in the annual minimal DO concentration was eutrophication. Hypoxic water enhanced the release of phosphorus from sediment (Figure 8.10). The increase in the concentration of DIP was predicted in Scenarios 2 and 3, and that led to enhanced primary production (Figure 8.11). The concentration of chlorophyll a increased by 5–15% for the period 2011–2110. Consequently, as dissolved oxygen declined with

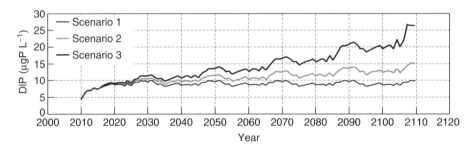

Figure 8.10 Prediction of DIP (dissolved inorganic phosphorus) concentration at Station B in Lake Biwa for the period 2010–2110 under three IPCC climate-warming scenarios.

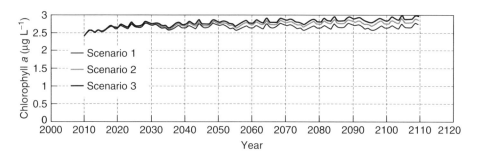

Figure 8.11   Prediction of chlorophyll *a* concentration at Station B in Lake Biwa for the period 2010–2110 under three IPCC climate-warming scenarios.

increased algal decomposition in the hypolimnion the increase in the internal phosphorus loading from sediment enhanced primary production in the photic zone. At the same time the higher sediment flux and mineralization resulted in still further consumption of DO at the lake bottom.

8.5   Conclusions

This chapter introduced a three-dimensional hydrodynamic and ecosystem coupled model to reproduce past change in Lake Biwa and Lake Ikeda and to predict future changes in the physical limnology and ecosystem indicators in Lake Biwa as impacted by global warming. Overturn occurred every year for the period 1982–1991. This was because the bottom water temperature increased during stratified seasons and both trends of the surface and bottom water temperatures intersect in winter. Overturn was predicted to continue for the period 2011–2110 in Lake Biwa. In contrast, the bottom water temperature did not reach the increase in surface water temperature in Lake Ikeda, resulting in a partly mixed water column. Surface areas of Lake Biwa and Lake Ikeda are 670 km^2 and 11 km^2 with maximal depths of 104 m and 233 m, respectively. Lake Ikeda has a relatively larger topographical inclination. Three-dimensional physical processes, such as wind-driven currents, density-driven currents, and internal waves, may hardly develop and consequently transport less heat downward. This hypothesis must be examined more precisely for generalized lakes in future studies. The increasing rate of bottom water temperature is a key parameter to predict the effects of global warming on physical limnology. This should be carefully observed and discussed in the context of vertical mixing, heat flux through the bottom, and the contribution of groundwater.

Warmer water temperature reduces the solubility of oxygen. Predicted DO concentrations decreased at a larger rate than expected from the solubility of oxygen alone. This may be attributed to prolonged stratification caused by the difference in the increasing rate of water temperature at the surface and bottom of Lake Biwa. Higher water temperature also increased biological processes such as algal growth and organic matter decomposition, which enhanced oxygen consumption at the bottom. The downward flux of dissolved oxygen will have been reduced by stronger stratification. Finally, the increase in the internal phosphorus loading from bottom sediment enhanced primary production at each overturn and then the resulting higher sediment flux and mineralization, resulted in further consumption of dissolved oxygen at the bottom.

A three-dimensional hydrodynamic and ecosystem coupled model can reproduce physical processes such as wind-driven currents, density-driven currents, and internal waves so that it can estimate not only vertical mixing and eddy diffusion but the advection of heat. The spread of the area of hypoxic water can also be predicted. The three-dimensional model is time consuming to construct but will be helpful for examining and discussing the effects of global warming on the future physical limnology and ecosystem characteristics of lake systems. The model is promising with the availability of advanced computers whose CPU speed of has increased by approximately 1000 times in 10 years. The accuracy of the model should be improved by using finer meshes, concise boundary conditions, sensitivity analysis and calibration of ecological parameters as well as validation with more and accurate environmental monitoring data.

8.6 Acknowledgements

The meteorological data at Hikone Meteorological Station and water quality in each river were provided by the Japan Meteorological Agency and National Institute for Environmental Studies, respectively. The present study was supported by the Global Environment Research Fund Fa-804 by the Ministry of the Environment, Japan, and the Ministry of Education, Science, Sports and Culture, Grant-in-Aid for Scientific Research (A), 20244079.

References

Akitomo, K., Kurogi, M., and Kumagai, M. (2004) Numerical study of a thermally induced gyre system in Lake Biwa. *Limnology*, **5**, 103–14

Danis, P.A., von Grafenstein, U., Masson-Delmotte, V., *et al*. (2004) Vulnerability of two European lakes in response to future climatic changes. *Geophys. Res. Lett.*, **31**, L21507.

DeStasio, B.T., Jr.,, Hill, D.K., Kleinhans, J.M., *et al*. (1996) Potential effects of global climate change on small north-temperate lakes: Physics, fish, and plankton. *Limnol. Oceanogr.*, **41**, 1136–49.

Droop, M.R. (1974) The nutrient status of algal cells in continuous culture. *J. Mar. Biol. Assoc. UK*, **54**, 825–55.

Fotonoff, N.P. and Millard, R.C. Jr. (1983) Algorithms for computation of fundamental properties of seawater. UNESCO Technical Papers in Marine Science, UNESCO, http://www.jodc.go.jp/info/ioc_doc/UNESCO_tech/059832eb.pdf. (accessed July 24, 2012).

George, D.G. (2007) The impact of the North Atlantic Oscillation on the development of ice on Lake Windermere. *Climatic Change*, **81**, 455–68.

Ikeda, S. and Adachi, N. (1978) A dynamic water quality model of Lake Biwa—A simulation study of the lake eutrophication. *Ecol. Model.*, **4**, 151–72.

IPCC (2007) *Climate change 2007: the physical science basis*. Contribution of working group I to the fourth assessment report of the Intergovernmental Panel on Climate Change, Cambridge University Press, Cambridge.

Kagoshima Prefecture (2001) The third plan for management of water quality in Lake Ikeda, Kagoshima Prefecture, Kagoshima. (In Japanese.)

Kitazawa, D., Kumagai, M., and Hasegawa, H. (2010) Effects of internal waves on dynamics of hypoxic waters in Lake Biwa. *J. Korean Soc. Mar. Environ. Engin.*, **13**, 30–42.

Kumagai, M., Vincent, W.F., Ishikawa, K., and Yasuaki Aota (2003) Lessons from Lake Biwa and other Asian lakes: global and local perspectives, in *Freshwater Management, Global Versus Local Perspectives* (eds. M. Kumagai, W. F. Vincent), Springer-Verlag, Tokyo.

Lehman, J.T., Botkin, D.B., and Likens, G.E. (1975) The assumptions and rationales of a computer model of phytoplankton population dynamics. *Limnol. Oceanogr.*, **20**, 343–65.

Livingstone, D.M. (2003) Impact of secular climate change on the thermal structure of a large temperate central European lake. *Climatic Change*, **57**, 205–25.

Livingstone, D.M. (2008) A change of climate provokes a change of paradigm: Taking leave of two tacit assumptions about physical lake forcing. *Internat. Rev. Hydrobiol.*, **93**, 404–14.

Matsuoka, Y., Goda, T., and Naito, N. (1986) An eutrophication model of Lake Kasumigaura. *Ecol. Model.*, **31**, 201–19.

Matzinger, A., Schmid, M., Veljanoska-Sarafiloska, E., *et al.* (2007) Eutrophication of ancient Lake Ohrid: Global warming amplifies detrimental effects of increased nutrient inputs. *Limnol. Oceanogr.*, **52**, 338–53.

Mellor, G..L. (1996) Introduction to physical oceanography. American Institute of Physics Press, New York.

Munk, W.H. and Anderson, E.R. (1948) Notes on a theory of the thermocline. *J. Mar. Res.*, **7**, 276–95

O'Reilly, G.M., Alin, S.R., Plisnier, P.D., *et al.* (2003) Climate change decreases aquatic ecosystem productivity of Lake Tanganyika, Africa. *Nature*, **424**, 766–8.

Peeters, F., Livingstone, D.M., Goudsmit, G.H., *et al.* (2002) Modeling 50 years of historical temperature profiles in a large central European lake. *Limnol. Oceanogr.*, **47**, 186–97.

Perroud, M., Goyette, S., Martynov, A., *et al.* (2009) Simulation of multiannual thermal profiles in deep Lake Geneva: A comparison of one-dimensional lake models. *Limnol. Oceanogr.*, **54**, 1574–94.

Quayle, W.C., Peck, L.S., Peat, H., *et al.* (2002) Extreme responses to climate change in Antarctic lakes. *Science*, **295**, 645.

Robarts, R.D., Waiser, M.J., Hadas, O., *et al.* (1998) Relaxation of phosphorus limitation due to typhoon-induced mixing in two morphologically distinct basins of Lake Biwa, *Japan*. *Limnol. Oceanogr.*, **43**, 1023–36.

Schindler, D.W., Bayley, S.E., Parker, B.R., *et al.* (1996) The effects of climatic warming on the properties of boreal lakes and streams at the Experimental Lakes Area, northwestern Ontario. *Limnol. Oceanogr.*, **41**, 1004–17.

Shiga Prefecture (2004) *Environmental White Paper*, Kyoeiinsatsu Corporation, Hikone. (In Japanese.)

Somiya, I. (2000) *Lake Biwa—Formation of Environment and Water Quality*, Gihodo-syuppan, Tokyo. (In Japanese.)

Straile, D., Johnk, K., and Rossknecht, H. (2003) Complex effects of winter warming on the physicochemical characteristics of a deep lake. *Limnol. Oceanogr.*, **48**, 1432–8.

Taguchi, K., and Nakata, K. (2009) Evaluation of biological water purification functions of inland lakes using an aquatic ecosystem model. *Ecol. Model.*, **220**, 2255–71.

Verburg, P., Hecky, R.E., and Kling, H. (2003) Ecological consequences of a century of warming in Lake Tanganyika. *Science*, **301**, 505–7.

Vollmer, M.K., Bootsma, H.A., Hecky, R.E., *et al.* (2005) Deep-water warming trend in Lake Malawi, East Africa. *Limnol. Oceanogr.*, **50**, 727–32.

Webb, E.K. (1970) Profile relationships: the log-linear range, and extension to strong stability. *Quart. J. Roy. Meteorol. Soc.*, **96**, 67–90.

Yoshimizu, C., Yoshiyama, K., Tayasu, I., *et al.* (2010) Vulnerability of a large monomictic lake (Lake Biwa) to warm winter event. *Limnology*, **11**, 233–9.

9

Model Development to Evaluate the Impacts of Climate Change on Total Phosphorus Concentrations in Lakes

Kohei Yoshiyama

River Basin Research Center, Gifu University, Japan

9.1 Introduction

Lakes that transform slowly under natural conditions have experienced two drastic human-induced changes since the latter part of the twentieth century. One is lake eutrophication (Vollenweider and Kerekes 1982) caused by anthropogenic nutrient loading, and the other is lake acidification (Schindler 1988), caused by industrial emission of nitrogen oxides and sulfur oxides. These changes act on lake ecosystems as stressors, resulting in such undesirable outcomes as the occurrence of harmful algal blooms, formation of anoxia, and the death of fish and other organisms. Following these two stressors, global warming poses still another threat to the world's lake ecosystems.

The effects of eutrophication and acidification on lake water quality operate on watershed spatial scales and atmosphere, respectively. Lakes are dynamic living systems and water quality can be recovered if they are managed appropriately by taking into account the time scale of their water renewal rates (Schindler 2006). On the other hand, climate warming directly operates on lake physics along with changes in wind, precipitation, humidity, and cloud cover patterns, and these effects propagate through a lake ecosystem. Climate warming is a global phenomenon and lakes all over the world are influenced by this new stress (Livingstone 2008). This impact is projected to continue for more than 100 years regardless of future emission scenarios

Climatic Change and Global Warming of Inland Waters: Impacts and Mitigation for Ecosystems and Societies, First Edition. Edited by Charles R. Goldman, Michio Kumagai and Richard D. Robarts.
© 2013 John Wiley & Sons, Ltd. Published 2013 by John Wiley & Sons, Ltd.

(Wigley 2005). Therefore, global warming is essentially different from the other two stressors in terms of its mode of operation and spatial and temporal scales.

A wide range of research has been undertaken to understand and reduce the impacts of eutrophication and acidification on lake water quality. As part of this process, constructing and testing models have played important roles (Vollenweider and Kerekes 1982; Jørgensen 1993). While detailed simulation models are necessary for prediction in each lake (see Chapter 8), empirical or semi-empirical models with fewer parameters and limited mechanistic details are applied effectively for water-quality management because of their clarity and generality. For example, the Vollenweider model (Vollenweider and Kerekes 1982) is arguably the most successful model in limnology. Yet, such simple models to account for effects of global warming are only now being developed (Blenckner 2005). In this chapter, I introduce a model framework to help understand the impacts of global warming on the phosphorus enrichment of lake ecosystems, and support environmental management decisions.

9.2   Model framework

Global warming influences diverse aspects of lake ecosystems and focusing on a single indicator does not capture the whole impact on these systems. Nevertheless, here I consider the effects of global warming on lake total phosphorus (TP) because it is a common variable of water quality models and can be related to other parameters (Chapra and Dobson 1981).

9.2.1   TP mass balance model

For a lake with the volume V (m^3), surface area A (m^2), and average depth $Z = V/A$ (m), TP mass balance (Vollenweider 1968) is expressed by

$$V \frac{dP}{dt} = M - QP - \sigma VP \qquad (9.1)$$

where P (mg m^{-3}) is the TP concentration, M (mg yr^{-1}) is the phosphorus loading, Q (m^3 yr^{-1}) is the outflow, and σ (yr^{-1}) is the sedimentation coefficient. At the steady state of Equation 9.1, P is written as

$$P = \frac{M}{Q + \sigma V} \qquad (9.2)$$

Redefining $L = M/A$ (mg m^{-2} yr^{-1}) as the areal loading and $\rho = Q/V$ (yr^{-1}) as the flushing rate, we obtain

$$P = \frac{L/Z}{\rho + \sigma} \qquad (9.3)$$

Equation 9.3 contains one unknown σ that needs to be determined from empirical relationships. According to Vollenweider (1976), σ and the square root of the flushing rate $\sqrt{\rho}$ show a good correlation. Substituting $\sigma = \sqrt{\rho}$ into Equation 9.3, the well known Vollenweider model is derived:

$$P = \frac{L}{q\left(1 + \sqrt{\tau}\right)} \qquad (9.4)$$

where $q = Q/A$ (m yr^{-1}) is the water discharge height and $\tau = 1/\rho$ (yr) is the water residence time. A more detailed account of the Vollenweider model can be found in Ahlgren, Frisk, and Kampnielsen (1988).

9.2.2 TP mass balance model with stratification and mixing periods

The Vollenweider model has been successful in capturing a general trend of lake responses to eutrophication. Obviously, the model cannot reflect responses of lakes to global warming because it does not consider the direct and indirect effects of water temperature increases. Global warming influences the sedimentation coefficient σ because it extends the summer stratification period where σ should be greater due to the shallower surface layer. Global warming may also influence phosphorus loadings because the internal loading increases with the formation of anoxia during the summer stratification period.

The TP mass balance model is extended by explicitly accounting for the summer stratification period and the winter mixing period (Figure 9.1). During the summer stratification period, a lake is separated into two layers, epilimnion and hypolimnion. The TP mass balances are considered in the two layers connected by a thermocline.

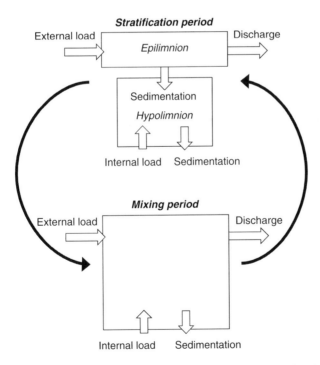

Figure 9.1 A schematic diagram of the mass balance model that account for the stratification and mixing periods. During the stratification period, a lake is separated into epilimnion and hypolimnion, and phosphorus mass balance in each layer is considered. During the mixing period, mass balance is considered for the whole lake.

The diffusive flux between the two layers is ignored because it is negligibly small (Yamazaki *et al.* 2010). I also assume that the external loading enters the epilimnion and that internal loading enters the hypolimnion. Let the stratification period of the first year start from $t = 0$ and end at $t = t_s$. TP mass balances in the epilimnion (thickness Z_e [m]; volume V_e [m^3]) and in the hypolimnion (thickness Z_h [m]; volume V_h [m^3]) for $0 < t \le t_s$ are respectively expressed by

$$V_e \frac{dP_e}{dt} = M_{ext} - QP_e - \frac{v}{Z_e} V_e P_e \tag{9.5}$$

$$V_h \frac{dP_h}{dt} = M_{int} - \frac{v}{Z_h} V_h P_h + \frac{v}{Z_e} V_e P_e \tag{9.6}$$

where P_e and P_h (mg m^{-3}) are TP concentrations in the epilimnion and hypolimnion, M_{ext} and M_{int} (mg year^{-1}) are the external and internal loadings, and v (m yr^{-1}) is the apparent sedimentation rate. The initial conditions are given by $P_e(0) = P_{e0}^1$ and $P_h(0) = P_{h0}^1$. Note that when a lake is homogeneous ($P_e = P_h$) and the lake surface area is close to the lake area at the thermocline ($V_h/Z_h = A$), the sum of Equations 9.5 and 9.6 is equivalent to Equation 9.1.

At the end of the stratification period of the first year, $t = t_s$, winter mixing homogenizes the lake and sets the initial condition of P for the mixing period,

$$P_0^1 = \frac{V_e P_e(t_s) + V_h P_h(t_s)}{V} \tag{9.7}$$

Phosphorus mass balance for $t_s < t \le 1$ is expressed by

$$V \frac{dP}{dt} = M_{ext} + M_{int} - QP - \frac{v}{Z} VP \tag{9.8}$$

where $V = V_e + V_h$ and $Z = (V_e + V_h)/A$. $P(1)$ in turn sets the initial conditions of P_{e0}^2 and P_{h0}^2 in the second year. TP mass balance equations for the consecutive years are formulated iteratively.

Equations 9.5, 9.6 and 9.8 are first-order linear differential equations and can be integrated explicitly. All model parameters can be time-dependent but, for simplicity, I assume that parameters are constant except for M_{int} that is influenced by seasonal formation of anoxia. For the stratification period of the first year ($0 < t \le t_s$), $P_e(t)$ and $P_h(t)$ are expressed by

$$P_e(t) = \left(P^0 - \frac{M_{ext}/V_e}{Q/V_e + v/Z_e} \right) e^{-\left(\frac{Q}{V_e} + \frac{v}{Z_e} \right)t} + \frac{M_{ext}/V_e}{Q/V_e + v/Z_e} \tag{9.9}$$

$$P_h(t) = P^0 e^{-\left(\frac{v}{Z_h} \right)t} + e^{-\left(\frac{v}{Z_h} \right)t} \int_0^t e^{\left(\frac{v}{Z_h} \right)s} \left[\frac{M_{int}(s)}{V_h} + \frac{v V_e P_e(s)}{Z_e V_h} \right] ds \tag{9.10}$$

where $P^0 = P_e^1 = P_h^1$ is the initial TP concentration. For the mixing period of the first year ($t_s < t \le 1$), the initial concentration P_0^1 is obtained by Equation 9.7 and $P(t)$ is written as

$$P(t) = P_0^1 e^{-\left(\frac{Q}{V} + \frac{v}{Z} \right)(t - t_s)} + e^{-\left(\frac{Q}{V} + \frac{v}{Z} \right)(t - t_s)} \int_0^{t - t_s} e^{\left(\frac{Q}{V} + \frac{v}{Z} \right)s} \left[\frac{M_{ext}}{V} + \frac{M_{int}(s)}{V} \right] ds \tag{9.11}$$

When the internal loading M_{int} is negligible during the winter mixing period, $P(t)$ is written as

$$P(t) = \left(P_0^1 - \frac{M_{ext}/V}{Q/V + v/Z}\right) e^{-\left(\frac{Q}{V} + \frac{v}{Z}\right)(t - t_s)} + \frac{M_{ext}/V}{Q/V + v/Z} \qquad (9.12)$$

$P_e(t)$, $P_h(t)$ and $P(t)$ are computed for the following years in the same way.

A numerical example of the model is given in Figure 9.2. Here I consider an ideal lake with a surface area $A = 10^6$ m^2 and maximum depth $Z_m = 30$ m. Assuming a parabolic shape (Carpenter 1983), the lake area $a(z)$ at depth z is expressed by

$$a(z) = \frac{A}{Z_m}(Z_m - z) = \frac{10^6}{30}(30 - z) \qquad (9.13)$$

where Z_m is the maximum depth. The thickness of the epilimnion Z_e can be expressed by a function of lake surface area A (Fee *et al.* 1996)

$$Z_e = 2.92 + 0.0607A^{0.25} = 4.84 \qquad (9.14)$$

Lake volumes, the mean lake depth, and the hypolimnion thickness are obtained accordingly: $V = 15 \times 10^6$ m^3, $V_e = 4.45 \times 10^6$ m^3, $V_h = 10.55 \times 10^6$ m^3, $Z = 15$ m, $Z_h = V_h/a(Z_e) = 12.58$ m. The external loading is $M_{ext} = 1$ tonne year$^{-1} = 10^9$ mg yr^{-1}, a typical value for mesotrophic lakes of this size (Saijo and Hayashi 2001). For this example, the internal loading is not considered ($M_{ext} = 0$). Assuming that the water-residence time is one year, Q is set to 15×10^6 m^3 year^{-1}. The sedimentation rate $v = 10$ m year^{-1} is taken from Vollenweider (1976). The length of the stratification period is set to $t_s = 0.7$ assuming a temperate monomictic lake.

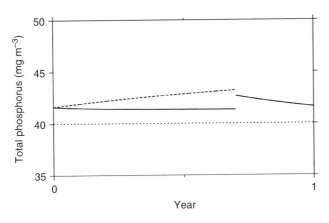

Figure 9.2 A numerical example of the model described in Equations 9.5, 9.6 and 9.8. Periodic solution of total phosphorus concentrations during a year is shown. One year consists of a stratification period from $t = 0$ to $t = t_s$ and a mixing period from $t = t_s$ to $t = 1$. During the stratification period, the lake is separated into two layers, epilimnion and hypolimnion. Solid and broken lines respectively show the total phosphorus in the epilimnion and hypolimnion. At the onset of the mixing period, the lake is homogenized (solid line). The dotted line shows steady state total phosphorus concentration of the corresponding whole-lake mass balance model (Equation 9.1).

During the stratification period, TP in the epilimnion decreases exponentially while TP in the hypolimnion increases due to sedimentation from the epilimnion. Note that TP in the hypolimnion may increase further due to internal loading if oxygen is depleted at the sediment-water interface. The lake is homogenized at the onset of the mixing period, and TP decreases during the period. The TP concentrations are generally higher than steady state of the original mass balance model of $40\,\mathrm{mg\,m^{-3}}$ (Equation 9.2) and of the Vollenweider model, $33\,\mathrm{mg\,m^{-3}}$ (Equation 9.4). In the model with stratification and mixing periods, phosphorus accumulates in the hypolimnion without flushing out from the lake during the stratification period, resulting in higher TP concentrations.

9.2.3   Stratification period model

In the previous section, a mass balance model that considers the stratification and mixing periods was introduced. In this section, I consider effects of global warming on length of the stratification period t_s.

Global warming directly affects the onset of spring stratification. Demers and Kalff (1993) compiled data from lakes in temperate and subtropical regions and found relationships between mean annual air temperature and date of the onset of stratification. According to the study, a $1\,^{\circ}\mathrm{C}$ increase in air temperature advances the onset of stratification by 3.4–6.1 days.

On the other hand, global warming also affects the onset of the winter mixing period (fall turnover), but the influence is not so straightforward. While spring stratification is a process near the surface boundary, winter mixing is a whole lake process. Intuitively, a warm winter keeps the epilimnion warmer and delays the date of fall turnover by stabilizing a water column. However, water temperature in the hypolimnion is also higher in a warmer climate, which reduces the stability of a water column. So water temperature inside a lake may counteract the increasing air temperature, either advancing or delaying the date of fall turnover. Nürnberg (1988) analyzed data taken from lakes worldwide and found that summer hypolimnetic temperature can predict nearly 80% of the variations of the fall turnover date. The empirical model suggested that a $1\,^{\circ}\mathrm{C}$ increase in hypolimnetic temperature advances the date by 6.8 days in southern Ontario lakes. Nürnberg (1988) also considered other parameters and constructed an empirical model:

$$\log(X) = 2.62 - 0.116\log(T_h) + 0042\log(Z) - 0.002\,\mathrm{lat}_{ad} \qquad (9.15)$$

where X is the date of fall turnover, T_h is the summer hypolimnetic temperature, Z is the mean lake depth, and lat_{ad} is the latitude adjusted for the elevation. According to Nürnberg (1988), latitude (lat, in decimal degree) is adjusted by altitude (alt, m) as:

$$\mathrm{lat}_{ad} = \mathrm{lat} + \frac{c}{100}\mathrm{alt} \qquad (9.16)$$

where c is a coefficient that depends on the latitude (Lewis 1983). For example, $c = 0.46$ at 40° latitude. Because air temperature decreases by $0.65\,^{\circ}\mathrm{C}$ with each $100\,\mathrm{m}$ increase in elevation, a $1\,^{\circ}\mathrm{C}$ temperature increase corresponds to a $154\,\mathrm{m}$ lower elevation. If we apply $c = 0.46$, the $154\,\mathrm{m}$ lower elevation results in 0.7° decrease in adjusted latitude. In this way, the effect of warming can be incorporated in the empirical model of fall turnover (Equation 9.15).

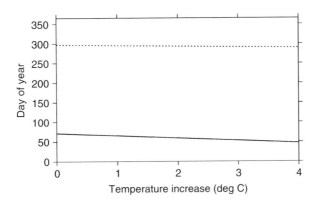

Figure 9.3 Effect of air temperature increase on the onset of stratification and fall turnover. A lake located at 40° latitude with current annual mean air temperature of 15°C is considered. The lake average depth is set to 15 m, and current summer hypolimnetic temperature, 10°C. With a temperature increase, both stratification (solid line) and fall turnover (dotted line) advance. The stratification period is prolonged due to the stronger response of the onset of stratification period than that of fall turnover.

Figure 9.3 shows responses of dates of the onset of stratification and fall turnover obtained from the empirical models with assumptions: (1) a lake is located at 40° latitude; (2) current mean annual air temperature and summer hypolimnetic temperature of 15 °C and 10 °C, respectively; and (3) the mean air and hypolimnetic temperatures increase in parallel. While the onset of stratification was advanced as expected, the date of fall turnover was advanced by global warming as well. Yet the response of the latter was relatively weak and the stratification period is prolonged by 4.14 days per 1 °C.

9.2.4 Hypolimnetic oxygen consumption model

As a result of a prolonged stratification period by warming, the hypolimnion tends to be anoxic. Here we adopt a model of Livingstone and Imboden (1996) that distinguishes oxygen consumption in a water column (volumetric) and in sediment (areal). The model assumes that: (i) no oxygen production occurs in the hypolimnion, (ii) volumetric (J_V mg O_2 m^{-3} d^{-1}) and areal (J_A mg O_2 m^{-2} d^{-1}) oxygen consumption rates are constant spatially and temporally during a summer stratification period, (iii) vertical mixing is negligible in the hypolimnion, and (iv) oxygen concentrations are homogeneous in each stratum due to horizontal mixing. Under these assumptions, oxygen depletion rate J at each depth z is expressed by

$$J(z) = J_V + \alpha(z)J_A \tag{9.17}$$

$\alpha(z)$ (m^{-1}) is the ratio of sediment surface area to water volume at z, given by

$$\alpha(z) = -\frac{1}{a(z)} \frac{da(z)}{dz} \tag{9.18}$$

For our idealized lake (Equation 9.13), we have

$$\alpha(z) = \frac{1}{30 - z} \qquad (9.19)$$

Oxygen concentration at the end of the stratification period (DO_s) for each depth z is calculated by:

$$DO_s(z) = DO_i - 365 \, t_s J(z) \qquad (9.20)$$

where DO_i is oxygen concentration at the onset of stratification. Note that DO_i is also affected by warming because oxygen saturation concentration is lower at higher water temperature.

Oxygen consumption rate in a water column J_V can be related to TP and hypolimnetic temperature (Pace and Prairie 2005). Oxygen consumption in sediment J_A, in contrast, is difficult to determine, and numerous factors including chemical (Rippey and McSorley 2009) and physical (Bryant et al. 2010) ones influence the process.

9.2.5   Internal loading model

From Equation 9.20, the extent of anoxia at time t, $a_X(t)$ (m^2) is calculated, which is related to internal phosphorus loading M_{int} (Nürnberg 1984):

$$M_{int} = (\text{Maximum P release rate } [mg \, m^{-2} \, yr^{-1}]) \times a_X(t) \qquad (9.21)$$

or it may be practical to use averaged internal loading through the stratification period,

$$\bar{M}_{int} = \frac{1}{t_s} \int_0^{t_s} M_{int} dt$$

$$= (\text{Maximum P release rate } [mg \, m^{-2} \, yr^{-1}]) \times \frac{1}{t_s} \int_0^{t_s} a_X(t) dt \qquad (9.22)$$

The maximum phosphorus release rate varies between 2 and $10 \, g \, m^{-2} \, year^{-1}$ (Nürnberg 1984).

9.3   Concluding remarks

I have introduced, here, a model framework that consists of four submodels: TP mass balance, stratification period, hypolimnetic oxygen consumption, and internal phosphorus loading. The causal mechanisms and the effects of warming are depicted in Figure 9.4. There is a well-known positive feedback mechanism between trophic state and anoxia formation in lakes (Carpenter, Ludwig, and Brock, 1999). Global warming fuels the feedback loop by reducing the initial oxygen concentration, extending the stratification period, and increasing oxygen consumption. The model framework presented here is not absolute at all, but such mechanisms should be clearly included in models that predict and evaluate the impact of global warming on lake ecosystems.

In order to develop the model further, several challenges must be addressed. The stratification period model needs to be evaluated with a larger data set because the lakes considered in the studies are rather concentrated in Canada. Models of hypolimnetic

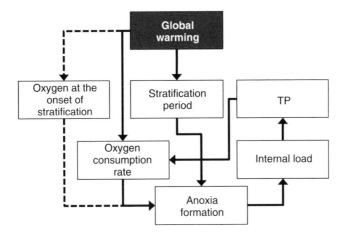

Figure 9.4 A diagram that describes the causal relationship between global warming and total phosphorus concentration. Solid arrows denote a positive effect and broken arrows, a negative effect. There is a positive feedback loop, TP → oxygen consumption rate → anoxia formation → internal load → TP that is fueled by global warming.

oxygen consumption and internal loading have several uncertainties. Among them, data of sediment oxygen consumption are still limited and empirical relationships are yet to be investigated. Studies suggest that phosphorus can be released from oxic sediments (Gachter, Meyer, and Mares, 1988); however, most models assume that phosphorus is released only from anoxic sediments. Whether phosphorus release from oxic sediments is a significant or a negligibly small part of internal loading is still to be determined. Though the TP mass balance model assumed all external loadings enter the epilimnion, river waters occasionally flow into the hypolimnion and may load significant amounts of nutrients. In particular, cold and turbid snowmelt waters enter the hypolimnion at the onset of stratification in the early spring (Kumagai and Fushimi 1995). Elucidating these processes may contribute to our understanding of the impacts of global warming on lakes and thereby help produce the methodologies and understanding necessary to reduce the negative impacts of global climate change.

References

Ahlgren, I., Frisk, T., and Kampnielsen, L. (1988) Empirical and theoretical models of phosphorus loading, retention and concentration vs. lake trophic state. *Hydrobiologia*, **170**, 285–303.

Blenckner, T. (2005) A conceptual model of climate-related effects on lake ecosystems. *Hydrobiologia*, **533**, 1–14.

Bryant, L.D., Lorrai, C., McGinnis, D.F., *et al.* (2010) Variable sediment oxygen uptake in response to dynamic forcing. *Limnol. Oceanogr.*, **55**, 950–64.

Carpenter, S. (1983) Lake geometry: implication for production and sediment accretion rates. *J. Theoret. Biol.*, **105**, 273–86.

Carpenter, S., Ludwig, D., and Brock, W. (1999) Management of eutrophication for lakes subject to potentially irreversible change. *Ecological Applications*, **9**, 751–71.

Chapra, S.C. and Dobson, F.H. (1981) Quantification of the lake trophic typologies of Naumann (surface quality) and Thienemann (oxygen) with special reference to the Great Lakes. *J. Great Lakes Res.*, **7**, 182–93.

Demers, E. and Kalff, J. (1993) A simple model for predicting the date of spring stratification in temperate and subtropical lakes. *Limnol. Oceanogr.*, **38**, 1077–81.

Fee, E., Hecky, R., Kasian, S., and Cruikshank, D. (1996) Effects of lake size, water clarity, and climatic variability on mixing depths in Canadian Shield lakes. *Limnol. Oceanogr.*, **41**, 912–20.

Gachter, R., Meyer, J., and Mares, A. (1988) Contribution of bacteria to release and fixation of phosphorus in lake-sediments. *Limnol. Oceanogr.*, **33**, 1542–58.

Jørgensen, S.E. 1993. Mangement and modelling of lake acidification. *UNEP/ILEC Guidelines of Lake Management Series* **5**, 79–104.

Kumagai, M. and Fushimi, H. (1995) Inflows due to snowmelt, in *Physical Processes in a Large Lake: Lake Biwa, Japan* (ed. S. Okuda, J. Imberger, and M. Kumagai), AGU, Washington, D.C., pp. 129–39.

Lewis, W. (1983) A revised classification of lakes based on mixing. *Can. J. Fish. Aquat. Sci.*, **40**, 1779–87.

Livingstone, D.M. (2008) A change of climate provokes a change of paradigm: taking leave of two tacit assumptions about physical lake forcing. *Internat. Rev. Hydrobiol.*, **93**, 404–14.

Livingstone, D. and Imboden, D. (1996) The prediction of hypolimnetic oxygen profiles: A plea for a deductive approach. *Can. J. Fish.Aquat. Sci.*, **53**, 924–32.

Nürnberg, G.K. (1984) The prediction of internal phosphorus load in lakes with anoxic hypolimnia. *Limnol. Oceanogr.*, **29**, 111–24.

Nürnberg, G.K. (1988) A simple model for predicting the data of fall turnover in thermally stratified lakes. *Limnol. Oceanogr.*, **33**, 1190–5.

Pace, M.L. and Prairie, Y.T. (2005) *Respiration in Aquatic Ecosystems*, *Respiration in Lakes*, Oxford University Press, Oxford, pp. 103–21.

Rippey, B. and McSorley, C. (2009) Oxygen depletion in lake hypolimnia. *Limnol. Oceanogr.*, **54**, 905–16.

Saijo, Y. and Hayashi, H. (eds.) (2001) *Lake Kizaki, Limnology and Ecology of a Japanese Lake*, Backhuys Publishers, Leiden.

Schindler, D.W. (1988) Effects of acid rain on freshwater ecosystems. *Science*, **239**, 149–57.

Schindler, D.W. (2006) Recent advances in the understanding and management of eutrophication. *Limnol. Oceanogr.*, **51**, 356–63.

Vollenweider, R.A. (1968) Scientific fundamentals of the eutrophication of lakes and flowing waters, with particular reference to nitrogen and phosphorus as factors in eutrophication. Technical report, OECD. Paris.

Vollenweider, R.A. (1976) Advances in defining critical loading levels for phosphorus in lake eutrophication. *Memorie Istituto Italiano di Idrobiologia.* **33**, 53–83.

Vollenweider, R.A. and Kerekes, J. (1982) Eutrophication of waters: monitoring, assessment and control. OECD, Paris.

Wigley, T. (2005) The climate change commitment. *Science*, **307**, 1766–9.

Yamazaki, H., Honma, H., Nagai, T., *et al.* (2010) Multilayer biological structure and mixing in the upper water column of Lake Biwa during summer 2008. *Limnology*, **11**, 63–70.

10

Recent Climate-Induced Changes in Freshwaters in Denmark

Erik Jeppesen[1,2,3], Brian Kronvang[1], Torben B. Jørgensen[1,4],
Søren E. Larsen[1], Hans E. Andersen[1], Martin Søndergaard[1],
Lone Liboriussen[1], Rikke Bjerring[1], Liselotte S. Johansson[1],
Dennis Trolle[1], and Torben L. Lauridsen[1,2]

[1]*Department of Bioscience, Aarhus University, Denmark*
[2]*Sino-Danish Centre for Education and Research (SDC), Beijing, China*
[3]*Greenland Climate Research Centre (GCRC), Greenland Institute of Natural Resources, Nuuk, Greenland*
[4]*Limfjordssekretariatet, Nørresundby, Denmark*

10.1   Introduction

As a result of the Intergovernmental Panel on Climate Change reports, it is now widely accepted that global warming is a fact. It is also recognized that the extent of the effects will differ at different locations, being stronger at high latitudes and less strong at lower latitudes (IPCC 2007). According to a set of regional climate models, Denmark can expect an overall warmer climate and particularly higher temperatures in late summer and winter (van Roosmalen *et al.* 2010). An overall wetter climate can be expected, with more precipitation during winter, but likely lower precipitation during summer (van Roosmalen *et al.* 2010). Such changes will have profound effects on discharge, nutrient input, temperature, water column stability (in lakes), nutrient dynamics as well as biological structure and dynamics. In this chapter we show that Danish freshwaters are already facing changes that can be attributed to the recent warming during the past two decades. We use long-term data from 18 streams and between 20 and 250 lakes.

Climatic Change and Global Warming of Inland Waters: Impacts and Mitigation for Ecosystems and Societies,
First Edition. Edited by Charles R. Goldman, Michio Kumagai and Richard D. Robarts.
© 2013 John Wiley & Sons, Ltd. Published 2013 by John Wiley & Sons, Ltd.

10.2   Trends in water flow in streams and diffuse nutrient losses

Since the early 1880s annual precipitation in Denmark has increased by, on average, 100 mm or about 15%. To elucidate the effect on runoff in streams we analyzed the daily mean water flow from 18 streams selected to cover Denmark geographically from north to south and from east to west, covering contrasting precipitation conditions. These streams have the longest recorded time series of water flows in Denmark with historical mean daily water flow since 1950. The time series (1950–2006) was analyzed for trends by using Kendall's trend test (Sen 1968). For the annual minimum and maximum values we fitted a generalized extreme value distribution (GEVD) (Kite 1978). For minimum and maximum values the empirical distributions were tested and compared between the two periods 1950–1977 and 1978–2006 using Kolmogorov–Smirnov's "Goodness-of-fit" test (Conover 1980). A Baseflow Index (BFI assuming values between 0 and 1) was calculated for each station, representing the ratio of mean annual base flow (low flow) to mean annual flow.

The River Ribe, draining coarse sandy soils in the western part of Jutland, Denmark, is given as an example of the trend in annual maximum and minimum water flow between the periods 1950–1977 and 1978–2006 (Figure 10.1). A significant (10% level used) increase appeared in the distributions of both maximum and minimum daily water flows between the two 28-year periods (Table 10.1). The River Harrested draining loamy soils on Zealand, eastern Denmark, a region receiving much less precipitation, shows another result (Figure 10.1). Here, the distribution of the annual minimum daily water flow values during 1950–1977 and 1978–2006 decreased significantly between the two periods, whereas the distribution of maximum daily water flow did not exhibit any significant change (Table 10.1).

For each station two threshold values were calculated. The first threshold value (t_{max}) is the 90% percentile in the distribution of all mean daily water flows and the second threshold value (t_{min}) is the 10% percentile for the same period. For each year the number of mean daily water flows larger than t_{max} was identified as well as the number smaller than t_{min}.

In River Ribe a significant trend in the annual t_{min} was detected, and the test probability was <0.10 for the number of transgressions of the t_{max} and the standard deviation of the annual water flows, the latter being a clear sign of more pronounced seasonal variations (Table 10.1). River Harrested showed no significant trends in annual t_{min} and t_{max}, whereas a significant downward trend was observed in BFI, indicating an increase in the proportion of water discharged from the upper soil horizons, tile drainage water being more enriched in nutrients than groundwater discharging to the stream. Twelve of the 18 investigated stations analyzed showed an increase in annual maximum daily water flow between the two periods. Likewise, the 12 stations exhibited an increase in annual minimum daily water flow, while six stations showed a decrease. The BFI analyses produced significant trends for eight stations of which five experienced a positive trend.

Using export coefficient models for nutrients (Andersen *et al.* 2006) and the water flow changes given above, we calculated the changes in catchment losses of nutrients from diffuse sources between the periods 1950–1977 and 1978–2005. In River Ribe total nitrogen (TN) loss from diffuse sources was estimated to have increased from

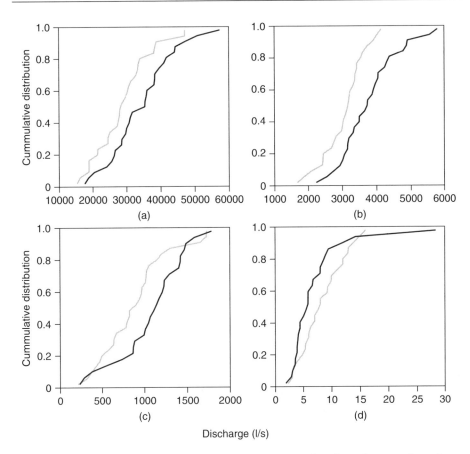

Figure 10.1 Comparison between cumulative distribution functions for annual maximum (a) and annual minimum (b) for River Ribe, annual maximum (c) and annual minimum (d) for River Harrested. The dark line represents the period 1950–1977, the light line the period 1978–2006.

22.7 kg N ha^{-1} during the first period (1950–1977) to 24.5 kg N ha^{-1} during the second period (1978–2005), equaling an 8% increase in catchment TN loss. For River Harrested the calculated increase in TN loss from diffuse sources was estimated to range from 14.3 to 15.7 kg N ha^{-1} (9% increase). The corresponding figures for P increases were 0.34 to 0.36 kg P ha^{-1} (5% increase) in River Ribe and 0.26 to 0.31 kg P ha^{-1} (19% increase) in River Harrested.

10.3 Trends in stream temperature

The air temperature in Denmark has increased by approximately 1.5 °C since the early 1880s and almost 1 °C since 1970. We screened for trends in annual extremes of water temperatures in 16 streams located in three different regions of Denmark. For the longest and most complex data series for each individual month (1–12) a Kendall's trend test (Kendall 1975; Hirsch, Slack, and Smith 1982) was conducted to trace monotonous development in water temperatures. Kendall's τ is a measure of the

Table 10.1 Sign of trend, absolute value for trend ($\hat{\beta}$) and significance level (P) for different runoff characteristics for River Ribe at Stavnager Bridge and River Harrested at Kramsvadgård.

Stream	Annual daily min.	Annual daily max.	Annual mean	Standard deviation of annual diurnal water flow	Baseflow Index (BFI)	Annual t_{min} threshold number of transgressions (10%)	Annual t_{max} threshold number of transgressions (90%)
River Ribe	+ $\hat{\beta}=16.4\,l/s$ P = 0.019	+ $\hat{\beta}=106\,l/s$ P = 0.017	+ $\hat{\beta}=29.4\,l/s$ P = 0.011	+ $\hat{\beta}=23.8\,l/s$ P = 0.092	+ $\hat{\beta}=0.0003$ P = 0.56	− $\hat{\beta}=0$ P = 0.10	+ $\hat{\beta}=0.460$ P = 0.089
River Harrested	− $\hat{\beta}=-0.076\,l/s$ P = 0.013	+ $\hat{\beta}=3.71\,l/s$ P = 0.35	+ $\hat{\beta}=0.37\,l/s$ P = 0.37	+ $\hat{\beta}=0.39\,l/s$ P = 0.52	− $\hat{\beta}=-0.0014$ P = 0.04	+ $\hat{\beta}=0$ P = 0.11	+ $\hat{\beta}=0.130$ P = 0.57

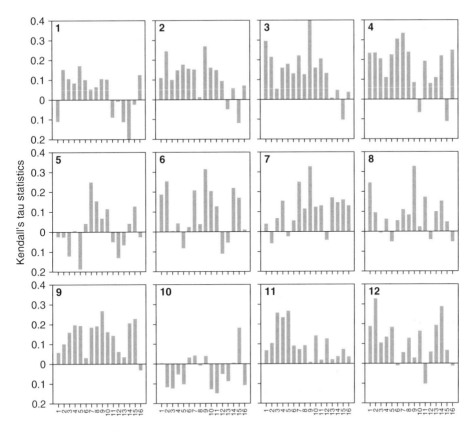

Figure 10.2 Kendall test for monotonous development trend in water temperatures in 16 streams calculated for each month (1 = January, 2 = February etc.). The data series behind the analysis ranges from minimum pre-1980 until 2006.

direction and strength of the trend, the higher the τ the stronger the trend. Kendall's τ was mainly positive for all streams during all months (Figure 10.2). For October, though, a tendency to a constant or slightly negative τ was observed, i.e., declining temperatures. The months exhibiting the highest general increase in water temperatures were March, April and September. Local variations occurred, which could be ascribed to differences in hydrological conditions, such as variation in groundwater buffering of the surface water temperature.

10.4 Trends in lake temperature

Temperature data from lake surface waters in 20 lakes followed as part of the Danish national monitoring program and sampled 19 times per year since 1989 showed an increase during summer and autumn (Figure 10.3). Interpolated daily values were tested for trends in the selected analysis period: 1989–2006 (Kendall's τ) for slope coefficients tested for significance at the 5% level, and these showed that the most profound changes occurred in summer (third quarter), showing an approximately 2 °C

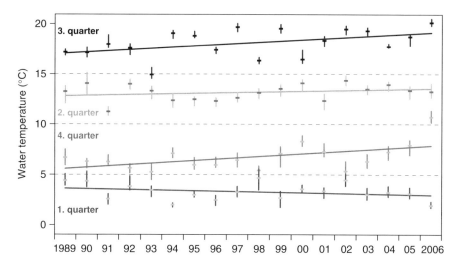

Figure 10.3   General temperature development in the surface water of the 20 intensively monitored lakes from 1989 to 2006 described by the 25% (the lowest point of the vertical line), 50% and 75% percentiles (the upper point of the vertical line) for the first, second, third and fourth quarter.

increase during 1989–2006. Also in autumn (fourth quarter) the lakes became warmer, whereas the temperature actually dropped during winter.

Lake temperatures were influenced by the North Atlantic Oscillation (NAO). A low NAO in winter means less impact by Atlantic winds during the winter period. In this case the Danish climate is regulated by the continental climate, leading to lower temperatures. A high NAO in winter reflects stronger western winds and higher temperatures in winter and thus a milder and wetter climate in Denmark. Generally, the NAO winter index has been high (positive values) since 1879. For a period around 1990 the index was particularly high and the Danish winters were very warm. The NAO winter index has shown a declining tendency since 1990, and our analyses demonstrated a significant relationship with declining water temperatures in Danish lakes during the first quarter since 1989, both at the surface and at the bottom (Hansen, Nedergaard and Skov 2008). In the third quarter winds are usually calm and temperatures high, the lake water column is relatively undisturbed and mixing is at its lowest, irrespective of whether the lakes are deep with a thermocline or shallow without stratification. Since the mid-1990s the summer NAO index has been lower than average, and the Danish weather has been influenced to a larger extent by the climate of the European continent with warmer summers (http://www.cru.uea.ac.uk/~timo/index.htm, accessed June 20, 2012). This stronger influence from the continent in the summer months with higher air temperatures has triggered an increase in water temperature as evidenced in the lakes included in the test data, particularly during the third but also during the fourth quarter (Figure 10.3).

During the first quarter, lake water temperatures declined in both surface and bottom water irrespective of depth (Z), area (A) and water column stability (Figure 10.4). During this season the lakes were fully mixed, though minor differences in temperature development between surface and bottom appeared in the individual lakes.

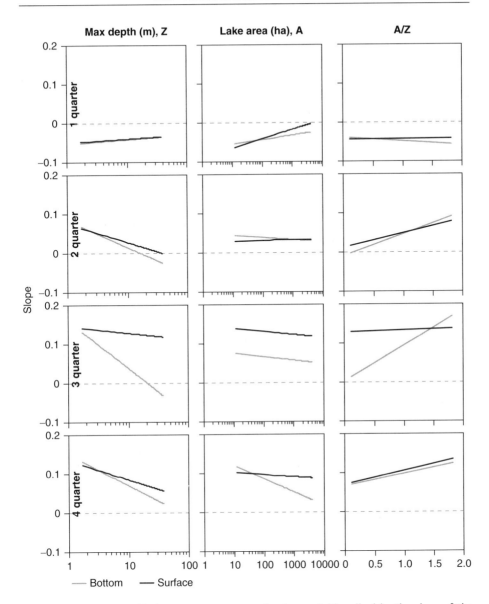

Figure 10.4 Relationship between temperature development (described by the slope of the regression line) and depth (log-transformed), area (log-transformed), and the stability of the water column ($A^{0.25}/Z_{max}$) in surface and bottom water (two colors) based on data from 20 Danish lakes covering the period 1989–2006. Positive values indicate increasing temperatures and negative values declining temperatures. The closer the value is to zero, the more negligible the temperature development. Water column instability increases with increasing A/Z values. (See insert for color representation.)

During the second quarter deep lakes with a large water volume (small A/Z value) exhibited largely unchanged temperatures, while the temperature in shallow more "mixed" lakes demonstrated a minor increase. In the third quarter, surface water temperatures increased markedly in all lakes irrespective of depth, area and water column stability. In the bottom water, significant differences occurred as well. In small shallow lakes with frequent mixing, temperatures increased as in the surface, whereas temperatures in bottom waters in deep lakes with little mixing had only a minor increase—in the deepest lakes temperatures actually declined. During the fourth quarter water temperatures increased both at the surface and at the bottom, most pronouncedly in shallow lakes.

Overall, differences in temperatures between surface and bottom water increased substantially in the lakes studied, reflecting that (i) the stratification of the lakes had become more stable, (ii) the thermocline was not as frequently broken during summer and lasted longer, and (iii) the stratification period was prolonged (Hansen, Nedergaard, and Skov 2008). This applied to both medium deep and deep lakes. Even in the medium deep lakes, where the wind impact was strongest and mixing highest, stratification periods have become longer. Thus, the increasing lake temperatures implied a more stable and longer lasting thermocline. Wind effects tended to partly counteract this impact, and in the lakes most affected by wind the temperature effect on the thermocline change will obviously be lowest.

It appeared that the temperature fluctuations were not as significant in deep lakes with a large water volume and little or no mixing of water (large stability/small A/Z value) as in small lakes with more frequent mixing. Likewise, there was a weak tendency for smaller temperature increases in large lakes than in small lakes. In the deep lakes an earlier and/or faster formation of the stable thermocline led to lower temperatures of the bottom water at earlier stratification (Hansen, Nedergaard, and Skov 2008). Accordingly, the temperatures of the bottom water of the two lakes with stable stratification declined in both the second and the third quarters of 1989–2006.

10.5 Trends in the formation and duration of the thermocline in lakes

The stability of the water column is strongly regulated by a lake's volume, maximum depth and surface area. We used the $A^{0.25}/Z_{max}$ value as an indicator as in Gorham and Boyce (1989). A stable and unbroken thermocline during the whole summer season is central for the chemical and biological conditions of lakes. In the available data, two lakes exhibited permanent stratification, two lakes were fully mixed and the remaining 16 were temporarily stratified where the thermocline was established at a gradually increasing depth during the test period; the depth of the thermocline increased with lake depth ($R^2 = 0.6$, $P<0.05$). Correspondingly, there was a positive correlation between the development of stratification depth and the stability of the water column (A/Z value) ($R^2 = 0.4$, $P<0.05$), the thermocline becoming more stable with increasing depth (Figure 10.5).

There were large differences to the timing of the onset of stratification, both from year to year and from lake to lake. Of the 18 lakes in which a form of thermocline was established during the season, the date of formation moved forward in nine lakes,

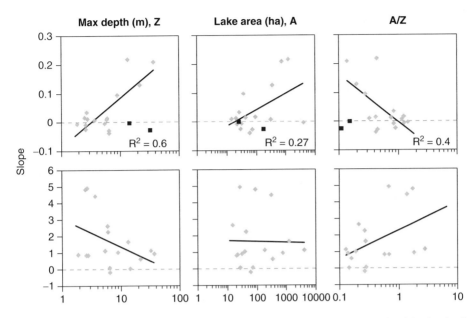

Figure 10.5 Upper panel: changes (described by the slope of the regression line) in depth of thermocline in the summer months, June, July and August, as a function of maximum depth, area and A/Z value (all log-transformed) based on data from 17 Danish lakes covering the period 1989–2006. The black squares represent lakes with stable stratification in summer; the rest are temporarily stratified lakes. When values are positive the thermocline depth has increased. Lower panel: changes in the date for breakdown of the thermocline (time for full mixing) in the 17 investigated lakes. Positive values: longer stratification period; negative values: shorter stratification period.

whereas a tendency to later formation was observed in another nine lakes (Hansen *et al.* 2008). In contrast, there was a clear tendency of later breakdown of the thermocline. Only in one of the lakes did full mixing occur earlier in the year. Primarily due to the varying time of thermocline formation, there were large variations in the total duration of the thermocline period (data not shown), but generally the thermocline period became longer. Apparently, there was no correlation between the time of thermocline breakdown and lake area; however, thermocline breakdown was postponed the longest in shallow lakes with unstable water layers (Figure 10.5).

Faster warming of the surface water and calmer wind conditions led to formation of a thermocline higher in the water column. A contrasting tendency was, however, reflected by the reduced eutrophication of Danish lakes during the past 18 years (Søndergaard, Jensen, and Jeppesen 2005), which led to reduced algal abundance and higher transparency. When water becomes clearer, light can penetrate deeper in a water column with a concomitant temperature increase. This means that a potential thermocline will be established deeper in a lake. Thus, changes in thermocline depth are likely caused by a combination of thermal development and weather conditions and the development of eutrophication and changes in water transparency.

Increasing water temperatures will influence oxygen concentrations and turnover rates. Oxygen consumption increases, and with reduced oxygen levels, P (phosphorus) release from sediments is likely to increase. Changes in the internal processes of a lake due to climate change may thus have a eutrophication effect via internal P release. The location of the thermocline in a water column influences the extension of the potential oxygen-free zone below the thermocline. The stability of the thermocline also influences the extent of oxygen depletion. Judged from the test data, there was a tendency toward having a deeper thermocline now than before and, with it, a smaller oxygen-free zone (Hansen, Nedergaard, and Skov 2008). Simultaneously, however, the thermoclines had become more stable, reducing the transfer of oxygen-containing water to the bottom (hypolimnion), and with this increasing the probability of oxygen depletion in the hypolimnion.

An analysis of data from five summer stratified Danish lakes, monitored biweekly during summer 1989–2003, showed large interannual variations in oxygen and nutrients in the hypolimnion, depending on summer temperature (Jeppesen *et al*. 2009, 2011). In years with high summer temperatures the depth where the minimum oxygen concentration in the hypolimnion passed below 3 mg L^{-1} during the stratification period moved upwards compared to the average for the study period, but was overall much deeper in colder years. This also meant that areas with low oxygen concentrations often, depending on lake morphometry, were larger in warm than in cold years, implying impoverished living conditions for benthic invertebrates and deteriorated foraging conditions for fish with increasing temperature. Maximum ammonium concentrations were often much higher in warm years as a result of higher mineralization and longer stratification periods. By contrast, nitrate concentrations were higher in cold years, due to lower loss of nitrate (lower denitrification during a shorter stratification period) (Jeppesen *et al*. 2011). However, total inorganic N and TN in the hypolimnion varied only negligibly between years and no clear effects were seen in surface water TN (Jeppesen *et al*. 2011). Maximum orthophosphate concentrations in the hypolimnion were overall higher in warm years, likely as a response to the lower oxygen concentrations leading to release of iron-bound phosphate. However, no clear effects of these changes were seen in the surface waters in subsequent years (Jeppesen *et al*. 2009).

10.6   Changes in trophic structure and ecological state

Multiple regression analyses of data from 250 Danish lakes sampled in August showed higher dominance of cyanobacteria in terms of biovolume, most notably potential N-fixing forms and also of dinophytes at higher temperatures. There was also a tendency to increasing chlorophyll *a* concentration and phytoplankton biovolume, while concomitantly diatoms became less important (Figure 10.6). Thus, the risk of dominance by potential toxic cyanobacteria will increase with warming and the period when blooming cyanobacteria occur will likely be longer (Romo *et al*. 2005; Blenckner *et al*. 2007; slso see Chapter 11) (Figure 10.7).

Changes in lake trophic structure may also indirectly enhance the risk of turbid conditions and dominance of cyanobacteria by affecting the fish community (Jeppesen *et al*. 2007). A multiple regression analysis made for August data from Danish lakes demonstrated a decrease in the average size of cladocerans and copepods with increasing temperature (Figure 10.6). This usually suggests enhanced predation by fish.

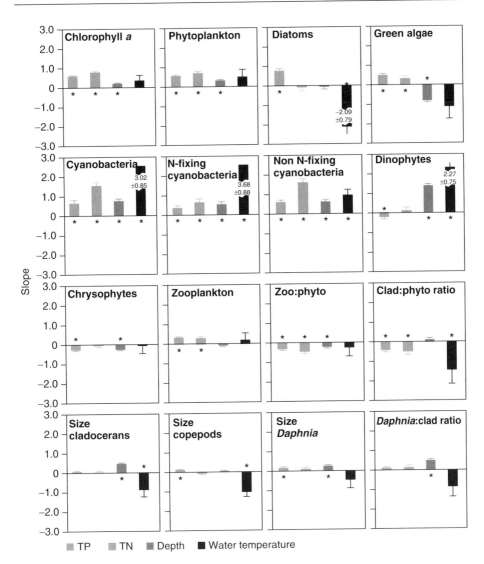

Figure 10.6 Multiple regressions between different plankton variables ($\log_e$-transformed) and total phosphorus, total nitrogen and water temperature of the surface layer and lake mean depth—all $\log_e$-transformed. When the value is positive, there is a positive effect of a given variable (when including also the other variables) and the opposite, if negative. All data are from August. Significant values are marked with an *. Modified from Jeppesen *et al.* (2009). (See insert for color representation.)

A tendency for a decrease in the zooplankton:phytoplankton biomass ratio and the proportion of *Daphnia* among the cladocerans provided further evidence of higher fish predation. With a lower proportion of large-sized *Daphnia* and a lower average zooplankton size, grazing on large-bodied phytoplankton is likely to decline, which will further enhance the risk of dominance by filamentous cyanobacteria. This shift in zooplankton size was likely due to changes in the composition of fish stocks with higher

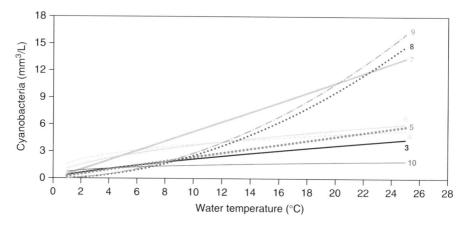

Figure 10.7   Calculated biovolume of cyanobacteria in the individual months (1–12) for a lake with a TP concentration of 100 μg l^{-1} (year round) and a mean depth of 3 m using a multiple regression model (see Jeppesen *et al.* 2009). Particularly pronounced changes in temperature occur in July, August and September.

dominance of zooplanktivorous and omnivorous fish, implying increased predation on zooplankton.

We analyzed monitoring data from 200 Danish lakes collected in late summer (15 August to 15 September) from 1989 to 2006. A significant increase in the proportion of small (<10 cm) perch (*Perca fluviatilis*) and bream (*Abramis brama*) with increasing temperature in May–July was revealed, while no temperature effect was found for roach (*Rutilus rutilus*) (Jeppesen *et al.* 2010). This increase in the proportion of small fish occurred despite an overall major reduction in nutrient levels following a reduction in external nutrient loading, which, according to monitoring data from numerous Danish lakes, should have resulted in a lower proportion of small fish (Jeppesen *et al.* 2000). The evidence thus indicated a fast response of bream and perch but not of roach populations to changes in temperature. Such fast responses to external changes have been seen for fish in lakes undergoing oligotrophication following nutrient loading reduction (Jeppesen *et al.* 2005), suggesting that fish are sensitive indicators of environmental change. Moreover, a long-term study in shallow Lake Søbygaard has shown a major decline in the average size of both roach and perch from 1989 to 2010, leading to a significant reduction in average size of roach, perch and rudd (*Scardinius erythrophthalmus*) when pooled together (E. Jeppesen *et al.*, unpublished). The decline coincided well with the change in average summer air temperature, indicating that the changes might have reflected an increase in ambient temperature. Thus, in a multiple regression relating fish size to phytoplankton chlorophyll *a*, TP and temperature, only temperature was retained in the final model, further emphasizing the key role of temperature for the size change (E. Jeppesen *et al.* unpublished).

Fish abundance, and thus fish predation on zooplankton and benthos, may also be influenced by changes in the duration of ice cover in areas where ice cover occurs. Thus, reduced ice cover in winter should enhance fish survival, which might cascade down the food web and also reinforce symptoms of eutrophication. In lakes covered by ice for up to five months a year, fish abundance was very low due to frequent intense fish kills likely due to anoxia under ice (Jackson *et al.* 2007). Comparative studies of Danish coastal lakes and continental Canadian lakes with similar summer temperatures, but major temperature differences during winter, have shown fourfold lower chlorophyll a:TP ratios and much higher zooplankton:phytoplankton biomass ratios in the winter-cold Canadian lakes, likely due to a lower winter survival of zooplanktivorous fish under ice (Jackson *et al*. 2007). Monitoring data from Danish lakes show clear indications of reduced fish predation in 1996, following the only cold winter with prolonged ice cover (c. 60–90 days) in the monitoring period 1989–2006. However, the effect was modest compared with observations from shallow Canadian prairie lakes (Balayla *et al*. 2010). The size structure of the main microcrustaceans was displaced towards larger size classes in the summer of 1996, resulting in a significantly greater grazing capacity on phytoplankton. Accordingly, phytoplankton biomass (as chlorophyll *a*) was lower and grazing apparently higher. All these effects were stronger in shallow than in deep lakes (Balayla *et al*. 2010). In a Finnish lake, Ruuhijärvi *et al*. (2010) recorded similar results of winter fish kills. In combination, the results from the long-tem study of Lake Søbygård and the less frequent samplings from numerous Danish lakes strongly indicate that despite a reduction in external nutrient loading and a subsequent reduction in the overall biomass of fish, fish density is increasing and the average body size decreasing, with potentially strong cascading effects.

The results from Danish lakes concur with a latitude gradient study by Gyllström *et al*. (2005) of shallow European lakes that showed that the ratio of fish biomass (expressed as catches per net in multi-mesh sized gillnets) to zooplankton biomass increased from north to south and that the zooplankton:phytoplankton biomass ratio decreased, both substantially. Moreover, higher latitude fish species are typically larger, grow more slowly, mature later, have longer lifespans and allocate more energy to reproduction than populations at lower latitudes (Blanck and Lammouroux 2007). Even within species such changes are seen along a latitude gradient (Blanck and Lammouroux 2007; Jeppesen *et al*. 2010).

10.7 Predictions of the future

A precipitation-runoff model coupled with empirical export coefficients run during the 30-year control period (1961–1990) and a 30-year scenario period (2071–2100) using the A2 scenario (IPCC 2007) showed an increase in the losses of TP from diffuse sources (mainly agriculture and background) between 3.3–16.5% in the nine regions of Denmark (Jeppesen *et al*. 2009). The seasonal pattern of changes in diffuse TP loss was more pronounced in the scenario period than in the control period over nine months. Particularly, the seasonal changes in TP losses are expected to be highest in late winter (February and March) and early summer (May and June), whereas the TP loss would decrease in autumn (September and October). For N, enhanced diffuse TN loading was predicted during winter and spring (2.7–25%), being highest in February. Simulated diffuse TN loading decreased during the late summer and autumn period (August to November) with the most marked decrease in TN loading occurring in September (−37%). However, the annual loading increased.

Higher nutrient loading will lead to eutrophication in lakes, a process that is already stimulated by the warming-induced changes in trophic structure, which can lead toward dominance by small-sized fish, lower zooplankton grazing, stronger cyanobacterial

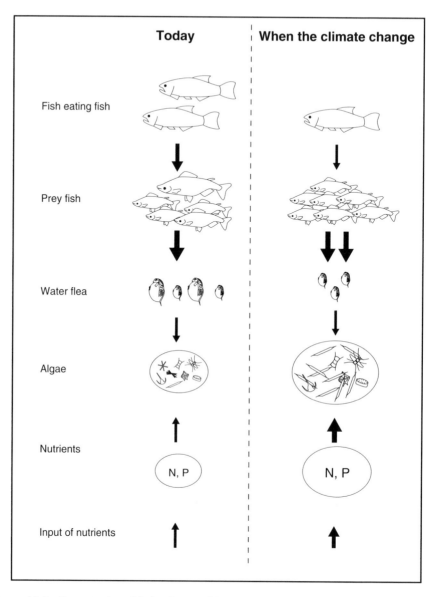

Figure 10.8 Conceptual model showing trophic structure in meso-eutrophic Danish lakes and the suggested changes in a climate change perspective. Today top-down control by piscivorous fish is medium to low depending on nutrient level and cyprinids feeding on large-bodied zooplankton is, conversely, medium to high. Accordingly, there are moderate to few *Daphnia* and moderate-to-low grazing on phytoplankton. Such lakes are sensitive to additional nutrient loading but somewhat buffered by grazing by zooplankton, if the nutrient level is not high. When the lakes become warmer they will, to a higher degree, be dominated by small fish, and zooplankton will be less abundant and smaller. Accordingly, the lakes will be more sensitive to addition of nutrients. Moreover, cyanobacteria will be stimulated by both higher temperature and higher external and internal nutrient loading.

dominance and higher internal loading. We are dealing with a series of allied attacks (Moss *et al*. 2011).

10.8   Adaptation needed

Our results clearly indicate that climate change in Denmark has lead—and will lead even more in the future—to enhanced runoff and higher variability in runoff in rivers, higher nutrient loading of rivers and lakes, higher eutrophication of lakes—both as a consequence of increased external nutrient loading, but also higher temperature induced stimulation of cyanobacteria, and to a shift in the food web dynamics that will reinforce eutrophication (Figure 10.8).

With increasing warming and winter precipitation it will therefore become more difficult to meet the present-day targets set for the ecological state of northern European lakes according to the European Water Framework Directive. Additional efforts must be initiated to reduce the external nutrient loading to levels lower than present-day recommendations in order to maintain or improve the ecological state of lakes (see, for example, Trolle *et al*. 2011). This calls for adaptation measures (Jeppesen *et al*. 2010), which in the north temperate zone should include less intensive land use in catchments with sensitive freshwaters to reduce diffuse nutrient inputs, re-establishment of riparian vegetation to buffer nutrient transfers to streams and rivers; improved land management to reduce sediment and nutrient export from catchments; improved design of sewage works to cope with the consequences of flood events and low flows in receiving waters and, where appropriate, re-establishment of lost wetlands, riparian buffer zones and remeandering of channelized streams to increase retention of organic matter and nutrients. Planting trees along the streams may further help reduce the temperature increase and also provide additional terrestrial-derived food sources and woody debris to the benefit of the stream ecosystems suffering more extreme runoff and washout of organic material in the future warmer climate.

10.9   Acknowledgments

We are grateful to A.M. Poulsen for manuscript editing. The project was supported by the EU REFRESH project (GOCE-CT-2003-505540) and EU-WISER, CLEAR (a Villum Kann Rasmussen Centre of Excellence project), the BUFFALO-P project funded by the Danish Ministry of Food, Agriculture and Fishery under the programme "Animal Husbandry, the Neighbors and the Environment," The Research Council for Nature and Universe (272-08-0406), and the MONITECH, CRES and IGLOO projects.

References

Andersen, H.E., Kronvang, B., Larsen, S.E., *et al*. (2006) Climate-change impacts on hydrology and nutrients in a Danish lowland river basin. *Sci. Total Environ.*, **365**, 223–37.

Balayla, D. Lauridsen, J.T.L., Søndergaard, M., and Jeppesen, E. (2010) Winter fish kills and zooplankton in future scenarios of climate change. *Hydrobiologia*, **646**, 159–72.

Blanck, A., and Lammouroux, N. (2007) Large-scale intraspecific variation in life-history traits of 44 European freshwater fish. *J. Biogeogr.*, **34**, 862–75.

Blenckner, T., Adrian, R., and Livingstone, D.M., *et al*. (2007) Large-scale climatic signatures in lakes across Europe: A meta-analysis. *Glob. Chang. Biol.*, **13**, 1314–26.

Conover, W.J. (1980) *Practical Nonparametric Statistics*, 2nd edn., John Wiley & Sons, Inc., New York.

Gorham, E. and Boyce, M. (1989) Influence of lake surface area and depth upon thermal stratification and the depth of the summer thermocline. *J. Great Lakes Res.*, **15**, 233–45.

Gyllström, M., Hansson, L.-A., Jeppesen, E., *et al*. (2005) Zooplankton community structure in shallow lakes: interaction between climate and productivity. *Limnol. Oceanogr.*, **50**, 2008–21.

Hansen, J.W., Nedergaard, M., and Skov, F. (eds.) (2008) *IGLOO- Indikatorer for den globale opvarmning i overvågningen*, By- og Landskabsstyrelsen, Miljøministeriet, København [in Danish].

Hirsch, R.M., Slack, J.R., and Smith, R.A. (1982) Techniques of trend analysis for monthly water quality data. *Wat. Resourc. Res.*, **27**, 803–13.

IPCC (2007) *Fourth Assessment Report, Climate Change 2007*, Cambridge University Press, Cambridge.

Jackson, L.J., Søndergaard, M., Lauridsen, T.L., and Jeppesen, E. (2007) A comparison of shallow Danish and Canadian lakes and implications of climate change. *Freshwat. Biol.*, **52**, 1782–92.

Jeppesen, E., Jensen, J.P., Søndergaard, M., and Lauridsen, T. (2005) Response of fish and plankton to nutrient loading reduction in 8 shallow Danish lakes with special emphasis on seasonal dynamics. *Freshwat. Biol.*, **50**, 1616–27.

Jeppesen, E., Jensen, J.P., Søndergaard, M., *et al*. (2000) Trophic structure, species richness and biodiversity in Danish Lakes: changes along a phosphorus gradient. *Freshwat. Biol.*, **45**, 201–18.

Jeppesen, E., Kronvang, B., Meerhoff, M., *et al*. (2009) Climate change effects on runoff, catchment phosphorus loading and lake ecological state, and potential adaptations. *J. Envir. Qual.*, **38**, 1930–41.

Jeppesen, E., Kronvang, B., Olesen, J.E., *et al*. (2011) Climate change effect on nitrogen loading from catchment in Europe: implications for nitrogen retention and ecological state of lakes and adaptations. *Hydrobiologia*, **663**, 1–21.

Jeppesen, E., Meerhoff, M., Holmgren, K., *et al*. (2010) Impacts of climate warming on lake fish community structure and potential ecosystem effects. *Hydrobiologia*, **646**, 73–90.

Jeppesen, E., Søndergaard, M., Meerhoff, M., *et al*. (2007) Shallow lake restoration by nutrient loading reduction—some recent findings and challenges ahead. *Hydrobiologia*, **584**, 239–52.

Kendall, M.G. (1975) *Rank Correlation Methods*, Charles Griffin & Co., London.

Kite, G.W. (1978) *Frequency and Risk Analysis in Hydrology*, Water Resources Publications, Fort Collins CO.

Moss, B., Kosten, S., Meerhoff, M., *et al*. (2011) Allied attack: climate change and nutrient pollution. *Inland Wat.*, **1**, 101–5.

Romo, S., Villena, M.-J., Sahuquillo, M., *et al*. (2005) Response of a shallow Mediterranean lake to nutrient diversion: does it follow similar patterns as in northern shallow lakes? *Freshwat. Biol.*, **50**, 1706–17.

Ruuhijärvi, J., Rask, M., Vesala, S., *et al*. (2010) Recovery of the fish community and changes in the lower trophic levels in a eutrophic lake after a winter kill of fish. *Hydrobiologia*, **646**, 145–58.

Sen, P.K. (1968) Estimates of the regression coefficient based on Kendall's tau. *J. Amer. Stat. Assoc.*, **63**, 1379–89.

Søndergaard, M., Jensen, J.P., and Jeppesen, E. (2005) Seasonal response of nutrients to reduced phosphorus loading in 12 Danish lakes. *Freshwat. Biol.*, **50**, 1605–15.

Trolle, D., Hamilton, D.P., Pilditch, C.A., *et al*. (2011) Predicting the effects of climate change on trophic status of three morphologically varying lakes: Implications for lake restoration and management. *Environ. Modell. Softw.*, **26**, 354–70.

van Roosmalen, L., Christensen, H.J., Butts, M.B., and Refsgaard, J.C. (2010) An intercomparison of regional climate model data for hydrological impact studies in Denmark. *J. Hydrol.*, **380**, 406–19.

11

Lake Phytoplankton Responses to Global Climate Changes

Kirsten Olrik[1], Gertrud Cronberg[2], and Heléne Annadotter[3]

[1]*Laboratory of Environmental Biology Aps, Hellerup, Denmark*
[2]*Department of Ecology, University of Lund, Sweden*
[3]*Regito Research Center on Water and Health, Vittsjö, Sweden*

11.1 Introduction

On a global scale, ongoing climatic changes have demonstrated the importance of the underlying physical environment as a driving force in pelagic ecosystems. The physical lake environment is strongly affected by warming and changes in wind pattern. These changes exert a variety of selective pressures on phytoplankton, from changes in the mixing pattern of a lake, the buoyancy stabilization, and the ratio between the photic zone and the mixing zone ($Z_{euphotic}/Z_{mixing}$), which determines the separation between available light and available nutrients for phytoplankton. These physical factors influence the shape, size, and physiology of phytoplankton (see, e.g., Reynolds 1992; Olrik 1994). Temperature increase speeds up all processes in a lake, chemical as well as physiological. Thus, photosynthesis, nutrient uptake, N_2-fixation, and cell division will all proceed at higher speed as lakes warm (Eppley 1972).

Prolonged stagnation due to warming aggravates oxygen depletion in lake sediments, and results in enhanced release of P from the sediment. This has cascading effects on the chemical lake environment and pelagic life at all levels.

A rise in lake temperature also enhances the degradation of humic N-compounds into dissolved inorganic nitrogen (DIN) and further degradation through denitrification of the DIN to free N_2. Over the last decades, NO_3-N-depleted (nitrate-nitrogen) conditions due to temperature-enhanced denitrification have increased significantly in shallow north European lakes (Weyhenmeyer *et al.* 2007) and this may well be the case worldwide as global warming proceeds.

Climatic Change and Global Warming of Inland Waters: Impacts and Mitigation for Ecosystems and Societies,
First Edition. Edited by Charles R. Goldman, Michio Kumagai and Richard D. Robarts.
© 2013 John Wiley & Sons, Ltd. Published 2013 by John Wiley & Sons, Ltd.

Phytoplankton organisms that have a competitive advantage in warming lakes are indeed dependent on individual lake characteristics, but there are some general tendencies that can be identified. The best adapted species in pelagic ecosystems must be able to adjust to some of the following climate-related in-lake conditions or a combination of them, which include elevated temperature, longer periods of stagnation, prolonged DIN-depletion, and oscillating oxygen regimes in the photic zone. There are also likely to be prolonged hypolimnetic anoxia, DOM- (dissolved organic matter) and P-repletion due to internal loadings from the sediment, and in the warmest regions, rising salinity.

This chapter will focus on the spreading of phytoplankton with energy-consuming survival strategies in warming and increasingly stagnant lakes. Unlike organisms in the temperate spring and autumn overturn, the food web in stagnant waters goes through bacteria, heterotrophic and mixotrophic flagellates, and cyanobacteria, all of which produce less oxygen than they consume. This is the reason for the increasing anoxia in strongly stagnating and stratified waters (Cushing 1989).

11.2   Changes in phytoplankton of the circumpolar zones

In circumpolar zones the growing season is short due to snow-cover and lack of sunlight. Air temperature has risen dramatically over the last two decades, especially in the Arctic. Earlier melting of the ice has prolonged the period when solar radiation is available. The effects are visible in the physical environment, as well as in the phytoplankton growth-period, although cold summers still occur where the ice breakup is late due to prolonged snowfalls. In these years the snow cover still limits the growing season for most organisms (see Chapter 2; see also Christoffersen *et al.* 2008).

The effect of increased loadings of melt water to glacial lakes is a reduction of slow-growing phytoplankton biovolumes in the lakes through dilution and increased runoff. Reduced light conditions, combined with earlier warming in spring and earlier depletion of DIN, will suppress the growth of obligate autotrophic phytoplankton (Christoffersen *et al.* 2008). This will take place both in glacial lakes due to increasing amounts of inorganic and organic material, and in melting tundra-lakes that partially sink into the groundwater, while contributing to the release of the climate regulating greenhouse gasses, CO_2 and CH_4 (Vincent *et al.* 2012).

These conditions favor heterotrophic carbon degrading bacteria, ciliates, and rotatories, and mixotrophic chrysophyte flagellates, such as *Dinobryon bavaricum* (Figure 11.1a), *Chromulina*, and *Ochromonas*; mixotrophic dinoflagellates, *Peridinium willei* (Figure 11.1b); and a few facultative planktic chlorophytes, *Koliella longiseta, Oocystis*, and desmids (Figures 11.1c, d).

Due to few fish predators, circumpolar and northern boreal zones have a rich herbivorous zooplankton fauna, especially *Daphnia*. In many lakes Arctic char is the only fish present, and in many shallow lakes there are no fish. In these lakes zooplankton control on phytoplankton can be very effective.

Phytoplankton in zooplankton-controlled lakes will consist of only a few dinoflagellate and desmid individuals. In fish-free lakes Arctic tadpole shrimp,

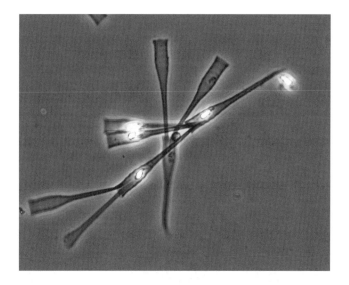

Figure 11.1a *Dinobryon bavaricum*. Photocredit: Gertrud Cronberg.

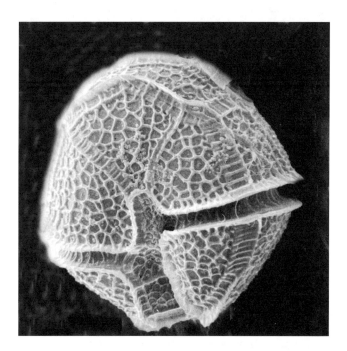

Figure 11.1b *Peridinium willei* SEM. Photocredit: Kirsten Olrik.

Lepidurus arcticus, primarily a benthic invertebrate, can thrive on *Daphnia* (Christoffersen *et al*. 2008; Jackson *et al*. 2007). This predation substantially reduces grazing pressure on phytoplankton. In a warming climate, small mixotrophic chrysophyte flagellates should thrive in these lakes.

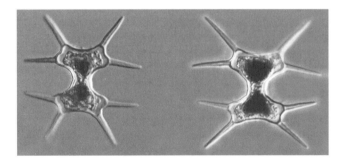

Figure 11.1c   *Octacanthium octocorne (bas. Xanthidium o.)*. Photocredit: Kirsten Olrik.

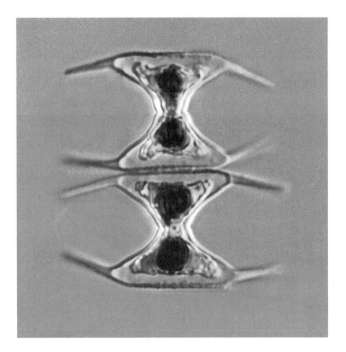

Figure 11.1d   *Staurodesmus triangularis*. Photocredit: Kirsten Olrik.

11.3   Changes in phytoplankton of the boreal and temperate climatic zones

The contemporary climate prognosis for the northern temperate zone is warmer winters with higher precipitation, drier summers, and greater frequency of extreme climatic events such as storms and heat waves. The warming of lakes in these zones is already approximately 2 °C since the late 1980s (IPCC 2007; see also Chapter 10). Up to now, mean summer precipitation in northern Europe has remained fairly constant, whereas the rain pattern in summer has changed to longer hot and dry periods, alternating with spells of sudden thunderstorms and heavy rainfalls that enhance runoff. Consequently, fertilizers from agricultural areas and sewage water from flooded sewers to a greater extent are washed into lakes. Apart from changes in the seasonal precipitation pattern and nutrient pulses, these events mimic the adverse pattern in lakes that undergo eutrophication.

In northern Europe the temperature is driven by the North Atlantic Oscillation (NAO), which is the difference in pressure at sea level between the Icelandic Low and the Azores High (Chen and Hellström 1999). The temperature rise in lakes takes place in both summer and winter and causes an earlier ice break-up. Prolonged warming over the last two decades has gradually lengthened the phytoplankton growth season from March/April–October to February/March–December/January. During the next decades a strong increase in winter temperature is expected (Weyhenmeyer 2009).

In order to detect climatically induced regional trends in Finnish lakes, Arvola, Järvinen, and Tulonen (2011) presented a long-term data series from 1982 to the present from 19 large lakes situated between 61° and 69° N. In 40% of the lakes TP (total phosphorus) decreased significantly during the study period, and in >25% of the lakes reductions of TN (total nitrogen) were also observed. During the same period, phytoplankton biomass declined in only 15% of the lakes. In contrast, the long-term trend of chlorophyll a was more often an increase than a decrease. In nutrient-poor lakes in the north, the east, and south of Finland, nutrient limitation shifted from N to P (Weyhenmeyer et al. 2007). This could be due to temperature-induced faster immobilization of phytoplankton P, when phytoplankton sinks to the lake bottom after P-depletion in the epilimnion. In combination with a worldwide increase in airborne N-deposition, the temperature-induced degradation of humus could also be a part of the explanation of a probable trend for an increase in phytoplankton biomass in these oligotrophic lakes. The Finnish lakes were clustered into three groups: A: eutrophic lakes with a relative biomass of cyanobacteria >20%, B: meso-eutrophic lakes with a relative biomass of diatoms >35%, and C: oligo-mesotrophic lakes with a relative biomass of chrysophytes >40%.

The prognosis for the phytoplankton in warming nutrient-poor Finnish lakes is twofold: accelerated P-sorption at the sediment surface due to warming will favor phytoplankton species that live at the detection limit of dissolved reactive phosphorus (DRP). Also due to warming, accelerated bacterial degradation of humic substances will continue. Bacteria have a much higher affinity to DRP in low concentrations than phytoplankton (Veen 1991; Olrik 1998), and the surviving phytoplankton strategists will be mixotrophic chrysophyte flagellates, which fit into Arvola's group C (Arvola, Järvinen, and Tulonen 2011).

Among the warming, nutrient-poor Swedish lakes, numerous similar changes in physical and chemical conditions are occurring: increasing surface water temperature and intensity of thermal stratification, and increasing concentration of sulfate (SO_4^{2-}) (Bloch 2010 and recent studies at Swedish University of Agriculture Sciences—SLU). Bloch found that the only species in Swedish lakes that seemed to take advantage of changing climatic conditions was the flagellate *Gonyostomum semen* (Figure 11.2a) that is spreading in southern Swedish humic-acid forest lakes. These lakes have steep thermal and oxygen stratification, a progressive summer anoxia and high concentrations of SO_4^{2-} and H_2S rising from the bottom. *G. semen* is a large mixotrophic flagellate, an effective swimmer, and a nuisance alga due to its excessive production of mucilage. It can migrate through the steep thermo- and oxyclines, which is not possible for obligate aerobic organisms. With its tolerance for SO_4^{2-} and H_2S,

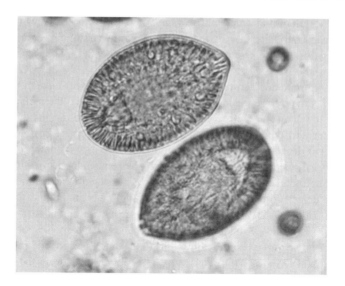

Figure 11.2a *Gonyostomum semen*. Photocredit: Gertrud Cronberg. Mixotrophic, SO_4^{2-} and H_2S tolerant flagellate with the unusual ability to switch between aerobic and anaerobic metabolism. It forms massive summer blooms in acid-humic stratified lakes.

and a rare ability to switch from aerobic to anaerobic metabolism, *G. semen* can use internally released PO_4-P in the anoxic hypolimnion. Here, it may also find a refuge from zooplankton grazing (Salonen and Rosenberg 2000). *G. semen* can proliferate for several months in humic-acid lakes, provided water temperature exceeds $10\,^{\circ}C$. An illustrative example is Danish Lake Gribsø (Figure 11.2b; Olrik and Sørensen 2006a).

The prognosis for phytoplankton in warming humic-acid stratifying lakes is that *Gonyostomum semen* will remain the main survivor, accompanied by mixo- and heterotrophic dinoflagellates.

During the 1950s through the 1980s, many Danish lakes deteriorated due to discharge of urban sewage to watercourses and lakes (see Chapter 10). In the first years after construction of secondary and tertiary sewage treatment plants in all towns, the lakes underwent an oligotrophication where TP, DIN, DOC (dissolved organic carbon), and biomass of cyanobacteria and chlorophytes declined markedly (Figures 11.2c, d; Olrik 2008a, 2008b). Nevertheless, during the warming period 1990–2010, a climatically induced re-eutrophication took place. A typical example of this development in a shallow alkaline lake is Lake Arreskov, where small centric diatoms began to develop large populations in November–December, and again in February–March (Figures 11.2e, f). This early diatom growth resulted in an early depletion of DIN and P in spring. Together with large herbivorous *Daphnia*, the DIN-depletion prolonged the clear-water phase to last through April–June (Olrik and Sørensen 2006b; Olrik 2008a). The early start of DIN-depletion due to warming in winter and spring aggravated the depletion of oxygen in the lake sediment, which turned anoxic earlier than before the warming. This was followed by release of loosely bound P from the upper sediment. In early July the sudden rise in P in the water induced a dramatic development of small and fast growing autotrophic phytoplankton (diatoms and chlorophytes). This species-rich assembly then vanished due to a combination of N-limitation and zooplankton

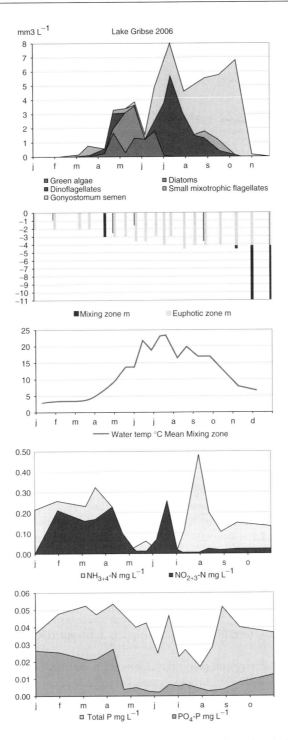

Figure 11.2b Lake Gribsø (Denmark), 2006. Diagrams of yearly cycle of phytoplankton and environmental variables. During summer, the euphotic zone exceeds the very shallow epilimnion layer, which favours *Gonyostomum semen* and small mixo- and heterotrophic dinoflagellates.

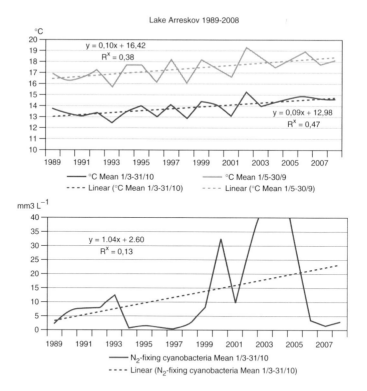

Figure 11.2c Lake Arreskov 1989–2008: Totally mixed, mean depth $<3.5\,$m; alkalinity $>2.0\,$meq$\,$L^{-1}; TP $<0.200\,$mg$\,$L^{-1}. Biovolumes and dominance of N_2-fixing cyanobacteria has risen significantly with warming.

grazing. It was replaced by large species of short-lived N_2-fixing cyanobacteria of the genus *Dolichospermum* (bas. *Anabaena*) (Figures 11.2h–k) and repeating *Aphanizomenon* blooms that reduced light in the photic zone by up to 80% (Figure 11.2g; Olrik 2009a). Like photosynthesis, N_2-fixation is a light-demanding process, and the species involved may either rapidly decline from lack of light due to self-shading, or sink to the bottom where they absorb new PO_4-P. During their decay, the N_2-fixing cyanobacteria enrich the water with organic N and P, and may also release neuro- and hepatotoxic substances. In Lake Arreskov they were succeeded by *Microcystis* spp. Figures 11.2l–o) and other chroococcalean cyanobacteria with a high demand for P and the ability to utilize dissolved organic nitrogen (DON), i.e., *Aphanothece* and *Cyanodictyon*.

The prognosis for phytoplankton development in warming shallow alkaline lakes of temperate zones is that autotrophic species will be further stressed by temperature enhanced DIN-depletion, and mainly be restricted to shade adapted diatoms, which bloom in winter and early spring, where DIN is not limiting. Due to a combination of increasing temperature, temperature induced DIN-depletion, and P-repletion from internal loadings during summer; blooms of harmful N_2-fixing cyanobacteria will thrive, succeeded by blooms of chroococcalean cyanobacteria, for example, *Microcystis*. As warming continues, current dominating cyanobacterial species (e.g.,

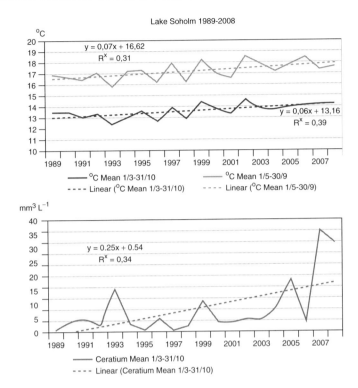

Figure 11.2d Lake Søholm 1989–2008: Summer stratified, mean depth 6–7 m, max. depth 16 m; alkalinity >2 meq L^{-1}; TP <0.090 mg L^{-1}. Biovolumes and dominance of the mixotrophic dinoflagellate *Ceratium* assembly has risen significantly with warming.

Figures 11.2g–o) will presumably be partly replaced by the invasive N_2-fixing *Cylindrospermopsis raciborskii* from even warmer climatic zones (Figure 11.3a).

Long-term investigations from 1989–2008 in the Danish Lake Søholm showed that through the last two decades, the temperature of the lake water has risen 2 °C as a mean from March–October. Since the early 2000s, summer stratification in the lake has intensified due to the temperature rise. Every summer the epilimnion became N-limited for long periods and the water column developed a steep vertical gradient of temperature and oxygen. At the same time, a progressive anoxia developed from the lake sediment under release of loosely bound P. The 1% light depth was 3-4 m, the mixing epilimnion depth 2–7 m, and the anoxic, phosphorus-rich hypolimnion was separated by a thermocline from the photic zone from early May through early November. During this period, it moved downwards from 2–3 m to full circulation at 16 m. In the thermocline, PO_4-P was in excess, as well as nutrient-rich organisms for the mixotrophic flagellates that could access it (Olrik 2008b, 2009b).

This environment was the niche of the large, mixotrophic, swimming dinoflagellates, *Ceratium* (Figure 11.2p), with *C. hirundinella* as the dominant species (Figure 11.2q) and other *Ceratium*-species as subdominants. Over most of the period since the early 1990s, this low-diversity species assembly has dominated with increasing biomass during the summer stratification period (Figures 11.2d and r). When warm winters preceded the recirculation of the water column in November, species with a large

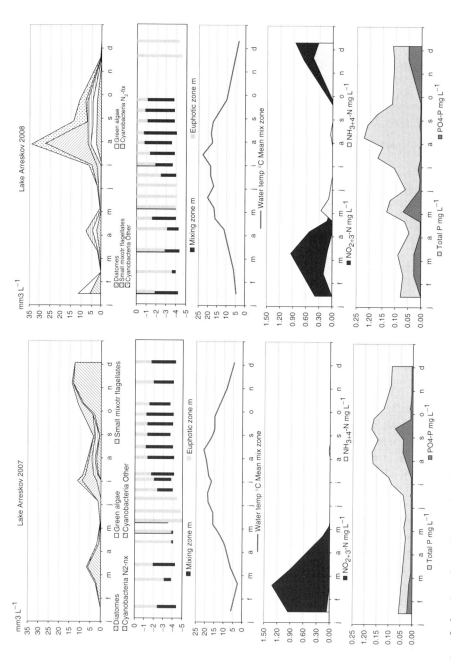

Figure 11.2e Lake Arreskov (Denmark). Two successive yearly cycles of phytoplankton and environmental variables. In the first year, high winter temperature favours diatoms in November–December. In the second year, high summer temperature, SIN-depletion, and internal

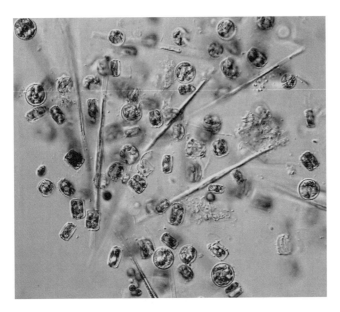

Figure 11.2f Bloom of centric diatoms (*Cyclotella* and *Synedra*) in December. Photocredit: Kirsten Olrik.

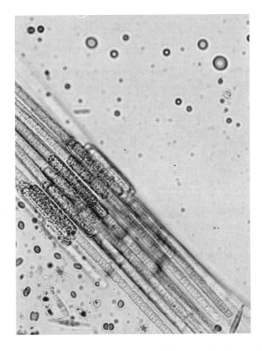

Figure 11.2g *Aphanizomenon klebahnii*. Photocredit: Gertrud Cronberg.

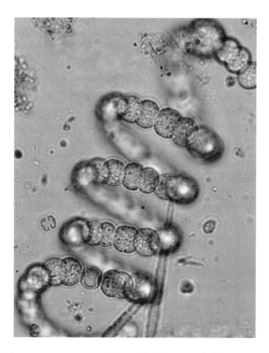

Figure 11.2h *Dolichospermum crassum (bas. Anabaena spiroides var. crassa).* Photocredit: Gertrud Cronberg.

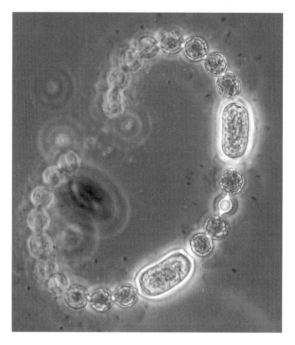

Figure 11.2i *Dolichospermum ellipsoides* (bas. *Anabaena ellipsoidea).* Photocredit: Gertrud Cronberg.

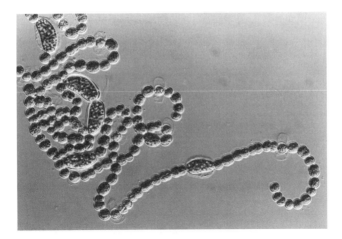

Figure 11.2j *Dolichospermum flos-aquae* (syn. and bas. *Anabaena fl.-a.*). Photocredit: Kirsten Olrik.

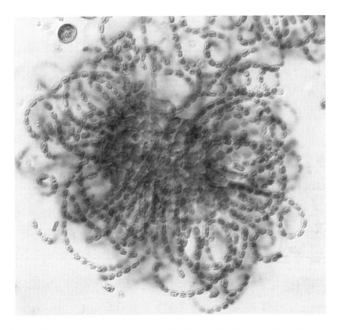

Figure 11.2k *Dolichospermum lemmermannii* (bas. *Anabaena l.*). Photocredit: Gertrud Cronberg.

light-catching capacity and tolerance to deep circulation into darkness prevailed until the light increased in March, which favored large centric diatoms and a long slender desmid, *Closterium aciculare* (Olrik 2008b, 2009b).

   Hot weather followed by atmospheric depressions and heavy rainfalls sometimes releases loosely bound P and noxious gasses (CH_4, NH_3, and H_2S) from the anoxic sediment in stratified waters. These gasses turn the water column completely anoxic and kill all living organisms in the water phase, with the exception of anaerobic bacteria. Evidence of this condition has been observed recently at Mariager Fiord in

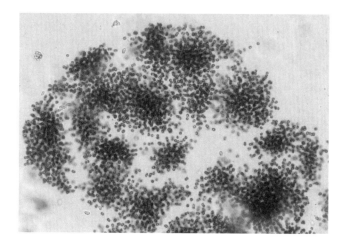

Figure 11.2l *Microcystis botrys*. Photocredit: Gertrud Cronberg.

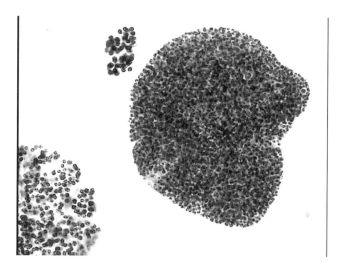

Figure 11.2m *Microcystis flos-aquae*. Photocredit: Gertrud Cronberg.

Denmark (Sørensen and Wiggers 1997). As there likely will be a further increase in temperature, this situation can be expected to occur more frequently.

The prognosis for phytoplankton in warming stratified alkaline eutrophic lakes is that the stratification period will be further prolonged and DIN-depletion in the epilimnion will last longer. The large swimming dinoflagellates, *Ceratium* (Figure 11.2q), will continue to increase biovolumes during the stratification period. If the thermocline is within a depth of 3 m from the water surface for some time, N_2-fixing cyanobacteria may become dominant.

Lake Biwa, the largest freshwater lake in Japan, has undergone eutrophication due to human activities since the 1950s and accelerated during the rapid economic growth period of the 1970s (see Chapter 6). A number of control measures were instigated

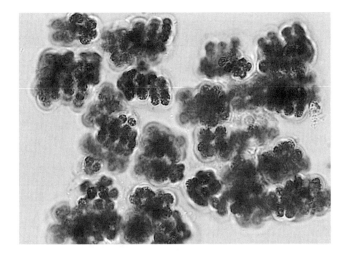

Figure 11.2n   *Microcystis viridis*. Photocredit: Gertrud Cronberg.

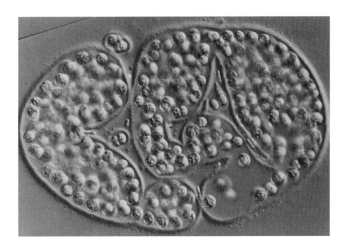

Figure 11.2o   *Microcystis wesenbergii*. Photocredit: Kirsten Olrik.

including sewage treatment in urban areas. An improvement followed this period when TP in the photic zone was reduced to $<10\,\mu g\,L^{-1}$ and the mean yearly phytoplankton biovolume became as low as $<0.07\,mm^3\,L^{-1}$. Still, the former phytoplankton assemblies of deep circulating small shade-adapted diatoms succeeded by chlorococcal green algae and desmids did not return. Instead, cyanobacteria have taken over the dominance, accompanied by the dinoflagellate *Ceratium*, and the chrysophyte, *Uroglena americana* in May and June (Hsieh *et al.* 2010). Since 1977 this species has caused taste and odor problems (Ichise *et al.* 2001). Reduced oxygen conditions have increased near the lake bottom. This oxygen decline is in part due to warming, which changes the production cycle from a winter overturn succeeded by weaker and shorter summer stagnation (see Chapters 6 and 8). Since the water has warmed, the period of overturn in winter has been reduced and the period of stagnation has been

Figure 11.2p   Bloom of a dinoflagellate *Ceratium* assembly. Photocredit: Gertrud Cronberg.

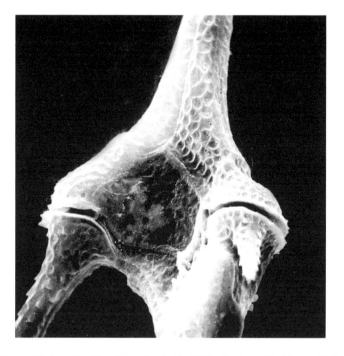

Figure 11.2q   *Ceratium hirundinella* (in SEM). Photocredit: Kirsten Olrik.

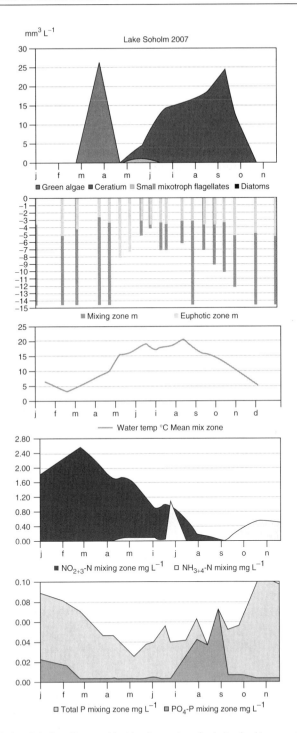

Figure 11.2r Lake Søholm (Denmark). Yearly cycle of phytoplankton and environmental variables with SIN-depletion in epilimnion during summer, and where the mixing depth of the epilimnion exceeds that of the photic zone. These conditions favour the dinoflagellate *Ceratium* assembly.

prolonged. So even though this lake is now essentially mesotrophic, the warming effect on the phytoplankton probably tends to mimic that of eutrophic lakes.

11.4 Changes in phytoplankton assemblies of the semi-arid subtropical climatic zone

Ongoing climatic changes in the subtropical zone are increased warming and long periods of drought that result in a negative balance between precipitation and evaporation. This development contributes to the eutrophication and salination of lakes. Freshwater for drinking water and irrigation is a limited resource so many lakes have been transformed into dammed reservoirs with devastating effects on their aquatic ecosystems (Zohary and Ostrovsky 2011).

Unlike the case in shallow temperate lakes, submerged plant covers in shallow warm lakes do not protect zooplankton from predation by small fish and shrimp (Meerhoff *et al.* 2007). This results in disappearance of large *Daphnia* and predominance of smaller cladocera, such as *Bosmina*, which are less effective grazers on phytoplankton than *Daphnia*. The invertebrate predator *Chaeoborus* is more abundant here than in temperate lakes, and is a much more effective predator on zooplankton. Consequently, the higher impact of fish on the trophic cascade in warm lakes may result in a stronger "bottom-up" effect and a higher sensitivity of warm lakes to increased external and internal nutrient loadings (Meerhoff *et al.* 2007). With increasing warming, these conditions may spread to temperate zones.

A bloom-forming filamentous and N_2-fixing cyanobacterium, *Cylindrospermopsis raciborskii*, (Figure 11.3a) has recently spread from lakes in tropical Africa to lakes in the Mediterranean and Central Europe. In its original tropical environment, it grows all the year round, whereas in Central Europe with hot summers and cold winters, it

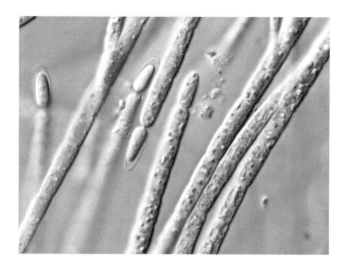

Figure 11.3a *Cylindrospermopsis raciborskii*. Photocredit: Gertrud Cronberg. A tropical invasive species in South and Central Europe.

blooms only in summer. Ten strains from various latitudes were cultivated to detect whether or not it had developed new subspecies, but this seemed unlikely, as the maximum growth of all of the strains was at $30\,°C$ and a light intensity $80\,\mu mol$ photons $m^{-2}\,s^{-1}$ (Briand *et al*. 2004). It is hard to imagine any other explanation for the spreading of this species than global warming that has induced a shift in especially shallow lakes from P- to N-limitation as a possible result of enhanced denitrification (Briand *et al*. 2004). At increasing temperature in both summer and winter, it is likely that the growing season of this harmful species will be prolonged and will continue spreading north.

Until the mid-1990s, Lake Kinneret in northern Israel had a predictable yearly succession of phytoplankton with four stages. The most obvious was the presence of a winter diatom bloom of *Aulacoseira granulata*, and a spring bloom of the dinoflagellate *Peridinium gatunense* (Figure 11.3b) in the early stratification period March–May (Pollingher and Hickel 1991). It was described by Reynolds (2002) as "one of the best known examples of year-to-year similarity in abundance, distribution and composition of phytoplankton." Since 1996 this is no longer the case and warming, desiccation and lowering of the water level are some of the reasons why this key-species has almost disappeared and been replaced by blooms of N_2-fixing cyanobacteria (*Aphanizomenon ovalisporum* and *Cylindrospermopsis raciborskii*) (Zohary 2004; see also Chapter 16).

As far as evaporation exceeds precipitation due to warming and changes in precipitation pattern in the watershed, the prognosis for phytoplankton in Lake Kinneret and other lakes in semi-arid zones is gradual disappearance of the traditional phytoplankton, zooplankton, and fish, and increase in harmful summer blooms of cyanobacteria.

Fish-killing *Prymnesium parvum* blooms are frequent in brackish eutrophied coastal waters of the temperate and subtropical zones. As a consequence of warming, blooms of this species have now been found in inland lakes that turned brackish due to

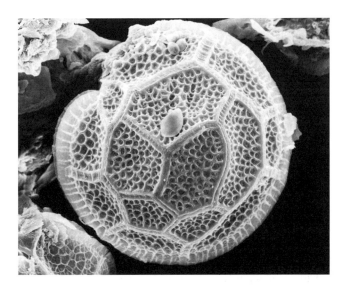

Figure 11.3b *Peridinium gatunense*. Photocredit: Gertrud Cronberg. The former ultimately dominant species during the spring stratification period in Lake Kinneret, Israel. Now it only blooms occasionally. Some of the changes are due to warming.

increased evaporation. This has happened, for example, in Texas, and in the artificial Lake Texoma in Oklahoma, USA where it kills both fish and their main food supply, *Daphnia*. Consequently, it has a dual negative effect on the fish population in these lakes (Kohmescher 2007; Hambright *et al*. 2010).

The prognosis is that as long as water temperature increases and evaporation exceeds precipitation, *Prymnesium parvum* will spread further in inland lakes of the semi-arid subtropical climatic zone, and result in further fish-kills.

11.5   Changes in phytoplankton assemblies of the tropical climatic zone

Tropical freshwater lakes are under the ultimate stress from warming and strong anthropogenic stress due to fishing, aquaculture in fish cages, and construction of artificial reservoirs for drinking water reclamation, irrigation, and energy production. Furthermore, discharge of untreated sewage from towns and settlements, as well as from outlets and seepage of manure, herbicides and pesticides from large greenhouses, are serious stressing factors. Hence, most lakes suffer from strong eutrophication and are anthropogenically affected to a degree beyond sustainability. Rising temperature, evaporation and salination, harmful storm events, increasing frequency of damaging flooding and soil erosion contribute to a deterioration of light conditions in the lakes, and alter in-lake chemical and biological processes. Anoxia in the sediment and bottom water and depletion of NO_3-N are common (Krienitz and Kotut 2010).

In the eastern arm of the Great Rift Valley in Kenya, a series of saline-alkaline closed basin lakes have undergone strong eutrophication due to a combination of warming and the growth of settlements, intensive cultivation, urbanization, and industrialization (Raini 2009). The consequences for the lakes of these developments are increased external P-loadings and the occurrence of massive blooms of harmful, potentially toxic N_2-fixing cyanobacteria, *Dolichospermum (syn. Anabaena) flos-aquae* (Figure 11.2j), *Anabaenopsis arnoldii* and *Anabaenopsis abijatae* (Figure 11.4a). The latter, with a lumpy filamentous colony form, has been misinterpreted as colonies of *Microcystis aeruginosa*, which is a true freshwater species that cannot exist in the high salinity soda lakes, although it is a frequent species in the inland freshwater and low-saline lakes of East Africa (see, for example, Wood and Talling 1988; Kotut, Ballot, and Krienitz 2006; Kotut and Krienitz 2011). Through the last two decades, increasing mass death of the lesser flamingo (*Phoenicopterus minor*, Figure 11.4b) in the soda-lakes Nakuru, Bogoria, and Oloidien, has been associated with periods of low biomass of the lesser flamingo's preferred algal food diet, the nontoxic cyanobacterium *Arthrospira fusiformis* (Figure 11.4c; Krienitz and Kotut 2010). The occurrence of this species was interrupted at irregular intervals in each lake and partly replaced by populations of *Dolichospermum flos-aquae*, *Anabaenopsis abijatae*, and *Anabaenopsis arnoldii*, or by picoplanktic cyanobacteria and chlorophytes, especially *Picocystis salinarum* (Krienitz *et al*. 2012). Non-toxic *Anabaenopsis* filaments can block the flamingo's food filtration system with mucilage. Together with cyanobacteria other than *A. fusiformis*, it can weaken the lesser flamingo to the point where the bird's cyanobacterial toxin elimination system is weakened (the toxins are stored in the feathers). Then they become more susceptible to attacks of infectious diseases (Krienitz

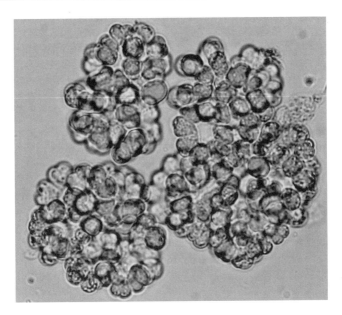

Figure 11.4a   *Anabaenopsis abijatae*. Photocredit: Gertrud Cronberg. A cyanobacteria, which stresses the lesser flamingo by blocking it's filtering system. It has invaded the GRV lakes after salination due to warming and anthropogenic disturbance and fertilization.

Figure 11.4b   Lake Nakuru. The lesser flamingo. Photocredit: Johan Forssblad.

and Kotut 2010). These are the recent environmental challenges that affect the lesser flamingo's food and water supply.

In the tropical zone, both top-down and bottom-up regulation strongly influences lake ecosystems. Herbivorous zooplankton is partly replaced by omnivorous fish, of which the Nile tilapia (*Oreochromis niloticus*) has been a successful invader in many South American and African lakes with disastrous consequences for the original life. An example is Lake Victoria. Together with the carnivorous Nile perch

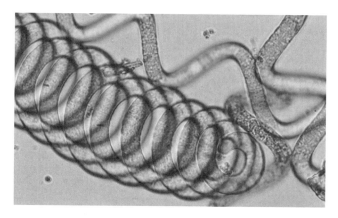

Figure 11.4c *Arthrospira fusiformis*, the preferred diet of the lesser flamingo. Photocredit: Gertrud Cronberg.

(*Lates niloticus*), it has outcompeted the original fish population (Nsinda and Mrosso 1999; Njiru *et al*. 2004).

In shallow hypertrophic freshwater lakes, the hot climate and heavy rainstorms in combination with the Nile tilapia's excretion and respiration aggravate oxygen instability. Thus, frequent oscillations take place between an oxic turbid stage and an anoxic fish-free stage. The oxic stage is with Nile tilapia and the potentially neuro- and hepatotoxic cyanobacteria, for example, *Microcystis flos-aquae* (Figure 11.2i), *M. viridis* (Figure 11.2j), and the non-toxic *M. wesenbergii* (Figure 11.2k). In the anoxic phytoplankton- and fish-free stage, the concentrations of PO_4-P and (NH_3+NH_4)-N are extremely high. In this stage, the lake water will slowly recover and start over again in an oxic stage when the Nile tilapia and cyanobacteria return (Mhlanga *et al*. 2006).

Figueredo and Giani (2005) described the interactions between Nile tilapia and phytoplankton from several aquatic environments in Brazil, where Nile tilapia had been introduced in many aquaculture programs due to its rapid growth rate. It has caused adverse changes in water quality and physical water conditions of the lakes and their phytoplankton assemblies, for example, in the large tropical Furnas Reservoir in southeastern Brazil. The authors described how the presence of Nile tilapia increased N and P availability in the lakes via excretion. In this way nutrient recycling by fish significantly affected nutrient dynamics of the reservoir. As a result of a positive bottom-up effect on the phytoplankton, fish excretion contributed to maintenance of a large phytoplankton biovolume and a high chlorophyll *a* concentration in the lake.

In warm lakes, the top-down regulating effect of zooplankton on phytoplankton is weakened, partly due to smaller zooplankton body size, so that zooplankton are less effective grazing on phytoplankton in tropical lakes than in temperate lakes (Meerhoff *et al*. 2007).

Another reason for the weakened grazing pressure on small phytoplankton in tropical lakes is that Nile tilapia predominantly feed on larger algae (mainly cyanobacteria and diatoms). They are replaced by small-sized or mucilaginous colony-forming cyanobacteria and chlorophytes. Thus, Nile tilapia contributes to eutrophication of lake water by both top-down and bottom-up forces, especially by grazing and excretion of considerable amounts of nutrients. In this way it promotes the increase of the small fast growing phytoplankton. Consequently, Nile tilapia must be used with great caution in aquaculture to avoid unexpected environmental degradation.

Without question, temperature plays a dominant role in pelagic ecosystems in the hot climate of tropical zones, which may warm even more in the future. But to differentiate between what exactly is temperature-related degradations, temperature induced acceleration of chemical and biological processes, and degradations caused by human activities, for example, introductions of invasive fish, is often difficult. For this reason, the interactions between fast-growing Nile tilapia and its harmful impact on tropical lakes and their phytoplankton was highlighted here.

11.6 Concluding remarks

What can be done to reduce negative impacts of global warming?

In this chapter we have shown how climatic changes in lakes influence phytoplankton growth conditions and select species strategies in the different climatic zones. This study has also evidenced how difficult, if not impossible, it is to entirely separate climate impacts from eutrophication impacts on lakes and their phytoplankton. Ongoing warming amplifies eutrophication impacts.

The most important message from phytoplankton developments in warming lakes discussed here is that the more anthropogenically disturbed lakes are, the more vulnerable is their phytoplankton to warming through accelerated DIN-depletion, anoxia, and internal P-loading from lake sediment, all of which support the growth and development of harmful phytoplankton blooms. By the same token, the warmer the lakes, the more vulnerable are their ecosystems to anthropogenic disturbances through sewage pollution, water level fluctuations, salination, and introduction of invasive species.

In nutrient-poor polar and boreal lakes, continuous warming will accelerate phytoplankton growth rate and the transformation of inaccessible humic N into NO_3-N. In the long run, warming of nutrient-poor lakes will probably result in oligotrophication due to immobilized particulate P at the sediment surface. Melting tundra lakes present a special problem as they partly disappear into the groundwater while contributing to the release of climate regulating gasses CO_2 and CH_4 through accelerated decomposition of peat.

In temperate lowland lakes, warming accelerates denitrification and depletes DIN due to transformation to atmospheric N_2. This denitrification happens both in shallow lakes and in the epilimnion of stratified deeper lakes through most of the summer. Despite great efforts to reduce external loadings of nutrients to temperate lakes, many of these still retain deposits of P in their sediment. Sediment oxygen depletion during summer is more rapid and lasts longer with continuous warming, and loosely bound P is increasingly released into lake water. Limitation by DIN reduces or eliminates true autotrophic species of diatoms and chlorophytes, which are replaced by less oxygen-productive species with alternative survival strategies. These include N_2-fixing cyanobacteria and mixotrophic flagellates that maintain lakes in an unhealthy condition. In shallow lakes, cyanobacteria with a luxury uptake of P, build up huge biomasses in summer, starting with filamentous N_2-fixing cyanobacteria, which are later succeeded by colony building cyanobacteria. Unfortunately, most of these species are potentially toxic.

In deeper alkaline stratifying lakes with internal P-loading in summer, swimming mixotrophic dinoflagellates of the genus *Ceratium* are able to build up huge biomasses during the stratification period.

Autotrophic diatoms are shade adapted, and due to warmer winters, their growth period in both shallow polymictic and deeper stratifying alkaline lakes have shifted from spring and autumn to winter-early spring, where essential DIN and PO_4-P are available throughout the water column. This change in growth periods for diatoms implies an early depletion of DIN in spring. In this respect, phytoplankton succession in warming temperate lakes resembles that of subtropical lakes.

The swimming flagellate *Gonyostomum semen* in temperate humic-acid stratified lakes with anoxic conditions in most of the water column has a parallel strategy to that of *Ceratium* in stratified alkaline lakes. But *Gonyostomum* is specialized in a different way than dinoflagellates, as it can shift from aerobic to anaerobic metabolism, and thereby explore the entire water column in lakes with an anoxic hypolimnion from where all true aerobic phytoplankton species are excluded.

There is no reason to assume that temperature-enhanced denitrification and depletion of DIN is restricted to the temperate zone. Presumably it happens all over the world, and is the reason why increasing numbers of harmful N_2-fixing cyanobacteria blooms now occur in so many inland waters, and why increasing numbers of harmful flagellates bloom along stratified eutrophic coastlines.

In the face of ongoing climate changes, the current urgent task is to reduce human-induced negative impacts on inland waters as well as coastal marine waters. It is not an easy task, and continues to require a thorough knowledge in several disciplines, and an awareness of how and when anthropogenic stress factors on the environment push climate change impacts to further harmful outcomes.

In order to elucidate interactions between warming and eutrophication, long-term data series of biological and environmental variables (see also Chapters 13 and 23) remain a valuable tool to understand what is actually occurring. In light of that, it is sad that monitoring programs are being reduced all over the world (see Chapter 23).

The question, "What can be done to reduce negative impacts of global warming?" can partly be answered by the same methods as those used to restore lakes from eutrophication. The method, however, varies with lake size and climate (Jeppesen *et al.* 2007; Søndergaard *et al.* 2007). The difference from earlier restoration attempts is that methods now must be restricted to those that are climate sustainable and focus on making lakes more resistant and resilient to warming and stagnation. First and foremost, external loadings of P to surface waters from agriculture and sewers must be greatly reduced. This requires modern agricultural practices that limit nutrient runoff and state-of-the-art treatment plants. In order to obtain the desired P-equilibrium in lake sediments with a high P-sorption capacity, lakes with internally recycling P may require more extensive measures than removal of external P-loading in order to immobilize the P-pool in the lake sediment (Annadotter *et al.* 1999).

An unlucky consequence of the efforts to keep freshwater lakes healthy in a warming world is that maintaining historical P in lake sediments still remains an important obstacle to success. Combined with the warming effect on the N-cycle in lakes with enhanced loss of free N_2, warming accelerates yearly recycling of historical P. Thus, autotrophic oxygen producing phytoplankton is repressed due to lack of DIN, and replaced by more energy demanding phytoplankton strategists. Lakes are thus maintained in an unhealthy oxygen-reduced condition.

As P is a limited resource worldwide, it is worth attempting its recovery from P-overloaded lakes by harvesting phytoplankton blooms, which circulate loosely bound sediment phosphorus every year. This would serve a twofold aim: to reduce the phosphorus content of lakes and reclaim a limited plant nutrient resource for agriculture.

11.7 Acknowledgements

We wish to thank the former staff of Miljøbiologisk Laboratorium for their contribution to processing the data on phytoplankton from Lake Arreskov and Lake Søholm through 1989–2008. The staff of the former County of Funen is thanked for the supply of the physical and chemical data from these two lakes. The staff of the former County of Frederiksborg is thanked for the supply of the physical and chemical data from Lake Gribsø 2007. Charles R. Goldman, Richard Robarts, and Eduardo Martins are thanked for their thorough reading of the manuscript; Tamar Zohary for interesting discussions and reading of the manuscript, proposals for new references, and for elucidation of questions about Lake Kinneret, and Lothar Krienitz for fruitful discussions and new references on the identity of the toxic cyanobacteria in East African soda lakes.

References

Annadotter, H., Cronberg, G., Aagren, R., *et al*. (1999) Multiple techniques for lake restoration. *Hydrobiologia*, **395/396**, 77–85.

Arvola, L., Järvinen, M., and Tulonen, T. (2011) Long-term trends and regional differences of phytoplankton in large Finnish lakes. *Hydrobiologia*, **660**, 125–34.

Bloch, I. (2010) Global Change Impacts on Phytoplankton Communities in Nutrient-poor Lakes. Licentiate Thesis, Swedish University of Agricultural Sciences, Uppsala.

Briand, J.F., Leboulanger, C., Humbert, J.F., *et al*. (2004) *Cylindrospermopsis raciborskii* (cyanobacterium) invasion at mid-latitudes: Selection, wide physiological tolerance, or global warming? *J. Phycol.*, **40**, 231–8.

Chen, D.L. and Hellström, C. (1999) The influence of the North Atlantic Oscillation on the regional temperature variability in Sweden: spatial and temporal variations. *Tellus Series A-Dyn. Meteorol. Oceanogr.* **51**, 505–16.

Christoffersen, K.S., Amsinck, S.L., Landkildehus, F., *et al*. (2008) Lake flora and fauna in relation to ice melt, water temperature and chemistry at Zackenberg. *Adv. Ecol. Res.*, **40**, 371–89.

Cushing, D.H. (1989) A difference in structure between ecosystems in strongly stratified waters and in those that are only weakly stratified. *J. Plank. Res.*, **11**, 1–13.

Eppley, R.W. (1972) Temperature and phytoplankton growth in the sea. *Fish. Bull.*, **70**, 1063–85.

Figueredo, C.C. and Giani, A. (2005) Ecological interactions between Nile tilapia (*Oreochromis niloticus* L.) and the phytoplanktonic community of the Furnas Reservoir (Brazil). *Freshwat. Biol.*, **50**, 1391–403.

Hambright, K.D., Zamor, R.M., Easton, J.D., *et al*. (2010) Dynamics of an invasive toxigenic protist in a subtropical reservoir. *Harmful Algae*, **9**, 568 77.

Hsieh, C.H, Ishikawa, K., Sakai, Y., *et al*. (2010) Phytoplankton community reorganization driven by eutrophication and warming in Lake Biwa. *Aquat. Sci.*, **72**, 467–83.

Ichise, S., Wakabayashi, T., Fujiwara, N., *et al*. (2001) Long-term changes of phytoplankton in Lake Biwa: 1978–2000. *Rep. Shiga Pref. Inst. Pub. Hlth. and Environ. Sci*., **36**, 29–35 [in Japanese].

IPCC (2007) *Summary for Policymakers, in Climate Change 2007*: The Physical Science Basis, Contribution of Working Group 1 to the Fourth Assessment Report of the Intergovernmental Panel on Climate Change (eds. S. Solomon, D. Qin, M. Manning, *et al*.), Cambridge University Press, Cambridge.

Jackson, A.R. (2009) Impact of land use changes on water resources and biodiversity of Lake Nakuru catchment basin, *Kenya. Afr. J. Ecol.*, **47**(Suppl. 1), 39–45.

Jackson, L.J., Lauridsen, T.L., Søndergaard, M., and Jeppesen, E. (2007) A comparison of shallow Danish and Canadian lakes and implications of climate change. *Freshwat. Biol.*, **52**, 1782–92.

Jeppesen, E., Meerhoff, M., Jacobsen, B.A., *et al*. (2007) Restoration of shallow lakes by nutrient control and biomanipulation—the successful strategy varies with lake size and climate. *Hydrobiologia*, **581**, 269–85.

Komescher, N. (2007) Experimental Analysis of Grazing and Life History Traits in Daphnia pulicaria Fed the Toxic Algae *Prymnesium parvum*. Master of Science thesis. Plankton Ecology and Limnology Laboratory, University of Oklahoma, Norman Oklahoma.

Kotut, K., Ballot, A., and Krienitz, L. (2006) Toxic cyanobacteria and their toxins in standing waters of Kenya: implications for water resource use. *J. Water Health*, **04.2**, 233–45.

Kotut, K. and Krienitz, L. (2011) Does the potentially toxic cyanobacterium *Microcystis* exist in the soda lakes of East Africa? *Hydrobiologia*, **664**, 219–25.

Krienitz, L., Bock, C., Kotut, K., and Luo, W. (2012) *Picocystis salinarum* (Chlorophyta) in saline lakes and hot springs of East Africa. *Phycologia*, **51**, (1), 22–32 .

Krienitz, L. and Kotut, K. (2010) Fluctuating algal food populations and the occurrence of lesser flamingos (*Phoeniconaias minor*) in three Kenyan Rift Valley Lakes. *J. Phycol.*, **46**, 1088–96.

Meerhoff, M., Clemente, J.M., De Mello, F.T., *et al*. (2007) Can warm climate-related structure of littoral predator assemblies weaken the clear water state in shallow lakes? *Global Change Biol.*, **13**, 1888–97.

Mhlanga, L., Day, J., Chimbari, M., *et al*. (2006) Observations on limnological conditions associated with a fish kill of *Oreoochromis niloticus* in Lake Chivero following collapse of an algal bloom. *Afr. J. Ecol.*, **44**, 199–208.

Njiru, M., Ojuk, J.E., Getabu, A., *et al*. (2004) Dominance of introduced Nile tilapia, *Oreochromis niloticus* (L.) in Lake Victoria: A case of changing biology and ecosystem. *Afr. J. Ecol.*, **42** (3), 163–70.

Nsinda, P.E. and Mrosso, H.D. (1999) Stock Assessment of *Lates niloticus* (L.), *Oreochromis niloticus* (L.), and *Rastrineobola argentea* (Pellegrin) using Fishery Dependent Data from the Tanzanian Waters of Lake Victoria. Jinjja Fisheries Data Working Group of the Lake Victoria Fisheries Research Project: 79–83. *LVFRP Technical Document No. 6*.

Olrik, K. (1994) Phytoplankton—Ecology. *Environmental Project No. 251*, Danish Ministry of the Environment and Danish Environmental Protection Agency, Copenhagen.

Olrik, K. (1998) Ecology of mixotrophic flagellates with special reference to Chrysophyceae in Danish lakes. *Hydrobiologia*, **369**/370, 329–38.

Olrik, K. (2008a) *Lake Arreskov 2007*. *Plante- og dyreplankton*. Rapport udført for Fyns Amt. Miljøbiologisk Laboratorium ApS, Humlebæk.

Olrik, K. (2008b) Lake Søholm 2007. *Plante- og dyreplankton*. Rapport udført for Fyns Amt.—Miljøbiologisk Laboratorium ApS, Humlebæk.

Olrik, K. (2009a) *Lake Arreskov 2008*. *Plante- og dyreplankton*. Rapport udført for Fyns Amt. Miljøbiologisk Laboratorium, Humlebæk.

Olrik, K. (2009b) *Lake Søholm 2008*. *Plante- og dyreplankton*. Rapport udført for Fyns Amt. Miljøbiologisk Laboratorium, Humlebæk.

Olrik, K. and Sørensen, A. (2006a) *Lake Store Gribsø 2005. Plante- og dyreplankton*. Rapport udført for Frederiksborg Amt. Miljøbiologisk Laboratorium ApS, Humlebæk.

Olrik, K. and Sørensen, A. (2006b) *Lake Maglesø at Brorfelde 2005. Plante- og dyreplankton*. Rapport udført for Vestsjællands Amt. Miljøbiologisk Laboratorium ApS, Humlebæk.

Pollingher, U. and Hickel, B. (1991) Dinoflagellate associations in a subtropical lake (Lake Kinneret, Israel). *Arch. Hydrobiol.*, **120**, 267–85.

Raini, J.A. (2009) Impact of land use changes on water resources and biodiversity of Lake Nakuru catchment basin, Kenya. *Afr. J. Ecol.* (special issue on ecosystem changes and implications for livelihoods of rural communities in Africa), **47** (suppl.), 39–45.

Reynolds, C.S. (1992) Eutrophication and the management of planktonic algae: what Vollenweider couldn't tell us, in *Eutrophication: Research and Application to Water Supply* (eds. D.W. Sutcliffe and J.G. Jones), Freshwater Biological Association, Ambleside, Cumbria, pp. 4–29.

Reynolds, C.S. (2002) On the interannual variability in phytoplankton production in freshwaters, in *Phytoplankton Productivity; Carbon Assimilation in Marine and Freshwater Ecosystems* (eds. P.J.L.B. Williams, D.N. Thomas, and C.S. Reynolds), Blackwell Science, Oxford, pp. 187–221.

Salonen, K. and Rosenberg, M. (2000) Advantages from dial vertical migration can explain the dominance of *Gonyostomum semen* (Raphidophyceae) in a small, steeply stratified humus lake. *J. Plank. Res.*, **22**, 1841–53.

Søndergaard, M., Jeppesen, E., Lauridsen, T.L., *et al*. (2007) Lake restoration: successes, failures, and long-term effects. *J. Appl. Ecol.*, **44**, 1095–105.

Sørensen, H.M. and Wiggers, L. (1997) Notat. Iltsvind i Mariager Fjord 1997. Status.—Århus Amt, Natur- og Miljøkontoret, Højbjerg, Danmark.

Veen, A. (1991) Ecophysiological studies on the phagotrophic phytoflagellate *Dinobryon divergens* Imhof. Doctoral dissertation, Department of Fundamental Applied Ecology, University of Amsterdam, Amsterdam.

Weyhenmeyer, G.A. (2009) Do warmer winters change variability patterns of physical and chemical lake conditions in Sweden? *Aquat. Ecol.*, **43**, 653–9.

Weyhenmeyer, G.A., Jeppesen, E., Adrian, R., *et al*. (2007) NO_3-N-depleted conditions on increase in shallow Northern European lakes. *Limnol. Oceanogr.*, **52**, 1346–53.

Wood, R.B. and Talling, J.F. (1988) Chemical and algal relationships in a salinity series of Ethiopian inland water. *Hydrobiologia*, **158**, 29–67.

Zohary, T. (2004) Changes to the phytoplankton assemblage of Lake Kinneret after decades of a predictable, repetitive pattern. *Freshwat. Biol.*, **49**, 1355–71.

Zohary, T. and Ostrovsky, I. (2011) Ecological impacts of excessive water level fluctuations in stratified freshwater lakes. *Inland Waters*, **1**, 47–59.

12

The Influence of Climate Change on Lake Geneva

Ulrich Lemmin and Adeline Amouroux
School of Architecture, Civil and Environmental Engineering (ENAC), Ecole Polytechnique Fédérale de Lausanne (EPFL), Switzerland

12.1 Introduction

The increase in human population, urbanization and industrialization continues to place mounting demands on the world's fresh water supply. Large, deep lakes already play a significant role in the supply system. In the future, they will be further stressed and their quantity and quality will diminish unless solid management strategies are applied to assure sustainable long-term development. Large, deep lakes form a set of lakes with specific characteristics. They have a long memory and thus a long recovery time, mainly due to physical factors (size, depth, climate), and are modulated by physical, chemical, and biological phenomena. Due to the large surface area, spatial heterogeneity increases, resulting in complex physical processes. As a consequence, they are highly sensitive ecosystems.

Processes in the hypolimnion of large, deep lakes such as the rare events of full water column mixing are of key importance for the dynamics of the whole system due to the vast storage volume. Changes in water and thermal regimes, as well as in the seasonal dynamics can influence the extent, timing and frequency of mixing with a strong impact on the trophic state of the lake and the functioning of the whole ecosystem (see also Chapter 3). In that context, it has to be considered whether the causes are reversible or irreversible.

In the past, eutrophication has been a prime management concern. However, due to progress in waste-water treatment, the trophic state of many lakes has greatly improved over the last few decades. More recently, a new threat has been posed by the well documented effect of climate change. The most significant features of climate change in the European mid-latitude region are a warming trend in the atmospheric boundary layer and an increasing tendency towards extreme weather events. Continuous warming may increase lake water temperature and extreme events may cause

Climatic Change and Global Warming of Inland Waters: Impacts and Mitigation for Ecosystems and Societies, First Edition. Edited by Charles R. Goldman, Michio Kumagai and Richard D. Robarts.
© 2013 John Wiley & Sons, Ltd. Published 2013 by John Wiley & Sons, Ltd.

strong fluctuations in lake water temperature. These phenomena are irreversible in the foreseeable timescales and may result in negative effects for water quality and biogeochemical cycles in lakes. It is therefore important to investigate the presently observable effects in lakes in order to determine the trends of the future development of lake systems under conditions of climate change.

An atmospheric warming trend has been observed over most of the western European continent. Since this is a large-scale phenomenon, it also has been linked to hemispheric forcing such as the North Atlantic Oscillation (NAO). It was shown in small Swedish lakes that the variation in lake chemistry can be related to NAO dynamics (Weyhenmeyer 2004). However, synchronous relationships were restricted to variables closely related to surface-water temperature. Livingstone and Dokulil (2001) came to a similar conclusion for the near surface layers of Austrian lakes. Straile, Jöhnk and Rossknecht (2003) indicated that the deep-water layers of Lake Constance may be weakly influenced by NAO dynamics. The change in the date of ice breakup on the lake has been used to determine climate change effects on lake dynamics (Weyhenmeyer, Meili and Livingstone, 2004). It was shown that the onset of ice breakup occurs earlier. The effect of temperature change on the reproductive cycle of the roach (*Rutilus rutilus* L.) in Lake Geneva was assessed by Gillet and Quetin (2006). They observed an advance in the onset of thermal stratification in the lake over twenty years. Due to this warming trend, roach spawning tended to occur two weeks earlier between 1983 and 2001. Rempfer *et al.* (2010), investigating the effect of an unusually warm winter on Lake Zürich, found high water temperatures and an increase in thermal stability. They conclude that in deep lakes, such as Lake Geneva, climate warming will inhibit complete mixing during winter. The strong warming of Lake Superior waters since the early 1980s, when compared to the previous seven decades, has led to an increase in the duration of the stratification period by 25 days (Austin and Colman 2008).

In this chapter, we will investigate some aspects of climate change on the water temperature dynamics in Lake Geneva (local name: Lac Léman). This large, deep, subalpine lake lies on the border between Switzerland and France (Figure 12.1), and is about 70 km long and 10 km wide with a surface area of 580 km^2. It has a maximum depth of 310 m on the central plateau in the eastern main basin (mean depth 153 m). Lake Geneva is the largest lake in Western Europe and is of considerable socioeconomic and ecological importance for the surrounding area. For this study, we will use meteorological data collected by the Federal Office of Meteorology and Climatology MeteoSwiss (FOMC) at stations near the shore of Lake Geneva for more than a century and water temperature and oxygen data collected by the CIPEL (Commision Internationale pour la Protection des Eaux du Léman; 2010) since 1957 at a station on the central plateau.

12.2   Atmospheric boundary layer temperatures

The thermal cycle in lakes is determined by the heat flux over the lake. However, over large lakes, it is difficult to measure all necessary atmospheric heat flux parameters in sufficient detail, because heat flux is dominated by the temperature difference between the lake and the atmospheric boundary layer and the wind field over the lake. Therefore, we will approximate the heat flux by using air temperatures taken

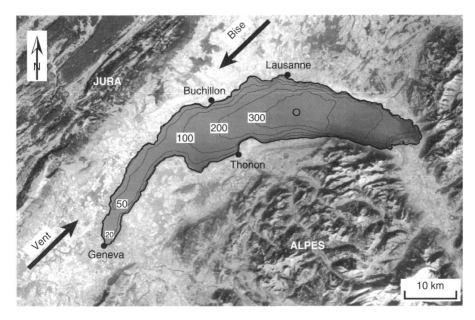

Figure 12.1 Map of Lake Geneva. Depth contours are indicated. The red dot in the center of the lake is the position where the data were taken and were analyzed in this study. The Rhone River, entering the lake from the east, is the main contributor. Arrows labeled "bise" and "vent" indicate the direction of the dominant winds over the lake.

at meteorological stations near the lake as an indicator. The FOMC has collected meteorological data in the region since 1864. A long-term trend analysis of the air temperature data by the FOMC (Bader and Bantle 2004) has shown that there has been a general warming trend between 1864 and 2000.

However, if air temperatures taken during summer and winter are analyzed separately, it becomes obvious that during the summer (April to September; Figure 12.2) a linear trend is not a good representation of the seasonal mean and the long-term mean temperature. Both of them strongly deviate from the linear trend in multi-decadal cycles. The linear trend in the winter data (October to March; Figure 12.3) is essentially governed by a decadal low at around 1890 and a continuous increase after 1970. During the period from 1910 to 1970, no significant trend is observed. Therefore, for the present analysis, we will consider the year 1970 as the beginning of the warming cycle related to climate change and limit the present analysis to the period after 1970. This corresponds closely to the onset of the climate change effect considered in most global scale studies (IPPC 2001, 2007).

A warming trend in a lake may be the result of either long-term increased warming in the atmospheric boundary layer typically taking place during the summer or reduced cooling in the atmospheric boundary layer during the winter. In order to determine whether the lake is affected by a climate-related change in heat gain during the summer or heat loss during the winter, we divided the air temperature data of the station at Geneva airport near the western end of the lake into four groups: summer (May to September) daytime, summer nighttime, winter (November to March) daytime and winter nighttime. The results are summarized in Table 12.1. Out of these

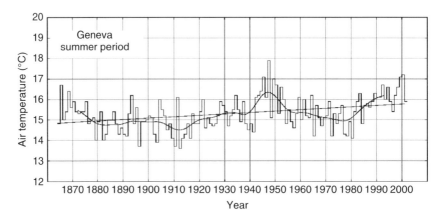

Figure 12.2   Time series of annual mean summer air temperatures calculated from monthly mean values for Geneva. The curve represents the 20-year mean (Gaussian low-pass filter). The straight line gives the linear trend over the whole period. Diagram modified after Bader and Bantle (2004); © MeteoSchweiz.

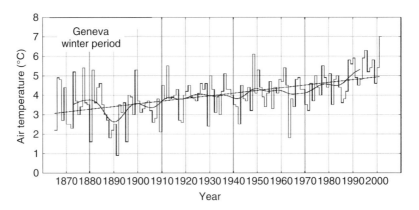

Figure 12.3   Time series of annual mean winter air temperatures calculated from monthly mean values for Geneva. The curve represents the 20-year mean (Gaussian low-pass filter). The straight line gives the linear trend over the whole period. Diagram modified after Bader and Bantle (2004); © MeteoSchweiz.

four groups, only the winter nighttime temperatures showed a statistically significant increase ($R = 0.8$) with a linear trend of about $0.07\,°C\ year^{-1}$. The other temperature groups also showed warming trends, but they were not as significant. Winter nighttime temperatures are normally the lowest air temperatures recorded around Lake Geneva, and the heat flux from the lake would be strongest during this time. The present results indicate that climate change may result in potentially less heat flux from the lake and thus less cooling during winter nights. Other parameters, such as sunshine hours and solar radiation, did not show any significant trend.

Seasonally averaged winter nighttime temperatures after 1970 have been plotted in Figure 12.3. A continuous increase but also interannual variability is evident.

Table 12.1 Air temperature trends.

			Gradient ($°C \ year^{-1}$)	Correlation coefficient
1970–2000	Day	Summer: May to Sept.	0.01	0.11
		Winter: Dec. to March	0.06	0.62
1970–2000	Night	Summer: May to Sept.	0.08	0.18
		Winter: Dec. to March	0.07	0.81

Livingstone (1997) suggested that, for Swiss lowland conditions, monthly winter mean temperatures can be considered as mild when above 0.2 °C. Keeping in mind that winter nighttime temperatures should be colder than monthly mean temperatures, it can be seen that for most of the period investigated, the temperatures in Figure 12.3 are above the value of 0.2 °C. This suggests that this period was dominated by mild winters.

As to winter cooling by extreme events, we investigated the number of ice days during each winter since 1970. We had previously observed that ice days lead to lake cooling by strongly developed cold density currents over the lateral slopes of the lake (Fer, Lemmin and Thorpe 2002). For the period 1970 to 2000, the number of ice days varied randomly between two and 32 with no clear trend. Thus, the number of ice days alone is not a suitable indicator for climate-affected lake cooling dynamics. The cumulated heat flux during the season would probably be needed instead. On the other hand, the high variability in the number of ice days indicates strong inter-annual climate dynamics which affect the heat content in the lake on a year-to-year scale.

12.3 Lake warming trends

The annual thermal cycle of lakes situated in Lake Geneva climate belt consists of a warming phase during the summer and a cooling phase during the winter. In many lakes, heat gain during the warming phase and heat loss during the cooling phase are nearly equal; thus, no long-term lake temperature trend develops. In most of these lakes, thermal stratification is eliminated during the cooling phase. This in turn resets the lake to almost the same starting conditions at the beginning of the warming phase in the following year. In deep lakes, however, such as Lake Geneva, where large quantities of heat have to be exchanged with the atmosphere during winter cooling, year-to-year variations in the rate of warming or cooling may be sufficiently strong to cause only partial destratification limited to the upper part of the water column during the winter season. Warming in the following year may then result in an overall shift to higher lake temperatures. If this happens for several consecutive years, a multiannual cycle of lake warming will occur, which is most evident in the deep-water layers. This cycle is typical for the long-term dynamics of large deep lakes and was observed well before the onset of climate change. It is therefore important to determine how this cycle may relate to climate change-induced trends.

12.3.1   Historical situation

The first systematic temperature profile measurements in Lake Geneva were reported by Forel (1895). These measurements, reproduced in Figure 12.5a, show that homogeneous water-column temperatures did not occur regularly during winter periods, thus indicating a multiannual warming cycle, even though mean winter air temperatures in those years were among the lowest within the recording period (Figure 12.3). Since 1956, CIPEL has carried out routine monthly temperature measurements. These have shown that the mean annual stratification cycle has not significantly changed since Forel's measurements (Figure 12.5b). The CIPEL data also confirmed the development of multiannual warming cycles already seen in Forel's observations, particularly in the deep-water layer below a 100 m depth (Figure 12.6). From 1956 to 2010, there were only five winter events during which stratification in the water column disappeared and the whole lake cooled significantly. Furthermore, during the 1950s and 1960s, mean temperatures, particularly in the deep-water layer, were in the same range as those reported by Forel (1895; Figure 12.5a), indicating that no significant long-term trend in heat gain occurred from 1890 to about 1970. This is in agreement with the meteorological observations by Bader and Bantle (2004; Figure 12.3). One finds two multiannual cycles between 1955 and 1970. Thus, multiannual warming cycles already occurred in Lake Geneva before the onset of climate change at around 1970. After 1970, deep-layer temperatures show a continuation of the occurrence of multiannual warming cycles superimposed on a continuous overall temperature increase. However, multiannual warming cycles are seen to be of irregular duration. In the present paper, the relationship between this "normally" occurring multiannual warming cycle and the trend resulting from climate change will be investigated. The temperatures in the less profound depths of 50 m and 100 m showed strong and random year-to-year fluctuations. This indicated that the dynamics of these intermediary layers were more strongly influenced by forcing from the atmosphere. This confirmed the observations of Zhang, Lemmin and Hopfinger (1994) that the upper 100 m of the water column had different dynamics than the deep-water layers. A trend of continuous warming was less obvious in these layers.

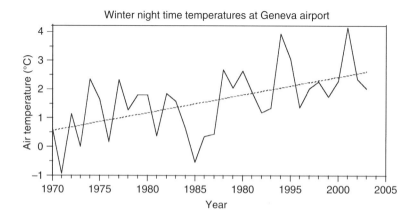

Figure 12.4   Mean winter nighttime air temperatures at Geneva.

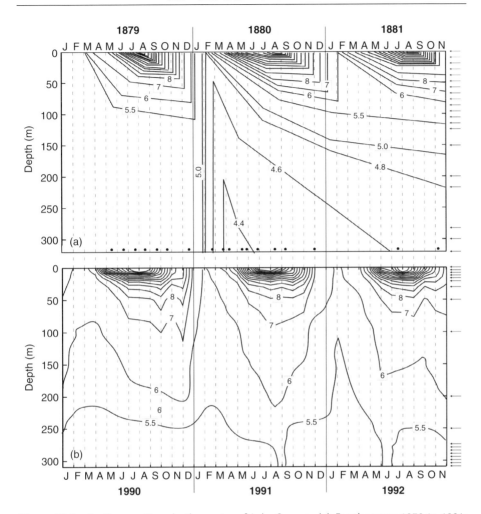

Figure 12.5 Isotherm pattern in the center of Lake Geneva: (a) For the years 1879 to 1881; this plot was reconstructed from data collected by F.A. Forel (1895). (b) For the years 1990 to 1992; this plot was reconstructed from data provided by CIPEL. Dots at the bottom indicate the times of profiling. Arrows on the right give the depths at which temperatures were measured.

12.3.2   Trends after 1970

CIPEL data (Commission Internationale pour la Protection des Eaux du Léman contre la pollution 2010) for the upper layer of the lake are reproduced in Figure 12.7. These data are taken at a depth of 5 m and are representative for the near surface layer of the lake. For these annual mean temperatures, CIPEL proposes a mean linear trend, which indicates a temperature increase of 1.5 °C over the 40-year period. This results in an increase of 0.0375 °C year^{-1}.

We analyzed the temperature data after 1970 for trends that may be related to the atmospheric temperature trend discussed above (Figures 12.3 and 12.4). For this study, we averaged the data over a winter (December to March) and a summer (June to September) period. For each of these periods, data were averaged over the epilimnion

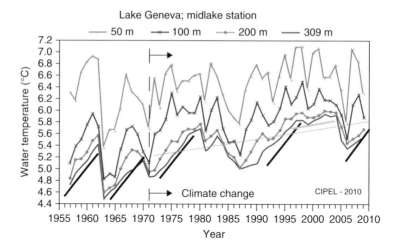

Figure 12.6 Annual mean values of the temperature at five selected depths in the deep-water layers of Lake Geneva. Data were collected at the center of the lake. All red lines have the same slope and indicate the rising leg of multiannual cycles at 309 m. The heavy green line is the mean slope for the years 1997 to 2003. The blue straight line indicates the mean long-term trend 1970 to 2010. Diagram modified after CIPEL (2010); reproduced with the permission of CIPEL. (See insert for color representation.)

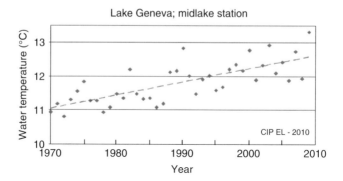

Figure 12.7 Annual mean temperatures in the upper layer of Lake Geneva. A linear trend is indicated. Diagram modified after CIPEL (2010); reproduced with the permission of CIPEL.

(0 to 20 m), the metalimnion (30 to 100 m), which includes the thermocline layer, and the deep hypolimnion (200 to 310 m). In a first step, we applied a linear trend to the data. Results for the two periods are shown in Figure 12.8. For the summer period (Figure 12.8a), a positive trend is seen in the epilimnion. However, year-to-year changes of temperature indicate a multiannual cyclic pattern and are nearly always greater than the annual linear trend. A similar observation can be made for the metalimnion. The trend lines for the two layers almost have the same slope. In the hypolimnion, the amplitude of the annual excursions around the trend line is greatly reduced with respect to the upper two layers and the slope of the trend is smaller than in those layers. Correlation coefficients for the linear trends are given in Table 12.2. The one for the hypolimnion may be considered significant.

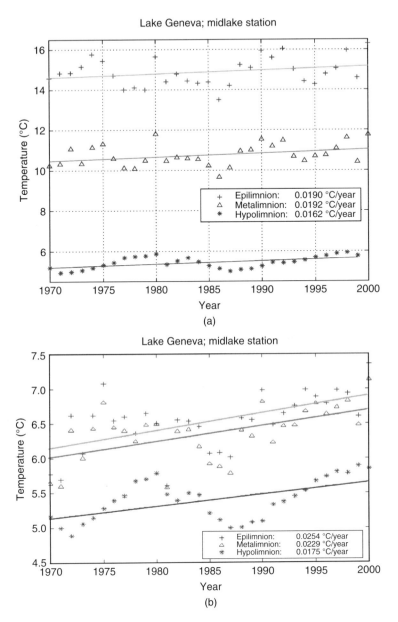

Figure 12.8 (a) Summertime mean temperature in the three layers of the water column; the linear trends are indicated and (b) Wintertime mean temperature in the three layers of the water column; the linear trends are indicated.

During winter (Figure 12.8b), the slopes of the trend lines in the epilimnion and the metalimnion are higher than during the summer. Amplitudes of the year-to-year variation and the extreme values in the two layers are similar to those in summer. A tendency towards a reduction in amplitude of the extreme values over the 30-year period, most obvious during the last 10 years, should be noted. This indicates a certain reduction in winter cooling variability and is different from the summer, where no such

Table 12.2   Rate of change ($^\circ$C year^{-1}) of linear trends; the correlation coefficients are given in parentheses.

		Epilimnion	Metalimnion	Hypolimnion
1970–2000	Summer : June to Sept.	0.019 (0.25)	0.019 (0.32)	0.016 (0.48)
	Winter: Dec. to March	0.025 (0.56)	0.023 (0.54)	0.018 (0.54)
1983 –2000	Summer: June to Sept.	0.068 (0.47)	0.052 (0.50)	0.039 (0.68)
	Winter : Dec. to March	0.050 (0.75)	0.051 (0.75)	0.043 (0.76)

tendency is seen. The mean values of the trend lines in the upper two layers are close to each other indicating winter cooling and convective downward mixing to at least 100 m, as would be expected even during relatively warm winters. In the hypolimnion, the winter situation is similar to that of the summer. Correlation coefficients for the linear trends are given in Table 12.2. They are higher than during the summer and are similar for all three layers.

Worldwide climate studies have indicated that human-induced climate change effects can be established after 1980 (IPPC 2001; Bader and Bantle 2004; CH2011 (2011)). From Figure 12.3, it appears that the rate of air temperature increase at Geneva is also greater and more continuous after this date. In order to investigate whether a change in the beginning of our lake temperature time series will affect the results for the linear trend, we repeated our analysis for the period 1983 to 2000. This is also the period over which Austin and Colman (2008) observed a climate change-related increase in temperatures in Lake Superior. As can be seen in Table 12.2, the results show a marked rise in all gradients, doubling them for the linear trends. In the epilimnion, the trend gradient during the summer is now greater than during the winter. At the same time, correlation coefficients in all layers increased significantly. For the winter period, they are almost equal in all layers.

In order to determine whether there are differences in the trends on time scales smaller than the summer-winter distribution studied above, we also calculated the linear trends for each month for all three layers. The results are given in Figure 12.9. It can be seen that as expected, the values generally fall close to those observed in the seasonal analysis above. However, much stronger trends are seen in the epilimnion in April and May during the onset of summer stratification. They also have the highest correlation coefficients (0.55 and 0.64, respectively). Correlation for the hypolimnion is above 0.5 for all months, except for February.

12.3.3   Inter-annual variability

As indicated above, the inter-annual variation for the summer and winter annual mean temperature time series is not completely random (Figure 12.8). At the same time, linear trends over different time periods show strong differences in the trend gradients. In order to investigate these points further, we smoothed the time series over three years and over 15 years by a running mean. Results for the epilimnion during the summer period are presented in Figure 12.10. It can be seen that the 15-year mean is quite different from a linear trend. During the first 15 years, there is actually a decrease in temperature, whereas for the remaining time, the mean temperatures increase almost

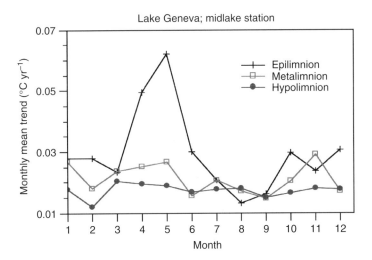

Figure 12.9 Mean monthly trends (1970 to 2000) for the three layers, as indicated.

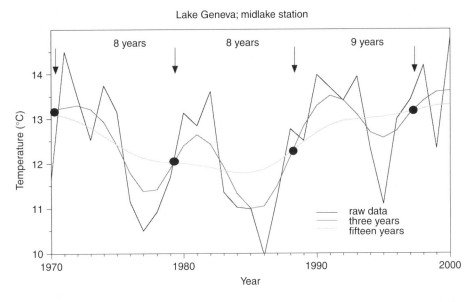

Figure 12.10 Mean summer temperatures in the epilimnion, smoothed over different periods, as indicated. The raw data are identical to those shown in Figure 12.8a.

continuously. The shape of this mean curve is similar to the one observed for the air temperature in Geneva (Figure 12.2). The three-year mean value curve reflects the inter-annual variation of the temperature. It is close to sinusoidal. The crossover points between the 15-year mean and the three-year mean have been marked in Figure 12.10. A fairly regular periodicity between eight and nine years is observed. An analysis carried out for individual months of the summer season leads to comparable results. In the hypolimnion, an almost sinusoidal oscillation around the linear trend with a period of about 20 years is seen between 1975 and 1995 (Figure 12.8).

12.4   Discussion

From 1970 to 2010, a general warming trend was observed in and around Lake Geneva. The warming in the lake can be related to the warming in the atmospheric boundary layer which is well documented and is strongest during winter nights leading to a reduction in lake cooling (see also Chapter 14, this volume). The gradient of the atmospheric warming trend is larger than the ones observed in different layers of the lake. However, it should be noted that the heat flux over the lake and the heat content of the lake would be needed to determine the relationship between the dynamics of the two mediums.

The periods 1970 to 2000 and 1983 to 2000 were analyzed for warming trends and different gradients were found for the two periods (Table 12.2). However, in both cases the gradients are similar for the epilimnion and the metalimnion. These two layers cover the water column down to a 100 m depth. The results indicate that climate change effects may penetrate into this part of the water column. We had previously observed that surface-controlled processes directly affect the dynamics of the water column down to about 100 m depth (Zhang, Lemmin and Hopfinger 1994; Michalski and Lemmin 1995) mainly by vertical turbulent mixing. In the context of climate change effects on lake dynamics, it is therefore important to understand well the small-scale aspects of mixing and the interaction of mechanisms that may be affected by climate change induced shifts in turbulent mixing regimes. When the data during the summer months are smoothed over a certain period (Figure 12.10), nonlinear behavior becomes obvious. Warming during the more recent years appears to have slowed down, even though air temperatures continue to rise. This observation, however, has to be taken with caution because of the short length of the time series.

Based on the above observations, at this time climate warming appears not to have a direct effect on the dynamics of the deep layers below a 100 m depth. The normal rate of warming in these layers during multiannual cycles can be estimated from Figure 12.6 for the periods 1973 to 1979 and 1988 to 1997. During these periods, a gradient of close to $0.5\,°C\ year^{-1}$ is found. Similar multiannual gradients were observed in the 1950s and 1960s when no climate change effect existed. During all multiannual cycles, the vertical temperature gradient of the warming leg in the deep layers remained nearly constant. This indicates that mixing caused by internal processes controls the multiannual warming cycle, independent of climate change. Poincaré waves, which were found to be active in this layer during the stratification period (Lemmin, Mortimer and Baeuerle, 2005), may contribute to this mixing.

It was noted by Bader and Bantle (2004) that winter temperatures during the years 1988 to 2004 were generally above the long-term mean. From 1997 to 2003, several of the warmest annual mean air temperatures were recorded and the summer of 2003 brought an unprecedented heat wave (Schär et al. 2004). However, the rate of warming in the hypolimnion during this time period slowed down (Figure 12.6; green line), decreasing to about $0.07\,°C\ year^{-1}$. This implies that lake warming in the deep layers is not directly related to climate change, and that internal processes in the lake may have changed the rate of vertical downward mixing of heat, with a tendency towards saturation. Climate change-induced warming cannot explain this change in trend because mixing is produced by the input of kinetic energy, forced by the wind field over the lake. The mean vertical gradient remains nearly unchanged; thus, thermal stability did not considerably change. This shows that in addition to turbulent mixing, other processes act at the same time to maintain a vertical temperature gradient. All multiannual gradients in the deep layers of Lake Geneva are much larger than the long-term (1970 to 2000) linear trend (around $0.02\,°C\ year^{-1}$) which may be related to climate change. No saturation is found in the near surface layers during the same period (Figure 12.7). However, interannual variation in this layer is again strong. Thus, climate change may shift the temperatures to a higher level, but it does not directly affect the processes by which the heat is distributed within the deep layers of the lake.

The temperatures in the hypolimnion at several depths were continuously measured over 33 months from February 2001 to December 2003, recording every 10 min. with a resolution of 1 mK. The data are presented in Figure 12.11. We found a linear increase in temperature with a gradient of about $0.05\,°C\ year^{-1}$. During the recording period, we observed two winter cooling events. The first one occurred in 2002, and only briefly affected the water column down to 250 m, but did not penetrate to the bottom (300 m). It did not alter the long-term rate of warming in those layers. The second cooling event occurred in 2003. It was more severe and lasted longer, resulting in a cooling of the whole water column below 150 m. This cooling event brought the temperature level in the deep layers back to those observed at around 1995, setting back the deep-water warming trend by about eight years. During both cooling events, the temperatures fell well below those observed before and after the events in those layers. This may be a documentation of deep-water cooling by density currents descending from the lateral slopes into the bottom layers (Fer, Lemmin and Thorpe 2002), rather than by vertical mixing. From Figure 12.10 it can also be seen that temperatures in the deep hypolimnion continually increase throughout the year after the winter cooling event. This is in contrast to the often stated assumption (Livingstone 1997, 2003) that the deep-water temperatures in deep lakes remain constant after winter cooling.

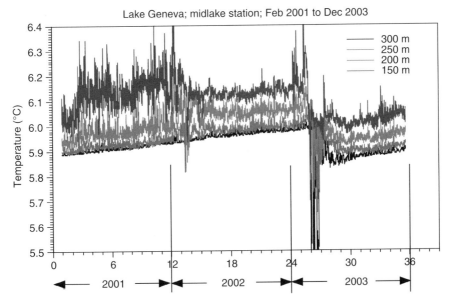

Figure 12.11 Time series of temperatures recorded at different depths in the deep layers in the center of Lake Geneva. Recording interval is 10 min. (See insert for color representation.)

A periodic pattern of temperature variation in the epilimnion of the lake with a period of about eight to nine years was suggested by the data smoothing (Figure 12.10). The gradients in this pattern are again stronger than those of the climate change induced trend. No atmospheric forcing with a comparable periodicity seems to exist. A certain correlation with an apparent signal of the 11-year sunspot cycle can be imagined (Labitzke and van Loon 1999). Maxima (1980, 1990) and minima (1986, 1995) in the monthly averaged solar flux correspond to those seen in the data in Figure 12.9. Unfortunately, the time series of the data is too short to determine the significance of this pattern. A below-average winter 2004/2005 has led to lake cooling and a reduction of epilimnion temperatures that falls close to this eight- to nine-year cyclic pattern.

In addition to an increase in the overall temperature level of the lake, climate change may produce a change in the stratification pattern and the thermal stability of the water column. This was seen in Figure 12.9 where a much stronger rate of change in the epilimnion temperature was found in the spring, which leads to an earlier onset of thermocline formation and may explain the advance in spawning reported by Gillet and Quetin (2006). At the same time, changes in plankton population dynamics with an advance of the spring bloom can be expected and may affect the timing of the clear-water phase in the lake. Thus, a disturbance of the ecosystem functioning and a decoupling of trophic relationships due to temporal mismatches of trophic interactions can be anticipated. An earlier onset of thermocline formation will also shorten the duration of the winter mixing regime and increase the length of the season for internal seiche and internal wave activity (Thorpe *et al.* 1996; Lemmin, Mortimer and Baeuerle 2005). This, in turn, will lead to a longer season of turbulent mixing induced by breaking internal waves and, thus, may more rapidly increase deep-water temperatures.

It may be argued that climate change effects on rivers flowing into the lake may affect the temperature pattern in the lake. It was stressed by Jakob (1997, 2000) that due to increased ice melt resulting from climate change, the temperature of rivers coming from glacier-covered high mountains might actually decrease instead of increase. The author found cooling mainly during the summer, and warming during the rest of the year. For the Rhone River which originates in the high mountains and is the principal river entering the lake, we analyzed the temperatures from 1970 to 2000 and found a slightly positive trend of about $0.02\,^{\circ}\text{C year}^{-1}$ during the summer (R = 0.12) and the winter (R = 0.35). This development may also be influenced by the strong hydroelectric activity in the Rhone River basin. The gradient of the trend that we found for the Rhone River is similar to the long-term trend observed in the lake. Therefore, the Rhone River is not likely to significantly change the warming trend detected in the lake.

The development of the climate in the alpine region is sensitive to the conditions over the North Atlantic Ocean (Quandrelli *et al.* 2001) and the effect of the North Atlantic Oscillation (NAO) on European climate is well established. We therefore analyzed the lake data for effects of NAO. Using the same statistical tools applied in the analysis above, we were not able to find a correlation between our data and the NAO index. Higher order statistical concepts such as those applied by Straile, Jöhnk and Rossknecht (2003) may bring out NAO effects. We have seen that the mean (over 15 years) water temperatures in the epilimnion (Figure 12.10) closely follow the mean air temperatures at Geneva (Figure 12.2). Therefore, a certain effect of the NAO on the water dynamics similar to the one observed by Weyhenmeyer, Meili and Livingstone (2004) can be expected. However, this effect may be limited to surface waters and may not penetrate into the deep layers. Due to the size of Lake Geneva and the shape of the surrounding topography, local climate effects are well developed and may mask the NAO effect on lake dynamics. Therefore, when compared to other climate change effects discussed above, the NAO effect appears not to be important for the dynamics in Lake Geneva, since it cannot be evidenced by the same first order statistics. Support for this conclusion comes from the analysis by Hilmer and Jung (2000). These authors found that for the period 1958 to 1997, the squared correlation coefficient between the NAO index and the winter (December to March) air temperatures was below 20% for Lake Geneva region and should therefore not be significant. Furthermore, it has to be recalled that several phenomena contribute to the low frequency variability of the North Atlantic climate signal affecting the alpine climate: the North Atlantic Oscillation (NAO), the Arctic Oscillation (AO), the East Atlantic pattern (EA), the Gulf Stream/Northern Current Index (GSI), the East Atlantic Western Russian (EA/WRUS) and the Northern Hemisphere Temperature (NHT).

An estimate for the long-term future development of the temperature pattern in Lake Geneva may be based on the predictions presented by Schär *et al.* (2004). Using IPCC specified scenarios, the authors show that for the forecasting period 2071 to 2100 during summer months, the mean air temperature may increase by about 4.6 °C and that, at the same time, air temperature variability may increase by 100% in Lake Geneva region. Although this analysis focused on summer dynamics, similar changes may be expected for the winter months. Therefore a certain continuation of lake warming is likely. However, as was shown above, the actual warming in the lake will also depend on how winter night temperatures develop.

From 1997 to 2003, we observed a certain "saturation" tendency in the increase in deep-water temperatures. This may suggest that an increase in deep-water lake temperatures comparable to atmospheric warming is not likely. Furthermore, the strong increase in climate variability predicted by Schär *et al.* (2004) indicates that relatively cold winters, which may offset the warming trend, could become more frequent. Unfortunately, no studies seem to exist that predict a trend for the probability and the amplitude of cold spells related to climate change-induced variability. Combined with the observations from the winter 2004/2005, discussed above, it would be difficult to argue that the mean long-term development of the lake temperature structure will strictly follow the mean air temperature trend as suggested by the study of Schär *et al.* (2004).

12.5 Conclusion

It has been shown that over the past 30 years there has been a warming trend in Lake Geneva, which follows the climate-related warming trend in the atmospheric boundary layer. However, during this period, the amplitude of the inter-annual temperature variations was always greater than the climate change induced long-term mean trend. At the same time, the amplitude of the multiannual deep-water cycles was still stronger than the climate change related long-term trend. Multiannual cycles have been observed before climate change induced warming started and their dynamics have been the same throughout the period of observation. Thus, climate change may continue to shift the temperatures in the lake to a higher level, but at present, it is not likely to modify significantly the processes by which the heat is distributed within the lake volume.

As a result of the long-term trend, the probability of the occurrence of certain system modifications in the future can be envisioned. In that case, the gradual change in physical processes in lakes requires a good understanding of the interplay of processes that lead to the downward mixing of heat into deeper layers. The observed difference in the slope of the climate change-induced trend lines during summer and winter shows that seasonal variability in mixing dynamics remains important. It should be noted, however, that the observation period of this trend is still too short to allow a clear determination of the future long-term development resulting from climate change. With longer time series it may be found that a linear trend starting in 1970 may not be the best representation of the future effects of climate change. At present, many predictions and climate-change scenarios are based on linear trend assumptions. It has to be remembered that climate-change effects may only follow linear trends up to a certain threshold, and that sudden long-lasting abrupt and more severe climate changes may occur once such a threshold is passed. This has been observed in the past and may happen again in the future (Alley 2004).

Nevertheless, it can be expected that a warming trend will continue or increase in rate in the foreseeable future. This may stress the ecosystem and may risk modifying the existing system. In particular, it may pose a threat for certain species that need a cold water environment during certain stages of their development cycle. It may also be detrimental to water quality, including oxygen concentration, which in turn will have consequences for drinking water supplies to the ever increasing population around the lake.

12.6   Acknowledgements

The data for the present investigation were kindly provided by the Federal Office of Meteorology and Climatology MeteoSwiss and the CIPEL (Commision Internationale pour la Protection des Eaux du Léman). Part of this work was carried out within the framework of the project 'EUROLAKES' (EU-Contract.-No. EVK1-CT1999-00004) sponsored by the Swiss Federal Office of Science and Research (Contract Nr. 99.0190). We are grateful for the support. The help of C. Perrinjaquet in the field work and the data preparation is appreciated.

References

Alley, B.R. (2004) Abrupt climate change. *Sci. Am.*, **291**, 40–8.
Austin, J., and Colman, S. (2008) A century of temperature variability in Lake Superior. *Limnol. Oceanogr.*, **53**, 2724–2730.
Bader, S. and Bantle, H. (2004) Das Schweizer Klima im Trend. Temperatur- und Niederschlagsentwicklung 1864–2001. Veröffentlichung der MeteoSchweiz Nr. 68, Zürich.
Commission Internationale pour la Protection des Eaux du Léman contre la pollution (CIPEL) (2010) *Rapport de la Commission internationale pour la protection des eaux du Léman contre la pollution, Campagne 2009*. Nyon, Switzerland
CH2011. 2011. Swiss climate change scenarios CH2011, published by C2SM, MeteoSwiss, ETH, NCCR Climate and OcCC, Zürich, Switzerland.

Fer, I., Lemmin, U., and Thorpe, S.A. (2002) Winter cascading of cold water in Lake Geneva. *J. Geophys. Res.*, **107**, C6,10.1029/2001JC000828.

Forel, F.A. (1895) *Le Léman, monographie limnologique*, Rouge, Lausanne.

Gerdeaux, D. (2011) Does global warming threaten the dynamics of Arctic char in Lake Geneva? *Hydrobiologia*, **600**, 69–78.

Gillet, C. and Quetin, P. (2006) Effect of temperature changes on the reproductive cycle of roach (*Rutilus rutilus*, L.) in Lake Geneva from 1983 to 2001. *J. Fish Biology*, **69**, 518–34.

Hilmer, M. and Jung, T. (2000) Evidence for a recent change in the link between the North Atlantic Oscillation and Arctic sea ice export. *Geophys. Res. Lett.*, **27**, 989–92.

IPPC (Intergovernmental Panel On Climate Change) (2001) *Climate Change 2001: The Scientific Basis*. Contribution of working group 1 to the third assessment report of the Intergovernmental Panel On Climate Change. Cambridge University Press, Cambridge, UK.

IPPC (Intergovernmental Panel On Climate Change) (2007) *Climate Change 2007: The Physical Science Basis*. Contribution of working group 1 to the fourth assessment report of the Intergovernmental Panel On Climate Change. Cambridge University Press, Cambridge, UK.

Jakob, A. (1997) Temperaturentwicklung in den Fliessgewässern. *Mitt. zur Fischerei*, **66**, 29–40.

Jakob, A. 2000. Evolution de la temperature des cours d'eau. *Informations concernant la pêche*, **66**, 29–40.

Labitzke, K., and Van Loon, H. (1999) The signal of the 11-year sunspot cycle in the upper troposphere-lower stratosphere. *Space Science Review*, **80**, 393–410.

Lemmin, U. (1998) *Courantologie Lémanique. Arch. Sci. Genève*, **51**, 103–20.

Lemmin, U., Mortimer, C.H., and Baeuerle, E. (2005) Internal seiches dynamics in Lake Geneva. *Limnol. Oceanogr.*, **50**, 207–16.

Livingstone, D.M. (1997) An example of the simultaneous occurrence of climate-driven "sawtooth" deep-water warming/cooling episodes in several Swiss lakes. *Verh. Internat. Verein. Limnol.*, **26**, 822–8.

Livingstone, D.M. (2003) Impact of secular climate change on the thermal structure of a large temperate central European lake. *Climatic Change*, **57**, 205–25.

Livingstone, D.M. and Dokulil, M.T. (2001) Eighty years of spatially coherent Austrian lake surface temperatures and their relationship to regional air temperature and North Atlantic Oscillation. *Limnol. Oceanogr.*, **46**, 1220–7.

Michalski, J. and Lemmin, U. (1995) Dynamics of vertical mixing in the hypolimnion of a deep lake: Lake Geneva. *Limnol. Oceanogr.*, **40**, 809–16.

Quandrelli, R., Lazzeri, M., Cacciamani, C., *et al.* (2001) Observed winter alpine precipitation variability and links with large-scale circulation patterns. *Clim. Res.* **17**, 275–84.

Räisäni, J. and Ylhäisi, J.S. (2011) Cold months in a warming climate. *Geophys. Res. Letters*, **38**, doi: 10.1029/2011GL049758.

Rempfer, J., Livingstone, D.M., Blodau, C., *et al.* (2010) The effect of the exceptional mild European winter of 2006–2007 on temperature and oxygen profiles in lakes in Switzerland: a foretaste of the future? *Limnol. Oceanogr.*, **55**, 2170–80.

Schär, C., Vidale, P.L., Lüthi, D., *et al.* (2004) The role of increasing temperature variability in European summer heatwaves. *Nature*, **427**, 332–6.

Straile, D., Jöhnk, K., and Rossknecht, H. (2003) Complex effects of winter warming on the physicochemical characteristics of a deep lake. *Limnol. Oceanogr.*, **48**, 1432–8.

Thorpe, S.A., Keen, J.M., Jiang, R., and Lemmin, U. (1996) High-frequency internal waves in Lake Geneva. *Phil. Trans. R. Soc. Lond. A*, **354**, 237–57.

Weyhenmeyer, G.A. (2004) Synchrony in relationships between the North Atlantic Oscillation and water chemistry among Sweden's largest lakes. *Limnol. Oceanogr.*, **49**, 1191–201.

Weyhenmeyer, G.A., Meili, M., and Livingstone, D.M. (2004) Nonlinear temperature response of lake ice breakup. *Geophys Res. Lett.*, **31**, C6,10.1029/2004L07203.

Zhang, S., Lemmin, U., and Hopfinger, E. (1994) The variability of vertical finescale temperature structures in Lake Geneva, Switzerland. Fourth International Symposium on Stratified Flows, Grenoble, Vol. 3.

13

Climate Change and Wetlands of the Prairie Pothole Region of North America: Effects, Management and Mitigation

Marley J. Waiser

Formerly of the Global Institute for Water Security, University of Saskatchewan, Canada

13.1 Introduction

Wetlands—scores of water bodies ranging in size from potholes to lakes that characterize the prairie landscape—are key ecological components of the vast prairie region of North America. A series of glaciation retreat events during the Pleistocene epoch some 12 000 years ago were instrumental in the creation of these wetlands. Glaciations reset the clock for soils, plants and animals by scraping the land surface clean (Beaudoin 2003). When the ice sheets retreated, they left glacial till deposits, largely clay rich glacial tills with interspersed deposits of clay, silt sand and gravel, tens to hundreds of meters thick (Conly and van der Kamp 2001). These deposits have produced not only the hummocky topography of the prairie landscape, but also the sulfate salts that characterize soils, and water in the region.

Millions of closed-basin wetlands, known locally as potholes or sloughs, have formed in depressions created by this rolling topography (Figure 13.1). The prairie pothole region (PPR) covers approximately 780 000 km², extending from north-central Iowa to Alberta (Mitsch and Gosselink 1993). This region is bounded in the north by the southern boundary of the boreal forest in Alberta, Saskatchewan and Manitoba, in the south by the limits of the Laurentide ice sheet of the Wisconsian glaciation, in the east by the prairie deciduous transition zone and in the west by the Missouri River in the United States and the foothills of the Rockies in Canada (Figure 13.2) (Millett *et al.* 2009). Wetlands in the PPR range from permanent to ephemeral and

Climatic Change and Global Warming of Inland Waters: Impacts and Mitigation for Ecosystems and Societies, First Edition. Edited by Charles R. Goldman, Michio Kumagai and Richard D. Robarts.
© 2013 John Wiley & Sons, Ltd. Published 2013 by John Wiley & Sons, Ltd.

Figure 13.1   Aerial photograph of the prairie pothole region of North America. Photocredit: Bill Waiser.

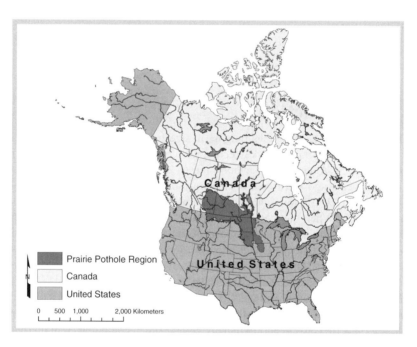

Figure 13.2   Map of the Prairie Pothole Region (PPR) of North America based on ecoregion classification. From Millett *et al.* (2009) with modifications by the author and used with permission.

freshwater to saline. They are highly dependent on precipitation for their water supply as most are refilled in spring from snow melt runoff on frozen or saturated soils (Covich *et al.* 1997).

Prairie wetlands are located within a highly variable semi-arid climatic area, where evaporation exceeds precipitation. In this climatic zone, annual evaporation is 30 cm more than the annual precipitation of 35 cm (LaBaugh *et al.* 1996), and this means that wetland water levels and vegetation fluctuate widely in response to seasonal wet/dry periods (Woo and Rowsell 1993; Poiani *et al*. 1996). In fact, within the prairie pothole region, most temporary wetlands are dry by mid-summer (Robarts and Waiser 1998).

The combination of unique climatic conditions (which initiates wet/dry cycles), depressional wetlands, and rich glacial till make for some of the most productive aquatic ecosystems in the world (Murkin 1989). In fact, prairie wetlands are the single most important breeding area for waterfowl on the North American continent (Poiani and Johnson 1991) producing 50–80% of the North American duck population in any given year (Batt *et al*. 1989). A critical habitat for amphibians, these wetlands also support a diverse community of insects and crustaceans including primitive crustaceans of the Orders Anostraca (fairy shrimp), Conchostraca (clam shrimp) and Notostraca (tadpole shrimp) (Donald *et al*. 1999). It is little wonder, then, that they have been dubbed "the foundation for biodiversity" on the prairies. Wetlands are also important in the surface and subsurface hydrology of the region (van der Kamp and Hayashi 1998).

Over the last century, these wetlands have undergone a number of anthropogenic stresses. Perhaps the most important of these has been drainage. From 1870 to 1930, for example, nearly all the wetlands in Iowa (89%), and south Minnesota (90%) were drained (Millett *et al*. 2009). More recently, increases in levels of atmospheric pollutants have caused atmospheric CO_2 levels to rise with concomitant decreases in stratospheric ozone levels. The ensuing climatic warming and increases in levels of UVB radiation reaching the earth's surface could both significantly affect prairie wetlands (Robarts and Waiser 1998). Due to the importance of the PPR, not only in terms of regional hydrology but also in supporting vast numbers of waterfowl, mitigation measures designed to offset effects of climate change on these systems will be needed in the coming years.

This chapter will first discuss the PPR climate, wetland hydrology and the influence of climate on wetland hydrology, chemistry, vegetation, and productivity. The many changes that have already occurred and others that are expected under various climate change scenarios will then be considered. Lastly, various mitigation measures designed to offset climate change effects on wetlands in the PPR will be examined.

13.2   Climate and its influence on wetlands in the PPR

Climate in the PPR is influenced by three major air masses: the Continental Polar, Maritime Tropical and the Maritime Polar. According to a recent paper "The complex interactions among these air masses in the middle of a large continent create one of the most extreme and dynamic climates on earth" (Millett *et al*. 2009). Here, drought to deluge conditions are common and temperatures regularly exceed 40 °C in summer and −40 °C in winter. High summer temperatures combined with low humidity mean that rates of evapotranspiration are elevated and in fact they exceed precipitation. This region is therefore classified climatically as semi- arid. Generally, the north and west PPR receive less precipitation than the south and east due to a NE to SE rise in elevation (200–1200 m). It is not surprising then that in the twentieth century, the mid- PPR wetlands (in the east Dakotas and SE Saskatchewan) experienced the most dynamic climate seen worldwide (Millett *et al*. 2009).

13.3   Wetland hydrology

Most wetlands in the PPR are shallow, usually <1 m in depth, and occur in hydro-
logically closed basins (no permanent stream inflow or outflow—Winter *et al.* 2001).
Because their watersheds are not connected by surface water drainage, water balance within individual wetlands is determined by precipitation, evapotranspiration,
and interaction with groundwater (LaBaugh *et al.* 1996). Subsurface flow in and out
of their catchments (groundwater), however, is minimal due to the impermanence
of clay-rich soils (Schindler and Donahue 2006) PPR hydrology is dominated by
cold weather processes including snow accumulation and subsequent spring melt over
frozen soils. Although snow is only one-third of the annual precipitation, it accounts
for 80% of annual runoff. Not surprisingly then most wetlands here are highly dependent on spring melt for their water supply (Covich *et al.* 1997). During spring and
summer, precipitation occurs mostly as highly convective storms which can deposit a
great deal of moisture within a short time period (Su, Stolte and Van der Kamp 2000).
Most of this precipitation, however, is absorbed by dry soils (little runoff) although
the more intense storms may result in flooding.

13.4   Linkages between climate and wetland processes

There is no doubt that climate is a driving factor and intimately linked to wetland
hydrology, vegetation, water chemistry and indeed wetland biology and productivity.
With regard to hydrology, in these topographically closed aquatic systems, response
to changes in the balance between precipitation and evaporation is much faster than
in those systems with distinct inflows and outflows (Fritz 1996). During summer,
wetland water levels decline sharply due to evaporation from open water as well as
evapotranspiration from the surrounding vegetation (Poiani *et al.* 1996; Conly and
van der Kamp 2001). As a result, smaller wetlands tend to dry out completely during
the summer months. As well, drought or deluge tends to shift wetland permanency
in opposite directions e.g., drought shifts permanent back to semi-permanent and
vice-versa (Johnson *et al.* 2004).

Just as wetland permanency is driven by climate, so too is the wetland vegetation
cover cycle. In fact, drought and deluge frequencies determine the speed with which
wetlands will undergo a complete cover cycle, which may be several decades or
longer, depending on climatic variability (Johnson *et al.* 2004).

Four wetland cover cycles associated with climate have been identified:

1. A dry stage with little or no standing water and characterized by heavy emergent
   vegetation and high nutrient sequestration.
2. A regenerating stage with reflooding and vegetation propagation. This phase
   exhibits high wetland productivity associated with periodic nutrient release during
   decomposition (dry marsh stage) and uptake during the regeneration phase.
3. A degenerating stage during which vegetative plants decline.
4. An open water, less productive lake stage with high water levels, little emergent
   vegetation and few nutrients.

Wetlands in the mid PPR, where climate is the most variable, are, on average, the most productive because they pass through all vegetative states noted above. Not surprisingly, then, productive temporary wetlands here are the most important for waterfowl. Historically, however, these are the wetlands which have been most affected by agricultural drainage (Johnson *et al*. 2004). Wetlands in the wetter east PPR are less productive because they are more or less in a permanent open water state (partially a result of intense ditch digging in the twentieth century, which connected wetland basins). These water bodies dry out less frequently and therefore experience fewer complete cover cycles per century. Finally, wetlands in the western PPR are less productive because they dry out more frequently (Johnson *et al*. 2004).

With regard to wetland biology, PPR aquatic invertebrate production has been shown to be driven by interannual fluctuations of water levels in response to climate wet/dry cycles (Anteau 2011). Wetland invertebrates provide protein and calcium for waterfowl energy and reproduction. Because survival, recruitment and migratory bird population size, in turn, are influenced by nutrition during spring migration and at breeding sites it is not surprising that a strong relationship exists between invertebrate density and juvenile bird survival (Anteau 2011). In another study, occupancy of prairie wetlands by tadpoles was found to be related to winter and spring precipitation, with 67% of the long term variation in occupancy related to snowfall from November to February (Donald *et al*. 2011).

Wetland water chemistry is closely tied to climate and hydrology as well with many wetlands exhibiting changes in water chemistry concomitant with seasonal declines in wetland water volume and depth (Fritz 1996; Waiser 2006). Prairie wetlands losing most of their water by evaporation and not infiltration to the pond margin tend to be saline and located in lowland areas. As water evaporates seasonally from these wetlands, salinity increases (Figure 13.3; Waiser 2006).

Along with increases in salinity, seasonal increases in dissolved organic carbon (DOC), sometimes by as much as three times springtime values, have also been observed (Figure 13.3). Although evapoconcentration likely explained this seasonal DOC increase, saline wetlands were also shown to be losing some DOC via infiltration to the wetland margin. Freshwater ponds which lost most of their water by infiltration to the pond margin, on the other hand, displayed less seasonal variation in DOC concentrations and no increase in salinity (Waiser 2006).

13.5 Existing climate change

According to tree rings, lake sediments and salinity, and shifting vegetation at least 40 droughts (including multi-decadal ones) have occurred on the Canadian prairies in the last 100 years. Although drought is a recurrent feature here, there is no doubt that PPR has experienced extensive and above normal warming since the early 1960s (Sauchyn 2010). From 1895–1991 temperatures in the Canadian PPR temperature rose by 1.1 °C with larger increases of up to 3 °C reported for parts of Montana and the Dakotas (Millett *et al*. 2009). With minimum temperatures warming more in winter than in summer, the PPR is in fact getting less cold (Millett *et al*. 2009). Such warming may, in fact, exceed the global average (Gan 1998). Also of note is the fact that during winter, spring and summer there are fewer days here with extremely low temperatures. This fact means that wetlands are freezing up later in winter and losing

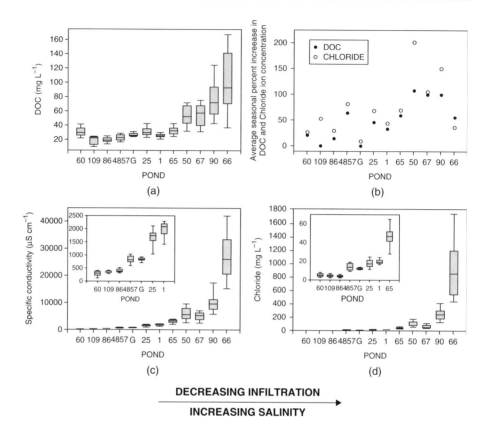

Figure 13.3 Box whisker plots indicating the median and variation in (a) DOC, (c) specific conductivity, and (d) and chloride ion for each of the 12 study ponds at the St. Denis National Wildlife Refuge. The boundary of the box closest to zero indicates the 25th percentile, the line within the box marks the median, and the boundary farthest from zero indicates the 75th percentile. Whiskers above and below the box (error bars) represent the 90th and 10th percentiles, respectively. (b) Average seasonal increases in DOC and chloride ion for each study pond (expressed as a percentage of the spring value). From left to right, ponds increase in salinity and decrease in percentage of water loss due to infiltration. Ponds in which the range in concentrations are obscured by the scale of the larger plot have been replotted on a smaller scale in the insets in Figures 13.3c and 13.3d. From Waiser (2006).

their ice earlier in spring. Along with the warming trend, significant drying has also occurred, with a steepening of the east west moisture gradient during the twentieth century. In fact, over the last 100 years, precipitation in eastern Montana and North Dakota declined by 10% while increasing by the same percentage in the eastern PPR.

13.6 Predicted climate change and its effects

Over long time scales, disturbance has clearly been a hallmark of the PPR and wetland flora and fauna have adjusted accordingly. And as noted earlier, the productivity of these wetlands is highly dependent on the frequency with which they experience the four vegetative cover cycles, which in turn are tied to climatic changes. Although the ecological reality here has been one of dynamic and random change, some scientists fear that the climate change currently being experienced across the PPR may be a tipping point with ecosystems here undergoing what is known as a regime shift. Under these circumstances, a "new normal" condition for prairie wetlands may occur. According to some scientists, "Future ecosystems that result from climate change in the Prairie Provinces will be unprecedented."

Under current scenarios of global warming, the Polar air masses which dominate prairie winters are predicted to weaken thereby allowing hot dry air from the American southwest and hot moist air from the American southeast to more routinely affect PPR. Drought interspersed with unusually wet conditions is also predicted to occur. Rainfall that does fall will be more concentrated in time with larger amounts falling in fewer storms. More flooding than normal may occur. An increase in the number of extreme weather events is also expected. Other impacts include a greater frost-free period; in fact from 2080–2099, 20 fewer frost free days on the Great Plains are predicted (Millett *et al.* 2009). Longer warmer winters will reduce snowpack accumulation resulting in less spring runoff and therefore shallower spring time wetland water depths and fewer wetlands.

Researchers have used the Palmer Drought Index (PDI) to tie decline in wetland numbers to climate. The PDI uses temperature, precipitation, available soil moisture and Thornthwaite's method to calculate potential evapotranspiration. PDI values above +4 and below −4 represent extreme deluge and drought, respectively (Millett *et al.* 2009). A statistical comparison of PDI to spring wetland counts in the PPR indicated that up to 70% of the observed variation in wetland numbers was explained by the PDI. This observation confirmed that year to year variation in wetland numbers was clearly related to climatic variation (Sorenson *et al.* 1998). Using the PDI, scientists have shown that a combined $3\,°C$ temperature increase and 10% decrease in precipitation could reduce the number of wetland basins by 74% in the Canadian parkland, by 31% in the Canadian grasslands and by 56% in the American grasslands (Larson *et al.* 1994).

A concomitant decline in waterfowl numbers associated with wetland loss has been predicted. Although waterfowl have historically adapted to the dynamic climate, fewer options may exist in the future (Millett *et al.* 2009). In fact, one scientist estimates that the numbers of breeding ducks could be reduced by as much as 50% (Sorenson *et al.* 1998). The loss of wetland basins will also mean a change in habitat for small mammals, reptiles and amphibians with some rare species facing possible extirpation. Invasion by exotic species, better adapted to the drier conditions, is also a possibility. According to some authors the consequences for prairie wetlands will be severe if future climate does not provide supplemental moisture to offset higher evapotranspiration.

An increasing frequency of dry basins will have a number of effects. Firstly, it will expand the area of wetland sediments exposed to air and turn wetlands into strong greenhouse gas emitters (Crooks *et al.* 2011). Wetlands sediments that are still moist but where overlying water is gone will see a great increase in methane (a greenhouse gas) emissions to the atmosphere. This methanogenesis will be favored by sediment warming due to air exposure and the fact that sediments for a time will still be saturated and anoxic. Disturbingly, such dramatic increases in CH_4 emissions have already been noted at sites in Alberta and Saskatchewan (Pennock *et al.* 2010; Badiou *et al.* 2011).

Wetland sediments, due to their anoxic nature, also contain a relatively high percentage of organic carbon. If exposed sediments are rewetted during deluge situations, this process may create a "champagne" effect—literally bubbling CO_2, another greenhouse gas, into the atmosphere. CO_2 fertilization in turn will favor woody vegetation and shrubs. Their increased growth will be especially favored with decreased soil moisture accompanying increased warming. Such increases are already occurring in a wide variety of semi-arid ecosystems in a process known as vegetation thickening (Walden *et al*. 2004).

Other possible effects include a resetting of the timing of certain wetland ecological processes due to earlier ice out and spring runoff. In saline wetlands, increasing salinity as a result of increasing temperatures and evaporative stress may favor mercury methylation. Wetlands are important sites of carbon sequestration, with greater carbon accumulation known to occur in freshwater as opposed to saline ecosystems. Under a drier climate scenario, increasing numbers of saline wetlands may decrease the amount of carbon being sequestered (Crooks *et al*. 2011).

A warmer future climate and steepening of the existing east–west moisture gradient in the PPR could also shift optimal bird-breeding conditions away from areas that are historically productive (the mid-PPR), eastward. In the eastern PPR, however, most wetlands have been drained and the remaining are in the less productive open-water phase (less suitable as waterfowl habitat—Millett *et al*. 2009). Finally, decreasing wetland water levels as a result of increasing temperatures may result in diminished aesthetics, which in turn may have an impact on tourism, hunting, and bird watching.

13.7 Mitigation/management of climate change

Due to the extreme climatic variability that already exists within the PPR, mitigation and conservation challenges presented by climatic change will be great. According to the Ramsar agreement, management of a wetland and its resources should be done in a manner that is consistent with maintenance of its ecological character. Ecological character, in turn, is defined as the sum of biological, physical and chemical components and their interaction, which maintains a wetland and its products, functions and attributes. Change in ecological character is defined as impairment or imbalance in any biological, physical or chemical component of a wetland ecosystem or in their interactions, which maintain productivity and functionality (as cited in Walden *et al*. 2004).

Changes in land-use practices could offer hope as a mitigation strategy to off-set climate change effects in the PPR. Using a modeling approach (WETSIM 3.2) and a semi-permanent wetland in eastern South Dakota, differing land-use practices (unmanaged grassland, grassland managed with moderately heavy grazing, and cultivated crops) and climate scenarios were used to predict wetland water levels (Poiani and Johnson 1991; Poiani *et al*. 1996). Climate scenarios were developed by adjusting the historical climate in combinations of 2 °C and 4 °C air temperature and ±10% precipitation. According to the model output, water levels in wetlands surrounded by managed grasslands were significantly greater than those surrounded by unmanaged grassland. Management reduced both the proportion of years the wetland went dry and the frequency of dry periods. This scenario also produced the most dynamic vegetation cycle of any scenario, which in turn would bode well in terms of waterfowl production.

Another proposed strategy for the mitigation/management of climate change on PPR wetlands has been proposed by Henderson and Thorpe (2010) and would involve field assessments of wetlands at risk, identification of vulnerable wetlands and monitoring to identify the scope and extent of the problem, and the likelihood of adverse change. Monitoring and field assessments will help to gain an appreciation of the risk presented and the existing knowledge gaps. Clearly there is an urgent need for long-term monitoring of wetland variables. Long-term data sets are essential for testing climate change models (Johnson *et al.* 2004), especially given the extreme variability of water levels, vegetation cover and composition that characterize the PPR (Johnson *et al.* 2010). Models (like the PDI) can identify possible risks. Once the risk is identified there are a number of ways in which it might be managed. The easiest way, the authors note, is passive management (Johnson *et al.* 2010). The benefits of this traditional, laissez faire approach are that it is relatively inexpensive, painless to manage, and avoids risks associated with active management (Henderson and Thorpe 2010).

Another way of risk management is embodied by the active "managed change model," a strategy currently being utilized in Britain. This strategy advocates acceptance of the fact that landscape modification will be driven by climate change and that differing wetlands (e.g., differing in their hydrology) will require differing management strategies. In the managed change model, the creation of "substitute ecosystems" will delay, direct or ameliorate ecosystem change. The advantages of this mitigation scheme include the preservation of ecosystem diversity, maintenance of a valued landscape, and a reduction in the risk of catastrophic change. On the negative side, the model's outcomes are unpredictable, it tends to be more expensive than passive management, and careful ecosystem monitoring is required. With increasing climatic change, the interventions required in order to maintain ecosystem integrity may become increasingly expensive. At this point, practicality becomes an important issue (Henderson and Thorpe 2010).

With specific reference to the PPR, active management practices might involve, for example, the introduction of drought-resistant vegetation (although there are concerns regarding introduction of exotic or foreign species), assisted species migration and re-establishment of wetlands in the wetter east (where the most wetland drainage has occurred but where breeding waterfowl populations are predicted to move under climate change scenarios—Millett *et al.* 2009). Widespread restoration of seasonal and semi-permanent wetlands across the prairies could also help mitigate GHG emissions in the short term and could sequester about 3.25 mg CO_2 equivalents ha^{-1} yr^{-1} (Badiou *et al.* 2011).

Another management/mitigation scheme is the so-called frozen landscape model. This strategy assumes that a particular landscape at a particular time and in a particular place is the correct one. As such, it seeks to maintain a particular landscape as it is by resisting anthropogenic change or reversing such change by landscape restoration. According to Henderson and Thorpe (2010), the advantage of this management strategy is that the goals are well understood. The landscapes to be restored existed before and therefore can be recreated.

Once a management strategy has been chosen, the last step in the mitigation process is ongoing monitoring. This step involves the use of indicators to verify effectiveness of the chosen management strategy. Within the PPR this would require the establishment of an intensive monitoring network to collect quantitative data within a few wetland complexes of differing hydrologies (areas $<10\,km^2$) (Sorenson *et al.* 1998).

Landsat imagery, which can discriminate between flooded and dry basins, might be a useful tool in the monitoring process. In addition to intensive monitoring, collecting data on a regional scale with more wetlands sampled less frequently and fewer parameters measured (Sorenson *et al*. 1998) has also been proposed as an important part of this last step.

13.8   Action required

Adapting the proposed long term wetland monitoring scheme necessary to identify risks to wetlands under various climate change scenarios will require not only upgrading the long term intensive monitoring sites on the PPR (e.g., equipped for snow sampling, new weather stations, and so forth) but the addition of more sites (as noted above). Such upgrading and additions for monitoring hydrology, vegetation, biological, and chemical variables, as well as weather, will assist in establishing current trends, forecasting future change as well as filling the many existing data gaps with respect to water quantity and quality, water use, and climate (Johnson *et al*. 2004; Millett, Johnson, and Gutenspergen 2009). It is, however, unclear who is responsible for collecting this data and sharing it, and who is going to pay for it.

Clearly, states and provinces across the PPR will have to adopt long-term effective climate, environment and water policies that not only address drought but prepare for extreme events predicted to occur as a result of climate change. Such policies will have to be comprehensive and integrate federal, provincial and local water strategies. Public awareness of the importance and values of wetlands through communication and education will also be an important part of the process. Expenditures to sustain wetland health through establishment of well equipped long term monitoring stations will only be accepted if the benefits are perceived, understood and valued (Hurlbert, Corkall, and Diaz 2010).

13.9   What is currently being done?

The North American Wetlands Conservation Act (NAWCA) of 1989 provides matching grants to organizations and individuals who have developed partnerships to carry out wetland conservation projects in the United States, Canada, and Mexico for the benefit of wetlands-associated migratory birds and other wildlife. The Act was passed, in part, to support activities under the North American Waterfowl Management Plan, an international agreement that provides a strategy for the long-term protection of wetlands and associated upland habitats needed by waterfowl and other migratory birds in North America. In December 2002, the American Congress expanded its scope to include the conservation of all habitats and birds associated with wetland ecosystems. In 2006, Congress once again reauthorized the Act, which extended its appropriation authorization up to $75 million per year until 2012 (http://www.phjv.ca/programs.html, accessed July 11, 2012). On June 7, 2012, the Migratory Bird Conservation Commission approved more than $22.3 million for the North American Wetlands Conservation Act (NAWCA) in Canada as well as $2.2 million in small grants to researchers in the USA (http://www.fws.gov/birdhabitat/Grants/NAWCA/index.shtm, accessed July 22, 2012).

The Prairie Habitat Joint Venture (PHJV) is a partnership organization tasked to deliver the North American Waterfowl management Plan. The PHJV Plan goal is to sustain waterfowl populations at the levels of the 1970s via four basic habitat goals: stop further wetland loss; stop further loss of native uplands, especially native grass-lands; restore lost wetlands, especially small basins; and restore upland habitat function in landscapes conducive for maintenance of bird populations. The PHJV is currently targeting restoration of 112 600 hectares of wetlands across Alberta, Saskatchewan and Manitoba. Achieving this objective will not only provide more wildlife habitat but could sequester 122 076 350 mg CO_2 equivalents ha^{-1} or about 0.3% of the 2008 annual emissions from industry in these three provinces (Badiou *et al.* 2011).

13.10 Acknowledgements

N.H. Euliss, USGS, for comments regarding the "champagne effect" when wetlands are rewetted following drought.

References

Anteau, M. (2011) Do interactions of land use and climate affect productivity of waterbirds and prairie-pothole wetlands? *Wetlands*. doi: 10.1007/s13157-011-0206-3.

Badiou, P., McDougal, R., Pennock, D., and Clark, B. (2011) Greenhouse gas emissions and carbon sequestration potential in restored wetlands of the Canadian prairie pothole region. *Wetlands Ecol. Manage.*, **19**, 237–56.

Batt, B.D.J., Anderson, M.G., Anderson, C.D., and Caldwell, F.D. (1989) The use of prairie potholes by North American ducks, in *Northern Prairie Wetlands* (ed. A. van der Valk), Iowa State University Press, Ames IA, pp. 204–27.

Beaudoin, A.B. (2003) Climate and landscape of the last 2000 years in Alberta, in *Archaeology in Alberta: A View from the New Millennium* (eds. J.W. Brink and J.F. Dormaar), Archeology Society of Alberta, Medicine Hat.

Conly, F.M. and Van der Kamp, G. (2001) Monitoring the hydrology of Canadian prairie wetlands to detect the effects of climate change and land use changes. *J. Environ. Monit. Assess.*, **67**, 195–215.

Covich, A.P., Fritz, S.C., Lamb, P.J., *et al.* (1997) Potential effects of climate change on aquatic ecosystems of the Great Plains of North America. *Hydrol. Processes*, **11**, 993–1021.

Crooks, S., Herr, D., Tamelander, J., *et al.* (2011) *Mitigating Climate Change through Restoration and Management of Coastal Wetlands and Near-shore Marine Ecosystems: Challenges and Opportunities,* Environment Department Paper 121, World Bank, Washington, DC.

Donald, D.B., Aitken, W.T., Paquette, C., and Wulff, S.S. (2011) Winter snowfall determines the occupancy of northern prairie wetlands by tadpoles of the wood frog (*Lithobates sylvaticus*). *Can. J. Zool.*, **89**, 1063–73.

Donald, D.B., Syrgiannis, J., Hunter, F., and Weiss, G. (1999) Agricultural pesticides threaten the ecological integrity of northern prairie wetlands. *Sci. Total Environ.*, **231**, 173–81.

Fritz, S. (1996) Paleolimnological records of climate change in North America. *Limnol. Oceanogr.*, **41**, 882–9.

Gan, T.Y. (1998) Hydroclimatic trends and possible climatic warming in the Canadian Prairies. *Wat. Resour. Res.*, **34**, 3009–15.

Henderson, N. and Thorpe. J. (2010) Ecosystems and biodiversity, in *The New Normal* (eds. D. Sauchyn, H. Diaz and S. Kulshreshtha), Canadian Plains Research Press, Regina, pp. 80–116.

Hurlbert, M., Corkall, D.R., and Diaz, H.P. (2010) Government institutions and water policy, in *The New Normal* (eds. D. Sauchyn, H. Diaz and S. Kulshreshtha), Canadian Plains Research Press, Regina, pp. 284–90.

Johnson, W.C., Boettcher, S.E., Poiani, K.A., and Guntenspergen, G. (2004) Influence of weather extremes on the water levels of glaciated prairie wetlands. *Wetlands*, **24**, 385–98.

Johnson, W.C., Werner, B., Guntenspergen, G.R., *et al.* (2010) Prairie wetland complexes as landscape functional units in a changing climate. *BioScience*, **60**, 128–40.

LaBaugh, J.W., Winter, T.C., Swanson, G.A., *et al.* (1996) Changes in atmospheric circulation patterns affect mid-continent wetlands sensitive to climate. *Limnol. Oceanogr.*, **41**, 864–70.

Larson, D.L. (1994) Potential effects of anthropogenic greenhouse gases on avian habitats and populations in the northern Great Plains. *Am. Midl. Nat.*, **131**, 330–46.

Millett, B., Johnson, W.C., and Guntenspergen, G. (2009) Climate trends of the North American prairie pothole region 1906–2000. *Climatic Change*, **93**, 243–67

Mitsch, W.J. and Gosselink, J.G. (1993) *Wetlands*, Van Nostrand Reinhold, New York.

Murkin, H.R. (1989) The basis for food chains in prairie wetlands, in *Northern Prairie Wetlands* (ed. A. van der Valk), Iowa State University Press, Ames IA, pp. 316–38.

Pennock, D., Yates, T., Bedard-Haughn, A., *et al.* (2010) Landscape controls on N_2O and CH_4 emissions from freshwater mineral soil wetlands of the Canadian Prairie Pothole region. *Geoderma*, **55**, 308–19.

Poiani, K.A. and Johnson, W.C. (1991) Global warming and prairie wetlands. *BioScience*, **41**, 611–18.

Poiani, K.A., Johnson, W.C., Swanson, G.A., and Winter, T.C. (1996) Climate change and northern prairie wetlands: Simulations of long term dynamics. *Limnol. Oceanogr.*, **41**, 871–81.

Robarts, R.D. and Waiser, M.J. (1998) Effects of atmospheric change and agriculture on the biogeochemistry and microbial ecology of prairie wetlands. *Great Plains Res.*, **8**, 113–36.

Sauchyn, D. (2010) Prairie climate trends and variability, in *The New Normal* (eds. D. Sauchyn, H. Diaz and S. Kulshreshtha, Canadian Plains Research Press, Regina, pp. 32–40.

Schindler, D.W. and Donahue, W.F. (2006) An impending water crisis in Canada's western prairies. *Proc. National Acad. Sci. of US* http://www.pnas.org/cgi/doi/10.1073/pnas.0601568103 (accessed June 29, 2012).

Sorenson, L.G., Goldberg, R., Root, T.L., and Anderson, M.G. (1998) Potential effects of global warming on waterfowl populations in the Northern Great Plains. *Climatic Change*, **40**, 343–69.

Su, M., W. J. Stolte, and Van der Kamp, G. (2000) Modeling Canadian prairie wetland hydrology using a semi-distributed streamflow model. *Hydrol. Processes*, **14**, 2405–22.

Van der Kamp, G., and Hayashi, M. (1998) The groundwater recharge function of small wetlands in the semi-arid northern prairies. *Great Plains Research*, **8**, 39–56.

Voldseth, R.A., Johnson, W.C, Guntenspergen, G.R., *et al.* (2009) Adaptation of farming practices could buffer effects of climate change on northern prairie wetlands. *Wetlands*, **29**, 635–47.

Waiser, M.J. (2006) The relationship between hydrological characteristics and DOC concentrations in prairie wetlands using a conservative tracer approach. *J. Geophys. Res.*, **111**, 1–15

Walden, D., Van Dam, R., Finlayson, M., *et al.* (2004) A risk assessment of the tropical wetland weed *Mimosa pigra* in northern Australia. Supervising Scientist Report 177, Supervising Scientist, Darwin NT.

Winter, T.C., Rosenberry, D.O., Buso, D.C., *et al.* (2001) Water source to four US wetlands: Implications for wetland management. *Wetlands*, **21**, 462–73.

Woo, M.K. and Rowsell, R.D. (1993) Hydrology of a prairie slough. *J. Hydrol.*, **146**, 175–207.

14

Historic and Likely Future Impacts of Climate Change on Lake Tahoe, California-Nevada, USA

Robert Coats[1], Goloka Sahoo[2], John Riverson[3], Mariza Costa-Cabral[4], Michael Dettinger[5], Brent Wolfe[6], John Reuter[1], Geoffrey Schladow[2], and Charles R. Goldman[1]

[1]*Department of Environmental Science and Policy, University of California, Davis, USA*
[2]*Department of Civil and Environmental Engineering, University of California, Davis, USA*
[3]*Tetra Tech, Inc., Fairfax VA, USA*
[4]*Hydrology Futures LLC, Seattle WA, USA*
[5]*US Geological Survey and Scripps Institute of Oceanography, La Jolla CA, USA*
[6]*Northwest Hydraulic Consultants, Sacramento CA, USA*

14.1 Introduction and background

Lake Tahoe is a large ultra-oligotrophic lake lying at an elevation of 1898 m in the central Sierra Nevada on the California–Nevada border (Figure 14.1). The lake is renowned for its deep cobalt blue color and clarity. Due to concerns about progressive eutrophication and loss of clarity, the lake has been studied intensively since the mid-1960s, and has been the focus of major efforts to halt the trends in clarity and trophic status (Goldman 1981; TERC 2011). Previous work on the effects of climate change on the lake (Coats *et al.* 2006) showed (i) that the lake is warming at an average rate of about 0.013 °C year^{-1}; (ii) the warming trend in the lake is driven primarily by increasing air temperature, and secondarily by increased downward long-wave radiation; (iii) the warming trend on monthly and annual timescales is correlated with the Pacific Decadal Oscillation (PDO) and (to a lesser extent) with El Niño-Southern

Climatic Change and Global Warming of Inland Waters: Impacts and Mitigation for Ecosystems and Societies, First Edition. Edited by Charles R. Goldman, Michio Kumagai and Richard D. Robarts.
© 2013 John Wiley & Sons, Ltd. Published 2013 by John Wiley & Sons, Ltd.

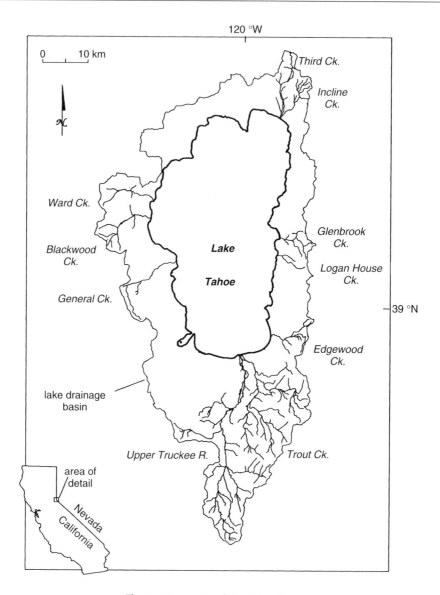

Figure 14.1 Map of the Tahoe Basin.

Oscillation (ENSO); and (iv) the warming of the lake is modifying its thermal structure, and increasing its resistance to deep mixing. Table 14.1 summarizes additional information about Lake Tahoe.

Warming trends and changes in thermal structure have been identified in lakes of Europe, Africa, and North America. Analysis of a 52-year record of monthly temperature profiles in Lake Zurich showed a secular temperature increase at all depths, resulting in a 20% increase in thermal stability. A temperature model of that lake showed that the warming trend in the lake is most likely explained by increasing night-time air temperature, concomitant with reduced night-time rates of latent and

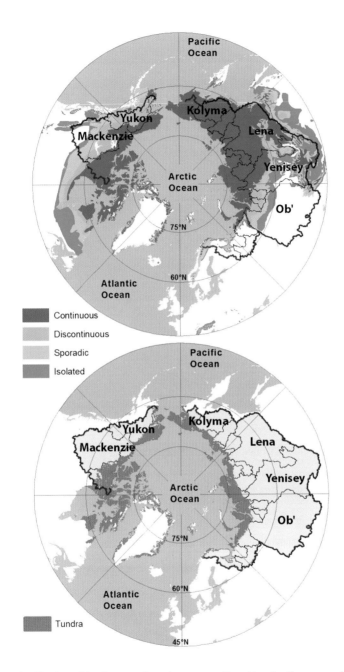

Plate 1.2 Distribution of Arctic permafrost (upper panel) and tundra (lower panel). Continuous permafrost indicates that >90% of the land surface is underlain by permafrost, discontinuous indicates 50–90%, sporadic indicates 10–50%, and isolate indicates <10% of the region contains permafrost.

Climatic Change and Global Warming of Inland Waters: Impacts and Mitigation for Ecosystems and Societies, First Edition. Edited by Charles R. Goldman, Michio Kumagai and Richard D. Robarts. © 2013 John Wiley & Sons, Ltd. Published 2013 by John Wiley & Sons, Ltd.

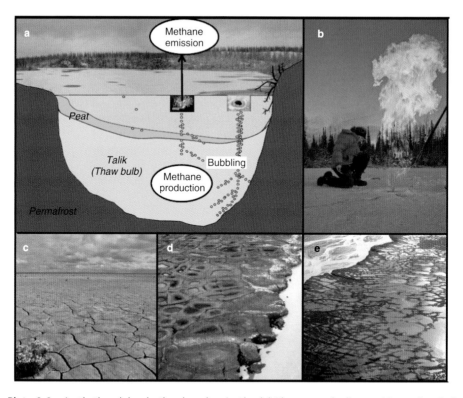

Plate 2.2 Arctic thaw lakes in the changing Arctic. (a) These waterbodies are biogeochemical hotspots on the tundra in which soil and lake organic matter is broken down by microbial activity in the thaw zone beneath the lake, resulting in the liberation of methane and carbon dioxide. Large quantities of these gases are released to the atmosphere via bubbling, which can produce and maintain holes in the ice. Modified from Walter *et al.* (2007). (b) The methane can accumulate as gas pockets beneath the ice, such as here in an Alaska lake where the gas has been vented through a hole made in the ice and then ignited. Photocredit: Todd Paris, November 2009; from Walter Anthony *et al.* (2010). Reproduced with permission. (c) In parts of the Arctic, thaw lakes are expanding in number and size, while in other areas, such as here in the Nettilling Lake region of Baffin Island, landscape erosion has resulted in complete drainage of some waterbodies. Photocredit: Reinhard Pienitz, August 2010. (d) Long-term as well as interannual variations in climate strongly affect the water balance and persistence of lakes on the permafrost. Many of these polygon ponds on Bylot Island, Canada evaporated to dryness in a warm, low precipitation year. Photocredit: Isabelle Laurion, July 2007. (e) The Bylot Island polygon ponds were numerous and extensive during a preceding cool, wet year. Photocredit: Isabelle Laurion, July 2005.

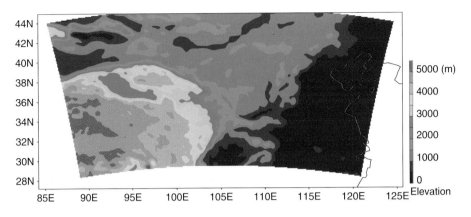

Plate 4.3 Domain of the WRF model for the Yellow River basin.

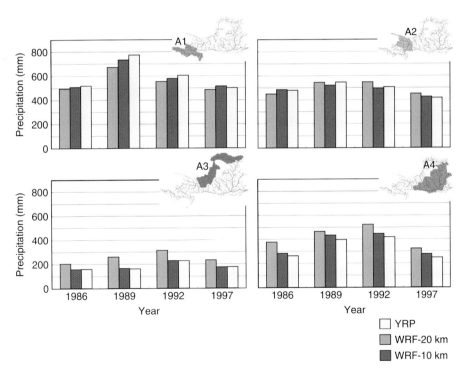

Plate 4.4 Sensitivity experiment for different horizontal resolutions (10 and 20 km), compared to observations (YRP).

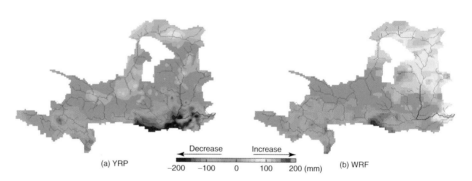

Plate 4.6 Comparison of the decadal annual precipitation difference between the 1990s and 1980s between (a) the YRP observed dataset and (b) the WRF simulation.

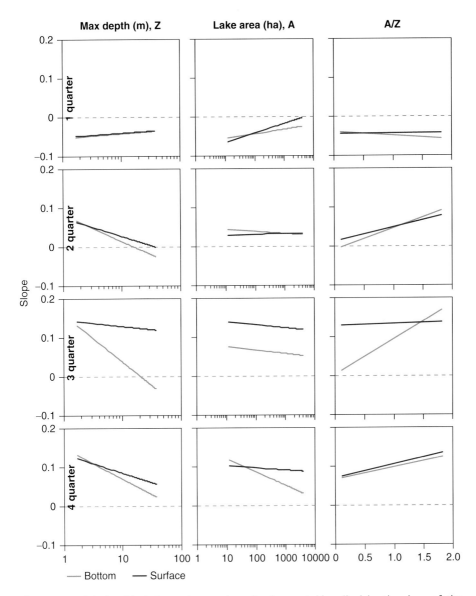

Plate 10.4 Relationship between temperature development (described by the slope of the regression line) and depth (log-transformed), area (log-transformed), and the stability of the water column ($A^{0.25}/Z_{max}$) in surface and bottom water (two colors) based on data from 20 Danish lakes covering the period 1989–2006. Positive values indicate increasing temperatures and negative values declining temperatures. The closer the value is to zero, the more negligible the temperature development. Water column instability increases with increasing A/Z values.

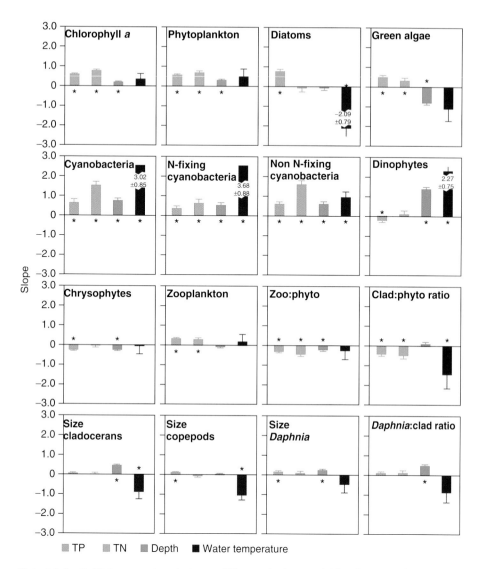

Plate 10.6 Multiple regressions between different plankton variables (log$_e$-transformed) and total phosphorus, total nitrogen and water temperature of the surface layer and lake mean depth—all log$_e$-transformed. When the value is positive, there is a positive effect of a given variable (when including also the other variables) and the opposite, if negative. All data are from August. Significant values are marked with an *. Modified from Jeppesen *et al.* (2009).

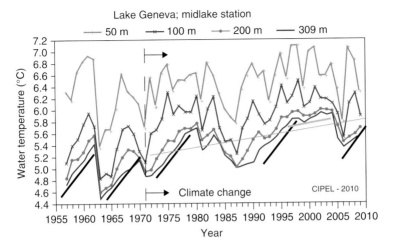

Plate 12.6 Annual mean values of the temperature at five selected depths in the deep-water layers of Lake Geneva. Data were collected at the center of the lake. All red lines have the same slope and indicate the rising leg of multiannual cycles at 309 m. The heavy green line is the mean slope for the years 1997 to 2003. The blue straight line indicates the mean long-term trend 1970 to 2010. Diagram modified after CIPEL (2010); reproduced with the permission of CIPEL.

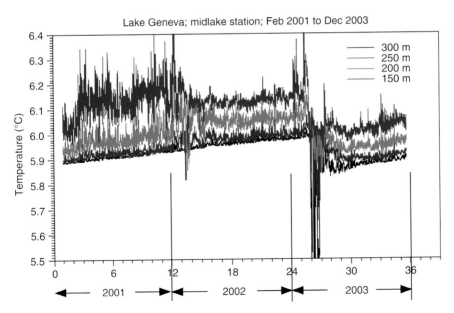

Plate 12.11 Time series of temperatures recorded at different depths in the deep layers in the center of Lake Geneva. Recording interval is 10 min.

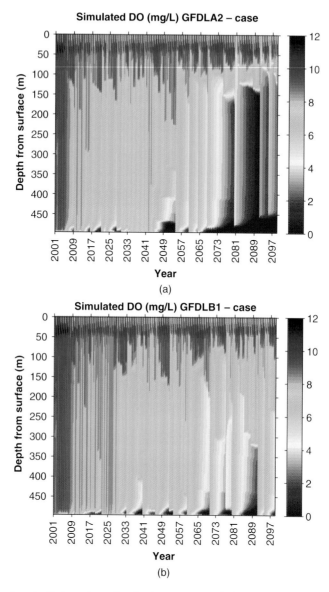

Plate 14.14 Twenty-first century dissolved oxygen trends in Lake Tahoe for two emission scenarios.

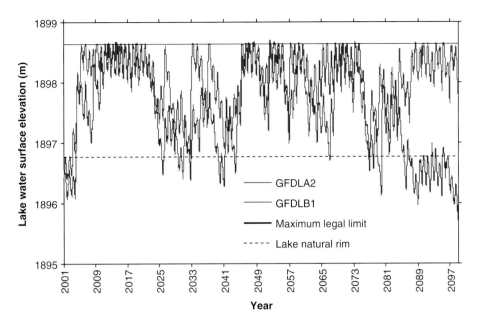

Plate 14.16 Modeled elevation of Lake Tahoe for two emission scenarios. When the lake level falls below the rim elevation, outflow ceases.

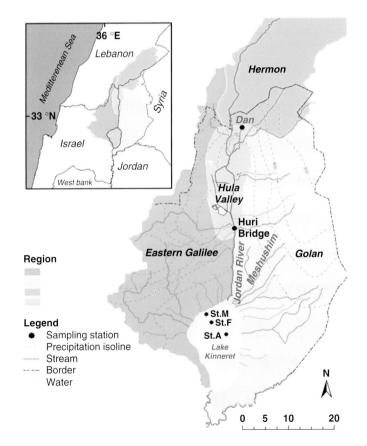

Plate 16.1 Lake Kinneret and it's watershed in northern Israel. Filled circles show locations of sampling stations. Modified from the original design by http://www.stav-gis.com/index_eng.htm (accessed July 11, 2012).

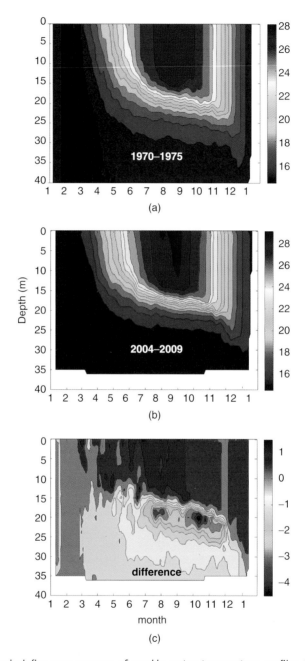

Plate 16.3 Typical five-year average of weekly water temperature profiles for the periods (a) 1970–1975 and (b) 2004–2009 at Stn. A (Figure 16.1); (c) Temperature differences throughout the water column are computed as the difference between (b) and (a). Monitoring and measurements of water-profile thermal structure were described by Rimmer *et al.* (2011a).

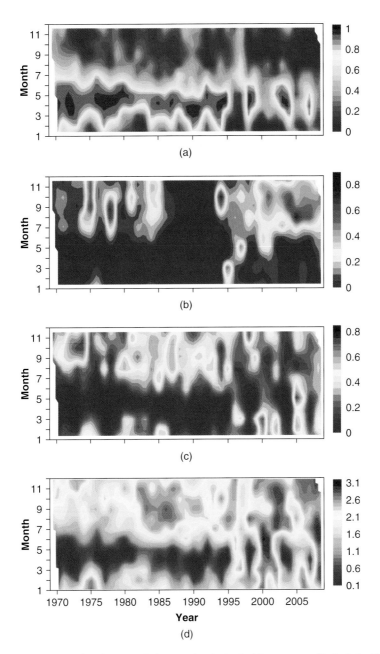

Plate 16.4 Interannual and seasonal changes in phytoplankton community in Lake Kinneret. (a–c) Relative biomass of three major groups of phytoplankton (dimensionless units). The relative biomass of each group is calculated as its proportion of the total biomass. (a) Dinoflagellates. (b) Cyanobacteria. (c) Chlorophytes. (d) Shannon diversity index. Data show recurrent spring-summer blooms of dinoflagellates (mainly *Peridinium gatunense*) until the mid-1990s. Since 1996 the bloom failed to develop in most years and the diversity index in spring and summer has increased. Intensive recurrent blooms of cyanobacteria commenced in the mid-1990s in summer and fall. Monitoring and determination of phytoplankton biomass was carried out at St. A and are described elsewhere (Zohary 2004; Zohary and Ostrovsky 2011).

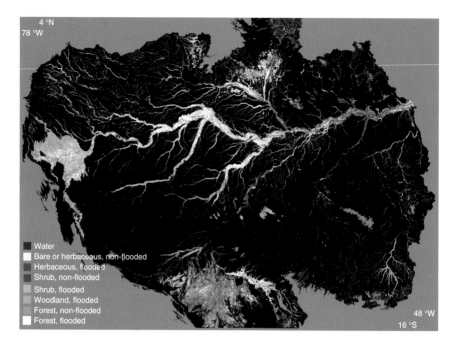

Plate 17.1 Wetland classification of lowland Amazon basin (area less than 500 m a.s.l) for high-water period in the central basin based on analysis of synthetic aperture radar data obtain from the JERS-1 satellite using algorithms described in Hess *et al.* (2003) and validated as described in Hess *et al.* (2002).

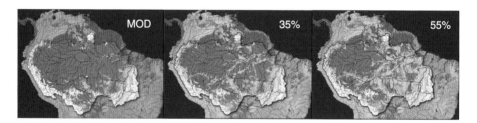

Plate 17.3 Vegetative cover of the Amazon basin. Forest (green) savanna (beige) and agriculture (orange) vegetation types are shown for the 6.7 million km^2 Amazon basin for the year 2000 (left) as defined by Eva *et al.* (2004), a scenario with 35% of the basin deforested (middle) and a scenario with 55% deforested (right). Scenarios of deforestation are from Soares-Filho *et al.* (2006).

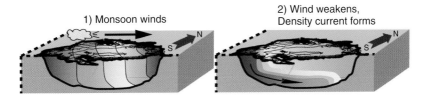

Plate 18.8 Schematic of north to south temperature differences created in Lake Victoria during the southeast monsoon. On relaxation of the winds, either gravitational readjustment leads to a gravity current that contributes to stratification to the north, or, following Song (2000), the gyre structure in the south is relaxed allowing northward flow of cool water. These currents appear to form when cooling is greater to the south during the southeast monsoon. Lower winds and higher relative humidity and a reduction in the cross-basin net heat losses, as occurred around 1980, would reduce the north–south temperature contrasts, suppress the formation of such density currents, and decrease the overall stratification allowing a larger volume of anoxic water.

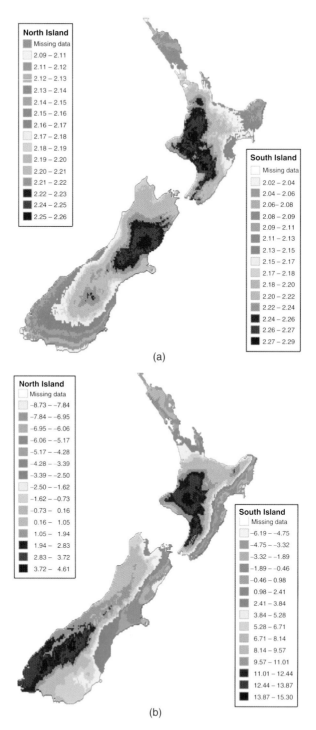

Plate 19.2 Change in (a) temperature (°C) and (b) percentage precipitation in 2091–2099 compared with baseline of 1965–1995, based on ensemble output and statistical downscaling from 12 GCM models.

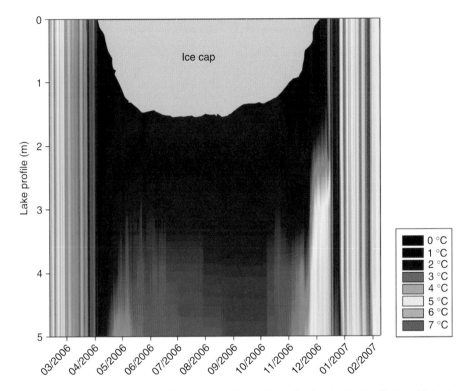

Plate 20.1 Thermal structure of Limnopolar Lake, Byers Peninsula, South Shetland Islands, from February 2006 to February 2007. Data were gathered at <1 m space resolution and at 30 min time resolution by means of a number of Hobo temperature loggers (Onset) hanging from a buoy.

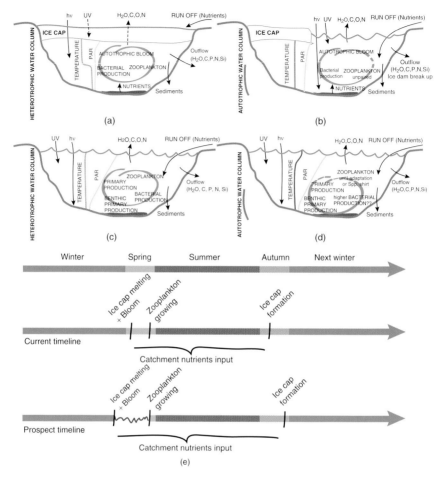

Plate 20.5 Schematic process diagram for a proposed maritime Antarctic lake's model. Mass flows: Grey arrows indicate terrestrial flux exchange of water and nutrients, the green loop represents the relationship between functional groups, the black arrows represent the flux exchange of chemical constituents of the lake with the atmosphere and benthos. (a) Descriptions of current relationships during early spring. (b) Scenario proposed for early spring under a global change scenario. (c) Current growing season scenario. (d) Scenario proposed for a growing season using current global change prognoses. (e) Timeline of seasonal shifts, where the main points of lake dynamics are addressed and the proposal for maritime Antarctic lakes in Antarctica. The time lapse that will cause a mismatch between a primary producer's bloom and zooplankton growth is marked here.

Table 14.1   Basic facts about Lake Tahoe.

Volume	156 km^3
Surface area	498 km^2
Average depth	313 m
Maximum depth	505 m
Depth of photic zone	To ca. 100 m
Basin area	1310 km^2
Maximum lake elevation	1899.54 m NAVD
Mean hydraulic residence time	ca. 650 years
Mean annual lake discharge	2 × 10^8 m^3
Mixing frequency below 450 m	ca. 1 year in 4
Rate of decline in Secchi depth	ca. 25 cm year^{-1}
Limiting nutrient	N until early 1980s; now P

sensible heat loss from the lake surface (Peeters *et al*. 2002; Livingstone 2003). Warming of Lake Tanganyika between 1913 and 2000, associated with increasing air temperatures, has increased the vertical density gradient and thus decreased both the depth of oxygen penetration and the nutrient supply in the upper mixed layer (Verburg, Hecky, and Kling 2003).

In North America, modeling and statistical analysis at Lake Mendota, Wisconsin, showed that increasing air temperature is related to higher epilimnion temperatures, earlier and more persistent thermal stratification, and decreasing thermocline depths in late summer and fall (Robertson and Ragotzkie 1990). Twenty years of temperature records from the Experimental Lakes Area in Northwestern Ontario showed an increase in both air and lake temperatures of 2 °C, and an increase in length of the ice-free season by three weeks (Schindler *et al*. 1996). Modeling studies of the effects of climate warming on the temperature regime of boreal and temperate-zone lakes are consistent with these observations, predicting increased summer stratification, greater temperature increases in the epilimnion than in the hypolimnion, and increased length of the ice-free season (Elo *et al*. 1998; Stefan, Fang, and Hondzo, 1998). From 1979 to 2006, Lake Superior warmed at an average rate of 0.11 °C year^{-1}, about twice the rate of regional air temperature, due in part to a positive feedback from reduced ice cover and reduced albedo (Austin and Colman 2007). At Lake Michigan, application of a one-dimensional temperature model showed that climate warming will decrease the summer thermocline depth, increase resistance to mixing, and could lead to permanent stratification (McCormick and Fahnenstiel 1999). Closer to Lake Tahoe, Arhonditsis *et al*. (2004) found a warming trend in Lake Washington, and showed that the relationship between lake temperature and climate may be different during lake cooling than during lake warming, though the ENSO and PDO indices were positively correlated with lake temperature during both the warming and cooling phases.

Recently, the trends in night-time summer surface temperatures of six lakes in northern California and Nevada—Tahoe, Almanor, Clear, Pyramid, Walker and Mono—were measured using ATSR and MODIS satellite sensor data (Schneider *et al*. 2009). The upward temperature trend (1992–2008) averaged 0.11 °C year^{-1}, about twice the trend in surface air temperatures, though none of the lakes freeze in winter. The trend of temperature rise in the surface waters of Lake Tahoe was 0.13 °C year^{-1} (1970–2007), ten times the rate for the entire lake volume.

14.2   Historic climate trends in the Tahoe Basin

Previous studies of the historic trends in hydroclimatology in the basin and nearby locations (1910–2007) indicated strong upward trends in air temperature (especially night-time temperature), a shift from snowfall to rain, a shift in snowmelt timing to earlier dates, increased rainfall intensity, and increased interannual variability (Coats 2010). The strongest warming trends for the period 1956–2005 were in T_{min} at Tahoe City (on the northwest side of the Lake), especially in summer months. At Truckee and Boca ($\sim$20–25 km north of Tahoe City and outside the Tahoe basin), both T_{max} and T_{min} also trended upward in winter and spring.

As might be expected, the trend in air temperature in the Tahoe basin is reflected in both the form of precipitation, and the timing of snowmelt. Figure 14.2 shows the fraction of annual precipitation at Tahoe City that fell as snow (that is, precipitation on days with average temperature $<0\,°C$) from 2010 to 2007. Since total annual snowfall affects the timing of snowmelt peak discharge, the timing of the latter was examined by testing the slope of the time trend in residuals after removal of the effect of total annual snowfall. On average, the timing of the spring snowmelt peak discharge (1961–2005) for five monitored streams in the Basin has shifted toward earlier dates at a rate of about 0.4 days year^{-1}. No such trend was observed for the streams outside of the basin. This was consistent with the observation of Johnson, Dozier, and Michaelsen (1999) from 30 years of snow survey data from 260 snow survey courses in the Sierra Nevada that the Tahoe basin experienced the highest loss—54%—in May snow-water equivalent (SWE) of any of the 21 river basins studied.

Good temperature profiles for the lake are available from late 1969 to the present. These records allow us to examine the time trends in temperature at a given depth, and in the volume-averaged temperature for the entire lake (Coats *et al.* 2006). Figure 14.3 shows the temperature at 400 m depth for the period 1969–2007. The great volume of

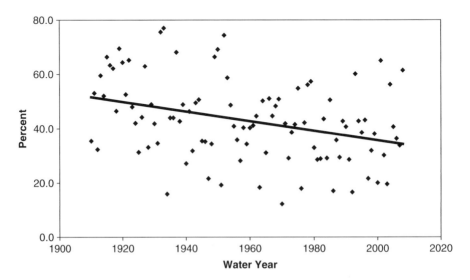

Figure 14.2 Percentage of annual precipitation falling as snow (defined as precipitation on days with mean temperature $<0\,°C$).

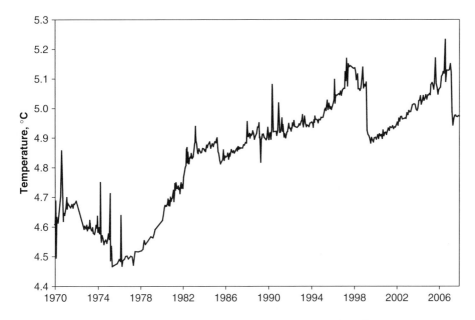

Figure 14.3 Temperature measured at 400 m in Lake Tahoe.

the lake filters out the short-term noise and reveals the long-term trend in temperature. Note the saw-tooth pattern, which is typical of warming temperate-region lakes (Livingstone 1997). During periods without deep mixing, heat is gradually transported downward by eddy diffusion. Then when the lake mixes to the bottom and becomes isothermal (typically during the late winter), the temperature drops sharply.

Figure 14.4 shows the trend in the deseasonalized volume-averaged lake temperature, 1970–2007. An interesting feature of this plot is the anomalous drop in temperature in 1982–1983. This was a period of a strong ENSO event, which is typically associated with warmer conditions. The short cooling trend was likely related to the March 1982 eruption of El Chichón in Chiapas, Mexico, which is known to have cooled air temperatures world-wide (Kerr 1982). The Seasonal Kendall Test (Salmi *et al*. 2002; Helsel and Frans 2006) for trend in monthly average lake temperature found a tau correlation coefficient of 0.54 and a Sen's slope of $0.013\,°C\ year^{-1}$, with $P < 5 \times 10^{-5}$.

The slope of the trend in volume-averaged temperature of Lake Tahoe is highest in September-November, and least in January-February, suggesting that increased summer warming plays a more important role that suppressed winter cooling (Coats *et al*. 2006). This contrasts with Lake Geneva, where Lemmin and Amouroux (Chapter 12) found that reduced winter cooling in the eplimnion and metalimnion played a larger role in the upward temperature trend than summer warming.

14.3 Modeling the impacts of future climate change

Understanding the historic and likely future conditions of Lake Tahoe's water quality and famed optical transparency requires consideration of the input of water, nutrients,

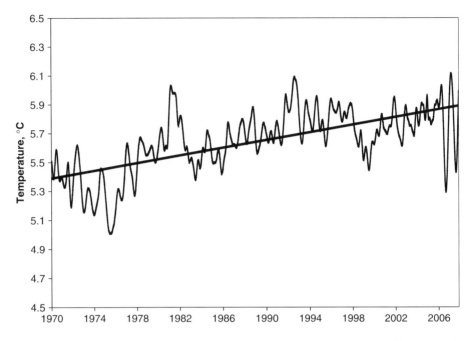

Figure 14.4 Volume-averaged daily temperature of Lake Tahoe, with seasonal effect removed.

sediment and energy from the lake's watershed and from the atmosphere (Jassby, Reuter and Goldman 2003; Reuter *et al*. 2003; Sahoo, Schladow and Reuter 2010). With continued change toward a warmer climate (IPCC 2007; Hansen *et al*. 2008), both research scientists and resource managers in the Tahoe basin would like to know: (i) How fast will the air temperature in the basin rise? (ii) How will the form, timing and annual amount of precipitation change? (iii) How will the changes in temperature and precipitation affect drought conditions, which in turn will affect vegetation, fire frequency and soil erosion? (iv) How will changes in precipitation affect streamflow regimes, especially high- and low-flow frequency-magnitude relationships? (v) How will continued warming of the lake affect its thermal stability, biogeochemical cycling and primary productivity?

Answering these questions is a daunting task, since the General Circulation Models (GCMs) used in global change research typically produce output at a 2.5° grid scale, too coarse to resolve the topographic complexity that drives climate in mountainous regions. Furthermore, to address extreme hydrologic events (precipitation and floods), an hourly time scale is needed, whereas GCM outputs are typically at daily or longer time scales.

To address these challenges, we used the twenty-first century downscaled daily temperature and precipitation (Hidalgo, Dettinger and Cayan, 2008; Dettinger, in press) from two GCMs (the Geophysical Fluid Dynamics Laboratory, or GFDL CM2.1 (Delworth *et al*. 2006) and the Parallel Climate Model, or PCM1 (Washington *et al*. 2000)) and two emissions scenarios (A2 and B1; Nakicenovic *et al*. 2000). For the GFDL, we also had wind (bias corrected), radiation and relative humidity data. The precipitation and wind data were corrected for bias by quantile mapping (Wood *et al*. 2001, 2002), to ensure that the empirical probability distribution functions matched the real station

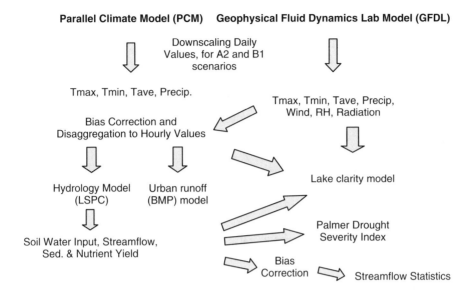

Figure 14.5 Information flowchart for project to model future impacts of climate change in the Tahoe basin.

data. The daily data were also adjusted using meteorological data for 12 SNOTEL stations (see http://www.wcc.nrcs.usda.gov/snotel/California/california.html, accessed July 11, 2012) in the Basin, and then disaggregated statistically to hourly values and used to drive a distributed hydrologic/water quality model—the Load Simulation Program in C++, or LSPC—for streams in the Tahoe basin (Riverson, *et al.* in press). The output from that model, along with the meteorological data, was used as input to a one-dimensional hydrodynamic/clarity model of the lake (Sahoo, Schladow and Reuter, 2010; Sahoo *et al.*, in press), and (with additional bias correction) to calculate stream-flow statistics for the Upper Truckee River (UTR) including flood frequency and runoff timing. It was also used to calculate soil water input as the sum of rainfall and snowmelt, for calculation of the weekly Palmer Drought Severity Index (PDSI) for selected sites (Palmer 1965; Wells, Goddard and Hayes 2004). The hydrology data were used in a stormwater routing and water-quality model to investigate the implications of climate change for design of Best Management Practices (BMPs) in the Tahoe basin (Wolfe 2010). Figure 14.5 shows the information flow in the study design (see Costa-Cabral *et al.* in press).

14.4 Impacts on the watershed

14.4.1 Precipitation and drought

Climate modeling studies of likely future precipitation in northern California reach no consensus on the direction of trends, possibly because that region is near the transition between the mid- to high-latitude zone of increasing precipitation and a band of drying conditions over the subtropical north Pacific and Mexico (Dettinger *et al.* 2004). Figure 14.6 shows the projected trends in total annual precipitation averaged over the

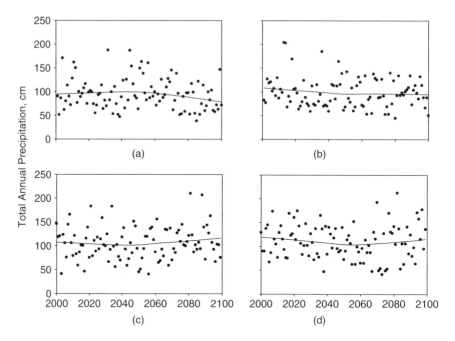

Figure 14.6   Total annual precipitation averaged over the Tahoe basin, for two models and two emissions scenarios. Curves are from a LOWESS smoothing (a) GFDL A2, (b) GFDL B1, (c) PCM A2 and (d) PCM B1.

Tahoe basin, from the two GCM models for each emissions scenario. For the GFDL A2 case, the downward trends in annual precipitation were significant (P < 0.01) for the months of October, December, April, May and June as well as for the annual total. For the GFDL B1 case, the downward trends were significant for January and December, as well as for the annual total. As shown in Figure 14.6, precipitation was projected (for the GFDL A2 case) to drop most sharply toward the end of the century. The 20-year running average for that case reached a maximum of 114 cm year^{-1} in 2060, and then dropped to a minimum of 79 cm year^{-1} in 2098. The PCM results, however, did not show major changes in future precipitation in the basin (see Coats *et al.* in press for more detail).

Changes in the form of precipitation, however, may be more important than changes in total amount. Figure 14.7 shows the expected trend in percent of precipitation falling as snow, averaged over the Basin, for the most contrasting scenario-model combinations. The loss of snowpack will have an effect on available soil moisture that is greater than the effect of precipitation change alone. Figure 14.8 shows the projected trends in the PDSI (expressed as the annual driest week), for the GFDL results at sites on the west (Tahoe City) and east (Glenbrook) sides of the basin. Note the sharp downturn for Glenbrook, where the annual snowpack is already thin and ephemeral in some years. At that site, the percent of annual soil water input as snowmelt explained 26% of the variance in the modeled maximum annual drought severity (minimum annual weekly PDSI value), with P < 10^{-7}.

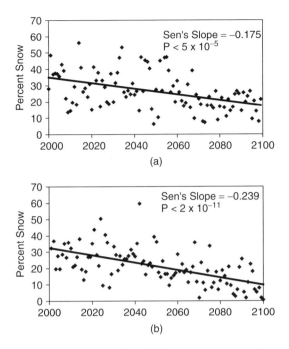

Figure 14.7   Percent of annual precipitation falling as snow averaged over the Tahoe basin. The B1 emissions scenario with the PCM (a) represents the least shift from snow to rain of the four cases modeled, and the A2 scenario with the GFDL model (b) represents the greatest shift.

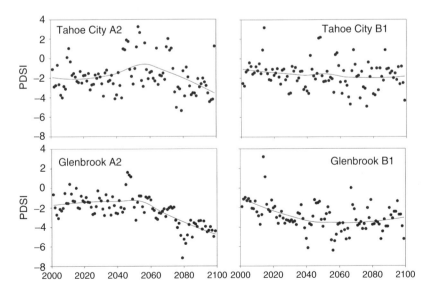

Figure 14.8   Trends in the modeled annual minimum weekly Palmer Drought Severity Index, for two locations (east and west sides of the Tahoe basin) and two emissions scenarios. Soil moisture input is calculated as daily rain plus snowmelt using a distributed hydrologic model for the basin.

14.4.2 Runoff

A number of studies have shown how climate change is shifting the timing of stream runoff in the western US, toward earlier dates, both historically (Lettenmaier and Gan 1990; Aguado *et al*. 1992; Wahl 1992; Pupacko 1993; Brown 2000; Stewart, Cayan and Dettinger 2005; Coats 2010) and in modeled future scenarios of climate change (Dettinger and Cayan 1995; Dettinger *et al*. 2004; Maurer 2007). Figure 14.9 shows the projected trends in the timing of the annual hydrograph centroid (also called the Center timing, or CT), for the Upper Truckee River, the lake's largest tributary. This measure takes account of the changes in precipitation form and amount throughout the water year, whereas the historic shift in snowmelt timing is more sensitive to warming trends in the spring. For the last third of the twenty-third century, the projected flow duration curve for the Upper Truckee shows a distinct downward shift for the A2 scenario (consistent with the decreasing precipitation and increasing drought), but not for the B1 scenario.

   The projected changes in runoff timing point toward reduced summer low-flow in basin streams, at least for the A2 scenario, but the statistical analysis does not show the full effects. Most of the streams in the Tahoe basin streams flow through coarse alluvium in their lower reaches, so even a small drop in summer stream flow can cause a shift from surface to subsurface flow, and thus destroy aquatic habitat.

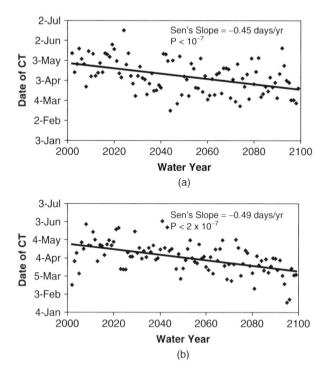

Figure 14.9 Modeled trends in the date of Center Timing of annual runoff, Upper Truckee River, for A1 and B2 scenarios, by GFDL (a) GFDL B1 and (b) GFDL A2.

14.4.3 Floods

Although climate change in the Tahoe basin will have a chronic effect on stream low-flow, the most dramatic effects may be on the flood frequency-magnitude relationship. Our future flood frequency curves created from downscaled GFDL/LSPC output were adjusted downward, using an equation that mapped the modeled GFDL flood frequency curves for the 1972–1999 period onto the actual measured curves for the same period. Figure 14.10 shows the modeled changes in flood frequency for the Upper Truckee River for three thirds of the twenty-first century compared with the historic curves from USGS data (USGS 1981). Note that the greatest increases for both the A2 and B1 scenarios are for the middle third of the century, perhaps because precipitation (for the A2 case) declines in the latter part of the century, and a reduced snowpack reduces the likelihood of a large contribution of snowmelt during a rainstorm. For the B1 scenario, the 100-year expected flood for the middle third of this century is projected to be 2.5 times the currently accepted value. These changes in flood frequency have major implications for stream morphology and habitat in the Tahoe basin, as well as for infrastructure, such as roads and bridges, and they support the assertion of Milly *et al.* (2008) that "stationarity is dead."

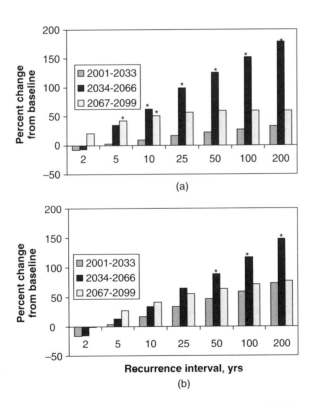

Figure 14.10 Percentage change in the modeled and adjusted GFDL/LSPC B1 and A2 flood frequency curves from the gauge record (1972–2008) (a) GFDL B1 and (b) GFDL A2.

14.4.4 Lotic ecology

Changes in stream flow regimes, especially an earlier and shorter snowmelt recession period, may have major effects on stream ecology. Possible effects (Yarnell, Viers and Mount, 2010) include: (i) a longer duration of the warm low-flow season; (ii) reduced suitable habitat and recruitment success for riparian woody vegetation; (iii) reduced arthropod diversity and changes in the macroinvertebrate community; (iv) flashier winter floods, with concomitant changes in sediment sorting and routing; and (v) greater habitat homogeneity and lower biodiversity.

14.5 Impact on the lake

Climate change and increased water temperature can have numerous effects on lake processes and biological communities. For example, Lake Tahoe has experienced a rapid increase in aquatic invasive species over the past decade, including bluegill sunfish (*Lepomis macrochirus*), largemouth bass (*Micropterus salmoides*), Eurasian water milfoil (*Myriophyllum spicatum*), curly leaf pondweed (*Potamogeton crispus*) and the Asian clam (*Corbicula fluminea*) (Kamerath, Chandra and Allen, 2008; Ngai 2008; Wittmann *et al*. 2010; TERC 2011; see also Chapter 15). These species inhabit the near shore region that is particularly vulnerable to temperature increase due to its shallow bathymetry.

The most important impacts of climate change on the lake, however, will be the direct result of lake warming on lake mixing, and in turn on internal nutrient loading and the lake's ability to support life in the hypolimnion. Figure 14.11 shows the modeled trends in volume-averaged, whole-lake temperature over the twenty-first century. For both scenarios, the warming trend was highly significant ($P < 10^{-6}$, by the Mann-Kendall test) but it did not really begin until well into the century (2039 for the A2 case and 2028 for the B1 case). In fact, the lake temperature under the A2 scenario dropped slightly ($P < 6 \times 10^{-4}$) from 2001 to 2039 before beginning a long upward trend.

The warming trend, especially under the A2 emissions scenario, will increase the lake's thermal stability and resistance to mixing, for two reasons. First, during summer, the lake warms from the surface downward, which increases the slope of the vertical temperature gradient; and second, the decrease in density of water with increasing temperature is highly non-linear. The work required to mix layered water masses at 24 and 25 °C is 30 times that required to mix the same masses at 4 and 5 °C (Wetzel 2001). The lake modeling results show that the lake's increased thermal stability will increase during the twenty-first century, suppressing complete lake turnover for prolonged periods of time (Sahoo *et al*. in press).

The Schmidt Stability (S) is a measure of the work required to mix a thermally stratified lake to an isothermal state without loss or gain of heat. It can be calculated from the lake's vertical density profile and bathymetry (Idso 1973). Figure 14.12 shows the lake's modeled minimum annual Schmidt Stability S (in $kJ\,m^{-2}$). As with the temperature trends, the upward trends in annual minimum S for both scenarios were highly significant ($P < 0.002$), but were strongest over the last 60–70 years of the twenty-first century. Note in Figure 14.12 that the Schmidt Stability showed some persistence. If the value of S in a given year remains above $2.5\,kJ\,m^2$, then the following year it never reaches 0. This persistence could be the result of the lake's thermal inertia, persistence in the modeled climate system, or both.

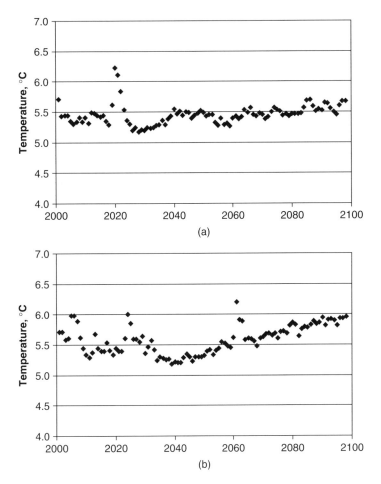

Figure 14.11 Modeled time trends of volume-averaged lake temperature (a) GFDL B1 and (b) GFDL A2.

Figure 14.13 shows the maximum mixing depth for the twenty-first century. Under the A2 scenario, the lake. mixed to the bottom only three times between 2069 and 2100. Under the B1, deep mixing occurred less frequently during the second half of the century, but did not cease altogether in this century.

The late spring mixing, however, was determined not just by the lake's thermal structure, but also by wind. The modeled GFDL A2 wind trends downward over the twenty-first century at an average of $-0.067 +/- 0.045 \, cm \, sec^{-1} \, year^{-1}$, according to the Mann–Kendall trend test. Sensitivity tests with the lake model showed that an increase of 10–15% in the modeled average daily wind speed over the twenty-first century would be sufficient to compensate for the effects of increasing temperature on thermal stability, in spite of the downward trend (Sahoo *et al.* in press). An evaluation of the modeled wind trends by month indicated increases in summer winds (which will tend to deepen the epilimnion) and decreases in the late fall and winter winds.

The current frequency of deep mixing—about one year in four—was sufficient to keep the lake well ventilated, and maintain oxidized conditions in the surficial

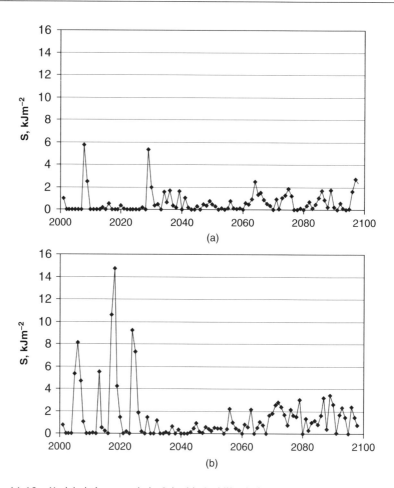

Figure 14.12   Modeled time trends in Schmidt Stability index (a) GFDL B1 and (b) GFDL A2.

bottom sediments. The loss of deep mixing would shut down the deep water ventilation, causing intense and prolonged periods of anoxia (Figure 14.14). This in turn
will trigger large releases of soluble reactive phosphorus (SRP; Figure 14.15) and
ammonium-nitrogen from the sediment into the bottom waters.

Beutel (2000) measured the release of available N and P from Lake Tahoe sediments,
and his experiments provide a basis for calculating internal nutrient loading under
anoxic conditions. This internal loading can then be compared to current estimates
of external loading, which comprise chiefly watershed runoff and direct atmospheric
deposition (LWRQCB and NDEP 2010). According to these calculations, by the end
of the century under the A2 scenario, the internal load of SRP will be about twice
the present external load, and the internal load of dissolved inorganic nitrogen (DIN)
will be about one-third of the present external load.

The rate at which the released nutrients will be transported into the photic zone is
currently unknown because active deep-mixing will cease as a result of the thermal
stability in the water column. Initially the released nutrients will be trapped below the
photic zone by density stratification, but eventually may be entrained into the upper

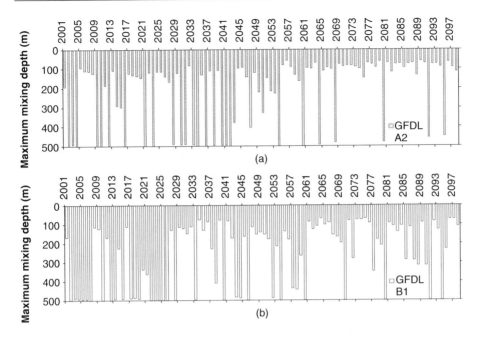

Figure 14.13 Maximum mixing depth of Lake Tahoe for two emission scenarios.

waters by vertical eddy diffusion. The net result would be a rapid increase in the lake's primary productivity and possibly a shift back to nitrogen-limitation because of the increased internal P-loading. These changes would most likely be irreversible, since the increased algal growth would increase the organic matter deposition in deep water, adding to the BOD of bottom waters and initiating, through a positive feedback mechanism, a "death-spiral of anoxia."

The development of anoxia and release of phosphorus from the lake sediments would trigger a cascade of microbial and biogeochemical changes (see also Chapter 6), though these are hard to predict. For example, denitrification in the sediments and water column could reduce available nitrogen, and hasten the shift back to nitrogen limitation. Blooms of cyanobacteria, however, could provide a new source of available nitrogen. While the pathways and rates of biogeochemical changes are uncertain, it is likely that Lake Tahoe will lose its unique ultra-oligotrophic status.

Release of methyl mercury from lake sediments and its incorporation into the food chain is another possible effect of anoxia (Watras 2010). In a study of heavy metal concentrations in the sediment of Lake Tahoe, Heyvaert *et al*. (2000) found that the Hg concentration in modern (post-1980) lake sediments was five times the pre-1850 concentration, and that the modern flux (mostly from direct atmospheric deposition) was 28 times the pre-industrial flux. Drevnick *et al*. (2009) found a lower modern to preindustrial flux ratio of 7.5 to ten, which was still relatively high compared with lakes world-wide. The concentrations of mercury in the tissue of lake trout (*Salvelinus namaycush*) and crayfish (*Pacifastacus leniusculus*) from Lake Tahoe showed a positive relationship between mercury concentration and weight, suggesting bioaccumulation. Consumption of one serving from the largest (10 kg) fish sampled (with a total Hg muscle tissue concentration of $0.24 \, \mu g \, kg^{-1}$) could easily exceed EPA's recommended

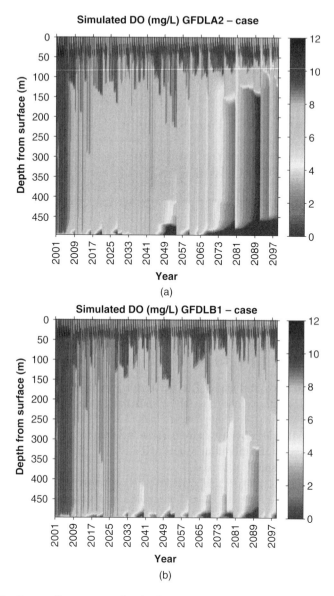

Figure 14.14   Twenty-first century dissolved oxygen trends in Lake Tahoe for two emission scenarios. (See insert for color representation.)

upper limit for daily intake of methylmercury (http://www.epa.gov/iris/subst/0073.htm, accessed July 11, 2012).

The change in the lake's thermal structure will affect the input of fine sediment and nutrients directly to the photic zone by modifying the "insertion depth" of streams flowing into the lake. This depth is determined largely by the relative densities of the inflowing streams and the receiving waters. Under current conditions, most of the inflow from January to March plunges into deep water below the photic zone, and has little immediate impact on primary productivity and Secchi depth. Future stream

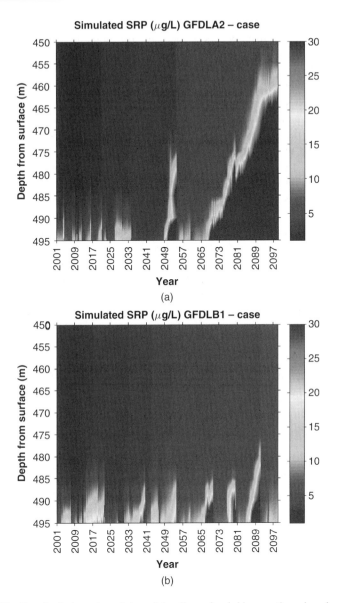

Figure 14.15 Twenty-first century internal loading of soluble reactive phosphorus in Lake Tahoe for two emission scenarios. The modeled pattern on NH_4-N release is the same as that of SRP, but the scale runs from $0-50 \, \mu g \, L^{-1}$ instead of $0-30 \, \mu g \, L^{-1}$.

temperatures were modeled from air temperature and solar radiation. Comparison of the modeled stream temperature with lake temperature indicated that with the A2 scenario, the insertion depth of the Upper Truckee river will remain below 200 m, and out of the photic zone during the latter quarter of this century, whereas under the B1 scenario, the river discharge will more frequently deliver its sediment and nutrient loads directly to the photic zone (Sahoo *et al.* in press).

The suppression of deep mixing and consequent anoxia will have profound biological effects, making the deep waters of the lake uninhabitable for salmonids and many invertebrates. More subtle biological changes, however, are already under way at the lowest and highest trophic levels. Winder, Reuter and Schladow (2008) showed how warming of the surface waters and increased stability is shifting the phytoplankton population toward small-bodied more buoyant species and away from larger bodied species that sink more readily out of the photic zone and are more dependent on vertical mixing. As the lake warms and its clarity declines, it is becoming increasingly hospitable for the warm-water fish mentioned above.

Lake Tahoe is not only a priceless scenic, recreational and biological resource. In most years, it drains into the Truckee River, supplying water to municipal and agricultural users downstream in the Reno, NV vicinity and helping to maintain Pyramid Lake at the river's terminus. From 1910 to 2010, the Tahoe basin contributed on average 23% (ranging from 0–40%) of the annual yield of the Truckee River at Farad, near the CA-NV border. In 20 out of the last 110 years, however, the lake level has fallen below the natural rim. Under both the A2 and B1 scenarios, the lake level will fall below the rim sporadically for much of the twenty-first century, but recover during years of heavy precipitation. Under the A2 scenario (Figure 14.16), the lake level will drop below the rim in about 2085, and plunge to an elevation of 1894.1 m by the end of the century, a level 1.5 m below the historic (since 1900) low stand of November 1992. Once the lake drops below the rim, evaporation from the lake is cumulative from one year to the next. Combined with earlier snowmelt and the shift from snow to rainfall, the declining lake level will present a serious challenge to water resource managers in the Truckee-Carson basin.

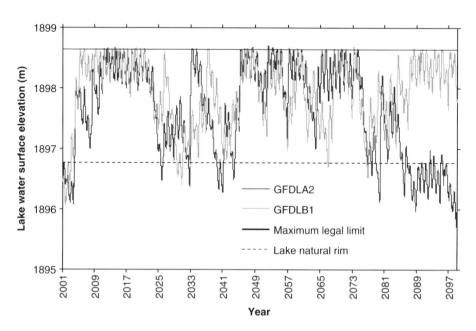

Figure 14.16 Modeled elevation of Lake Tahoe for two emission scenarios. When the lake level falls below the rim elevation, outflow ceases. (See insert for color representation.)

14.6 Reducing negative impacts

Climate change over this century threatens to shut off the deep mixing of Lake Tahoe, inducing anoxia and triggering a large release of algae-stimulating nitrogen and phosphorus from the bottom sediments. Changes in the watershed (described above) may increase the input of nutrients and fine sediment, reducing water transparency and promoting the reproductive success of invasive warm-water fish. At the regional and local levels, there is no way to halt the warming of the lake and to prevent the increase of internal nutrient loading—that solution must be found at the national and international levels. But it is possible to reduce the external nutrient loading to at least partially compensate for the future increase in internal loading, and watershed changes associated with climate change.

In 2010, the Lahontan Regional Water Quality Control Board and the Nevada Division of Environmental Protection completed their work on the "Lake Tahoe Total Maximum Daily Load Report," or TMDL (LRWQCB and NDEP 2010). Using a data base of 160 station years of stream gauging and sampling records, intensive sampling (450 events) of urban runoff, atmospheric deposition data, modeling of groundwater nutrient load from well data, stream channel erosion data, together with LSPC (Riverson 2010; Riverson *et al.* in press), the TMDL project calculated total loads of nitrogen, phosphorus and fine sediment from six source categories: non-urban uplands, urban uplands, atmospheric deposition (wet and dry), stream channel erosion, groundwater and shoreline erosion. Multiple regression was used to attribute nutrient and sediment loads to land use/land cover class (Coats *et al.* 2008).

Table 14.2 shows the final estimates of nutrients and sediment loads by source area. Note that direct atmospheric deposition is the most important source of total nitrogen, and upland urban areas are the most important sources of total phosphorus and fine sediment. The agencies have established a water transparency standard of 24 m Secchi depth after 20 years, and 29.7 m (the average transparency from 1967 to 1971) after 65 years. Using the Lake Clarity Model (Sahoo, Schladow and Reuter 2010), they found that the transparency standard could be met after 65 years by reducing the total loads of fine sediment, phosphorus and nitrogen by 65, 35 and 10%, respectively, from the currently estimated annual loads. If we assume that all of the current external load of phosphorus to the lake will ultimately be stored in the sediment and become available for release under anoxic conditions, then achieving the TMDL goal of a 35% reduction

Table 14.2 Pollutant loading estimates by source category, from Lahontan and NDEP (2010).

Source category		Total N, metric tonnes year^{-1}	Total P, metric tonnes year^{-1}	No. of fine sediment particles ($<16\,\mu$m), 10^{18}
Upland	Urban	63	18	348
	Non-urban	62	12	41
Atmosph. deposition (wet+dry)		218	7	75
Stream channel erosion		2	<1	17
Groundwater		50	7	NA
Shoreline erosion		2	2	1
Total		397	46	481

in load over the next 65 years could significantly reduce the internal load when anoxic conditions develop.

To achieve these goals, the TMDL project developed a broad array of Best Management Practices (BMPs) that could be used to reduce pollutant loads. Included are measures such as stabilizing and revegetating road shoulders, installing and maintaining storm water treatment vaults, detention basins, and treatment wetlands, putting to bed poorly-maintained dirt roads, and so forth. As with programs to reduce carbon emissions, no single measure can contribute a large load reduction, but each contributes a thin wedge, and taken together, the wedges add up to meet the program goals. The total estimated cost of construction and maintenance needed to achieve the 15-year transparency goal was estimated to be $100 million per year (LRWQCB and NDEP 2010).

The effectiveness (or design requirements) of some BMPs, however, may be affected by climate change, especially increases in rainfall intensity and rain-on-snow events. Wolfe (2010) used a combined water quality-runoff model—the Pollutant Load Reduction Model—together with the downscaled and bias-corrected twenty-first century precipitation (from GFDL with A2 and B1 scenarios) to investigate this problem. They found that increases in storm water runoff were reflected in a 10% decline in the treatment performance of facilities (such as treatment vaults and detention basins) designed for current hydrologic conditions. However, the designs based on current criteria would still be able to detain about 80% of the average annual runoff from developed areas. In general, the water quality benefits of implementing the BMPs in the proposed TMDL program would far outweigh the slight loss in functionality associated with climate change.

14.7 Summary and conclusions

Analysis of historic climate data in the Tahoe basin dating back from the early twentieth century indicated strong upward trends in temperature, especially for minimum daily temperature during spring and summer months. The upward trend was associated with a shift from snow to rainfall, a decrease in the number of days with average temperature below freezing, and a shift in the timing of the spring snowmelt peak runoff to earlier dates, and (since 1970) a warming trend in the lake itself. Comparisons of the rates of warming and the shift in snowmelt timing with stations nearby but outside of the basin suggested that the Tahoe basin is warming faster than the surrounding region.

The downscaled GCM output for both the B1 and A2 emissions scenarios indicated that the warming of the basin will continue. When we used the bias-corrected output for the GFDL model to force a hydrologic model over the twenty-first century, the results indicated a loss of snowpack, shift of the annual runoff hydrograph toward earlier dates, a large increase in flood risk in the middle third of the twenty-first century, and increased drought severity, due in part to the declining contribution of snowmelt to soil moisture storage. Changes in precipitation were less certain, though with the higher emission scenario (the A2), there was indication of decreasing precipitation in the last quarter of the century.

To evaluate the impacts of climate change on the lake, we used the output of the hydrology model together with downscaled climate data to drive a combined hydrodynamic/water quality model for the lake. The results indicated that, with the higher emission scenario (A2), the lake will continue to warm, especially in the latter two-thirds of the century, and become more thermally stable. Deep mixing will cease for long periods and the deep water of the lake will become anoxic, triggering a large release of available phosphorus and nitrogen, and a significant loss of well-oxygenated lake volume that now serves as salmonid habitat. By the end of the century, the lake may drop below the rim to an elevation not seen in historic time. Although these changes cannot be prevented without a large global reduction in future greenhouse gas emissions, the strict application of water quality Best Management Practices, though expensive, offers an opportunity to mitigate and perhaps forestall the potentially disastrous impacts of climate change on Lake Tahoe.

14.8   Acknowledgements

We thank Patricia Arneson for years of careful data stewardship; Scott Hackley and Robert Richards for field data collection; Michelle Brecker and Kelly Redmond of the Western Regional Climate Center for providing historic climate data; Sudeep Chandra for helpful suggestions and discussion; and Ayaka Tawada for diligent editing. This research was supported in part by grant #08-DG-11272170-101 from the USDA Forest Service Pacific Southwest Research Station using funds provided by the Bureau of Land Management through the sale of public lands as authorized by the Southern Nevada Public Land Management Act.

References

Aguado, E., Cayan, D., Riddle, L., and Roos, M. (1992) Climatic fluctuations and the timing of west coast streamflow. *J. Clim.*, **5**, 1468–83.

Arhonditsis, G. B., Brett, M.T., DeGasperi, C.L., and Schindler, D.E. (2004) Effects of climatic variability on the thermal properties of Lake Washington. *Limnol. Oceanogr.*, **49**, 256–70.

Austin, J.A. and Colman, S.M. (2007) Lake Superior summer water temperatures are increasing more rapidly than regional air temperatures: A positive ice-albedo feedback. *Geophys. Res. Lett.*, **34**, L06604, doi:10.1029/2006GL029021.

Beutel, M. (2000) Dynamics and Control of Nutrient, Metal and Oxygen Fluxes at the Profundal Sediment-Water Interface of Lakes and Reservoirs. Ph.D. thesis. University of California , Department of Environmental Science and Policy, Davis, CA.

Brown, R.D. (2000) Northern Hemisphere snow cover variability and change, 1915–97. *J. Clim.*, **13**, 2339–55.

Coats, R. (2010) Climate change in the Tahoe Basin: regional trends, impacts and drivers. *Climatic Change*, **102**, 435–66, doi:10.1007/s10584-010-9828-3.

Coats, R., Larsen, M., Thomas, H.J., *et al.* (2008) Nutrient and sediment production, watershed characteristics, and land use in the Tahoe Basin, California-Nevada. *J. Am. Water Resour. Assoc.*, **44**, 754–70.

Coats, R., Perez-Losada, J., Schladow, G., *et al.* 2006. The Warming of Lake Tahoe. *Climatic Change*, **76**, 121–148, doi:10.1007/s10584-005-9006-1.

Coats, R.N., Costa-Cabral, M. Riverson, J., *et al.* (in press) Projected twenty-first century trends in hydroclimatology of the Tahoe basin. *Climatic Change*.

Costa-Cabral, M., Coats, R., Reuter, J., *et al.* (in press) Climate variability and change in mountain environments: some implications for water resources and water quality in the eastern Sierra Nevada (USA). *Climatic Change*.

Delworth, T.L., Broccoli, A.J., Rosati, A., *et al*. (2006) GFDL's CM2 global coupled climate models. Part 1, Formulation and simulation characteristics. *J. Clim.*, **19**, 643–74.

Dettinger, M. (in press) Climate-change projections and downscaling of 21st Century temperatures, precipitation, radiative fluxes and winds for the Western US, with particular attention to the Lake Tahoe Basin. *Climatic Change*.

Dettinger, M.D. and Cayan, D.R. (1995) Large-scale atmospheric forcing of recent trends toward early snowmelt runoff in *California*. *J. Clim.*, **8**, 606–23.

Dettinger, M.D., Cayan, D.R., Meyer, M.K., and Jeton, A.E. (2004) Simulated hydrologic responses to climate variations and change in the Merced, Carson, and American River Basins, Sierra Nevada, California, 1900–2009. *Climatic Change*, **62**, 283–317.

Drevnick, P.E., Shinneman, A.L.C., Lamborg, C.H., *et al*. (2009) Mercury Flux to Sediments of Lake Tahoe, California–Nevada. *Water, Air, Soil Pollution*, **210**, 399–407, doi: 10.1007/s11270-009-0262-y.

Elo, A., Huttula, T., Peltonen, A., and Virta, J. (1998) The effects of climate change on the temperature conditions of lakes. *Boreal Env. Res.*, **3**, 137–50.

Goldman, C.R. (1981) Lake Tahoe: two decades of change in a nitrogen-deficient oligotrophic lake. *Verh. Internat. Verein. Limnol.*, **21**, 45–70.

Hansen, J., Sato, M., Kharecha, P., *et al*. (2008) Target atmospheric CO2: Where should humanity aim? *Open Atmos. Sci. J.*, **2**, 217–31, doi:10.2174/1874282300802010217.

Helsel, D.R., and Frans, L.M. (2006) Regional Kendall Test for Trend. *Environ. Sci. Tech.*, **40**, 4066–73, doi:10.1021/es051650b.

Heyvaert, A.C., Reuter, J.E., Slotton, D.G., and Goldman, C.R. (2000) Paleolimnological Reconstruction of Historical Atmospheric Lead and Mercury Deposition at Lake Tahoe, California–Nevada. *Environ. Sci. Technol.*, **34**, 3588–97, doi:10.1021/es991309p.

Hidalgo, H.G., Dettinger, M.D., and Cayan, D.R. (2008) Downscaling with constructed analogues: daily precipitation and temperature fields over the United States. California Climate Change Center Public Interest Energy Research Program 2007–027.

Idso, S.B. (1973) On the concept of lake stability. *Limnol. Oceanogr.*, **18**, 681–3.

IPCC (2007) *Climate Change 2007, the Fourth Assessment Report (AR4) of the United Nations Intergovernmental Panel on Climate Change (IPCC)*, United Nations Environment Programme and World Meteorological Organization, Geneva.

Jassby, A.D., Reuter, J.E., and Goldman, C.R. (2003) Determining long-term water quality change in the presence of climate variability, Lake Tahoe (USA). *Can. J. Fish. Aquat. Sci.*, **60**, 1452–61, doi:10.1139/f03-127.

Johnson, T., Dozier, J., and Michaelsen, J. (1999) Climate change and Sierra Nevada snowpack, in *Interactions Between the Cryosphere, Climate and Greenhouse Gases* (eds. M. Tranter, R. Armstrong, E. Brun, *et al*.) IAHS Press and Institute of Hydrology, Wallingford, p. 63–70.

Kamerath, M., Chandra, S., and Allen, B. (2008) Distribution and impacts of warm water invasive fish in Lake Tahoe. *Aquatic Invasions*, **3**, 35–41, doi:10.3391/ai.2008.1.1.7.

Kerr, R.A. (1982) El Chichón forebodes climate change. *Science*, **217**, 1023, doi:10.1126/science.217.4564.1023.

Lettenmaier, D.P., and Gan, T.Y. (1990) Hydrologic sensitivities of the Sacramento–San Joaquin River Basin, California, to global warming. *Water Resour. Res.*, **26**, 69–86, doi:10.1029/WR026i001p00069.

Livingstone, D.M. (1997) An example of the simultaneous occurrence of climate-driven "sawtooth" deep-water warming/cooling episodes in several Swiss Lakes. *Verh. Internat. Verein. Limnol.*, **26**, 822–8.

Livingstone, D.M. (2003) Impact of secular climate change on the thermal structure of a large temperate central European lake. *Climatic Change*, **57**, 205–25, doi:10.1023/A:1022119503144.

LRWQCB, and NDEP (2010) *Lake Tahoe Total Maximum Daily Load Final Report,* Lahontan Regional Water Quality Control Board, South Lake Tahoe, California, and Nevada Division of Environmental Protection, Carson City, Nevada.

Maurer, E.P. (2007) Uncertainty in hydrologic impacts of climate change in the Sierra Nevada, California under two emissions scenarios. *Climatic Change,* **82,** 309–25, doi:10.1007/s10584-006-9180-9.

McCormick, M.J. and Fahnenstiel, G.L. (1999) Recent climatic trends in nearshore water temperatures in the St. Lawrence Great Lakes. *Limnol. Oceanogr.,* **44,** 530–40.

Milly, P.C.D., Betancourt, J., Falkenmark, M., *et al.* (2008) Stationarity is dead: whither water management? *Science,* **319,** 573–574.

Nakicenovic, N., Alcamo, J., Davis, G., *et al.* (2000) *Special Report on Emissions Scenarios,* Cambridge University Press, Cambridge.

Ngai, K.L.C. (2008) Potential Effects of Climate Change on the Invasion of Largemouth Bass *(Micropterus salmoides)* in Lake Tahoe, California-Nevada. M.S. thesis. University of Toronto.

Palmer, W.C. (1965) *Meteorological Drought,* US Weather Bureau, Washington, DC.

Peeters, F., Livingstone, D.M., Goudsmit, G.-H., *et al.* (2002) Modeling 50 years of historical temperature profiles in a large central European lake. *Limnol. Oceanogr.,* **47,** 186–97.

Pupacko, A. (1993) Variations in northern Sierra Nevada streamflow, implications of climate change. *Water Resour. Bull.,* **29,** 283–90.

Reuter, J.E., Cahill, T.A., Cliff, S.S., *et al.* (2003) An integrated watershed approach to studying ecosystem health at Lake Tahoe, CA-NV, in *Managing for Healthy Ecosystems* (eds. D.J. Rapport, W.L. Lasley, D.E. Rolston, *et al.*), Lewis Publishers, Boca Raton FL, pp. 1283–98.

Riverson, J. (2010) Projected flow, nutrient and sediment loads based on climate change using output from the Lake Tahoe watershed model, in *The Effects of Climate Change on Lake Tahoe in the Twenty-First Century, Meteorology, Hydrology, Loading and Lake Response* (eds. R. Coats and J. Reuter), University of California, Davis, Tahoe Environmental Research Center and Department of Environmental Science and Policy, Davis CA, pp. 38–65.

Riverson, J., Coats, R., Costa-Cabral, M., *et al.* (in press) Modeling the impacts of climate change on streamflow, nutrient and sediment loads in the Tahoe Basin. *Climatic Change.*

Robertson, D.M., and Ragotzkie, R.A. (1990) Changes in the thermal structure of moderate to large sized lakes in response to changes in air temperature. *Aquat. Sci.,* **52,** 360–380, doi:10.1007/BF00879763.

Sahoo, G.B., Schladow, S.G., and Reuter, J.E. (2010) Effect of sediment and nutrient loading on Lake Tahoe optical conditions and restoration opportunities using a newly developed lake clarity model. *Water Resour. Res.,* **46,** doi:10.1029/2009WR008447.

Sahoo, G.B., Schladow, S.G., Reuter, *et al.* (in press) The response of Lake Tahoe to climate change. *Climatic Change.*

Salmi, T., Määttä, A., Anttila, P., *et al.* (2002) *Detecting Trends of Annual Values of Atmospheric Pollutants by the Mann-Kendall Test and Sen's Slope Estimates—The Excel Template Application MAKESENS,* Finnish Meteorological Institute, Helsinki.

Schindler, D.W., Bayley, S.E., and Parker, B. (1996) The effects of climatic warming on the properties of boreal lakes and streams at the Experimental Lakes Area, northwestern Ontario. *Limnol. Oceanogr.,* **41,** 1004–17.

Schneider, P., Hook, S.J., Radocinski, R.G., *et al.* (2009) Satellite observations indicate rapid warming trend for lakes in California and Nevada. *Geophys. Res. Lett.,* **36,** doi: 10.1029/2009GL040846.

Stefan, H.G., Fang, X., and Hondzo, M. (1998) Simulated climate change effects on year-round water temperatures in temperate zone lakes. *Climatic Change,* **40,** 547–76, doi:10.1023/A:1005371600527.

Stewart, I.T., Cayan, D.R., and Dettinger, M.D. (2005) Changes toward earlier streamflow timing across western North America. *J. Clim.,* **18,** 1136–55.

TERC (2011) *Tahoe: State of the Lake Report 2010*, Tahoe Environmental Research Center, University of California at Davis, Davis CA.

USGS (1981) *Guidelines for Determining Flood Flow Frequency*. Bulletin #17B of the Hydrology Subcommittee, US Department of the Interior Geological Survey, Office of Water Data Coordination, Reston VA, p. 193.

Verburg, P., Hecky, R.E., and Kling, H. (2003) Ecological consequences of a century of warming in Lake Tanganyika. *Science*, **301**, 505–7, doi:10.1126/science.1084846.

Wahl, K.L. (1992) Evaluation of trends in runoff in the Western United States, in *Managing Water Resources during Global Change* (ed. R. Herrmann), American Water Resources Association 28th Annual Conference and Symposium, American Water Resource Association, Reno NV, pp. 701–10.

Washington, W.M., Weatherly, J.W., Meehl, G.A., *et al.* (2000) Parallel climate model (PCM) control and transient simulations. *Climate Dynamics*, **16**, 755–74, doi:10.1007/s003820000079.

Watras, C.J. (2010) Mercury pollution in remote fresh waters, in *Biogeochemistry of Inland Waters* (ed. G.E. Likens), Academic Press, New York, pp. 648–57.

Wells, N., Goddard, S., and Hayes, M.J. (2004) A self-calibrating Palmer drought severity index. *J. Clim.*, **17**, 2335–51.

Wetzel, R. (2001) *Limnology: Lake and River Ecosystems*, 3rd edn., Academic Press, New York.

Winder, M., Reuter, J.E., and Schladow, S.G. (2008) Lake warming favors small-sized planktonic diatom species. *Proc. Royal Soc. B.*, **276**, 427–35, doi:10.1098/rspb.2008.1200.

Wittmann, M.W., Chandra, S., Reuter, J.E. *et al.* (2010) Final Report for Asian Clam Pilot Project. Technical Report—Tahoe Environmental Research Center, University of California-Davis, One Shields Ave., Davis, CA. p. 89.

Wolfe, B. (2010) Implications of Climate Change for Design of BMPs in the Lake Tahoe Basin, in *The Effects of Climate Change on Lake Tahoe in the Twenty-First Century, Meteorology, Hydrology, Loading and Lake Response* (eds. R. Coats and J. Reuter), Tahoe Environmental Research Center, and Department of Environmental Science and Policy, University of California, Davis CA, pp. 66–80.

Wood, A.W., Leung, L.R., Sridhar, V., and Lettenmaier, D.P. (2004) Hydrologic implications of dynamical and statistical approaches to downscaling climate model outputs. *Climatic Change*, **62**, 189–216, doi:10.1023/B:CLIM.0000013685.99609.9e.

Wood, A.W., Maurer, E.P., Kumar, A., and Lettenmaier, D.P. (2002) Long-range experimental hydrologic forecasting for the eastern United States. *J. Geophys. Res.*, **107**, 4429, doi:10.1029/2001JD000659.

Yarnell, S.M., Viers, J.H., and Mount, J.F. (2010) Ecology and management of the spring snowmelt recession. *BioScience*, **60**, 114–27.

15

Our New Biological Future? The Influence of Climate Change on the Vulnerability of Lakes to Invasion by Non-Native Species

Marion E. Wittmann[1], Ka Lai Ngai[2], and Sudeep Chandra[2]

[1]Department of Biological Sciences, University of Notre Dame, USA
[2]Aquatic Ecosystems Analysis Laboratory, Department of Natural Resources and Environmental Science, University of Nevada, Reno, USA

15.1 Introduction

Freshwater aquatic ecosystems have been observed to respond rapidly to climate change (Carpenter *et al*. 1992; Rosenzweig *et al*. 2008; Adrian *et al*. 2009), and these responses have impacts on ecosystem structure and function (Schindler *et al*. 1996; Magnuson *et al*. 2000; Verburg, Hecky, and Kling 2003). The disturbance resulting from climate change can occur indirectly at the watershed level, with subsequent impacts on the downstream aquatic environment, or directly influence the physical characteristics of lake water. For example, changes in temperature and precipitation patterns can alter stream flows and water levels, reduce winter ice cover in mid- and high latitude lakes and alter downstream water chemistry (quality) and transparency in lake systems (Poff, Brinson, and Day 2002; Mooij *et al*. 2005; Austin and Colman 2007; Murdoch, Baron, and Miller 2000). Both long-term (species composition) and short-term (growth rates, abundances) shifts in biological communities have been correlated with changes in temperature, precipitation and hydrologic regimes (Adrian *et al*. 2006; Rühland, Paterson, and Smol 2008; Winder and Hunter 2008).

Climatic Change and Global Warming of Inland Waters: Impacts and Mitigation for Ecosystems and Societies,
First Edition. Edited by Charles R. Goldman, Michio Kumagai and Richard D. Robarts.
© 2013 John Wiley & Sons, Ltd. Published 2013 by John Wiley & Sons, Ltd.

At the most fundamental level shifts to biological communities are a function of the impacts to basic biological processes (such as metabolism or reproduction) that are caused by environmental change. With climate change understanding the differential response of native and non-native communities to these changes is vital for the design of meaningful ecological research programs and the efficient management of freshwater systems.

In this chapter we explore some observed and proposed responses of non-native species to two climate change induced stressors to lakes—altered water transparency resulting from increases in run off and thermal regime shifts. We propose that these shifts can impact the establishment of non-native species and promote their expansion within a lake ecosystem as well as across landscapes (Figure 15.1). We consider Lake Tahoe (CA-NV, USA) as a case study to investigate the impacts of these stressors. Because of its long-term water quality monitoring record, Lake Tahoe provides an ideal system to look at the impacts of climate change and other human stressors such as the impacts of urban development. Additionally, Lake Tahoe has a significant history of both intentional (to benefit sport fisheries or other economic activity) and unintentional (hitchhikers on boats or equipment, or aquarium or bait bucket release) invasions that have impacted both the near-shore and offshore regions of the lake. We present a history of climate change and invasive species impacts on Lake Tahoe and describe the potential for climate change to impact two species, bluegill sunfish (*Lepomis macrochirus*) and Asian clam (*Corbicula fluminea*), which are non-native to and are currently established in Lake Tahoe. Finally, we provide suggestions and motivation for the integration of climate change and invasive-species research—and outline the importance of their synthesis to sustainable and long-term lake management.

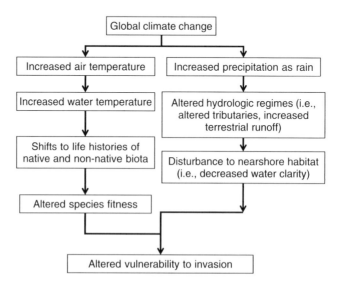

Figure 15.1 Mechanisms by which the Lake Tahoe ecosystem can be affected by climate change, and how these changes can impact the invasibility of Lake Tahoe.

15.2 Lake Tahoe: a history of change attributed to cultural eutrophication, thermal changes from climate, and the introduction of nonnative species

Lake Tahoe is a subalpine lake in the northern Sierra Nevada range spanning the California–Nevada border, in the United States (Figure 15.2). Tahoe is known for its

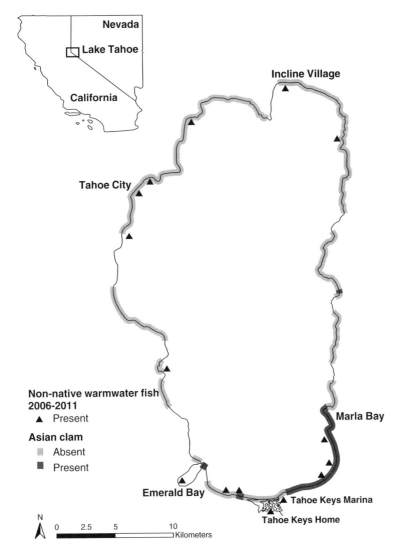

Figure 15.2 Lake Tahoe is situated on the border of California and Nevada in the Sierra Nevada mountain range, in the United States. The triangles on the map indicate the presence of non-native warmwater fish, the gray areas indicate Asian clam absence, and dark grey indicates Asian clam presence.

high transparency, which is mostly due to its small watershed, granitic geology, and large depth (Jassby *et al*. 1994; Goldman 2000). Two long-term (>50 year) effects that have been observed in Lake Tahoe include a general warming of lake temperatures (Coats *et al*. 2006; see also Chapter 14), and a decrease in transparency (Jassby *et al*. 1999; Goldman 2000; UCD 2011). The transparency changes are largely due to cultural eutrophication and increased loading of particles from the watershed, and interannual and intradecadal variability in transparency have been linked to climate (Goldman 1988; Jassby *et al*. 1999; Goldman 2000). The observed warming trend is greatest in the shallow and very deep pelagic-profundal waters, and was largely attributed to increased daily air temperatures and a slightly positive trend in downward long-wave radiation (Coats *et al*. 2006). This altered thermal regime measured in the open water has changed the lake's stability (measured as Schmidt's Index) and resulted in a shift in the phytoplankton community structure (Winder, Reuter, and Schladow 2009).

The first series of non-native species introductions to Lake Tahoe were intentional and occurred at the end of the nineteenth century (Figure 15.3). They included nine species of salmonids thought to be suited to Tahoe's environment (Dill and Cordone 1997; VanderZanden *et al*. 2003). Only rainbow (*Onchorynchus mykiss*), brown (*Salmo trutta*), brook (*Salvelinus fontinalis*) and lake trout (*Salvelinus namaycush*) persist in the lake today. Predatory impacts from lake trout combined with overfishing, hybridization with other trout, and siltation of spawning streams contributed to the extirpation of native Lahontan cutthroat trout (*Oncorhynchus clarki*

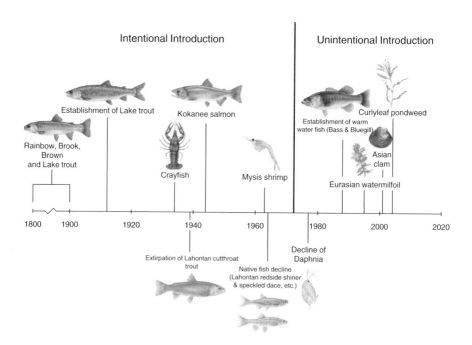

Figure 15.3   Lake Tahoe species invasion timeline. Prior to the 1970s, all introductions to Lake Tahoe were intentional, and intended to support the fishery. Beginning in the 1980s, a number of invasions occurred in Lake Tahoe's nearshore zone through unintentional introductions of various warmwater fishes, Asian clam, Eurasian milfoil and curlyleaf pondweed. Illustration courtesy of Sarah Adler.

henshawi) from Lake Tahoe by 1939. The establishment of non-native kokanee salmon (*Onchorynchus nerka*) occurred in 1945 and has since been supported through stocking. In the 1960s, state fish and game agencies introduced another invertebrate, *Mysis relicta* or the opossum shrimp, which corresponded with shifts in the feeding behavior and biomass of native chubs, non-native kokanee, and lake trout as well as a tenfold decrease in the abundance of forage fishes (Thiede 1997).

Another invertebrate, signal crayfish (*Pacifastacus leniusculus*), was introduced at least four times between 1888 and 1936 as forage for fishes and protein for the local population. There were approximately 55 million crayfish in Lake Tahoe in the 1960s (Abrahamsson and Goldman 1970), and by 2001 the population was estimated to be greater than 230 million. This increase was likely due to a variety of factors, including increased algal production in the nearshore and a lack of natural predators. Recent surveys of the bottom diversity of Lake Tahoe (surface to 501 m) suggested there may have been a 50–80% decline in native invertebrate densities for native invertebrate taxa since measurements were made in the 1960s. It was hypothesized that the large crayfish density was contributing to the decline of these invertebrates through predation and to the stimulation of near-shore production through nutrient recycling.

In the last 25 years there have also been a number of unintentional plant, invertebrate, and fish introductions to Lake Tahoe (Figure 15.3). These newer invasions differed from the game fish, *Mysis* and crayfish introductions in that they were inadvertent and not intended to support the sport fishery of the lake. Most unintentional introductions to Lake Tahoe were likely a result of "hitchhiking" or transport on recreational boats, or via bait bucket or aquarium release. The establishment of Eurasian watermilfoil (*Myriophyllum spicatum*) in Lake Tahoe was formally confirmed by experts in 1995, but was thought to have been introduced to south Lake Tahoe sometime after an early 1960's installation of a 740-acre residential development (the Tahoe Keys). Severe impacts from aquatic plants were observed in the Tahoe Keys by the 1980s, and at this time a mechanical harvesting program was initiated to remove nuisance plant growth and permit boater navigation within the Keys and out into the lake. In 2010, Eurasian watermilfoil was abundant throughout the entirety of the Tahoe Keys, and has since spread to over 30 locations lakewide. Another invasive macrophyte, curly leaf pondweed (*Potamogeton crispus*), was first observed in 2003 in a few small discrete locations along the south shore and has since rapidly increased its range to an approximate 20 km^2 area along the southern shoreline of Tahoe.

In the mid to late 1970s, and again in the late 1980s, a variety of non-native, warmwater fish species were also established in Lake Tahoe. Some observed taxa included largemouth bass (*Micropterus salmoides*), smallmouth bass (*Micropterus dolomieu*), and bluegill sunfish and were thought to be illegally introduced by recreational sports fishermen. In locations where warmwater fishes have established native forage fish populations have exhibited significant reductions (Kamerath, Chandra, and Allen 2008). Recent studies suggested that there was an expansion of warmwater fishes (largemouth bass and bluegill) in the nearshore areas of the lake (Kamerath, Chandra, and Allen 2008). This expansion was potentially due to increase in nearshore temperature (due to climate change), the creation of habitat due to establishment of invasive plants (see above), and a rich supply of food from a large crayfish population (see above).

The most recent nonnative species to establish in Lake Tahoe was the Asian clam (*Corbicula fluminea*). This invasive bivalve has been established in the nearby

Sacramento-San Joaquin Delta system since the 1960s (Prokopovich 1969), was first observed in Lake Tahoe in 2002 at low densities (2–20 individuals m^{-2}), and by 2008 high density populations (>2000 m^{-2}) were observed in the southeastern portion of the lake. By 2011, maximum population densities reached over 10 000 m^{-2}, and where the Asian clam was established, it was the dominant taxa in biomass and abundance. Limited by temperature and food availability, Asian clams in Lake Tahoe are smaller in size (maximum shell length = 25 mm) and have a reduced reproductive rate compared to observations from warmer, more eutrophic systems (Denton *et al.* 2012). Dense algal blooms of the green filamentous algal species *Cladophora glomerata, Zygnema* sp. and *Spirogyra* sp. were found to be associated with high density Asian clam populations (Forrest *et al.* 2012) suggesting their contribution to the eutrophication of the nearshore environment. Since 2008, the development of non-chemical control strategies for Asian clam in Tahoe (suction dredging and bottom barriers) has been a priority for scientists and managers collaborating in the basin to reduce the rapid spread of this species within the lake.

Given the success of both intentional and unintentional non-native species introduction to Lake Tahoe, the continued propagule pressure from nearby waterways (Dreissenid mussels in Lake Mead, AZ-NV or New Zealand mud snail in Lake Shasta, CA) and the resources spent to date to monitor, research and control them, it is of value to lake managers to explore whether the impacts of climate change will heighten the impacts of invasive species in Lake Tahoe and if so, what steps can be taken to mitigate them.

15.3   The effects of an altered thermal regime to Tahoe invasive species

Aquatic organisms are generally ectothermic and have specific temperature tolerance ranges and optimal temperature preferences for growth and reproduction. An altered thermal regime will therefore have direct impacts on aquatic organisms' fundamental biological processes, potentially affecting their fitness and competitiveness (Figure 15.1; Poff, Brinson, and Day 2002; Lockwood, Hoopes, and Marchetti 2007). For example, given the availability of sufficient resources, the growth and development of aquatic organisms with wide thermal tolerances could increase with a warmer climate, thus giving these species a strong competitive advantage over coldwater species (Hill and Magnuson 1990; Adrian *et al.* 2009).

One important conjecture for Asian clam establishment in new environments is temperature limitation. This conclusion is drawn from observations that continued exposure to temperatures below 1–2 °C correlated with mortality (Mattice and Dye 1976; Rodgers *et al.* 1977; Morgan *et al.* 2003; Werner and Rothhaupt 2008; Weitere *et al.* 2009), and that the occurrence of Asian clams in rivers such as the St. Clair River (Michigan) and the lower Connecticut River (New England) was restricted to the thermal refugia associated with power plant discharges during cold winter periods (French and Schlösser 1991; Morgan *et al.* 2003). Asian clams are also limited by high temperature extremes; massive die-offs were observed during a European heat wave in 2003 in the Rhine River with water temperatures in excess of 28 °C (Westermann and Wendling 2003; Cooper *et al.* 2005; Morgan *et al.* 2003). While water temperature has not limited the establishment of Asian clams in Lake Tahoe, it has impacted its reproduction and growth rate (Denton *et al*. 2012). Here we describe a scenario in which a nearshore warming of 1.5–2.0 °C, as predicted by Ngai (2008), increases the fecundity of Asian clam in Lake Tahoe, which could lead to its subsequent population increase. In particular, we look at the impact of this warming on its univoltine (a single brood per year) reproductive strategy—a mode that is typically associated with habitat limitation such as temperature or food availability and the subsequent successful recruitment for increased population growth.

15.3.1 Enhancement of reproductive cycle: Increased temperatures and univoltine reproduction

Asian clams are hermaphroditic, with adults able to carry eggs for the entirety of their life (Kraemer and Galloway 1986). Most published accounts of Asian clam reproduction indicate bivoltine (two broods per year) spawning—beginning with the release of D-shaped juveniles in the spring continuing into the summer, and once again in late summer continuing into the fall (McMahon and Bogan 2001; Sousa, Antunes, and Guilhermino 2008). Spermatogenesis and fertilization generally begin when temperatures rise in spring and exceed 10 and 15 °C, respectively (Rajagopal, Van der Velder, and Bij De Vaate (2000); Cataldo *et al*. 2001) with upper thermal limits of 30 °C observed (Aldridge and McMahon 1978). Thermal extremes observed in the mid-summer period likely cease spawning of Asian clam individuals, and hinder their metabolic and reproductive rates, leading to the observed bivoltine spawning patterns (Gillooly *et al*. 2001; Enquist *et al*. 2003; Figure 15.4).

Nearshore temperatures in Lake Tahoe are above 10 °C from early May to November and above 15 °C between late May through early September (Ngai 2008; Figure 15.5a). Given this temperature setting, the reproductive period for Asian clams in Lake Tahoe was estimated to occur from late spring through late autumn with spermatogenesis occurring at 10 °C in May, fertilization in June at 14–15 °C and juvenile release

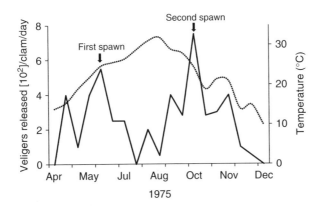

Figure 15.4 Asian clam bivoltine reproductive pattern as observed in Lake Arlington, Texas. Temperature extremes (>30 °C) limit Asian clam veliger release. Spawning once again resumes when temperatures decrease in autumn. Solid line represents veligers and dotted line represents temperature (°C). Figure after Aldridge and McMahon (1978) and Denton (2011).

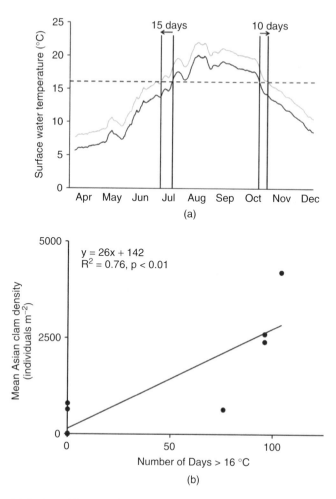

Figure 15.5 (a) Projected increase in the number of Asian clam spawning days as a result of a 2 °C warming in the Lake Tahoe nearshore zone. Current day temperatures (2010) shown in black and projected temperatures shown in gray. The dashed line indicates the minimum temperature required for pediveliger release (16 °C); (b) Asian clam abundance as a function of the number of days with temperatures >16 °C in Lake Tahoe. Each point on the graphic represents the mean abundance for three ponar grab samples collected from 2–50 m water depths in Marla Bay, Lake Tahoe.

at 16–18 °C in July–October (Figure 15.5a). Maximum temperatures in the Tahoe nearshore do not exceed 25 °C, which is well below the upper thermal limit (30 °C+) for Asian clam reproduction to occur. In contrast to most systems reported in the literature (Sousa, Antunes, and Guilhermino 2008), Lake Tahoe's Asian clam populations currently exhibit a univoltine reproductive strategy (Figure 15.6), with eggs present in gills continuously from May through November, and late stage (D-shaped) juveniles in August through October with a peak in mid-September (Denton *et al.* 2012). Low water temperature and food availability are the main drivers of this reduced fecundity.

Given a predicted 1.5–2.0 °C temperature range increase in the nearshore of Lake Tahoe over the next 100 years (Ngai 2008), Asian clam reproduction is likely to be

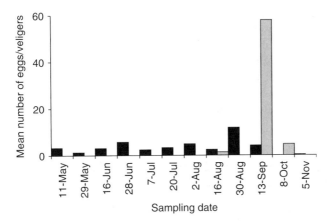

Figure 15.6 Mean number of eggs (black bars) and late stage veligers (gray bars) counted on Asian clam demibranchs. On each sampling date, clams (N = 40) were collected from 5 m water depth from Marla Bay, Lake Tahoe, 2010. After Denton *et al.* (2012).

enhanced. With a linear application of a 2 °C temperature increase, the number of days above the 16 °C threshold will increase by a total of 25 days (Figure 15.5a). This is to say that warming the lake based on a two degree warming of the measured 2010 nearshore temperature profile will lead to an estimated increase of 15 days in the spring/summer period and 10 in the late summer/autumnal period, in the number of days ≥16 °C. In addition, at no point in the predicted increased temperature scenario will water temperatures exceed the range appropriate for Asian clam reproduction.

Assuming that habitat variables other than temperature are equivalent at Tahoe's littoral zone (5–50 m), (as chlorophyll *a* and CDOM concentrations are similar and low and dissolved oxygen concentrations are high throughout most of the zone's water column—Coats *et al.* 2006), the population level implication for the increase in number of spawning degree days can then be deduced based on the current relationship between Asian clam density and degree days >16 °C observed at these water depths (Figure 15.5b). As stated above, Asian clams occur to a 50 m depth in Lake Tahoe, where water temperatures do not exceed 10 °C. From 5–10 m water depth there are over 80 days per year in which the water temperature is appropriate for reproduction, and less than 10 days at 30 m water depth. This relationship suggests that for each day over 16 °C the average abundance of Asian clam increases by 26 individuals (p<0.01, Figure 15.5b); an increase of 25 degree days will result in an increase of mean abundance by 650 clams m^{-2}.

While the coarse estimation from our proposed linkage between water temperature and Asian clam abundance does not include other important factors for reproduction and population growth such as food availability, mortality rates, competition or density dependence, these results show that on an annual time scale a 2 °C temperature increase can potentially result in greater abundance in Lake Tahoe, as has been observed elsewhere (Weitere *et al.* 2009). Additionally, increases in water temperatures are positively correlated with Asian clam filter-feeding rates (Lauritsen 1986). The combination of an increase in population size with accelerated filter feeding rates could present additional pressure to the abundance and composition of Lake Tahoe's algal community similar to those observed in Lake Michigan due to Dreissenid mussel filter feeding (Fahnenstiel *et al*. 2010; Vanderploeg *et al*. 2010). The increased fitness of Asian clams as a result of more suitable environmental conditions can lead to its expansion within the Tahoe system and potentially to other nearby waterways through increased risk of transport by recreational boaters or fishermen. Further research regarding the impact of increased invasive species fitness under climate change scenarios and the link between dispersal rates within and between managed systems could contribute to the reduction of spread of non-native species.

15.4   The effect of climate change to water transparency in western montane lakes and the impact on vulnerability to invasion

Climate-change research on western North American systems has focused on the shift in snowmelt timing toward earlier dates (Stewart, Cayan, and Dettinger 2005), the shift from snow to rain (Regonda *et al*. 2005; Knowles, Dettinger, and Cayan 2006), and an earlier onset of spring melt (Cayan *et al*. 2001). These hydrological shifts in response to altered precipitation regimes are likely to modify the timing and quantity of hydrological, chemical, and particle fluxes in montane catchments (Melack *et al*. 1997), which can impact transparency of montane lakes through altered inputs of nutrients and other particles (for a review see Adrian, Wilhelm, and Gerten 2009). Ultimately, transparency impacts the biota of lakes through exposure to ultraviolet light (UVR) (Williamson *et al*. 1994; Vehniainen, Hakkinen, and Oikari 2007), which can ultimately affect an aquatic ecosystem's vulnerability to invasion (Figure 15.1).

Lake Tahoe's clarity decline has been attributed to increases in both biological (phytoplankton and detritus) and inorganic (terrestrial sediment) particulate matter (Swift *et al*. 2006) resulting largely from human impacts (Goldman 1988) and increased loading of fine sediment particles (FSP) due to atmospheric deposition, particulates originating from roads, upland runoff and stream bank erosion (Byron and Goldman 1989). Although both eutrophication and FSP loading affects light penetration in Tahoe's water column, analyses indicated that FSP (particularly those $<16\,\mu m$) were more important than nutrients due to their light-scattering effect (Swift *et al*. 2006). Predicted shifts to precipitation regimes as a result of climate change will likely contribute to increased loading of particulate matter within the Tahoe catchment, causing further declines in transparency.

Here we summarize the important work of Tucker *et al*. (2010) as a case study in the effect of decreased water transparency on the establishment of a non-native warmwater species, bluegill sunfish in Lake Tahoe. Bluegill are very sensitive to environmental conditions (Lemly 1993; Olson *et al*. 2006; Essington, Sorensen, and Paron 1998; Huff, Grad, and Williamson 2004) and are typically limited to shallow waters of most lakes and rivers because of thermal requirements for reproduction (Kitchell *et al*. 1974). Bluegill were first observed in Lake Tahoe in the late 1980s and are currently limited to nearshore sites characterized by heavy urban development or modification (e.g., marinas, embayments) and some of the lake's largest tributaries (Kamerath, Chandra and Allen, 2008). These nearshore sites are particularly susceptible to terrestrial particulate loading, which has resulted in significant water clarity reductions compared to other undeveloped locations (which do not have established bluegill populations). Tucker *et al.* (2010) sought to test the hypothesis that UVR controls the suitability of nearshore habitats for the earliest life history stages of exotic bluegill, given its minimum temperature thresholds for spawning and high sensitivity of larval forms to UVR exposure (Williamson *et al.* 1999).

Through a combination of field measurements and laboratory experimentation, they found that larval bluegill experienced an 84% mortality rate when exposed to high UVR microcosms (those representing high water transparency in low urban development nearshore regions) compared to 11% mortality in low UVR microcosms (those representing decreased water transparency in high urban development nearshore conditions). Through field measurements, these authors found that both DOC and chlorophyll *a* were important regulators of variation in the UVR environment in nearshore areas of the lake. These authors suggested that avoidance of future declines in UVR transparency in Lake Tahoe could help decrease the potential for bluegill sunfish and other non-native species that are sensitive to UVR exposure to establish.

Lake Tahoe has a unique biological community that is adapted to its oligotrophic conditions such as cold water temperature, low nutrient availability, and high UVR penetration, which can limit the establishment of new species. These conditions provide resistance to species introductions, but in the face of a changing global climate, this resilience is likely to be compromised.

15.5 Complications for natural resource managers: understanding and mitigating the response of nonnative aquatic species to climate change

Until recently climate change models have dealt largely with broad-scale patterns assessing specific attributes of the environment (e.g., net precipitation, temperature) linked with watershed processes such as hydrological regime and water temperature. Increased monitoring, the growth of long term global climate datasets (e.g., World-Clim), as well as methodological and technological advances have produced higher resolution climate models, which have enabled climate-change scientists to reduce uncertainty associated with future scenarios of change. At the same time, biological invasion research has benefitted from great increases in the development of invasion theory. There has also been an increase in the number of peer-reviewed publications in the last decade, particularly with respect to interdisciplinary approaches to invasion science, heightened awareness by the public on invasive species issues, and improved collaborations between scientists and natural resource managers on preventing or controlling species invasions in public waterways (Lockwood, Hoopes, and Marchetti 2007).

Advances in climate modeling and biological invasion research have encouraged a field of research that examines the synergy between climate change and biological invasions and their mutually reinforcing effects to ecosystem change (Hellmann *et al.* 2008; Rahel and Olden 2008; Mainka and Howard 2010). Research using global climate change models to create and develop potential invasion scenarios in aquatic ecosystems, such as predicting range shifts of aquatic organisms under future climate scenarios (e.g., Sharma *et al.* 2007; Britton *et al.* 2010; Buisson *et al.* 2010), evaluating the synergistic impacts of climate change and biological invasion to recipient systems and resident species (e.g., Sharma, Jackson, and Minns 2009), have recently become more prevalent. Overall, however, publications remain limited (Smith *et al*. 2012). In particular, most of these studies have focused largely on large, regional landscapes; studies that quantify scenarios of invasion using downscaled climate models and smaller landscapes or within lake scales are generally lacking. Pyke *et al*. (2008) have urged policymakers to consider the interactions of climate change and species invasion. These authors suggested that climate policy inherently must consider assessment, mitigation, and adaptation. These are similar to attributes that should be considered by managers tackling the introduction of nonnative species. We suggest that for watershed managers to be effective they need an understanding of scenarios based on fundamental biological processes—the building blocks of population expansion or vulnerability to invasion, in order to produce environmental policies that provide secondary protection to aquatic ecosystems after broader federal and state policies are developed. We propose that understanding the linkage between climate change impact on basic biological processes such as growth, metabolism, and reproduction, and its subsequent implication for the primary and secondary spread of invasive species will provide for a more efficient management of our increasingly limited aquatic resources facing a future of climatic change.

15.6   Acknowledgements

We would like to thank the following for their assistance in collecting, analyzing, and discussing the information presented in this manuscript: Marianne Denton, Annie Caires, Joe Sullivan, Marcy Kamerath (University of Nevada-Reno), Brant Allen, Katie Webb, Raph Townsend, Dr. John Reuter, Dr. Geoffrey Schladow, and Dr. Charles R. Goldman (University of California-Davis), Andrew Tucker, Dr. Craig Williamson, and Dr. James Oris (Miami University in Ohio), Lars Anderson (USDA ARS Laboratory), Dr. Brian Shuter (University of Toronto), Dr. Almo Cordone (California Department of Fish and Game-retired).

References

Abrahamsson, A.A. and Goldman, C.R. (1970) Distribution, density and production of the crayfish *Pacifastacus leniusculus* Dana in Lake Tahoe, California-Nevada. *Oikos*, **21**, 83–91.

Adrian, R.S., Wilhelm, S., and Gerten, D. (2006) Life-history traits of lake plankton species may govern their phenological response to climate warming. *Glob. Ch. Biol.*, **12**, 652–61.

Adrian, R.S., O'Reilly, C.M., Zagarese, H. *et al*. (2009) Lakes as sentinels of climate change. *Limnol. Oceanogr.*, **54**, 2283–97.

Aldridge, D.W. and Mcmahon, R.F. (1978) Growth, fecundity, and bioenergetics in a natural population of the Asiatic freshwater clam, *Corbicula manilensis Philippi*, from North Central Texas. *J. Molluscan Stud.*, **44**, 49–70.

Austin, J.A. and Colman, S.M. (2007) Lake Superior summer water temperatures are increasing more rapidly than regional air temperatures: A positive ice-albedo feedback. *Geophys. Res. Lett.*, **34**, L06604, doi:10.1029/2006GL029021.

Britton, J.R., Cucherousset, J., Davies, G.D., *et al*. (2010) Non-native fishes and climate change: Predicting species responses to warming temperatures in a temperate region. *Freshwat. Biol.*, **55**, 1130–41.

Buisson, L., Grenouillet, G., Casajus, N. and Lek, S. (2010) Predicting the potential impacts of climate change on stream fish assemblages. *Am. Fish. Soc. Symp.*, **73**, 327–46.

Byron, E.R., and Goldman, C.R. (1989) Land use and water quality in tributary streams of Lake Tahoe, CA/NV. *J. Environ. Qual.*, **18**, 84–8.

Carpenter, S.R., Fisher, S.G., Grimm, N.B., and Kitchell, J.F. (1992) Global change and freshwater ecosystems. *Annu. Rev. Ecol. Syst.*, **23**, 119–39.

Cataldo, D.H., Boltovskoy, D., Stripeikis, J., and Pose, M. (2001) Condition index and growth rates of field caged *Corbicula fluminea* (Bivalvia) as biomarkers of pollution gradients in the Paraná River Delta (Argentina). *Aquat. Ecos. Health Manage*, **4**,187–201.

Cayan, D.R., Kammerdiener, S.A., Dettinger, M.D., and Caprio, J.M. (2001) Changes in the onset of spring in the Western United States. *Bull. Am. Met. Soc.*, **82**, 399–415.

Coats, R., Perez-Losada, J., Schladow, G., *et al.* (2006) The warming of Lake Tahoe. *Clim. Change*, **76**, 121–48.

Cooper, N.L., Bidwell, J.R., and Cherry, D.S. (2005) Potential effects of the Asian clam (*Corbicula fluminea*) die-offs on native freshwater mussels (Unionidae) II: Porewater ammonia. *J. N. Am. Benthological Soc.*, **24**, 381–94.

Denton, M., Chandra, S., Wittmann, M.E., *et al.* (2012) Reproduction and population structure of *Corbicula fluminea* in an oligotrophic, subalpine Lake. *J. Shellfish Res.*, **31** (1), 145–52.

Dill, W.A., and Cordone, A.J. (1997) History and status of introduced fishes in California, 1871–1996. *California Department of Fish and Game Fish Bulletin*, **178**, http://content.cdlib .org/view?docId=kt8p30069f&brand=calisphere&doc.view=entire_text (accessed July 24, 2012).

Enquist, B.J., Economo, E.P., Hugman, T.E., *et al.* (2003) Scaling metabolism from organism to ecosystem. *Nature*, **423**, 639–42.

Essington, T.E., Sorensen, P.W., and Paron, D.G. (1998) High rate of red superimposition by brook trout (*Salvelinus fontinalis*) and brown trout (*Salmo trutta*) in a Minnesota stream cannot be explained by habitat availability alone. *Can. J. Fish. Aquat. Sci.*, **55**, 2310–16.

Fahnenstiel, G., Pothoven, S., Nalepa, T., *et al.* (2010) Recent changes in primary production in the offshore region of southeastern Lake Michigan. *J. Great Lakes Res.*, **36**, 20–9.

Forrest, A.L., Wittmann, M.E., Schmidt, V. *et al.* (2012) Quantitative assessment of invasive species in lacustrine environments through benthic imagery analysis. *Limnol. Oceanogr. Methods*, **10**, 65–74.

French, J.R.P. and Schlosser, D.W. (1991) Growth and overwinter survival of the Asiatic clam, *Corbicula fluminea*, in the St. Clair River, Michigan. *Hydrobiologia*, **219**, 165–70.

Gillooly, J.F., Brown, J.G., West, G.B., *et al.* (2001) Effects of size and temperature on metabolic rate. *Science*, **293**, 2248–51.

Goldman, C.R. (1988) Primary productivity, nutrients, and transparency during the early onset of eutrophication in ultra-oligotrophic Lake Tahoe, CA/NV. *Limnol. Oceanogr.*, **33**, 1321–33.

Goldman, C.R. (2000) Four decades of change in two subalpine lakes. Baldi lecture. *Verhandlungen des Internat. Ver. Limnol.*, **27**, 7–26.

Hellmann, J.J., Byers, J.E., Bierwagen, B.G., and Dukes, J.S. (2008) Five potential consequences of climate change for invasive species. *Conserv. Biol.*, **22**, 534–43.

Hill, D.K. and Magnuson, J.J. (1990) Potential effects of global climate warming on the growth and prey consumption of Great Lakes Fish. *Trans. Am. Fish. Soc.*, **119**, 265–75.

Huff, D.D., Grad, G.G., and Williamson, C.E. (2004) Environmental constraints on spawning depth of yellow perch: The roles of low temperature and high solar ultraviolet Radiation. *Trans. Am. Fish. Soc.*, **133**, 718–26.

Jassby, A.D., Goldman, C.R., Reuter, J.E., and Richards, R.C. (1999) Origins and scale dependence of temporal variability in the transparency of Lake Tahoe, California- Nevada. *Limnol. Oceanogr.*, **44**, 282–94.

Jassby, A.D., Reuter, J.E., Axler, R.P., *et al.* (1994) Atmospheric deposition of nitrogen and phosphorus in the annual nutrient load of Lake Tahoe (Calfornia-Nevada). *Water Resour. Res.*, **30**, 2207–16.

Kamerath, M., Chandra, S., and Allen, B.C. (2008) Distribution and impacts of warmwater fish in Lake Tahoe, *USA. Aquat. Invasions*, **3**, 35–41.

Kitchell, J.F., Koonce, J.F., Manguson, J.J., *et al.* (1974) Model of fish biomass dynamics. *Trans. Am. Fish. Soc.*, **103**, 786–98.

Knowles, N., Dettinger, M.D., and Cayan, D.R. (2006) Trends in snowfall versus rainfall in the Western United States. *J. Clim.*, **19**, 4545–59.

Kraemer, L.R., and Galloway, M.L. (1986) Larval development of *Corbicula fluminea* (Müller) (Bivalvia: Corbiculacea): An appraisal of its heterochrony. *Am. Malacol. Bull.*, **4**, 61–79.

Lauritsen, D.D. (1986) Filter-feeding in *Corbicula fluminea* and its effect on seston removal. *J. N. Am. Benth. Soc.*, **5**, 165–72.

Lemly, A.D. (1993) Metabolic stress during winter increases the toxicity of selenium to fish. *Aquat. Toxicol.*, **27**, 133–58.

Lockwood, J.L., Hoopes, M.F., and Marchetti, M.P. (2007) *Invasion Ecology*, Blackwell, Oxford.

Mainka, S.A. and Howard, G.W. (2010) Climate change and invasive species: Double jeopardy. *Int. Zool.*, **5**, 102–11.

Magnuson, J.J., and others (2000) Ice cover phenologies of lakes and rivers in the Northern hemisphere and climate warming. *Science*, **289**, 1743–6.

Mattice, J.S. and Dye, L.L. (1976) Thermal tolerance of adult Asiatic clam, in *Thermal Ecology II* (eds. E.W. Esch, and R.W. McFarlene), Environmental Sciences Division, Oak Ridge National Laboratory, Oak Ridge TN, pp. 130–5.

McMahon, R.F. and Bogan, A.E. (2001) Mollusca: Bivalvia, in *Ecology and Classification of North American Freshwater Invertebrates* (eds. J.H. Thorp, and A.P. Covich), 2nd edn. Academic Press, New York, pp. 331–429.

Melack, J.M., Dozier, J., Goldman, C.R., *et al.* (1997) Effects of climate change on inland waters of the Pacific coastal mountains and Western Great Basin of North America. *Hydro. Proc.*, **11**, 971–92.

Mooij, W.M., Hülsmann, S., De Senerpont Domis, L.N., *et al.* (2005) The impact of climate change on lakes in the Netherlands: A Review. *Aquat. Ecol.*, **39**, 381–400.

Morgan, D.E., Keser, M., Swenarton, J.T., and Foertch, J.F. (2003) Population dynamics of the Asiatic clam, *Corbicula fluminea* (Müller) in the Lower Connecticut River: Establishing a foothold in New England. *J. of Shell. Res.*, **22**, 193–203.

Murdoch, P.S., Baron, J.S., and Miller, T.L. (2000) Potential effects of climate change on surface-water quality in North America. *J. Am. Wat. Res. Assn.*, **36**, 347–66.

Ngai, K.L. (2008) Potential effects of climate change on the invasion of largemouth bass *(Micropterus salmoides)* in Lake Tahoe, California-Nevada. University of Toronto. Master Thesis. University of Toronto.

Olson, M.H., Colip, M.R., Gerlach, J.S., and Mitchell, D.L. (2006) Quantifying ultraviolet radiation mortality rish in bluegill larvae: Effects of nest location. *Ecol. App.*, **16**, 328–38.

Poff, N.L., Brinson, M.M., and Day, J.R. Jr., (2002) *Aquatic Ecosystems and Global Climate Change: Potential Impacts on Inland Freshwater and Coastal Wetland Ecosystems in the United States*, Pew Center on Global Climate Change, Arlington VA.

Prokopovich, N.P. (1969) Deposition of clastic sediments by clams. *J. Sed. Res.*, **39**, 891–901.

Pyke, C.R., Thomas, R., Porter, R.D., *et al.* (2008) Current practices and future opportunities for policy on climate change and invasive species. *Cons. Biol.*, **22**, 585–92.

Rahel, F.J. and Olden, J.D. (2008) Assessing the effects of climate change on aquatic invasive species. *Cons. Biol.*, **22**, 521–33.

Rajagopal, S., Van der Velde, G., and Bij De Vaate, A. (2000) Reproductive biology of the Asiatic clams *Corbicula fluminalis* and *Corbicula fluminea* in the River Rhine. *Archiv für Hydrobiol.*, **149**, 403–20.

Regonda, S.K., Rajagopalan, B., Clark, M., and Pitlick, J. (2005) Seasonal cycle shifts in hydroclimatology over the Western United States. *J. of Clim.*, **18**, 372–84.

Rodgers, J.H., Cherry, D.S., Clark, J.R., *et al*. (1977) The invasion of Asiatic clam *Corbicula manilensis* in the New River, Virginia. *Nautilus*, **91**, 43–6.

Rosenzweig, C., Karoly, D., Vicarelli, M., *et al*. (2008) Attributing physical and biological impacts to anthropogenic climate change. *Nature*, **453**, 353–7.

Ruhland, K.M., Paterson, A.M., and Smol, J.P. (2008) Hemispheric-scale patterns of climate-related shifts in plankton diatoms from North American and European lakes. *Global Change Biol.*, **14**, 2740–54.

Schindler, D.W., Bayley, S.E., Parker, B.R., *et al*. (1996) The effects of climatic warming on the properties of boreal lakes and streams at the Experimental Lakes Area, Northwestern Ontario. *Limnol. and Oceanograph.*, **41**, 1004–17.

Sharma, S., Jackson, D.A., and Minns, C.K. (2009) Quantifying the potential effects of climate change and the invasion of smallmouth bass on native lake trout populations across Canadian lakes. *Ecography*, **32**, 517–25.

Sharma, S., Jackson, D.A., Minns, C.K., and Shuter, B.J. (2007) Will northern fish populations be in hot water because of climate change. *Global Change Biol.*, **13**, 2052–64.

Smith, A.L., Hewitt, N., Klenk, N. *et al*. (2012) Effects of climate change on the distribution of invasive alien species in Canada: a knowledge synthesis of range change projections in a warming world. *Environ. Rev.*, **20**, 1–16.

Sousa, R., Antunes, C., and Guihermino, L. (2008) Ecology of the invasive Asian clam *Corbicula fluminea* (Müller, 1774) in aquatic ecosystems: An overview. *Ann. Limnol.*, **44**, 85–94.

Stewart, I.T., Cayan, D.R., and Dettinger, M.D. (2005) Changes toward earlier streamflow timing across Western North America. *J. of Clim.*, **18**, 1136–55.

Swift, T.J., Perez-Losada, J., Schladow, S.G., *et al*. (2006) Water clarity modeling in Lake Tahoe: Linking suspended matter characteristics to Secchi depth. *Aquat. Sci.*, **68**,1–15.

Thiede, G. (1997) Impact of lake Trout Predation on Prey Populations in Lake Tahoe: A Bioenergetics Assessment. Thesis, Utah State University.

Tucker, A.J., Williamson, C.E., Rose, K.C., *et al*. (2010) Ultraviolet radiation affects invasibility of lake ecosystems by warm-water fish. *Ecology*, **91**, 882–90.

UCD (2011) State of the Lake Report 2011. University of California Davis. http://terc.ucdavis.edu/stateofthelake/StateOfTheLake2011.pdf (accessed June 29, 2012).

Vanderploeg, H.A., Liebig, J.R., Nalepa, T.F., *et al*. (2010) Dreissena and the disappearance of the spring phytoplankton bloom in Lake Michigan. *J. Great Lakes Res.*, **36** (suppl. 3), 50–9.

VanderZanden, M.J., Chandra, S., Allen, B.C., *et al*. (2003) Historical food web structure and restoration of native aquatic communities in Lake Tahoe (CA-NV) basin. *Ecosystems*, **3**, 274–88.

Vehniainen, E.R., Hakkinen, J.M., and Oikari, A.O.J. (2007) Fluence rate or cumulative dose? Vulnerability of larval Northern pike (*Esox lucius*) to ultraviolet radiation. *Photochem. Photobiol.*, **83**, 444–9.

Verburg, P., Hecky, R.E., and Kling, H. (2003) Ecological consequences of a century of warming in Lake Tanganyika. *Science*, **301**, 505–7.

Weitere, M., Vohmann, A., Schulz, N., *et al*. (2009) Linking environmental warming to the fitness of the invasive clam *Corbicula fluminea*. *Global Change Biol.*, **15**, 2838–51.

Werner, S., and Rothhaupt, K.O. (2008) Mass mortality of the invasive bivalve *Corbicula fluminea* induced by a severe low-water event and associated low water temperatures. *Dev. Hydrobiol.*, **204**, 143–50.

Westermann, F. and Wendling, K. (2003) Aktuelles Muschelsterben im Rhein Was ist die (wahrscheinlichste) Ursache des "Muschelsterbens" im Rhein im Hochsommer 2003? Report: Landesamt für Wasserwirtschaft Rheinland Pfalz

Williamson, C.E., Hargreaves, B.R., Orr, P.S., and Lovera, P.A. (1999) Does UV play a pole in changes in predation and zooplankton community structure in acidified lakes? *Limnol. Oceanog.*, **44**, 774–83.

Williamson, C.E., Zagarese, H.E., Schulze, P.C., *et al*. (1994) The impact of short-term exposure to UV-B on zooplankton communities in north temperate lakes. *J. of Plankton Res.*, **16**, 205–18.

Winder, M. and Hunter, D.A. (2008) Temporal organization of phytoplankton communities linked to physical forcing. *Oecologia*, **156**, 179–92.

Winder, M., Reuter, J.E., and Schladow, S.G. (2009) Lake warming favours small-sized planktonic diatom species. *Proc. R. Soc. Lond., Ser. B: Biol. Sci*., **276**, 427–35.

16

Long-Term Changes in the Lake Kinneret Ecosystem: The Effects of Climate Change and Anthropogenic Factors

Ilia Ostrovsky, Alon Rimmer, Yosef Z. Yacobi, Ami Nishri, Assaf Sukenik, Ora Hadas, and Tamar Zohary

Israel Oceanographic and Limnological Research, Yigal Allon Kinneret Limnological Laboratory, Israel

16.1   Introduction

Global climate change may have direct and indirect impacts on aquatic ecosystems. Direct effects are due to changes in forcing factors such as temperature, long- and short-wave radiation, precipitation, and wind. Indirect effects are often caused by anthropogenic manipulations or management, such as modifications of the hydrological regime in the watershed, increased water withdrawal during droughts, intentional or unintentional introduction of alien species, and changes in agricultural practices in the watershed. In many cases, anthropogenic eutrophication and climate change impacts reinforce each other (Moss *et al.* 2011) such that it is difficult to separate their effects. Lack of knowledge about ecosystem functioning together with attempts to "improve" the adverse changes in ecosystems (e.g., deterioration of water quality, eutrophication) by "proper management measures" may exert more stress on ecosystems and lead to unpredictable consequences. Here we report on long-term changes in the large, deep, subtropical Lake Kinneret (L. Kinneret) and evaluate effects of climate change and anthropogenic factors on the lake ecosystem.

Israel, like other Middle East semi-arid countries, is vulnerable to climate change due to generally dry conditions and limited water availability. Recent IPCC global circulation models as well as recent runs of four different climate models agree on a drying scenario in the Middle East by the end of the twenty-first century (IPCC 2007; Krichak *et al.* 2011; Rimmer *et al.* 2011b). Future climate is predicted to show trends

Climatic Change and Global Warming of Inland Waters: Impacts and Mitigation for Ecosystems and Societies,
First Edition. Edited by Charles R. Goldman, Michio Kumagai and Richard D. Robarts.
© 2013 John Wiley & Sons, Ltd. Published 2013 by John Wiley & Sons, Ltd.

of decreasing precipitation, increasing temperatures and evaporation, concurrent with higher incidence of extreme events (Krichak *et al*. 2007; Samuels *et al*. 2009). The potential decrease in water sources and steadily growing freshwater demands already impose large pressure on all available water resources in Israel. Lake Kinneret supplies about 30–50% of the national water demands. Therefore the ecological sustainability of the lake and its water quality are of great national concern.

A long-term monitoring program (>40 years) on L. Kinneret revealed prominent multiannual changes in ecosystem parameters. These changes were related to exploitation of terrestrial and aquatic resources in the watershed or the consequence of anthropogenic modifications aimed to improve the impaired ecological situation in the lake and its watershed (Hambright *et al*. 2008; Nishri 2011; Zohary and Ostrovsky 2011). Since the 1980s and 1990s, increased demand for freshwater as well as climate- and anthropogenic decline of inflow volumes resulted in three evident cycles of water level declines followed by exceptional floods which rapidly refilled the lake. The excessive fluctuations of the water level are typical for reservoirs in semi-arid regions (Sánchez-Carrillo 2007). The outcome of increasing stress on water resources in semi-arid and arid regions in response to combined (and interrelated) processes of climatic changes, population growth, and industrial development are as yet not well investigated.

This chapter provides an overview of the multiannual responses of a subtropical lake to long-term environmental changes including direct or indirect climate effects, and anthropogenic activities that may appear like climate shifts. We consider changes in the watershed, lake management, nutrient loads, thermal structure, salinity regime, plankton and fish communities, methane emission, and sedimentation processes, to better understand how subtropical lakes respond to alterations of these forcing factors. We highlight the importance of anthropogenic impacts which may magnify climate-like ecosystem changes or even surpass them.

16.1.1   Study site

Lake Kinneret is a warm lake located at ∼210 m below sea level. Its watershed area is 2730 km^2 (Figure 16.1). The major inflow to the lake is the Jordan River, which drains ∼1700 km^2 of the watershed. At its maximal water level the lake holds 4200×10^6 m^3 and its surface area is 167 km^2. The main outflow from the lake is via Israel's National Water Carrier (NWC). By pumping ∼330×10^6 m^3 annually through the NWC to the coastal area of Israel, the lake water is used for drinking, irrigation and reuse in populated areas. The water level of the lake strongly depends on the balance between natural inflows and artificial withdrawal for water supply. The lake is thermally stratified during April–December. Anoxic conditions in the hypolimnion are usually developed by May. Lake Kinneret is meso-eutrophic with a mean annual primary production of ∼600 g C m^{-2}. Chlorophyll *a* concentrations and primary productivity gradual increase between February and April–May and decline in the summer and fall (Yacobi 2006). The epilimnion is largely deprived of nutrient supply from the hypolimnion such that the phytoplankton community becomes nutrient limited following the establishment of stratification. The intensity of phytoplankton build up during holomixis determines to a large extent the phytoplankton biomass during the period of stratification (Yacobi and Ostrovsky 2008) portraying the consequential cascading of the earlier accumulated energy through the food web over summer and fall.

16.2   Changes in the watershed and their long-term impacts on Lake Kinneret

The Jurassic karst of Mt. Hermon is the major source of the Jordan River. The lake is also fed by runoff from the Golan Heights and the eastern Galilee. The Hula Valley constitutes the center of the drainage basin (Figure 16.1). Until the early 1950s, the southern and central parts of this valley were covered by the shallow Lake Hula and adjacent marshes. These water bodies served as natural filters for downstream L. Kinneret, retaining part of the nutrients from the watershed. Lake Hula was drained in the 1950s in order to increase arable land, to eradicate malaria, and to reduce evaporation losses (Hambright *et al*. 2008). The drainage engineering activities required the removal of peat from the marsh-bed sediments, which were later scattered over the soil surface. Agro-engineering activities that followed resulted in exposure of peat to atmospheric oxygen, thereby accelerating its biodegradation (Shoham and Levin 1968), and creating heavy dust storms.

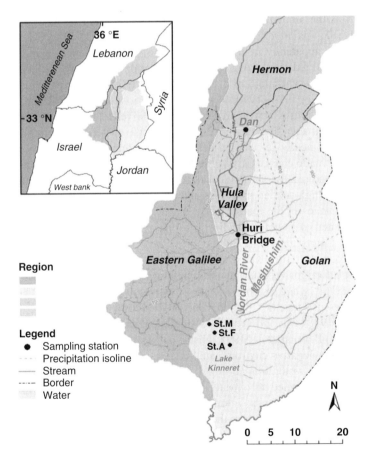

Figure 16.1 Lake Kinneret and it's watershed in northern Israel. Filled circles show locations of sampling stations. Modified from the original design by http://www.stav-gis.com/index_eng.htm (accessed July 11, 2012). (See insert for color representation.)

The long-term record of annual mean organic nitrogen (N_{org}) concentrations in Dan stream, a major source for the Jordan River and L. Kinneret, showed the highest concentrations in the mid-1970s (Figure 16.2a). Those concentrations declined in the following decade and remained at low levels throughout the 1990s and 2000s. A similar pattern was observed for organic carbon (C_{org}) in the Jordan River (Nishri 2011), where its mean annual values strongly correlated with N_{org} ($r = 0.93$, $P<0.001$). The long-term dynamics of N_{org} display large similarity in L. Kinneret, the Jordan River (downstream of the Hula Valley), Dan stream (the upper part of the watershed, upstream of the Hula Valley), and Meshoshim stream (external to the Jordan River catchment area, Figure 16.1). The coherent changes in N_{org} are confirmed by strong correlations of its concentrations in various water bodies around the Hula Valley (Figure 16.2b). These findings reflect the immense aeolian transport of nutrients from the degraded peat soils in the 1970s when the highest N_{org} concentrations were detected. The erosion of peat and its aeolian transport was also documented by finding its $\delta^{13}C$ signature in a sediment core retrieved from L. Kinneret (Dubowski, Erez, and Stiller 2003).

It is noticeable that in 1974 the import of C_{org} to L. Kinneret (ca. 10^5 tons C year^{-1}) was as high as the amount of carbon assimilated by primary producers. During the following decade this import decreased threefold (Nishri 2011). The declining load of organic material apparently influenced the rates of organic matter degradation in the lake. This is supported by a strong negative correlation ($r = -0.84$, $P<0.001$) between the annual mean concentrations of dissolved N_{org} and dissolved oxygen in L. Kinneret surface water during 1974–1990. Negative correlations between pH and N_{org} in the Jordan River and Dan stream (Figure 16.2c) may also reflect dependence of respiratory activity (organic matter degradation rate) on organic material concentrations in the riverine waters. A nearly 25-year lag between the beginning of the Hula drainage and the observed dramatic changes in N_{org} and C_{org} supplied to L. Kinneret was associated with the time required for the gradual decomposition of the peat soils, their erosion and transport through aeolian processes and by floods into L. Kinneret (Nishri 2011).

Temporal changes in the loads of N_{org} and C_{org} originating from the peat erosion were accompanied by a twofold decline in zooplankton biomass (Gophen, Serruya, and Spataru 1990; Hambright 2008) and the ratio of zooplankton metabolism (respiration) to primary production in L. Kinneret (Nishri et al. 1998). This suggested that organic material of peat origin was one of the major sources of C_{org} for the Kinneret food web in the 1970s, while its role essentially decreased by the 1990s. The lessening of grazing pressure together with the elevated inflows of bioavailable phosphorus (Markel et al. 1998) was followed by a significant rise in phytoplankton biomass in the early 1980s (Berman et al. 1995; Nishri and Hamilton 2010).

To alleviate the negative environmental consequences of the drainage of Lake Hula and its wetlands, a restoration program for the Hula Valley was realized in 1994. This project was designed to elevate the water table in the Hula Valley by partial reconstruction of the wetlands. In addition, to minimize the flow of polluted water from Hula Valley into L. Kinneret, these waters were diverted into a storage reservoir to be used locally for irrigation (Hambright et al. 2008). Such regulation of storm water runoff via the valley together with a general decrease in precipitation minimized the provision of water and chemical compounds from Hula Valley in dry and normal rainfall years. The later may have affected the bloom of a "keystone" phytoplankton species (see Section 16.5).

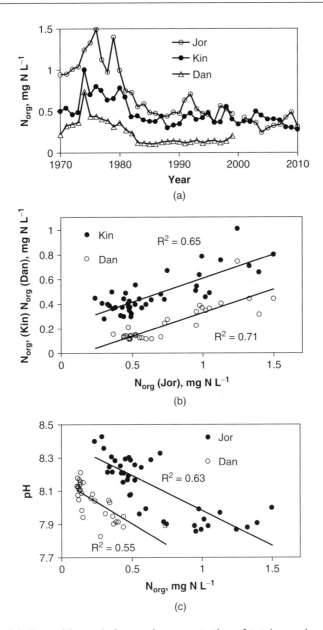

Figure 16.2   (a) The multiannual changes in concentration of total organic nitrogen, N_{org}, in surface waters (0–5 m) of Lake Kinneret (Kin), the Jordan River (Jor), and the Dan stream (Dan). (b) Correlations between concentrations of total organic nitrogen in the Jordan River, N_{org}(Jor), Lake Kinneret, N_{org}(Kin), and the Dan stream, N_{org}(Dan). (c) Correlations between concentrations of total organic nitrogen, N_{org}, and pH in the Jordan River and the Dan stream. Each point represents annual mean. Lake Kinneret data were collected in its center. Monitoring stations on rivers and Lake Kinneret are shown in Figure 16.1. Sampling and analytical measurements of chemical variables were described by Nishri (2011).

16.3   Long-term lake salinity changes

The salinity of L. Kinneret fluctuated between 190 and 280 mg Cl L^{-1} and was significantly higher than the salinity of the Jordan River ($\sim$30 mg Cl L^{-1}) and other surface streams. A lake salinity model (LSM) calculates the long-term salinity variations of the lake, based on the main components of the annual water and solute balance: (i) solute inflows to the lake from onshore and offshore saline springs; (ii) freshwater inflows from the Jordan River and the direct watershed; (iii) water pumped and released from the lake; and (iv) direct evaporation from the lake (Rimmer *et al.* 2006). The model helped to identify two dominant processes that determined the major salinity changes since 1965. The first process was a step reduction in solute inflow from $\sim$1.60 $\times$ 10^5 to $\sim$1.05 $\times$ 10^5 tons Cl$^-$ year^{-1}, caused by the operation of the Saline Water Carrier (SWC) in 1965 (SWC: a pipe that diverts $\sim$0.55 $\times$ 10^5 tons Cl$^-$ year^{-1}). As a result, the solute mass in the lake declined exponentially from 1965 to 1988 (Figure 8 in Rimmer *et al.* 2011b). The second process was the continuous reduction of freshwater inflows from $\sim$830 $\times$ 10^6 m^3 in the 1960s to $\sim$630 $\times$ 10^6 m^3 in the 2000s that resulted in a gradual linear increase of the lake residence time from $\sim$7 to $\sim$10 years, which, in turn caused a mild increase in lake salinity. The cumulative effect of both processes can be seen in the exponential decrease of solute mass in the lake from an average of $\sim$1.655 $\times$ 10^6 tons in 1965 to $\sim$0.8 $\times$ 10^6 tons during the end of the 1980s, and then a gradual increase back to $\sim$0.95 $\times$ 10^6 tons in 2010. The successful verification of past salinity trends revealed that our understanding of the combination of past salinization processes was good, and could be used for prediction of future salinity changes.

The diversion of saline water through the SWC was considered a successful operation to reduce lake salinity, which allowed, for many years, the use of the lake water for irrigation and domestic consumption. However, the significant reduction of fresh water inflows during the last decades and the expected further reduction, pose a threat of a significant salinity increase in the coming decades. On one hand the potential damage to the ecology of the lake is not obvious since the natural salinity of the lake was once much higher ($>$ 320 mg Cl L^{-1}) than it is today; on the other hand a salinity increase reduces water quality and creates a call for further changes in the Israeli water supply system.

16.4   Long-term lake stratification changes

Long-term increases in water temperature and changes to thermal stratification patterns have been observed in lakes across diverse climatic regions (Livingstone 2003), and systematically associated with the increase of air temperature (Livingstone *et al.* 2010). This increase seems to be a reflection of global warming (Williamson, Saros, and Schindler 2009), and is often related to documented regional variation of meteorological forcing (Dokulil *et al.* 2006). Long-term changes in thermal stratification in a lake can be caused by (i) gradually increasing air temperature (global climate change); (ii) local climatic and hydrological variations, such as gradually decreasing precipitation, followed by reduced stream flows and increased retention time (local causes), or (iii) increased exploitation of water upstream before it reaches the lake, overpumping from the lake, reduced lake level and reduced retention time.

Changes in L. Kinneret stratification since the 1970s can be seen by comparing five-year average water temperature contour maps (Figure 16.3a: 1970–1975 and Figure 16.3b: 2004–2009). An obvious difference in temperature regime is presented in Figure 16.3c. Analysis of the stratification change revealed that average epilimnion thickness decreased for all seasons by ~1–2 m over the entire period, especially during autumn; metalimnion thickness decreased by ~1 (autumn) to ~2 m (spring); and hypolimnion thickness slightly increased (~1 m) in time (Rimmer *et al.* 2011a). The average temperature of the epilimnion increased by ~1 °C (~0.025°C year^{-1}), the average hypolimnetic temperature remained constant (~15 °C), and the temperature gradient across the metalimnion increased by ~0.016 °C m^{-1}year^{-1} during the summer and fall. Rimmer *et al.* (2011a) concluded that the observed increase in water temperature and subsequent modification of thermal stratification were mainly imposed by the reduction in lake level, which was mainly an anthropogenic effect. Secondary importance to the stratification changes was attributed to the general increase of air temperature in the Jordan River catchment area and to the trend of declining inflows, especially during April and December. Analysis of the potential effect of two additional variables—water transparency and wind speed—did not result in a significant effect on the observed long-term changes in stratification. Most of these findings were supported by an analysis with DYRESM (Imberger and Patterson 1981), a 1D hydrodynamic model.

Thus, in contrast to the widespread observations on climate-related changes to the thermal regimes of various lakes, the changes in L. Kinneret were mainly due to local hydrological management, while only a minor effect should be attributed to global climate change.

16.5 Long-term changes to phytoplankton

The L. Kinneret phytoplankton assemblage was reported by Pollingher (1986) to have a typical annual succession pattern that was repeated from year to year. Its characteristic feature was an intense spring bloom of the large-celled dinoflagellate *Peridinium gatunense* (Figure 16.4a) with a biomass between 100–300 g wet weight m^{-2} at the bloom peak. The blooms disappeared early in the summer, after which phytoplankton biomass was low and composed of a diversity of nanoplanktonic chlorophytes, diatoms, and cyanophytes. Reynolds (2002) referred to this annual pattern: "Kinneret provides one of the best-known and best-attested examples of year-to-year similarity in the abundance, distribution and composition of the phytoplankton."

However, in the mid 1990s the Kinneret phytoplankton underwent a major change, resembling a regime shift (Zohary 2004). Since then *P. gatunense* blooms developed only after winters with high riverine inflows (Figure 16.5); a bloom failed to develop in 10 out of the 16 years between 1996 and 2011. Fungal epidemics attacking *P. gatunense* became prominent (Alster and Zohary 2007). Other deviations from the previous typical annual pattern included higher summer phytoplankton biomass with replacement of the species assemblage of mostly nanoplanktonic, palatable forms with filamentous or spiny, less palatable forms; new appearance and establishment of toxin-producing, nitrogen-fixing cyanobacteria in summer; and an increase in the absolute biomass and proportion of cyanobacteria to total biomass (Zohary 2004).

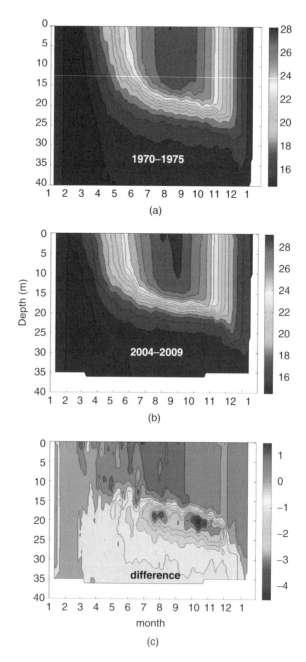

Figure 16.3 Typical five-year average of weekly water temperature profiles for the periods (a) 1970–1975 and (b) 2004–2009 at Stn. A (Figure 16.1); (c) Temperature differences throughout the water column are computed as the difference between (b) and (a). Monitoring and measurements of water-profile thermal structure were described by Rimmer *et al.* (2011a). (See insert for color representation.)

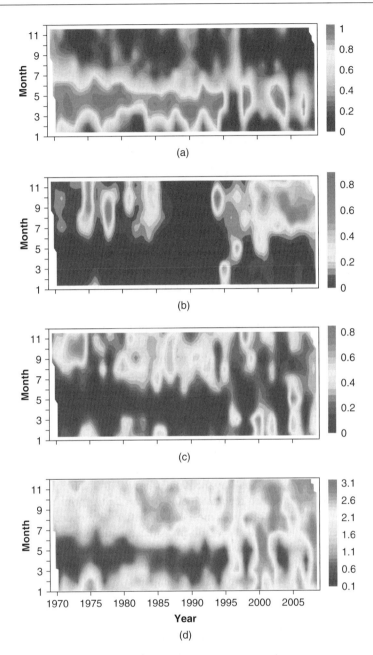

Figure 16.4 Interannual and seasonal changes in phytoplankton community in Lake Kinneret. (a–c) Relative biomass of three major groups of phytoplankton (dimensionless units). The relative biomass of each group is calculated as its proportion of the total biomass. (a) Dinoflagellates. (b) Cyanobacteria. (c) Chlorophytes. (d) Shannon diversity index. Data show recurrent spring-summer blooms of dinoflagellates (mainly *Peridinium gatunense*) until the mid-1990s. Since 1996 the bloom failed to develop in most years and the diversity index in spring and summer has increased. Intensive recurrent blooms of cyanobacteria commenced in the mid-1990s in summer and fall. Monitoring and determination of phytoplankton biomass was carried out at St. A and are described elsewhere (Zohary 2004; Zohary and Ostrovsky 2011). (See insert for color representation.)

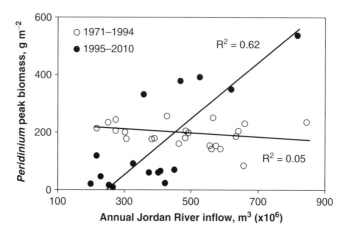

Figure 16.5   The relationship between the annual peak *Peridinium* biomass and annual Jordan River inflow volume over hydrological years (Oct–Sep) in Lake Kinneret for 1971–1994 (P>0.05, no significant relationship) and 1995–2010 (P<0.001, highly significant relationship).

The observed change in the phytoplankton was apparently associated with hydrological changes in the watershed following the 1994 Hula Restoration Project, when water draining the peat soils was redirected and no longer flowed into L. Kinneret on low- or normal-inflow years (see above). We noticed that since then massive *P. gatunense* blooms occurred only with rare extreme floods, when runoff from the peat soils did reach L. Kinneret. Such bloom dynamics highlight the main source of the limiting nutrient. One can suggest that selenium, known as an important micronutrient for *Peridinium* (Lindstrom and Rodhe 1978) may be a key agent regulating the blooming of this alga in L. Kinneret. Because selenium is efficiently removed from the water column by sedimentation (Oliver *et al*. 2009), its replenishment in the lake may depend on external supply. This explains why the regular provision of the peat-soil water with winter-spring floods stimulated regular blooms of *Peridinium* prior to 1996.

The lack of a *Peridinium* bloom in most of years after 1995 opened new ecological niches for other algal taxa, and successful invasions by alien species, including toxic cyanobacteria (Figure 16.4b, see below). Other manifestations included dominance by numerous nanoplanktonic algae in spring (Figure 16.4c) when an unusually high Shannon diversity index was recorded, for example, in 2000 (Figure 16.4d).

16.6   The invasion of nitrogen-fixing cyanobacteria (Nostocales)

The long-term stable pattern of a high-diversity, low-biomass phytoplankton assemblage in summer-fall ended in 1994 when the first-ever bloom of an invasive nitrogen fixing cyanobacterium *Aphanizomenon ovalisporum* (Nostocales) occurred (Pollingher *et al*. 1998; Hadas *et al*. 1999). Since then, this filamentous toxin-producing species has comprised a variable component of the summer-fall phytoplankton community and its toxin, cylindrospermopsin, was occasionally detected. An additional diazotrophic cyanobacterium, *Cylindrospermopsis raciborskii*, was observed for the first time in summer 2000 (Zohary and Shlichter 2009) forming a major summer bloom in 2005 and since then co-dominating the summer assemblage with *A. ovalisporum*. The local *C. raciborskii* is a non-toxic strain (Alster *et al*. 2010). This shift in the phytoplankton composition toward enhanced diazotrophic (N-fixing) activity in the summer may be driven by gradual changes in environmental variables (Hadas *et al*. 2002). It has been suggested that the proliferation of Nostocales in L. Kinneret is part of a worldwide phenomenon of the invasion of cyanobacteria into mesotrophic and oligotrophic lakes in the temperate zones (Wiedner *et al*. 2007; Mehnert *et al*. 2010). The high adaptability of Nostocales is attributed not only to their diazotrophic activity but also to their ability to form dormant cells known as akinetes. Akinetes reside in the sediment and resist adverse conditions, germinate to produce vegetative cells in response to improved environmental conditions, and are brought to the water column by resuspension and gas vacuole production (Kaplan-Levy *et al*. 2010).

We suggested that the persistence of Nostocales in L. Kinneret was associated with a set of environmental conditions that had been met since 1994, such as the concentrations of dissolved inorganic nitrogen reaching their annual minimum early enough in the summer, high water temperature (Figure 16.6), low wind-driven turbulence, and occasional pulses of phosphorus during the summer (Hadas *et al*. 1999). The decline in allochthonous N_{org} and the respective shift in the nitrogen-to-phosphorus ratio of the loads entering the lake in the 1980s (see above) may have favored the blooms of nitrogen-fixing cyanobacteria (Pollingher *et al*. 1998; Gophen *et al*. 1999). The sudden blooms of these cyanobacteria could be associated with rapid changes in the ecosystem and appearance of new ecological niches (see also Chapter 11).

16.7 Winter populations of toxic *Microcystis*

Following the interruption of *Peridinium* blooms after 1995 the toxic cyanobacterium *Microcystis* sp. appeared in the lake and formed thin surface scums. Unlike the Nostocales, *Microcystis* is not an invasive species; it was reported to bloom in L. Kinneret during the late 1960s and occasionally in the 1970s. In most cases when significant *Microcystis* populations were observed early in winter, the *Peridinium* bloom was delayed. Allelopathic interactions between *Microcystis* and *Peridinium* were demonstrated and a specific mechanism that inhibits the growth of *Peridinium* was postulated (Sukenik *et al*. 2002).

The disappearance of *Peridinium* opened the ecological system to other phytoplankton species (see also Chapter 11) including the toxic cyanobacterium *Microcystis* sp. Frequent blooms of *Microcystis* sp. could also be associated with reconstruction of the Hula wetlands, the alterations in nutrient supply from the watershed, and reduction in the ratio between nitrogen and bioavailable phosphorus due to a rise in fertilization of cultivated soils (Nishri 2011).

Based on the multiannual survey of microcystin derivatives in L. Kinneret (Table 16.1), we concluded that the blooming population frequently consisted of the same strains of *Microcystis*. Nevertheless, in 2009 microcystin-LF was observed, suggesting the presence and coexistence of various *Microcystis* chemotypes (strains that produce different sets of microcystins). Such variability is related to the strain response to changes in environmental conditions, imposed by anthropogenic activities and regional climate conditions.

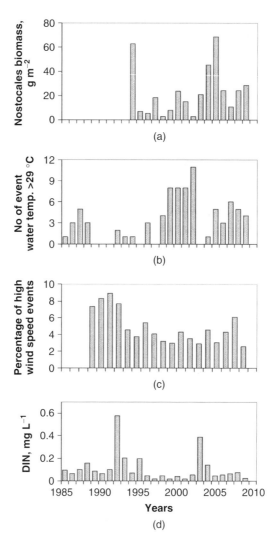

Figure 16.6   Annual variations in maximal Nostocales biomass in Lake Kinneret (a) presented together with number of high ($\geq 29.0\,^{\circ}$C) water temperature records observed between June and November each year (b), the ratio of high ($>6\,$m s^{-1}) wind speed records during that period (c), and the average dissolved inorganic nitrogen (DIN) in the upper 10 m water layer in June (d).

16.8   Long-term variation of chlorophyll and primary production

The mean chlorophyll a (Chl) concentrations in the second half of the year varied only slightly between the early 1970s and the mid-1990s; however, since then a trend of increasing Chl was noted (Yacobi 2006). At the same time we could not discern a long-term trend in primary productivity (PP) (Figure 16.7a). The different patterns of Chl and PP affected the interannual variation of biomass-based photosynthetic activity

Table 16.1 Abundance and concentration of microcystins in Lake Kinneret from 2005–2010. The most abundant micocystin derivatives are identified by their amino acid composition at position 2 and 4. L-Leu, R-Arg, Y-Tyr and F-Phe.

Date	Total microcystins ($\mu g\,L^{-1}$)	Microcystin content in biomass ($\mu g\,g^{-1}$)
Feb. 2010	2.6 (LR, RR)	61 (YR, LR, RR)
Jan. 2009	0.95 (LR, YR, RR)	710 (YR, LR, RR)
Feb. 2009	0.8 (LR, YR,RR)	1690 (LF, LR, YR, RR)
Mar. 2009	1.9 (LR, YR, RR)	1070 (LR, YR, RR)
Mar. 2008	<0.1	551 (LR, YR, RR)
Feb. 2007	<0.1	1300 (YR, LR, RR)
Mar. 2007	<0.1	210 (YR, RR, LR)
Feb. 2006	<0.1	156 (YR, RR, LR)

(assimilation number, A.N.). The A.N. did not show any trend from 1972 through 1993, but since then it displayed a clear declining trend (Figure 16.7b). The observed increase in Chl concentrations and respective decrease in A.N. were apparently related to the adaptation of the phytoplankton community to reduced light availability in the euphotic zone between 1993 and 2009 when a significant decrease ($r = -0.77$, $P<0.001$) in Secchi depth from ~4.0 m to 2.9 m was recorded.

The steadiness of PP since the 1970s, despite major changes in the phytoplankton community, suggests large functional adaptability of the Kinneret ecosystem to changing ambient conditions. Phosphorous availability in winter and spring, which is the main factor controlling PP, remained at a relatively stable level in the upper productive stratum of the lake despite large variability of external loads from the watershed (Nishri 2011). This suggests long-term stability of the phosphorus internal load. Thus, primary productivity is one of the most conservative ecological parameters that may not be easily affected without fundamental changes to the nutrient loads.

16.9 Long-term sedimentation flux changes

Multiannual dynamics of sedimentation rate measured with traps displayed immense differences at various locations (Figure 16.8). At the lake center (St. A, depth 40 m), where material resuspension was usually insignificant, the sedimentation rate displayed mainly a downward flux of particulate material from the epilimnion (Ostrovsky and Yacobi 2010). The observed strong positive correlation between sedimentation flux at St. A and the maximal annual water level ($r = 0.81$, $P<0.01$) and annual water inflow ($r = 0.75$, $P<0.01$) suggested that the loads from the watershed affected the sedimentation regime at this location. The amount of nutrients, which are imported with floods and affect the winter-spring bloom of large-celled algae, could also determine the amount of sedimenting newly-produced particles. A three- to fourfold decrease in sedimentation flux between 2003 and 2010 at the lake center reflected the essential reduction in export of particulates from the upper productive stratum, which could have been associated with superior recycling of organic material in the epilimnion.

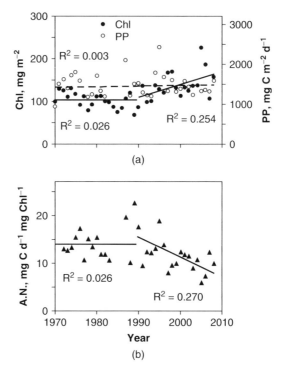

(a)

(b)

Figure 16.7 Multiannual changes in (a) the average chlorophyll *a* (Chl) density in the 0–15 m water layer and primary productivity (PP), and (b) assimilation number (A.N.) for the period from July to December. Note the lack of a temporal change in the values of PP (a), and the trend of increasing Chl density (a) and concomitant decreasing A.N. (b) from 1993 to 2008.

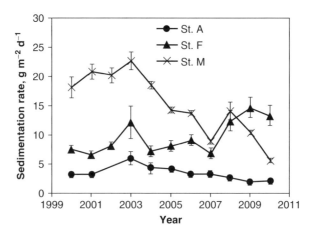

Figure 16.8 Multiannual variations of the mean sedimentation rate at three stations. St. A is a pelagic station located in the lake center (∼40 m depth), St. F is a deep peripheral station (∼20 m depth) and St. M is a littoral station (∼10 m depth). In 2003 a rapid rise of the water level occurred. Vertical bars represent standard error. Sedimentation rates were measured using sedimentation trap techniques (Ostrovsky and Yacobi 2010). Station locations are shown in Figure 16.1.

Resuspended material strongly affected the sedimentation rate at shallower areas (Ostrovsky and Yacobi 1999, 2010). A strong decrease in sedimentation rate since 2003 was observed at the littoral St. M (r = −0.96, P<0.001). This could have been associated with unusually large water-level fluctuations (WLF) during which bottom sediments in the littoral became periodically shallower and were exposed to energetic surface wave activity that removed fine particles and redeposited them to deeper locations. As a result, much coarser particles, which are harder to resuspend, became dominant in the littoral area.

The sedimentation rate at deep sublittoral St. F positioned close to the thermocline-bottom interface was strongly influenced by local resuspension and advective transport of recently settled particulate material (Ostrovsky and Yacobi 2010). These processes became the most intensive during the period of rapid thermocline deepening in fall when large areas of the lake bottom that were previously part of the hypolimnion and where fresh organic particles accumulated had now become part of the metalimnion. The interaction between internal waves, which are continuously present in the met-alimnion, with exposed sediments caused massive resuspension of recently deposited particles and their relocation toward the lake center. An increase in sedimentation rate at St. F (r = 0.73, P<0.01) over the last several years reflected enhancement of sediment transport in the sublittoral area associated with a large reduction of the water level and related shift in position of the thermocline-bottom interface.

16.10   Gaseous methane emission

In lakes and reservoirs gas ebullition could substantially envelop the total methane flux from bottom sediments to the water column and to the atmosphere (Ostrovsky *et al.* 2008). Methane ebullition from lakes is a much larger and globally signifi-cant source of atmospheric methane than formerly thought (Bastviken *et al.* 2011). Increase in temperature of organic-rich sediments may intensify methane production and ebullition (DelSontro *et al.* 2010). Thus, global warming, as well as anthropogenic factors causing increases in near-bottom temperature may intensify methane ebullition in water bodies.

Reduction of hydrostatic pressure at the bottom due to a decrease of water level is one of the most significant mechanisms causing large emissions of methane bubbles (Ostrovsky 2003). The effect of WLF on the amount of gas bubbles in the near-bottom stratum in L. Kinneret, was shown by the high correlation between bubble density in the water and the difference between the water level at the time of sampling and the lowest water level in the preceding year (Figure 16.9). Such a difference in water level determined the amount of free gas that remained in the sediment after a certain degree of its degassing in the previous year.

The predicted possible decrease in precipitation, on the one hand, and an increase in the frequency of extreme climatic events (for example, heavy rains) on the other hand (Krichak *et al.* 2007; Samuels *et al.* 2009) concurrently with an enhancement of anthropogenic pressure on the water resources, will increase the degree of WLF in lakes and reservoirs in arid and semi-arid regions. The latter will enlarge the direct flux of methane bubbles from sediments, and will intensify its delivery to the atmosphere thus contributing to global warming. In contrast, at smaller WLF, a higher propor-tion of methane will be exported from anoxic sediments via diffusive flux, which is

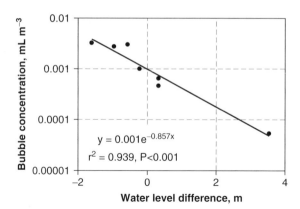

Figure 16.9 Temporal variability in near-bottom volumetric concentration of bubbles (mL m^{-3}) at the center of Lake Kinneret in relation to water-level fluctuation. Water-level difference is the difference between the water level at time of sampling and the lowest water level in the preceding year. Negative numbers reflect a decrease in water level and positive numbers reflect an increase in water level. Bubble concentration in the 3 m near-bottom water layer was measured using hydroacoustic techniques (Ostrovsky 2003; Ostrovsky *et al.* 2008) in 1998–2004.

largely depleted by methane-oxidizing bacteria in upper sediment layers and in water (Bastviken *et al.* 2011). Therefore, suppression of WLF in lakes and reservoirs is an obvious mitigating measure that will lessen delivery of methane to the atmosphere.

16.11   Long-term changes in fish populations and the fishery

Reproduction of the endemic bleak *Acanthobrama terraesanctae* (lavnun) has been evolutionarily adapted to natural water fluctuations in L. Kinneret. These fish attach their eggs to stony substrates prior to their colonization by periphyton. Such clean, hard substrates are available in the littoral zone at the time of year when water levels are increasing. This spawning behavior ensures high survival of the lavnun eggs (Gafny, Gasith, and Goren 1992). The gradual decrease in water level between 1988 and 1991 and from 1995 to 2002 was accompanied by the disappearance of hard substrates from the littoral. These water level decreases were followed by two abnormally rainy winters (1991/1992 and 2002/2003) resulting in a 4–5 m rise in water level. Such rises were extremely favorable for lavnun reproductive success, which resulted in a fewfold increase in fish density (Figure 16.10). These high spawning successes were followed by an increase in population biomass with one-two years lag time, since biomass of a lavnun cohort generally reaches its maximum during the second year of life (Ostrovsky and Walline 1999).

A huge increase in biomass of this zooplanktivorous fish led to a misbalance between their food requirements and zooplankton productivity (Ostrovsky and Walline 2001; Zohary and Ostrovsky 2011) and resulted in a crash of the zooplankton population (mainly copepods) due to an enormous increase in fish predation pressure. As a result of food shortage in such years, the fish were found in poor body condition

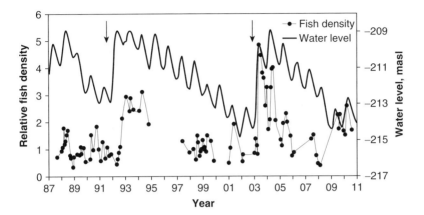

Figure 16.10 Time-series of water level and fish abundance in Lake Kinneret. Arrows mark the two events of a 4 m (winter 1991–1992) and a 4.7 m (winter 2002–2003) rise in water level, which were followed by a huge increase in the number of fish (mainly *Acanthobrama terraesanctae*), whose size exceeded −60 dB. The relative densities of fish were normalized based on the mean density of acoustic targets (>−60 dB) prior to the rapid rises in water level. The fish abundance data were collected with a 70 kHz Simrad single-beam echo sounder (model EY-M) during 1987–1994, a 120 kHz Biosonics dual beam echo sounder (model 102) during 1997–1999, and a 120 kHz Biosonics dual beam echo sounder (model DE5000) after 2001.

(Figure 16.11). An abnormal decrease in the body condition index (BCI) was found for post-spawning fish, which are more sensitive to lack of food than smaller fish. For the 2001–2005 period a strong inverse correlation ($r = 0.95$, $P < 0.05$) between the mean density of fish of −60 to −50 dB (mainly lavnun; see Figure 16.10. for method description) and zooplankton biomass was found. The latter suggests that the excessive biomass of planktivorous fish could reduce zooplankton biomass in critical years. Deteriorating body condition of large fish was followed by disappearance of these fish from the population due to increased post-spawning mortality (Ostrovsky and Walline 2001). The disappearance of large individuals of commercial size was the apparent reason for the lavnun fishery collapse in 1994–1995 and 2004–2005.

Multiannual change in catches of the Mango tilapia *Saratherodon galileus* (Figure 16.12), which was the most commercially valuable fish in L. Kinneret until 1999, is another example of the effect of WLF on the fish populations and the fishery. Strong negative correlation ($r = -0.84$, $P < 0.01$) between the tilapia catch and water level was observed from 1985 to 1998. It suggests that the catches significantly increased with declines in water level. With an increase in fishing pressure at lower water levels the average size of landed fish declined (Ostrovsky 2005). The decline in water level was accompanied by a notable decrease in stony littoral areas (Gasith and Gafny 1990), where a certain portion of the tilapia population could escape from being harvested. At exceptionally low water levels below −213.5 m.a.s.l these substrates vanished, such that the entire population of Mango tilapia became exposed to intense fishing pressure (Ostrovsky 2005). This also negatively affected fish reproduction capacity because large spawning fish became an easy target for fishermen. An extreme harvest of the unprotected fish since 1998 destroyed the ability of the tilapia population to reproduce itself due to an enormous decrease in

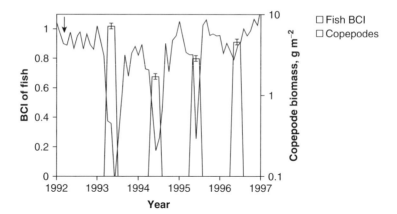

Figure 16.11   Changes in mean body condition index (BCI) of post-spawning (>13 cm long) fish, *Acanthobrama terraesanctae*, in May–July and dynamics of copepod density in 1993–1996. A dramatic decrease of the fish BCI was detected in 1994, about two years after an unusual success in fish reproduction. The lowest copepod abundance was observed in summer 1993, i.e., before the following decrease in BCI of the fish. BCI is a ratio of the actual fish weight to the mean weight of fish of the same length under good feeding conditions (Ostrovsky and Walline 2001). The arrow indicates the time of rapid water level rise. Vertical bars represent BCI standard error.

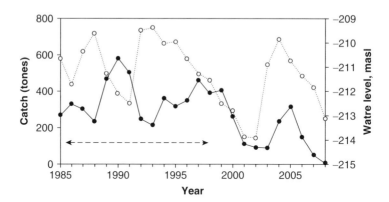

Figure 16.12   Changes in catches of Mango tilapia, *Saratherodon galileus* (thick line), in relation to the mean annual water level (circles and dotted line). A negative correlation between tilapia annual catch and water level was observed until 1998. Since 1999 catches co-varied with water level until the complete disappearance of these fish from the catches in 2008. Arrows indicate time interval of high negative correlation between fish catches and water level (see the text). The catch data are supplied from Snovsky and Shapiro (2001), Pisanty (2005), Baror and Mezner (2010).

the maternal stock. The reduced sustainability of the exploited tilapia stock in L. Kinneret during the last decade resulted in a lowered harvest with a decrease in water level. During the rapid rise in the water level in winter 2002–2003, large amounts of flooded vegetation appeared in the littoral zone all around the lake. This together with the appearance of stony littoral-formed fish refuges, improved breeding success, and resulted in an increase in harvest with a one- to two-year delay, which was needed for small fish to attain the minimal catch size. However, later an uncontrolled fishery under a long-term decrease in water level destroyed the tilapia population in 2008. Venturelli, Shuter, and Murphy (2009) showed a direct link between population age structure and reproductive rate that was consistent with strong effects of maternal quality on population dynamics. They concluded that a population of older, larger individuals had a higher maximum reproductive rate than an equivalent population of younger, smaller individuals, and that this difference increased with the reproductive lifespan of the population.

The vulnerability of fish reproduction and fisheries to extreme WLF can cause dramatic changes in fish population in lakes under conditions of global climate change. While strict fishery control is an obvious measure under such conditions, special attention should be given to restricting WLF to their natural regime that native fish had adapted evolutionarily. Creation of artificial aquatic habitats and spawning grounds, and development of captive breeding programs can be additional measures to protect fish. In contrast, the development of artificial fish stocking should be done with a large degree of caution as it can reduce genetic diversity of populations and thus affect the long-term sustainability of ecosystem functions.

16.12   Conclusions

Drainage of wetlands and concurrent anthropogenic changes in the watershed in the 1950s resulted in a multidecade deterioration of the Kinneret ecosystem. This case demonstrates the importance of upstream wetlands in lake ecological regimes and suggests that their disappearance due to decreasing precipitation may have significant impacts on lake functioning. Recent attempts to improve the ecological situation in the L. Kinneret watershed by wetland reconstruction and control of outflows from peat soils unpredictably affected the occurrence of "keystone" algal blooms in winter and spring, and thus adversely changed the entire lake ecosystem.

The anthropogenic increase in water demands and reduction of precipitation during the last decades altered the hydrological, thermal, and salinity regimes in L. Kinneret, emulating the effect of global warming. We demonstrated that such changes concurrently with increasing frequency of extreme events (such as floods, WLF) may have a conspicuous ecological impact on a lake in a semi-arid region. Particularly, the increase beyond the natural amplitude of WLF as well as long-term water level decline affected fish reproduction, trophic relationships, littoral communities, sedimentological regime, and enhanced methane ebullition. To diminish the negative effect of WLF, management policies should be directed to progressively decrease water consumption during drawdown periods. This requires developing a comprehensive strategy for economic fresh water usage and supply.

The L. Kinneret case demonstrates that superimposing anthropogenic impacts on milder climate-related changes on a lake can substantially intensify the effect of the latter. To mitigate the effect of rapidly altering environmental conditions on lake ecosystems, measures should be directed to better maintain their natural ecological regime to which these ecosystems have adapted evolutionarily. In contrast, attempts to actively "improve" complex aquatic ecosystems together with our insufficient knowledge of their functioning may lead to deep and possibly irreversible shifts.

16.13   Acknowledgements

We thank N. Koren, T. Fishbein, Y. Lechinsky, A. Rynskiy, S. Kaganovsky, and M. Diamant for technical assistance. Chemical analyses were performed by the Watershed Unit, Mekorot Water Company. Inflow data were provided by the Israel Hydrological Services. Fish catch data are courtesy of J. Shapiro and Z. Snovsky (Fisheries Dept., Ministry of Agriculture). Copepod biomass is courtesy of M. Gophen (Kinneret Limnological Laboratory database). Our research was supported by the Lake Kinneret Monitoring Program funded by the Israel Water Authority and by grants from the Israeli Science Foundation, German Ministry of Research and Technology, Israel Ministry of Science and Technology (ISF grants 211/02 and 627/07 1011/05; BMBF grant FKZ 02WT0985; MOST grant WR803).

References

Alster, A., Kaplan-Levy, R., Sukenik, A., and Zohary, T. (2010) Morphology and phylogeny of a non-toxic invasive *Cylindrospermopsis raciborskii* from a Mediterranean Lake. *Hydrobiologia*, **639**, 115–28.

Alster, A. and Zohary, T. (2007) Interaction between the bloom-forming dinoflagellate *Peridinium gatunense* and the chytrid fungus *Phlyctochytrium* sp. *Hydrobiologia*, **578**, 131–9.

Bastviken, D., Tranvik, L.J., Downing, J.A., *et al.* (2011) Freshwater methane emissions offset the continental carbon sink. *Science*, **331**, 50.

Behrenfeld, M. (2011) Biology: Uncertain future for ocean algae. *Nature Clim. Change*, **1**, 33–4.

Berman, T., Stone, L., Yacobi, Y.Z., *et al.* (1995) Primary production and Phytoplankton in Lake Kinneret: A long-term record (1972–1993). *Limnol. Oceanogr.*, **40**, 1064–76.

DelSontro, T., McGinnis, D.F., Sobek, S., *et al.* (2010) Extreme methane emissions from a Swiss hydropower reservoir: contribution from bubbling sediments. *Environ. Sci. Tech.*, **44**, 2419–25.

Dokulil, M.T., Jagsch, A., George, G.D., *et al.* (2006) Twenty years of spatially coherent deepwater warming in lakes across Europe related to the North Atlantic Oscillation. *Limnol. Oceanogr.*, **51**, 2787–93.

Dubowski, Y., Erez, J., and Stiller, M. (2003) Isotope paleolimnology of Lake Kinneret. *Limnol. Oceanogr.*, **48**, 68–78.

Gafny, S., Gasith, A., and Goren, M. (1992) Effect of water level fluctuation on shore spawning of *Mirogrex terraesanctae* (Steinitz), Cyprinidae in Lake Kinneret, *Israel. J. Fish Biol.*, **41**, 863–71.

Gasith, A. and Gafny, S. (1990) Effects of water level fluctuation on the structure and function of the littoral zone, in *Large lakes, Ecological Structure and Function* (eds. M.M. Tilzer and C. Serruya), Springer-Verlag, Berlin, pp. 156–71.

Gophen, M., Serruya, S., and Spataru, S. (1990) Zooplankton community changes in Lake Kinneret (Israel) during 1969–1985. *Hydrobiologia*, **191**, 39–46.

Gophen, M, Smith, V.H., Nishri, A., and Therkland, T. (1999) Nitrogen deficiency phosphorus sufficiency, and the invasion of Lake Kinneret, Israel, by N_2-fixing cyanobacterium *Aphanizomenon ovalisporum*. *Aquat. Sci.*, **61**, 293–306.

Hadas, O., Pinkas, R., Delphine, E., *et al.* (1999) Limnological and ecophysiological aspects of *Aphanizomenon ovalisporum* bloom in Lake Kinneret, *Israel. J. Plankton Res.*, **21**, 1439–53.

Hadas, O., Pinkas, R., Malinsky-Rushansky, N., *et al.* (2002) Physiological variables determined under laboratory conditions may explain the bloom of *Aphanizomenon ovalisporum* in Lake Kinneret. *European J. Phycol.*, **37**, 259–67.

Hambright, K.D. (2008) Long-term zooplankton body size and species changes in a subtropical lake: implications for lake management. *Fund. Appl, Limnol.*, **173**, 1–13.

Hambright, K.D., Zohary, T., Eckert, W., *et al.* (2008) Human engineered hydrological changes: exploitation and destabilization of the Sea of Galilee. *Ecol. Appl.*, **18**, 1591–603.

Imberger, J. and Patterson, J.C. (1981) A dynamic reservoir simulation model –DYRESM5, in *Transport Models for Inland and Coastal Waters* (ed. H.B. Fischer), Academic Press, New York, pp. 310–61.

IPCC (2007) *Climate Change 2007: Mitigation. Contribution of Working group III to the Fourth Assessment Report of the Intergovernmental Panel on Climate Change* (eds. B. Metz, O.R. Davidson, P.R. Bosch, R. Dave and L.A. Meyer), Cambridge University Press, Cambridge.

Kaplan-Levy, R.N., Hadas, O., Summers, M.L., *et al.* (2010) Akinetes—dormant cells of cyanobacteria, in *Dormancy and Resistance in Harsh Environments* (eds. E. Lubzens, J. Cerda and M.S. Clark), Springer, Berlin, pp. 109–32.

Krichak, S., Alpert, P., Bassat, K., and Kunin, P. (2007) The surface climatology of the eastern Mediterranean region obtained in a three-member ensemble climate change simulation experiment. *Adv. Geosci.*, **12**, 67–80.

Krichak, S., Breitgand, J., Samuels, R., and Alpert, P. (2011) A double-resolution transient RCM climate change simulation experiment for near-coastal eastern zone of the Eastern Mediterranean region. *Theor. Appl. Climat.*, **103**, 167–95.

Lindstrom, K. and Rodhe, W. (1978) Selenium as a micronutrient for the dinoflagellate *Peridinium cinctum fa westii*. *Ver. Internat. Verein..Theoret. Limnol.*, **21**, 168–73.

Livingstone, D.M. (2003) Impact of secular climate change on the thermal structure of a large temperate Central European lake. *Climatic Change*, **57**, 205–25.

Livingstone, D.M., Adrian, R., Arvola, L., *et al.* (2010) Regional and supra-regional coherence in limnological variables, in *The Impact of Climatic Change on European Lakes* (ed. D.G. George), Springer Science+Business Media B.V., Dordrecht, pp. 311–37.

Markel, D. (2012) Lake Kinneret fisheries as tools for better lake management. *Eretz Hakinneret*, **5**, 17–20. (In Hebrew.)

Markel, D., Saas, E., Lazar, B., and Bein, A. (1998) Biogeochemical evolution of sulfur rich aquatic system in a re-flooded wetland environment (Lake Agmon, Northern Israel). *Wetlands Ecology and Management*, **6**, 103–20.

Mehnert, G., Leunert, F., Cirés, S., *et al.* (2010) Competitiveness of invasive and native cyanobacteria from temperate freshwaters under various light and temperature conditions. *J. Plankton Res.*, **32**, 1009–21.

Moss, B., Kosten, S., Meerhoff, M., *et al.* (2011) Allied attack: climate change and eutrophication. *Inland Waters*, **1**, 101–5.

Nishri, A. (2011) Long-term impacts of draining a watershed wetland on a downstream lake, Lake Kinneret, Israel. *Air, Soil Water Res.*, **4**, 57–70.

Nishri, A. and Hamilton, D. (2010) A mass balance evaluation of the ecological significance of historical nitrogen fluxes in Lake Kinneret. *Hydrobiologia*, **655**, 109–19.

Nishri, A., Zohary, T., Gophen, M, and Wynne, D. (1998) Lake Kinneret dissolved oxygen regime reflects long term changes in ecosystem functioning. *Biogeochem.*, **42**, 253–83.

Oliver, W., Fuller, C., Naftz, D.L., *et al.* (2009) Estimating selenium removal by sedimentation from the Great Salt Lake, *Utah. J. Appl. Geochem.*, **24**, 936–49.

Ostrovsky, I. (2003) Methane bubbles in Lake Kinneret: Quantification, temporal, and spatial heterogeneity. *Limnol. Oceanogr.*, **48**, 1030–6.

Ostrovsky, I. (2005) Assessing mortality changes from size-frequency curves. *J. Fish Biol.*, **66**, 1624–32.

Ostrovsky, I., McGinnis, D.F., Lapidus, L., and Eckert, W. (2008) Quantifying gas ebullition with echo sounder: the role of methane transport with bubbles in a medium-sized lake. *Limnol. Oceanogr. Methods*, **6**, 105–18.

Ostrovsky, I. and Walline, P. (1999) Growth and production of the dominant pelagic fish, *Acanthobrama terraesanctae* (Cyprinidae), in subtropical Lake Kinneret (Israel). *J. Fish Biol.*, **54**, 18–32.

Ostrovsky, I. and Walline, P. (2001) Multiannual changes in the pelagic fish *Acanthobrama terraesanctae* in Lake Kinneret (Israel) in relation to food sources. *Ver. Internat. Verein. Theoret. Limnol.*, **27**, 2090–4.

Ostrovsky, I. and Yacobi, Y.Z. (1999) Organic matter and pigments in surface sediments: possible mechanisms of their horizontal distributions in a stratified lake. *Can. J. Fish. Aquat. Sci.*, **56**, 1001–10.

Ostrovsky, I. and Yacobi, Y.Z. (2010) Sedimentation flux in a large subtropical lake: spatio-temporal variations and relation to primary productivity. *Limnol. Oceanogr.*, **55**, 1918–31.

Pisanty, S. (2005) Quasi-cyclic fluctuation pattern of St. Peter's fish catches continue. *Fisheries and Fishbreeding in Israel*, **1**, 777–81 [in Hebrew with English abstract].

Pollingher, U. (1986) Phytoplankton periodicity in a subtropical lake (Lake Kinneret, Israel), *Hydrobiologia* **138**, 127–38.

Pollingher, U., Hadas, O., Yacobi, Y.Z., *et al.* (1998) *Aphanizomenon ovalisporum* (Forti) in Lake Kinneret, *Israel. J. Plankton Res.*, **20**, 1321–39.

Reynolds, C.S. (2002) On the interannual variability in phytoplankton production in freshwaters, in *Phytoplankton Productivity: Carbon Assimilation in Marine and Freshwater Ecosystems* (eds. P. J. le B. Williams, D.N. Thomas and C.S. Reynolds), Blackwell Science, Oxford, pp. 187–221.

Rimmer, A., Boger, M., Aota, Y., and Kumagai, M. (2006) A lake as a natural integrator of linear processes: Application to Lake Kinneret (Israel) and Lake Biwa (Japan). *J. Hydrol.*, **319**, 163–75.

Rimmer, A., Gal, G., Opher, T., *et al.* (2011a) Mechanisms of long-term variations of the thermal structure in a warm lake. *Limnol. Oceanogr.*, **56**, 974–88.

Rimmer A., Givati, A., Samuels, R., and Alpert, P. (2011b) Using ensemble of climate models to evaluate future water and solutes budgets in Lake Kinneret, *Israel. J. Hydrol.*, **410**, 248–59.

Samuels, R., Rimmer, A., and Alpert, P. (2009) Effect of extreme rainfall events on the water resources of the Jordan River, *Israel. J. Hydrol.*, **375**, 513–23.

Sánchez-Carrillo, S., Alatorre, L.C., Sánchez-Andrés, R., and Garatuza-Payán, J. (2007) Eutrophication and sedimentation patterns in complete exploitation of water resources scenarios: an example from northwestern semi-arid Mexico. *Environ. Monitor. Assess.*, **132**, 377–93.

Shoham, D. and Levin, I. (1968) Subsidence in the reclaimed Hula swamp area of Israel. *Israel J. Agricul. Res.*, **18**, 15–18.

Snovsky, G. and Shapiro, J. (eds.) (2001) *The Fisheries and Aquaculture of Israel, 1999,* State of Israel, Ministry of Agriculture and Rural Development. Department of Fisheries and Agriculture, Tiberias.

Sukenik, A., Eshkol, R., Livne, A., *et al.* (2002) Inhibition of growth and photosynthesis of the dinoflagellate *Peridinium gatunense* by *Microcystis* sp. (cyanobacteria): a novel allelopathic mechanism. *Limnol. Oceanogr.*, **47**, 1656–63.

Venturelli, P.A., Shuter, B.J., and Murphy, C.A. (2009) Evidence for harvest-induced maternal influences on the reproductive rates of fish populations. *Proc. Roy. Soc. B, Biol. Sci.*, B **276**, 919–24.

Wiedner, C., Rucker, J., Bruggemann, R., and Nixdorf, B. (2007) Climate change affects timing and size of populations of an invasive cyanobacterium in temperate regions. *Oecologia*, **152**, 473–84.

Williamson, C.E., Saros, J.E., and Schindler, D.W. (2009) Climate change: Sentinels of change. *Science*, **323**, 887–8.

Yacobi, Y.Z. (2006) Temporal and vertical variation of chlorophyll a concentration, phytoplankton photosynthetic activity and light attenuation in Lake Kinneret: possibilities and limitations for simulation by remote-sensing. *J. Plankton Res.*, **28**, 725–36.

Yacobi, Y.Z. and Ostrovsky, I. (2008) Downward flux of organic matter and pigments in Lake Kinneret (Israel): Relationships between phytoplankton and the material collected in sediment traps. *J. Plankton Res.*, **30**, 1189–202.

Zohary, T. (2004) Changes to the phytoplankton assemblage of Lake Kinneret after decades of predictable repetitive pattern. *Freshw. Biol.*, **49**, 1355–71.

Zohary, T. and Ostrovsky, I. (2011) Ecological impacts of excessive water level fluctuations in stratified freshwater lake. *Inland Waters*, **1**, 47–59.

Zohary, T. and Shlichter, M. (2009) Invasion of Lake Kinneret by the N_2-fixing cyanobacterium *Cylindrospermopsis cuspis* Komarek and Kling. *Ver. Internat. Verein..Theoret. Limnol.*, **30**, 1251–4.

17

Climate Change and the Floodplain Lakes of the Amazon Basin

John M. Melack[1] and Michael T. Coe[2]

[1]Bren School of Environmental Science and Management, University of California, Santa Barbara CA, USA
[2]The Woods Hole Research Center, Falmouth MA, USA

17.1   Introduction

Floodplains are important components of the biogeochemistry, ecology and hydrology of river systems throughout the world, and are under increasing threat from climate change and hydrological modifications. In the Amazon basin, floodplains contain thousands of lakes and associated wetlands linked to one another, and to the many rivers and streams of the basin. These floodplain lakes modify hydrology of the basin, influence carbon and nutrient biogeochemistry, emit carbon dioxide and methane to the atmosphere, and support highly diverse ecosystems and productive fisheries (Melack *et al.* 2009). Large seasonal and interannual variations in depth and extent of inundation are characteristic of Amazon floodplains, and, as water levels vary, the proportion of aquatic habitats changes considerably. Characterization of the areal extent and temporal changes of inundation and wetland vegetation on local, regional and basin-wide scales is now possible because of the availability of optical and microwave data from sensors on spacecraft and recently developed hydrological models. Projected changes in climate are likely to alter, directly and indirectly, floodplain inundation and related ecological conditions in the lakes. As background to the limnological and climatic conditions in the Amazon basin, we first describe current conditions, and then present results from simulations of changes in inundation as a result of deforestation and variations in rainfall followed by limnological implications of these changes.

Climatic Change and Global Warming of Inland Waters: Impacts and Mitigation for Ecosystems and Societies,
First Edition. Edited by Charles R. Goldman, Michio Kumagai and Richard D. Robarts.
© 2013 John Wiley & Sons, Ltd. Published 2013 by John Wiley & Sons, Ltd.

17.2   Current limnological conditions

Melack and Forsberg (2001) discuss the hydrology and limnology of floodplain lakes and biogeochemical aspects of carbon, nitrogen and phosphorus within the central Amazon basin with a focus on intensively studied lakes. Various facets of the ecology of Amazon floodplains are presented in Junk (1997). Biogeochemical and hydrological aspects of rivers and floodplains in the Amazon basin are discussed in Costa, Coe, and Guyot (2009), Melack *et al.* (2009), and Richey *et al.* (2009).

Lakes occur in seasonally flooded areas and include open water plus vegetated areas that vary in extent as water levels rise and fall. The limnological conditions in the open water portion of floodplain lakes are similar to those typical of shallow tropical lakes (Melack 1996; Talling and Lemoalle 1998; MacIntyre and Melack 2009). Productive beds of floating, emergent macrophytes, that develop annually (Engle *et al.* 2008; Silva, Costa, and Melack 2010), and seasonally inundated forests (Junk *et al.* 2010) are ecologically important and extensive. Floodplains along a 2600 km reach of the Amazon River from 52.5° W to 70.5° W and the lower 400 km of four major tributaries (Japurá, Purús, Negro and Madeira) contain about 8800 lakes larger than 250 m across (Sippel *et al.* 1992).

Underwater light attenuation in floodplain lakes is often high because of suspended particles and dissolved organic matter. Inflows of nutrient-rich water from the Amazon River, other tributaries and local watersheds help sustain the lakes' productivity (Melack and Forsberg 2001). Complex flow patterns (Alsdorf *et al.* 2007) and differences in the sources of water account, in part, for differences in productivity among lakes (Forsberg *et al.* 1988).

Floodplain lakes are significant in the organic carbon balance of the Amazon River system and are a major source of methane and carbon dioxide to the atmosphere (Richey *et al.* 2002; Melack *et al.* 2004; Melack and Engle 2009; Melack *et al.* 2009). The biogeochemistry of nitrogen and phosphorus involves numerous processes, pools and fluxes, and most of the requisite measurements are lacking for floodplain lakes, with the exception of Lake Calado for which the Amazon River is the major supplier of P while inputs of N predominantly are from local sources (summarized in Melack and Forsberg 2001).

17.3   Current climatic conditions

Characteristics of the complex dynamics of the climate system in the Amazon basin are summarized in Marengo and Nobre (2001); Betts *et al.* (2009); Costa *et al.* (2009); Marengo *et al.* (2009), and Nobre *et al.* (2009). Our review of current climatic conditions focuses on aspects related to factors that influence stratification and mixing in lakes and river discharge and inundation.

The lowland Amazon basin is warm and humid. In the central basin, mean annual temperature is approximately 27 °C, with August through November slightly warmer than the mean and January through April slightly cooler than the mean (Irion, Junk, and De Mello 1997). Diel variations can exceed 10 °C. Cool, southern air masses occasionally influence the central Amazon, and minima can fall below 20 °C for a few days during the austral winter (Melack and Forsberg 2001). Relative humidity is high year round, averaging about 76% (Ribeiro and Adis 1984). Meteorological records from lakes in the central Amazon floodplain are limited in time and space to a few sites for only a few years—for example, Lake Calado (Melack and Forsberg 2001) and Lake Curuai (Rudorff *et al.* 2011).

Long-term rates of temperature change are modest in the Amazon basin (Burrows *et al.* 2011). Based on surface temperatures measured from 1960 to 2009, these authors calculated the velocity and seasonal shifts of isotherm movement on a 1° by 1° grid. Within the lowland Amazon basin slight warming on the order of one to three tenths of a degree Centigrade per decade occurred except for regions of slight cooling in the far western and southwestern basin.

Considerable regional differences occur in amounts and timing of rainfall (Marengo 2004). While rainfall averages about 6 mm day^{-1} basin-wide, annual means in the Brazilian Amazon range from less than 2000 mm in the south, northern edge and east, except for higher values near the mouth of the Amazon River, to more than 3000 mm in the northwest (Liebmann and Marengo 2001). On the eastern slopes of the Andes, annual totals exceed 4000 mm (Costa and Foley 1998). Seasonality of rainfall is related to conditions in the tropical Atlantic and movement of the intertropical convergence zone leading to seasonal maxima in the northern Amazon from March to May and in the southern Amazon from January to April. Analyses using records from rain gauges for the period from 1949 to 1999 indicate a statistically significant (at the 5% level) negative trend in rainfall for the northern Amazon and a positive trend for the southern Amazon with alternating wetter and drier periods in both regions (Marengo 2004). Further time-series analysis revealed an interannual mode in variance about every five years in the northern Amazon and evidence of decadal scales of variability in both northern and southern regions (Marengo 2004).

Anomalously large or small amounts of rainfall, leading to especially high or low water level in rivers and lakes, usually occur about every 10 years. For example, the return period for high (2800 mm) and low (2000 mm) rainfall near Manaus in the central Amazon basin is a decade (Lesack and Melack 1991). Major droughts have been linked to intense El Niño/Southern Oscillation (ENSO) events and strong warming of surface waters in the tropical North Atlantic (Marengo *et al.* 2008). Warmer temperatures in the tropical North Atlantic lead to a northward shift of the intertropical convergence zone with increased convective rain over the North Atlantic and subsidence and decreased moisture transport in the Amazon (Zeng *et al.* 2008). Marengo *et al.* (2011) applied a criterion of maximum monthly rainfall of 100 mm to delineate dry periods from 1951 to 2010 and found that the dry season was longer in the 1950s and 1960s, became shorter in the mid 1970s and has become longer again in the last five years.

An increased frequency and severity of droughts are projected by some general circulation models as a result of anthropogenic climate warming (Malhi *et al.* 2008). Perhaps in line with such results, severe droughts occurred in 2005 and 2010, though exceptionally high water was observed in 2009. Based on a decade of satellite-derived dry season rainfall anomalies, Lewis *et al.* (2011) estimated that the 2010 drought was more extensive than the 2005 drought. Sea surface temperature anomalies in the tropical Atlantic were especially large in 2005 (about 0.9 °C) and 2010 (1.5 °C) (Marengo *et al.* 2011). The magnitude of 2005 drought was increased by the combination of the 2002–2003 ENSO event and 2005 tropical Atlantic warming (Zeng *et al.* 2008).

17.4   Current characteristics of discharge and inundation

Annual average runoff from the Amazon basin is approximately 1000 mm (Marengo 2005), about 40% of the annual rainfall. The annual hydrograph of the Amazon River has an amplitude from 2 m to 18 m depending on location and year and is about 10 m in the central basin with a maximum discharge there in June (Richey *et al.* 1986). A several year cycle of interannual variation in discharge measured in the central basin correlates with the El Ninõ-Southern Oscillation (Richey, Nobre, and Deser 1989). Minimum water levels of the Negro River at its confluence with the Amazon River and the Amazon River at Obidos from the 1970s to 2010 have detectable negative anomalies only at very low water levels (Marengo *et al.* 2011). During the severe drought in 2005, discharge at Obidos was 32% lower than the long-term average (Zeng *et al.* 2008).

Gravity anomalies associated with flooding in the Amazon detected by the Gravity Recovery and Climate Experiment (GRACE) indicated a basin-wide average deficiency of about 8 to 9 cm of water or approximately 515 km^3 in drought year 2005 compared to other years from 2002 to 2007 (Chen *et al.* 2009). In contrast, in 2009 a basin-wide surplus of 624 km^3 relative to the GRACE records from 2002 to 2009 was estimated (Chen, Wilson, and Tapley 2010). Xavier *et al.* (2010) applied GRACE data to examine interannual variability in total water storage (TWS) in combination with data on rainfall and river stage along the mainstem Amazon and four major tributaries (Madeira, Tapajós, Xingu and Negro). They detected large TWS in mid-2006 in eastern, northern and southern regions in strong contrast to the low TWS associated with the 2005 drought, and found that the derivative of basin-wide TWS to be highly correlated with the Southern Oscillation Index.

River water is stored in floodplain lakes during seasonal floods and returns when the river levels fall (Richey *et al.* 1989; Alsdorf *et al.* 2010). In some reaches, a net flux of water from the floodplain to the river can occur even when river levels are rising (Lesack and Melack 1995). Basin-wide, the annual variation calculated from GRACE data is approximately 1800 km^3 of water, and the phase and amplitude of this variation matches rainfall (Crowley *et al.* 2008).

Passive and active microwave systems on satellites have been used to detect the extent of inundation, and, in the case of synthetic aperture radar (SAR), to characterize vegetative structure on Amazon floodplains (Hamilton, Sippel, and Melack 2002; Hess *et al.* 2003; Melack 2004; Melack and Hess 2010). Using SAR, Hess *et al.* (2003) mapped the inundation extent and wetland vegetative cover under low and high water conditions at 100 m resolution for a 1.77 million km^2 quadrant in the central Amazon and reported that 17% of the quadrant could be identified as wetlands. Flooded forests occupy nearly 70% of the wetlands at high water, and aquatic habitats vary regionally. Applying the same approach to the entire lowland Amazon basin (the region less than 500 m above sea level) indicates a total floodable area of about 800000 km^2, or 14% of the area (Melack and Hess 2010). Basin wide, about three-fourths of wetlands are covered by forest, woodland or shrubland (Figure 17.1); coverage of aquatic habitats by river basin are summarized in Melack and Hess (2010).

Basin-scale hydrological models that include floodplain inundation operate at moderate resolution (e.g., ~9 km × 9 km, Coe *et al.* 2002; Coe *et al.* 2007). When forced with time varying observed climate data these models provide estimates of monthly

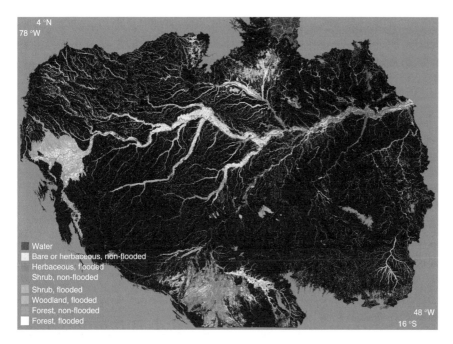

Figure 17.1 Wetland classification of lowland Amazon basin (area less than 500 m a.s.l) for high-water period in the central basin based on analysis of synthetic aperture radar data obtain from the JERS-1 satellite using algorithms described in Hess *et al.* (2003) and validated as described in Hess *et al.* (2002). (See insert for color representation.)

mean inundated area and water depth for relatively long time periods (e.g., >50 years). Model results illustrate important inter-annual and spatial variations in floodplain inundation associated with relatively short (e.g., ENSO) and long time-scale variations in precipitation that are difficult to deduce from the shorter record of satellite observations.

The results of simulations of flood height and extent in the twentieth century (Coe *et al.* 2007; Foley *et al.* 2002) illustrate the potential scale of response of the flood system to climate variability. In these simulations, exceptionally small flooded area during the dry season coincides with relatively warm tropical North Atlantic sea surface temperature (SST). For example, the simulated low-season flooded area in 1963 was 15% less than the mean low season flood for the period (1950–2000) in qualitative agreement with the observations of a record low flood wave height at Manaus (Marengo *et al.* 2008). ENSO strongly effects total simulated flooded area (Figure 17.2), with increased high-season flooding on the mainstem Amazon River during La Niña years and decreased high season flooding during El Niño in comparison to years not affected by ENSO (Coe *et al.* 2002; Foley *et al.* 2002). Low season flooded area tends to be greater in both phases of the ENSO. Variation of the simulated total flooded area in the high season during the average El Niño and La Niña years exceeds +/− 10% compared to the mean of years that are not affected by ENSO (Foley *et al.* 2002). The effect of ENSO is superimposed on the long timescale effect; exceptionally large floods occur when La Niña coincides with the wet phase in the 28-year cycle and exceptionally low floods occur when El Niño coincides with the dry phase

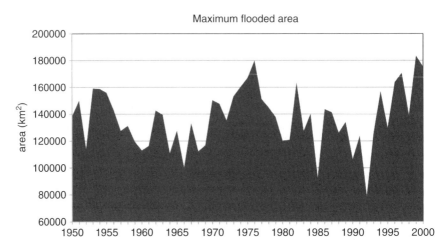

Figure 17.2   Simulated maximum flooded area of the main stem of the Amazon River. Maximum mean monthly area of inundation (km^2) is shown for each year 1950–2000 for the central Amazon basin. Results are from the simulations of Coe *et al.* (2007).

(Coe *et al.* 2002). For example, the maximum simulated flooded area in 1992 was about 45% less than the mean maximum flooded area, coincident with the dry period of the 1980s–1990s and the strong El Niño of 1992–1993 (Figure 17.2). The simulated flooded area in the middle 1970s was 10–50% greater than the average in both the high and low flood season consistent with several strong La Niña events during the peak of the wet period of the 1970s.

17.5   Simulations of inundation under altered climates and land uses

In addition to natural climate variability, land-cover change and anthropogenic climate changes may have a large impact on floodplain inundation in the twenty-first century. Locally, deforestation decreases evapotranspiration and increases discharge and flooded area (Bruijnzeel 1990; D'Almeida *et al.* 2006; Coe *et al.* 2011). However, reduced evapotranspiration and energy balance changes from deforestation can lead to decreased rainfall (see, for example, Butt, Oliveira, and Costa 2011; Knox *et al.* 2011), discharge and flooded area (D'Almeida *et al.* 2007; Coe, Costa, and Soares-Filho 2009). Thus, because the runoff and flooded area are the residual of the precipitation minus evapotranspiration, the net effect of deforestation on the flooded area is a complex combination of the local and non-local effects. Climate change is expected to decrease precipitation and increase temperature in much of the Amazon (Malhi *et al.* 2008), thus decreasing discharge and flooded area. However, the biophysical response of plants to increasing CO_2 and its impact on evapotranspiration are not well known and will complicate the water cycle and floodplain inundation response in the future. Although uncertainties are large, models give some insight into the scale of changes that could occur in the future.

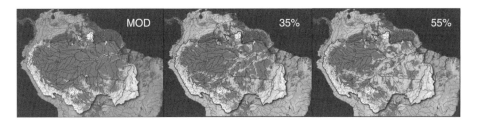

Figure 17.3 Vegetative cover of the Amazon basin. Forest (green) savanna (beige) and agriculture (orange) vegetation types are shown for the 6.7 million km² Amazon basin for the year 2000 (left) as defined by Eva *et al.* (2004), a scenario with 35% of the basin deforested (middle) and a scenario with 55% deforested (right). Scenarios of deforestation are from Soares-Filho *et al.* (2006). (See insert for color representation.)

Two simulations in which 35% and 55% of the Amazon were deforested (Figure 17.3) illustrate the complex ways in which deforestation may feed back to climate and influence floodplain inundation (Coe *et al.* 2009). A land-surface model was coupled to a global climate model and the impact of deforestation on the climate and water balance of the Amazon was quantified. In the simulation in which about 35% of the Amazon was deforested, precipitation was decreased in all years (and mostly in the southern Amazon), but the evapotranspiration from the reduced forest area was decreased even more. As a result, the average maximum flooded area for the period simulated (1950–2000) was increased by about 2% compared to the simulation with modern land cover distribution and the change in flooded area did not in any year exceed +/− 7% of that in the simulation with modern land cover. The annual minimum flooded area increased on average by 6%, with a few years being 10–15% greater than the simulation with modern land cover. Therefore, the net result of climate feedbacks from a 35% decrease in forest cover was a small increase in the flooded area along the mainstem Amazon River.

In the more drastic 55% deforested simulation the response was different. The annual mean precipitation decreased by 10–15% in the southern half of the Amazon basin compared to the simulation with modern vegetation. Evapotranspiration also decreased because of the reduced vegetation cover but not as much as the precipitation. As a result, the maximum and minimum flooded area of the mainstem Amazon River was decreased in all years (1950–2000) compared to the simulation with modern vegetation. The mean change for the period and both seasons was about −5% but was −10% or more in many years, particularly in El Niño years (i.e., 1992 with a −14% change in the maximum flood). Persistence of the flood was also affected when 55% of the basin was deforested (Figure 17.4). In the simulations with modern land cover, on average, there were only two months each year in which the total flooded area did not exceed 50% of the mean maximum flooded area and only nine years in the 51 year record had four or more months in which the flooded area was less than 50% of the maximum (Figure 17.4). The results were the same with 35% of the basin deforested. The average number of months below the 50% threshold increased modestly to three months per year with 55% of the basin deforested, but the number of years with greater than four months below the 50% threshold nearly doubled to 16 out of the 51 years. Therefore, climate feedbacks from a 55% reduction in forest

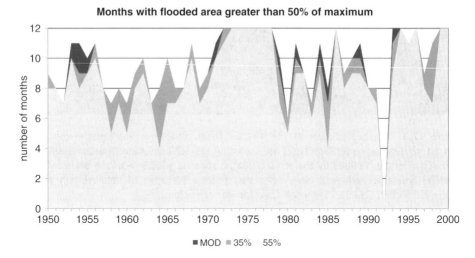

Figure 17.4   Number of months each year in which total flooded area is greater than 50% of the mean maximum flooded area. Results are shown for each year simulated (1950–2000) for modern climate and vegetation cover (MOD), 35% of the basin deforested (35%), and 55% of the basin deforested (55%). See Figure 17.3 for the deforestation distribution.

cover result in a strong reduction in flooded area in all seasons and accentuation of the variability, with very low flood levels being more common and longer lasting.

In summary, simulations suggested that relatively realistic scenarios of deforestation in the Amazon will result in increased water yield per unit precipitation due to a decrease in total evapotranspiration that accompanies deforestation. Increased water yield can result in increased flooded area in both low and high flood seasons when the deforestation is modest. However, extensive deforestation is likely to result in feedbacks to climate that reduce precipitation over large areas of the Amazon and subsequently to strong and persistent decreases in the flooded area on the main stem of the Amazon River.

Uncertainties in global climate model simulations of climate response to increasing greenhouse gases makes accurate prediction of future changes to the water balance of the Amazon quite difficult. However, analysis of the results of the climate models taking part in the IPCC AR4 simulations suggests that a decrease in precipitation and increase in evapotranspiration, and thus a decrease in discharge, is highly probable (Malhi *et al.* 2008). A series of simulations were performed in which precipitation was decreased by 10% and 25% for all years and throughout the Amazon Basin. These simplified simulations ignore important seasonal and spatial variations in the water balance but illustrate the potential sensitivity of the simulated flooded area to decreased water yield.

The annual mean maximum and minimum flooded areas for the period 1950–2000 decreased comparable to the prescribed change in precipitation. With a 10% reduction in rainfall the means are decreased by 10% compared to the simulation with no precipitation disturbance and vary interannually from −5% to −20%. With a 25% reduction in rainfall the mean maximum and minimum flooded area both decreased by 25% and varied inter-annually between −12% to −30%. The persistence of the

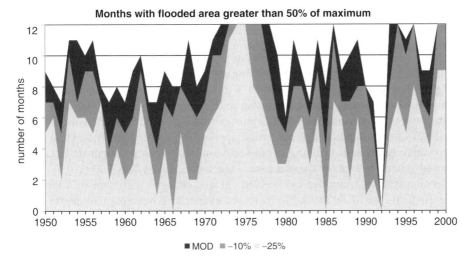

Figure 17.5 Number of months each year in which total flooded area is greater than 50% of the mean maximum flooded area. Results are shown for each year simulated (1950–2000) for the modern climate and vegetation cover (MOD), a 10% rainfall reduction (−10%), and a 25% rainfall reduction (−25%). See Figure 17.3 for the deforestation distribution.

flooded area was strongly influenced in both scenarios of decreasing precipitation (Figure 17.5). The average number of months each year in which total flooded area was less than 50% of the mean maximum flooded area increased from two months/year in the case with no precipitation change to four months/year with 10% less rainfall and to seven months/year with a 25% reduction (Figure 17.5). Furthermore, the total number of years in which the flooded area in four or more months was less than 50% of the maximum nearly tripled to 26 of the 51 years simulated in the case of −10% change in rainfall and to 44 of the 51 years in the series in the case of 25% reduction in rainfall. Therefore, even a modest reduction in rainfall with no change in other parameters (e.g., temperature and evapotranspiration) is sufficient to result in very large changes in the seasonal and inter-annual distribution of flooded area on the main stem of the Amazon River.

The results of simulations of both deforestation and prescribed climate changes suggested that the flooded area of the main stem of the Amazon River is sensitive to water balance changes. Discharge changes locally and remotely can result in changes to the magnitude and persistence of flooded area that may have important implications for ecosystem functioning.

17.6 Limnological implications of altered climate, land uses and inundation

17.6.1 Changes in stratification and mixing

Since few meteorological measurements have been made on floodplain lakes, and downscaled reanalysis data is too spatially coarse for local applications, modeling likely climate induced changes in vertical mixing of lakes is not feasible. Instead, inferences from observed mixing dynamics and likely responses to changes in climate based on mechanistic understanding of physical processes are offered.

Based on hundreds of vertical profiles of temperature obtained in a floodplain lake, patterns in the frequency of vertical mixing are evident (MacIntyre and Melack 1984, 1988): When water depths are less than 3 to 4 m, diel mixing from top to bottom is common. When depths exceed 5 m, a thermocline develops at about 3 m and mixing to the bottom is intermittent. When depths exceed 8 m, mixing to the bottom is rare. Analyses of heat inputs and losses, wind speeds, and thermal structure indicate that evaporative heat losses resulting in convective mixing often predominate over wind-induced mixing in shallow tropical lakes (MacIntyre and Melack 2009).

As temperatures warm due to expected climatic changes, the increased vapor pressure gradients are likely to lead to increased evaporative heat loss and deeper and/or more frequent convective mixing. On a regional scale, enhancement of the hydrological cycling of water may lead to augmented convective storms with associated intense rains and winds and concomitant impacts on vertical mixing. Cloud cover over Amazonian rivers and floodplains influences direct solar heating and short- and long-wave net radiative fluxes, but changes in cloud cover are difficult to forecast.

17.6.2   Other limnological changes

Floodplain lakes receive solutes and particulates from mainstem rivers and local sources including direct rainfall and dry deposition, passage of rain through vegetation, upland runoff, groundwater seepage, exchanges with neighboring lakes, and nitrogen fixation within the lake (reviewed in Melack and Forsberg 2001). All these processes are influenced, to varying degrees, by climatic conditions. As discussed earlier, the discharge and inundation of the floodplains is modified directly by changes in the amount and timing of rainfall. These variations, in turn, alter the relative importance of mainstem river inputs versus local runoff (Lesack and Melack 1995; Bonnet *et al.* 2008). The concomitant differences in nutrient and sediment concentrations from these two sources alter the supply of limiting nutrients and can cause alternations from nitrogen to phosphorus limitation (Setaro and Melack 1984; Forsberg *et al.* 1988).

Changes in extent of inundation lead to changes in the area of floodplain habitats and biogeochemical processes associated with these habitats. Floating macrophytes and flooded forests provide substratum for periphytic bacteria and algae as well as supplying organic carbon to the lakes. The extent of these two habitats varies seasonally and inter-annually as water level fluctuates. Nitrogen fixation occurs in association with periphytic bacteria (Doyle and Fisher 1994). Evasion of methane is higher in habitats associated with floating macrophytes and flooded forests than in open waters (Melack *et al.* 2004).

Carbon dioxide and methane are supersaturated in most Amazon waters and outgas to the atmosphere (Richey *et al.* 2002; Melack *et al.* 2004). The accumulation and air-water exchange of these two greenhouse gases depends on stratification and mixing of the water column, changes in water level and local meteorological conditions (Rudorff *et al.* 2011). If stratification is prolonged, high concentrations can develop in the hypolimnion and episodic mixing will then release large amounts (Engle and Melack 2000) as will changes in hydrostatic pressure as water levels decrease (Rosenqvist *et al.* 2002).

17.6.3 Changes related to deforestation

Extensive areas of the Amazon basin have been and are being deforested, including floodplain environments and neighboring upland catchments, and, as a result, hydrologic characteristics that determine transport of nutrients and other solutes are altered. These changes modulate influences of climate variations or trends in rainfall amounts and intensity on solute movements.

As discussed earlier, regional deforestation can decrease rainfall, which will alter the extent and depth of inundation. These changes will, in turn, modify the frequency of mixing to the bottom. One secondary influence of deforestation is biomass burning, which results in injection of aerosols into the atmosphere. These aerosols decrease solar radiation and alter the formation of clouds and rainfall (Longo *et al.* 2009).

Local deforestation in the catchment bordering floodplain lakes increases runoff of water and inputs of dissolved nutrients and suspended sediments (Williams, Fisher, and Melack 1997; Williams and Melack 1997). These changes increase flushing of portions of the lakes, increase growth of algae and floating macrophytes and decrease transparency. Decreased transparency will result in increased heating of the surficial waters and stronger diel stratification.

17.6.4 Changes related to dams

Large hydroelectric reservoirs have been built within the Amazon, including Tucuruí and Balbina, and several are under construction, for example Belo Monte on the Xingu River and Santo Antonio and Jirau on the Madeira River. Though construction of dams is occurring for economic reasons, climate changes may modify the motivation for and performance of hydroelectric and flood control dams. Further, the impacts of dams on inundation interact with climate-driven changes and may feedback to alter climate. Alterations of flows downstream of dams have major implications for floodplain ecosystems and organisms adapted to a natural flood regime. Considerable outgassing of greenhouse gases, especially methane, occur upstream and down-stream of hydroelectric dams (Kemenes, Forsberg, and Melack 2007, 2011). Hence, there is a feedback to climate warming associated with tropical reservoirs.

17.7 Management options and research directions

In an area as large as the Amazon basin with its thousands of lake and hundreds of thousand of km^2 of aquatic habitats, mitigation of changes wrought by climate changes are largely related to processes occurring at the regional or global scale. In light of the feedbacks between deforestation and associated burning and rainfall, reductions in deforestation will be valuable. Reservoirs are a relatively small source of methane compared to natural wetlands in the Amazon, but reductions in methane evasion from the reservoir surface and as water leaves the turbines would be a positive step. The largest benefit would result from global reductions in emission of greenhouse gases with participation by all countries and especially those with the largest rates of emission.

To improve projections of changes in Amazon floodplain lakes related to altered climatic conditions requires comparative analyses of the hydrology and limnology in a spectrum of floodplain lakes representative of the range of chemical, optical, morphometric, physical and biological conditions throughout the basin. While surveys have revealed the range of conditions, very few lakes have received focused studies of sufficient length to be able to evaluate and model their current status and variability. Advances in remote sensing systems are providing multi-year and synoptic spatial data on key conditions, such as inundation, sediment and chlorophyll levels and land cover alterations. *In situ* sensors are making it possible to monitor thermal structure, turbidity, chlorophyll and dissolved organic carbon and to profile water velocities. When combined with mechanistic models, these measurements are making it feasible to better understand temporal and spatial dynamics in Amazon lakes. As improved simulations of predicted climate changes become available at fine spatial resolution and are coupled with data-driven mechanistic models of limnological processes, quantitative forecasts of the future of the lakes in the Amazon will be possible.

17.8 Acknowledgements

Our research and perspectives on Amazon floodplain lakes have benefitted from collaborations with many individuals in Brazil and the United States, and we especially acknowledge B.R. Forsberg, M.H. Costa, S. MacIntyre, S.K. Hamilton, C. Rudorff, T.S Silva, E.M. Novo, L.L. Hess, and A. Kemenes. Funding from the US National Science Foundation and National Aeronautical and Space Administration has supported much of our research in the Amazon.

References

Alsdorf, D., Bates, P., Melack, J.M., *et al.* (2007) Spatial and temporal complexity of the Amazon flood measured from space. *Geophys. Res. Lett.*, **34**, L08402, doi: 10.1029/2007GL029447.

Alsdorf, D., Han, S.-C., Bates, P., and Melack, J. (2010) Seasonal water storage on the Amazon floodplain measured from satellites. *Remote Sens. Environ.*, **114**, 2448–56.

Betts, A.K., Fisch, G., Randow, C. von, *et al.* (2009) The Amazonian boundary layer and mesoscale circulations, in *Amazonia and Global Change* (eds. J. Gash, M. Keller and P. Silva-Dias), Geophysical Monograph Series 186, American Geophysical Union, Washington, D.C., pp. 163–81.

Bonnet, M.P., Barroux, G., Martinez, J.M., *et al.* (2008) Floodplain hydrology in an Amazon floodplain lake (Lago Grande de Curuai). *J. Hydrology*, **349**, 18–30.

Bruijnzeel, L.A. (1990) *Hydrology of Moist Tropical Forests and Effects of Conversion: A State of Knowledge Review*, UNESCO, Paris.

Burrows, M.T., Schoeman, D.S., Buckley, L.B., *et al.* (2011) The pace of shifting climate in marine and terrestrial ecosystems. *Science*, **334**, 652–5.

Butt, N., Oliveira, P.A., and Costa, M.H. (2011) Evidence that deforestation affects the onset of the rainy season in Rondonia, Brazil. *J. Geophys. Res.*, **116** (D11120), doi:10.1029/2010JD015174.

Chen, J.L., Wilson, C.R., and Tapley, B.D. (2010) The 2009 exceptional Amazon flood and interannual terrestrial water storage change observed by GRACE. *Water Resource Res.* **46**, W12526.

Chen, J.L., Wilson, C.R., Tapley, B.D., *et al.* (2009) 2005 drought event in the Amazon River basin as measured by GRACE and estimated by climate models. *J. Geophys. Res.*, **14**, B05404.

Coe, M.T., Costa, M.H., Botta, A., and Birkett, C. (2002) A long-term simulation of discharge and floods in the Amazon basin. *J. Geophys. Res.—Atmos.*, **107** (D20) 8044.

Coe, M.T., Costa, M.H., and Howard, E. (2007) Simulating the surface waters of the Amazon River basin: Impacts of new river geomorphic and dynamic flow parameterizations. *Hydrol. Processes*, **22**, 2542–53.

Coe, M.T., Costa, M.H., and Soares-Filho, B.S. (2009) The influence of historical and potential future deforestation on the stream flow of the Amazon River—Land surface processes and atmospheric feedback., *J. Hydrol.*, doi:10.1016/j.jhydrol.2009.02.043.

Coe, M.T., Latrubesse, E.M., Ferreira, M.E., and Amsler, M.L. (2011) The effects of deforestation and climate variability on the streamflow of the Araguaia River, Brazil. *biogeochemistry*, **105**, 119–131, doi:10.1007/s10533-011-9582-2.

Costa, M.H., Coe, M.T., and Guyot, J.L. (2009) Effects of climatic variability and deforestation on surface water regimes, in *Amazonia and Global Change* (eds. J. Gash, M. Keller and P. Silva-Dias), Geophysical Monograph Series 186, American Geophysical Union, Washington, D.C., pp. 543–53.

Costa, M.H., and Foley, J.A. (1998) A comparison of precipitation datasets for the Amazon Basin. *Geophys. Res. Lett.*, **25**, 155–8.

Crowley, J.W., Mitrovica, J.X., Bailey, R.C., *et al.* (2008) Annual variations in water storage and precipitation in the Amazon Basin. *J. Geod.*, **82**, 9–13.

D'Almeida, C., Vörösmarty, C.J., Marengo, J.A., *et al.* (2006) A water balance model to study the hydrological response to different scenarios of deforestation in Amazonia. *J. Hydrol.*, **331**, 125–36.

D'Almeida, C., Vörösmarty, C.J., Hurtt, G.C., *et al.* (2007) The effects of deforestation on the hydrological cycle in Amazonia: a review on scale and resolution. *Int. J. Climat.*, **27**, 633–47.

Doyle, R.D. and Fisher, T.R (1994) Nitrogen fixation by periphyton and plankton on the Amazon floodplain at Lake Calado. *Biogeochemistry*, **26**, 41–66.

Engle, D. and Melack, J.M. (2000) Methane emissions from the Amazon floodplain: enhanced release during episodic mixing of lakes. *Biogeochemistry*, **51**, 71–90.

Engle, D.L., Melack, J.M., Doyle, R.D., and Fisher, T.R. (2008) High rates of net primary productivity and turnover for floating grasses on the Amazon floodplain: Implications for aquatic respiration and regional CO_2 flux. *Global Change Biol.*, **14**, 369–81.

Eva, H.D., Belward, A.S., De Miranda, E.E., *et al.* (2004) A land cover map of South America. *Global Change Biol.*, **10**, 731–44.

Foley, J.A., Botta, A., Coe, M.T., and Costa, M.H. (2002) El Niño–Southern oscillation and the climate, ecosystems and rivers of Amazonia. *Global Biogeochem. Cycles*, doi: 10.1029/2002GB001872.

Forsberg, B.R., Devol, A.H., Richey, J.E., *et al.* (1988) Factors controlling nutrient concentrations in Amazon floodplain lakes. *Limnol. Oceanogr.*, **33**, 41–56.

Hamilton, S.K., Sippel, S.J., and Melack, J.M. (2002) Comparison of inundation patterns among major South American floodplains. *J. Geophys. Res.*, **107**, No. D20 1029/2000JD000306.

Hess, L.L., Melack, J.M., Novo, E.M.L.M., *et al.* (2003) Dual-season mapping of wetland inundation and vegetation for the central Amazon basin. *Remote Sens. Environ.*, **87**, 404–28.

Hess, L.L., Novo, E.M.L.M., Slaymaker, D.M., *et al.* (2002) Geocoded digital videography for validation of land cover mapping in the Amazon basin. *Int. J. Remote Sens.*, **7**, 1527–56.

Irion, G., Junk, W.J., and De Mello, J.A.S.N. (1997) The large central Amazonian River floodplains near Manaus: Geological, climatological, hydrological, geomorphological aspects, in *The Central Amazon Floodplain* (ed. W.J. Junk), Springer, Berlin, pp. 23–46.

Junk, W.J. (ed.) (1997) *The Central Amazon Floodplain*, Springer, Berlin.

Junk, W.J., Piedade, M., Wittmann, F., *et al.* (eds.) (2010) *Amazonian Floodplain Forests: Ecophysiology, Ecology, Biodiversity and Sustainable Management*, Springer, Dordrecht.

Kemenes, A., Forsberg, B.R., and Melack, J.M. (2007) Methane release below a hydroelectric dam. *Geophys. Res. Lett.*, **34**, L12809, doi:10.1029/2007GL029479.

Kemenes, A., Forsberg, B.R., and Melack, J.M. (2011) CO_2 emissions from a tropical hydroelectric reservoir (Balbina, Brazil). *J. Geophys. Res.—Biogeosci.*, **116**, G03004, doi:10.1029/2010JG001465.

Knox, R., Bisht, G., Wang, J., and Bras, R.L. (2011) Precipitation variability over the forest to non-forest transition in southwestern Amazonia. *J. Climate*, **24**, 2368–77. doi: 10.1175/2010JCLI3815.1

Lesack, L.F. and Melack, J.M. (1991) The deposition, composition, and potential sources of major ionic solutes in rain of the central Amazon basin. *Water Resource Res.*, **27**, 2953–77.

Lesack, L.F.W. and Melack, J.M. (1995) Flooding hydrology and mixture dynamics of lake water derived from multiple sources in an Amazon floodplain lake. *Water Resource Res.*, **31**, 329–45.

Lewis, S.L., Brando, P.M., Phillips, O.L., *et al.* (2011) The 2010 Amazon drought. *Science*, **331**, 554.

Liebmann, B. and Marengo, J.A. (2001) Interannual variability of the rainy season and rainfall in the Brazilian Amazon basin. *J. Clim.*, **14**, 4308–18.

Longo, K.M., Freitas, S.R., Andreae, M.O., *et al.* (2009) Biomass burning in Amazonia: Emissions, long-range transport of smoke and its regional and remote impacts, in *Amazonia and Global Change* (eds. J. Gash, M. Keller and P. Silva-Dias), Geophysical Monograph Series 186. American Geophysical Union, Washington, D.C., pp. 207–32.

MacIntyre, S. and Melack, J.M. (1984) Vertical mixing in Amazon floodplain lakes. *Verh. Int. Verein. Limnol.*, **22**,1283–7.

MacIntyre, S. and Melack, J.M. (1988) Frequency and depth of vertical mixing in an Amazon floodplain lake (L. Calado, Brazil). *Verh. Int. Verein. Limnol.*, **23**, 80–5.

MacIntyre, S. and Melack, J.M. (2009) Mixing dynamics in lakes across climatic zones. *Encyclopedia of Ecology of Inland Water*, Vol. **2**, Elsevier, New York, pp. 603–12.

Malhi, Y., Roberts, J.T., Betts, R.A., *et al.* (2008) Climate change, deforestation, and the fate of the Amazon. *Science*, **319**, 169–72.

Marengo, J.A. and Nobre, C.A. (2001) General characteristics and variability of climate in the Amazon basin and its links to the global climate system, in *The Biogeochemistry of the Amazon Basin and its Role in a Changing World* (eds. M.E. McClain, R.L. Victoria and J.E. Richey), Oxford University Press, Oxford, pp. 17–41.

Marengo, J.A. (2004) Interdecadal variability and trends of rainfall across the Amazon basin. *Theor. Appl. Climatol.*, **78**, 79–96.

Marengo, J.A. (2005) Characteristics and spatio-temporal variability of the Amazon River basin water budget. *Climate Dynamics*, **24**, 11–22.

Marengo, J., Nobre, C.A., Betts, R.A., *et al.* (2009) Global warming and climate change in Amazonia: Climate-vegetation feedback and impacts on water resources, in *Amazonia and Global Change* (eds. J. Gash, M. Keller and P. Silva-Dias), Geophysical Monograph Series 186, American Geophysical Union, Washington, D.C., pp. 273–92.

Marengo, J.A., Nobre, C.A., Tomasella, J., *et al.* (2008) The drought of Amazonia in 2005. *J. Clim.*, **21**, 495–516.

Marengo, J.A., Tomasella, J., Alves, L.M., *et al.* (2011) The drought of 2010 in the context of historical droughts in the Amazon region. *Geophys. Res. Lett.*, **38**, L12703.

Melack, J.M. (1996) Recent developments in tropical limnology. *Verh. Int. Verein. Limnol.*, **26**, 211–17.

Melack, J.M. (2004) Remote sensing of tropical wetlands, in *Manual of Remote Sensing. Remote Sensing for Natural Resources Management and Environmental Monitoring* (ed. S. Ustin), 3rd edn. John Wiley & Sons, Inc., New York, vol. **4**, pp. 319–43.

Melack, J.M. and Engle, D. (2009) An organic carbon budget for an Amazon floodplain lake. *Verh. Int. Verein. Limnol.*, **30**, 1179–82.

Melack, J.M. and Forsberg, B. (2001) Biogeochemistry of Amazon floodplain lakes and associated wetlands, in *The Biogeochemistry of the Amazon Basin and its Role in a Changing World* (eds. M.E. McClain, R.L. Victoria and J.E. Richey), Oxford University Press, Oxford, pp. 235–76.

Melack, J.M. and Hess, L.L. (2010) Remote sensing of the distribution and extent of wetlands in the Amazon basin, in *Amazonian Floodplain Forests: Ecophysiology, Ecology, Biodiversity and Sustainable Management* (eds. W.J. Junk, M. Piedade, F. Wittmann, *et al.*), Ecological Studies, Springer, Dordrecht, pp. 43–59.

Melack, J.M., Hess, L.L., Gastil, M., *et al.* (2004) Regionalization of methane emissions in the Amazon basin with microwave remote sensing. *Global Change Biol.*, **10**, 530–44.

Melack, J.M., Novo, E.M.L.M., Forsberg, B.R., Piedade, M.T.F., and Maurice, L. (2009) Floodplain ecosystem processes, in *Amazonia and Global Change* (eds. J. Gash, M. Keller and P. Silva-Dias), Geophysical Monograph Series 186, American Geophysical Union, Washington, D.C., pp. 525–41.

Nobre, C.A., Obregón, G.O., Marengo, J.A., *et al.* (2009) Characteristics of Amazonian climate: Main features, in *Amazonia and Global Change* (eds. J. Gash, M. Keller and P. Silva-Dias), Geophysical Monograph Series 186, American Geophysical Union, Washington, D.C., pp. 149–62.

Ribeiro, M.N.G. and Adis, J. (1984) Local rainfall variability—a potential bias for bioecological studies in the central Amazon. *Acta Amazonica*, **14**, 159–74.

Richey, J.E., Krusche, A.V., Johnson, M.S., *et al.* (2009) The role of rivers in the regional carbon balance, in *Amazonia and Global Change*, Geophysical Monograph Series 186 (eds. Gash, M. Keller and P. Silva-Dias), American Geophysical Union, Washington, D.C., pp. 489–504.

Richey, J.E. , Meade, R.H., Salati, E., *et al.* (1986) Water discharge and suspended sediment concentrations in the Amazon River, 1982–1984. *Water Resource Res.*, **22**, 756–64.

Richey, J.E., Melack, J.M., Aufdenkampe, A.K., *et al.* (2002) Outgassing from Amazonian rivers and wetlands as a large tropical source of atmospheric carbon dioxide. *Nature*, **416**, 617–20.

Richey, J.E., Mertes, L.A.K., Dunne, T., *et al.* (1989) Sources and routing of the Amazon River flood wave. *Global Biogeochem. Cycles*, **3**, 191–204.

Richey, J.E., Nobre, C., and Deser, C. (1989) Amazon River discharge and climatic variability: 1903–1985. *Science*, **246**, 101–3.

Rosenqvist, A., Forsberg, B.R., Pimentel, T., *et al.* (2002) The use of spaceborne radar data to model inundation patterns and trace gas emissions in the central Amazon floodplain. *Int. J. Remote Sens.*, **23**, 1303–28.

Rudorff, C.M., Melack, J.M., MacIntyre, S., *et al.* (2011) Seasonal and spatial variability in CO_2 emissions from a large floodplain lake in the lower Amazon. *J. Geophys. Res.—Biogeosci.*, **116**, G04007, doi:10.1029/2011JG001699.

Setaro, F.V., and Melack, J.M. (1984) Responses of phytoplankton to experimental nutrient enrichment in an Amazon lake. *Limnol. Oceanogr.*, **28**, 972–84.

Silva, T.S.F., Costa, M.P.F., and Melack, J.M. (2010) Spatio-temporal variability of macrophyte cover and productivity in the eastern Amazon floodplain: a remote sensing approach. *Remote Sens. Environ.*, **114**, 1998–2010.

Sippel, S.J., Hamilton, S.K., Melack, J.M. (1992) Inundation area and morphometry of lakes on the Amazon River floodplain, *Brazil. Arch. Hydrobiol.*, **123**, 385–400.

Soares-Filho, B.S., Nepstad, D.C., Curran, L.M., *et al.* (2006) Modelling conservation in the Amazon basin. *Nature*, **440**, 520–3.

Talling, J.F. and Lemoalle, J. (1998) *Ecological Dynamics of Tropical Inland Waters*, Cambridge University Press, Cambridge.

Williams, M.R., Fisher, T.R., and Melack, J.M. (1997) Solute dynamics in soil water and groundwater in a central Amazon catchment undergoing deforestation. *Biogeochemistry*, **38**, 303–35.

Williams, M.R. and Melack, J.M. (1997) Solute export from forested and partially deforested catchments in the central Amazon. *Biogeochemistry*, **38**, 67–102.

Xavier, L., Becker, M., Cazenave, A., *et al.* (2010) Interannual variability in water storage over 2003–2008 in the Amazon Basin from GRACE space gravimetry, *in situ* river level and precipitation data. *Remote Sens. Environ.*, **114**, 1626–37.

Zeng, N., Yoon, J.-H., Marengo, J.A., *et al.* (2008) Causes and impacts of the 2005 Amazon drought. *Environ. Res. Let.*, **3**, 014002.

18

Climatic Variability, Mixing Dynamics, and Ecological Consequences in the African Great Lakes

Sally MacIntyre

Department of Ecology, Evolution and Marine Biology, University of California, Santa Barbara, USA

18.1   Introduction

To understand how climate change influences the ecological functioning of the African Great Lakes, it is necessary to examine controls on mixing dynamics. Lake Tanganyika is the largest of these lakes, 670 km long and 1470 m deep. Lake Victoria, the second largest lake by surface area on Earth, 68 800 km^2, is located between the Eastern and Western Rift Valleys of East Africa and is the shallowest of the African Great Lakes with a maximum depth of about 80 m. The Eastern Rift Valley, which extends from northern Tanzania into the Danakil Depression in Ethiopia, is semi-arid. The majority of its lakes are within endorheic basins—that is, they receive water from incoming rivers and precipitation but have no outlet and are salty due to evaporative concentration of their waters. Lake Turkana (290 km long, 109 m deep) is the only great lake in the Eastern Rift Valley. Lakes in the Western Rift Valley experience a wetter climate, have outlets and, with the exception of Lake Kivu, which has deep hydrothermal saline springs, are fresh. From south to north, the Great Lakes in the western rift are Lake Malawi (560 km long, 706 m deep), Lake Tanganyika (670 km, 1470 m deep), Lake Kivu (90 km long, 480 m deep), Lake Edward (75 km long, 112 m deep), and Lake Albert (160 km long, 58 m deep). The lakes deeper than 100 m are permanently stratified and classified as meromictic and are anoxic in

Climatic Change and Global Warming of Inland Waters: Impacts and Mitigation for Ecosystems and Societies,
First Edition. Edited by Charles R. Goldman, Michio Kumagai and Richard D. Robarts.
© 2013 John Wiley & Sons, Ltd. Published 2013 by John Wiley & Sons, Ltd.

their lower water column. The shallower lakes are monomictic, and mix during the southeast monsoon. Due to their location near the equator and muted variations in solar radiation and air temperature, temperature differences between the upper and lower water columns in the shallower lakes are only a degree or two whereas in Lake Tanganyika they are $4\,^{\circ}$C at the warmest time of the year. Lake Kivu is in part fed by subterranean springs, and high concentrations of dissolved salts and CO_2 and CH_4 have accumulated below $100\,$m. Consequently, it is chemically stratified. The stratification and mixing of lakes in tropical Africa are influenced by their geological setting particularly as it influences their depth and by seasonality induced by the movement of the Intertropical Convergence Zone. Climate-related changes in their mixing will depend not only on increases in air temperatures but also on wind speed through its impacts on circulation, internal wave motions, and evaporation rates. Some aspects of the mixing dynamics of the African Great Lakes are reviewed in Beadle (1981), Spigel and Coulter (1996), and Talling and Lemoalle (1998) and a brief synopsis of mixing dynamics in East African Lakes is provided in MacIntyre (2012).

Climate and land use changes are occurring within the lake basins, and understanding climate and land use related influences on mixing dynamics and the links with ecosystem function is important for managing these lakes and their fisheries (Kolding *et al.* 2008). The sedimentary record indicates the African Great Lakes have undergone large changes over the last 140 000 years in response to solar and atmospheric cycles which influence the water budget on a planetary scale (Scholz *et al.* 2011). Heating in the last 90 years, based on paleolimnological proxies for temperature, is the most rapid in the last 1300 years and is occurring at a rate similar to that in the northern hemisphere (Tierney *et al.* 2010). Temperature measurements within the lakes taken over the last 100 years indicate that the deep African Great Lakes are warming (O'Reilly *et al.* 2003; Verburg, Hecky, and Kling 2003; Lorke *et al.* 2004; Vollmer *et al.* 2005). Warming is, in fact, occurring more rapidly in northern waters of Lake Tanganyika than global air temperatures (Verburg and Hecky 2009). Fully deciphering the implications of climate change—that is, how the changes in air temperature, wind speeds, rainfall, solar radiation and relative humidity influence stratification and the potential for mixing—requires understanding controls on meteorology and how these feedback and influence the surface energy budgets and consequent horizontal transport and mixing. With respect to consequences for ecosystem function, a number of aspects of mixing affect primary productivity. For instance, depth of mixing relative to depth of the euphotic zone determines whether phytoplankton are light limited; duration of diurnal stratification will affect degree of photoinhibition; frequency and depth of upwelling will affect nutrient supply; and riverine inflows and their flow paths, affect productivity nearshore and potentially offshore.

In the following, the physical processes important for structuring tropical lakes are briefly described. The mixing dynamics in Lakes Victoria and Tanganyika, as representative of a shallow and a deep Great Lake, are described in more detail and an evaluation is made of the consequences of climate change, through its interaction with these processes, on ecosystem function. Lake Kivu has been receiving considerable interest recently due to the high concentrations of carbon dioxide and methane in its lower water column, mixing within that lake will be described.

18.2   Controls on mixing dynamics in the African Great Lakes

Evaporation, also known as latent heat flux, and the relatively large changes in density that occur when water is heated above 20 °C exert a strong influence on the mixing dynamics in tropical lakes. Warming of the upper waters in the day creates a diurnal thermocline (Figures 18.1, 18.2). Nighttime cooling, largely driven by evaporation, erodes this feature. The minimum temperatures in lakes at all latitudes are set, in part, by minimum air temperatures. For the lakes closest to the equator, minimum air temperatures are only a few degrees cooler than maxima, and the temperature gradient between surface and bottom waters is small. Due to warm surface temperatures and the reduced vertical temperature gradient, mixing due to cooling from high evaporation rates can extend to deeper depths than in lakes outside the tropics, and conditions are less favorable for the establishment of seasonal stratification such as found in temperate and Arctic lakes. The shallow embayments of large lakes, such as Pilkington Bay and Winam Gulf, Lake Victoria, tend to stratify and mix to the bottom on a diel basis and

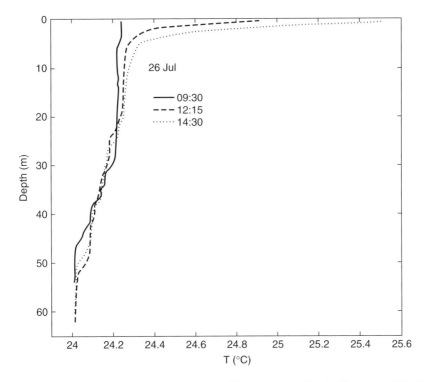

Figure 18.1   Temperature profiles in northern offshore waters of Lake Victoria, 1994, illustrating formation of the diurnal thermocline in the upper 5 m in response to morning heating during the southeast monsoon. Steppy structure is indicative of internal waves. Data courtesy of J.R. Romero and S. MacIntyre.

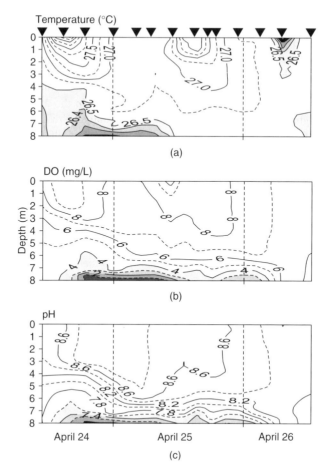

Figure 18.2 Time series of temperature (a), oxygen (b), and pH (c) in Pilkington Bay, Lake Victoria in April 1996 showing diurnal increases in temperature and nocturnal cooling. Cooling does not always lead to isothermy, or, as in the case here, cold water masses can be advected near the bottom (April 24; see MacIntyre *et al.* 2002). Hypoxic water with depressed values of pH, indicative of formation of CO_2, is associated with the persistent stratification. Even when the thermal stratification is reduced due to further cooling at night, patches of water with low concentrations of oxygen can persist. Data courtesy of J.R. Romero and S. MacIntyre.

are classified as polymictic (Figure 18.2). Seasonal stratification in shallow African Great Lakes is caused or enhanced by the offshore flow of cooler water produced in the shallow regions (Talling 1963, 1966). The depth of permanent thermoclines in very deep lakes, Malawi, Tanganyika, and Kivu, ~100 m to 200 m, is set by the depth of mixing during the windiest seasons when evaporation rates are highest. Stratification in the deep lakes may be influenced by river inflows or offshore flow of cool water from inshore locations (Patterson, Wooster, and Sear 1998; Vollmer *et al.* 2005), but the major factor is their deep depth relative to the depth of mixing set by evaporation and wind shear.

The annual cycle of stratification in tropical lakes is caused by seasonal changes in cloud cover as it affects solar radiation and net longwave radiation and seasonal changes in wind speed and relative humidity as they affect latent heat fluxes (Talling and Lemoalle 1998). The movement of the Intertropical Convergence Zone (ITCZ) is the driver for the changes in surface meteorology, which cause seasonality in mixing dynamics. The movement of the ITCZ determines the attributes of air masses which are advected over the lakes. As it moves north in the southern hemisphere, cool dry air sweeps across the lakes, wind speeds tend to be higher increasing evaporation, and, for lakes with a persistent thermocline, upwelling occurs primarily at the southern ends of the lakes (Figures 18.3, 18.4). Due to these processes, ventilation of the thermocline is greater towards the south and vertical fluxes of nutrients can be greatest there (Coulter 1963; Eccles 1974). Because the dry air accumulates moisture as it transits the lakes, relative humidity and cloud cover are higher at the northern ends of the lakes in the southern hemisphere and on Lake Victoria (Figure 18.3). Rainy seasons occur when the ITCZ passes over a lake, although there may be lags. Lakes Malawi and Tanganyika, which are farthest from the equator, experience one rainy season, whereas lakes which are closer or straddle the equator, such as Lake Victoria, have two, with these differences creating differences in the annual cycles of stratification (Figures 18.5, 18.6). Thus, the surface energy budgets at the northern ends of the lakes vary less than those at the southern ends, and evaporation and associated cooling, in particular, is enhanced towards the south. As will be discussed below, these differences contribute to horizontal transports and variations in stratification.

Internal wave mediated fluxes are important in the African Great Lakes. Coulter (1963) provided evidence for internal wave upwelling to the south and downwelling to the north in response to the southeast monsoon in Lake Tanganyika. His observations were followed by those of Eccles (1974) for Lake Malawi. Internal wave motions can induce oscillatory transports whose periodicity depends on the degree of stratification and length of the lake. For Lakes Tanganyika and Victoria, the computed period of the first vertical mode internal wave is about a month. Modeled seiche-induced transports for a mid-lake station in Lake Tanganyka have speeds up to $15\,\text{cm}\,\text{s}^{-1}$ (Figure 18.7). Due to the lakes' long fetches and weak stratification, the internal waves in Lakes Tanganyika, Malawi, and Victoria are expected to become nonlinear (Spigel and Coulter 1996). When waves become nonlinear, they break and turbulence results (MacIntyre et al. 1999). Further, non-oscillatory transports are then expected (Maderich, van'Heijst, and Brandt 2001). Time series measurements in Lake Tanganyika show that second vertical mode waves with amplitudes of 50 to 100 m occur in response to the high winds of the southeast monsoon, and higher frequency nonlinear waves are embedded in these larger scale motions (P. Verburg and S. MacIntyre, unpublished data). The mixing they cause will induce fluxes of nutrients from depths below the depth of upwelling. Thus, the nonlinear waves are important for nutrient supply and ventilation of the thermocline. Coulter and Spigel (1991) predicted they would cause north–south transports. Given the likely diel oscillations of the thermocline near the lake's boundaries from the diel wind patterns (Hamblin, Bootsma, and Hecky 2003) or Kelvin waves (Naithani and Deleersnijder 2004; Antenucci 2005), inshore-offshore exchange would occur with lake-wide transports of inshore waters resulting from the interactions of the different wave types. Fish (1957) and Talling (1966) inferred that lake-wide up- and downwelling occurred in response to strong wind events on Lake Victoria, and recent field studies have verified their inference (S. MacIntyre, J.R. Romero, and G.M. Silsbe, unpublished data). Decreases in thermal stratification with concomitant losses of anoxia, as noted in Hecky et al. (1994) and in

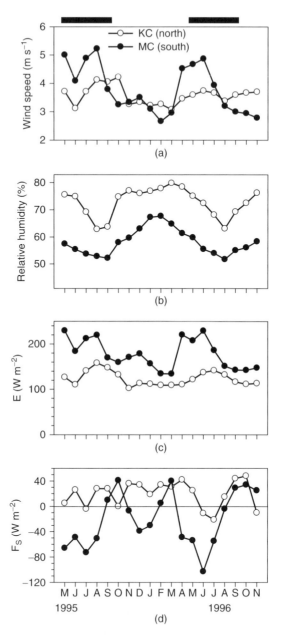

Figure 18.3 Time series of (a) wind speed, (b) relative humidity, (c) evaporation, and (d) total surface heat fluxes at the northern and southern stations on Lake Tanganyika in 1995 and 1996. Winds are stronger to the south during the southeast monsoon, relative humidity is always lower to the south, and evaporation is least to the north and greatest to the south during the southeast monsoon. The changes in surface heat fluxes are muted to the north, but cooling is appreciable to the south during the southeast monsoon. From Verburg *et al.* (2011), p. 913, Copyright (2000) by the Association for the Sciences of Limnology and Oceanography, Inc, used with permission.

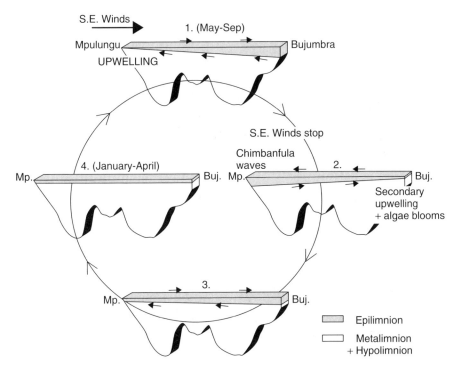

Figure 18.4 Schematic of thermocline tilting in Lake Tanganyika in response to the enhanced winds of the southeast monsoon. Internal waves are not illustrated. Reproduced from Hydrobiologia 407, 45–58, Plisnier *et al.* (1999) Limnological annual cycle inferred from physical-chemical fluctuations at three stations of Lake Tanganyika, with kind permission from Springer Science+Business Media B.V.

the recent field studies mentioned earlier, are likely indicative of the vertical mixing which results from instabilities associated with these large scale motions. Fish (1957) also showed internal waves cause considerable inshore-offshore transport. Thus, the ecosystem functioning of the African Great Lakes depends on the intensity of winds and the resulting up and downwelling of the thermocline, vertical mixing, and horizontal transport. Increased stratification, as caused by global warming, can reduce the amplitude of internal waves. At present, though, the stratification within the Great Lakes is so weak relative to wind forcing and their length that nonlinearity is likely to remain a persistent feature.

Density-driven flows can occur if there are spatial differences in rates of heating and cooling due to variable bottom slope, wind exposure, or, in the case of large lakes, differences in climate around the lake (Monismith, Imberger, and Morison 1990; Wells and Sherman 2001; Verburg, Antenucci, and Hecky 2011). The processes include differential heating and cooling, with the former causing buoyant overflows and the latter causing gravity currents which flow offshore at depth. With their weak stratification, the cool water introduced in the lower water column may substantially increase stratification in tropical lakes. Talling (1963) reported the stratification from differential cooling in Lake Albert. Flows due to differential cooling have been observed in the southern basin of Lake Malawi (Eccles 1974; Patterson, Wooster, and Sear 1998).

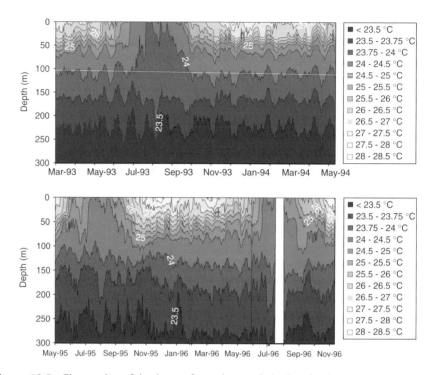

Figure 18.5 Time series of isotherms from the south basin of Lake Tanganyika showing upwelling to a depth between 100 and 150 m in 1993 and 1996 and 50 to 100 m in 1995. Internal waves were prevalent in all years, and the metalimnion was deeper and thicker in 1995 than in 1993. From Huttula (1997).

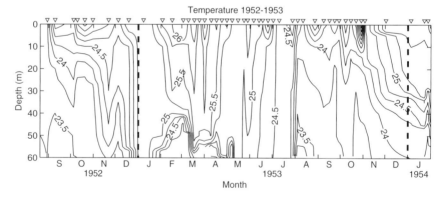

Figure 18.6 Time depth diagram of isotherms (lines of constant temperature) for Lake Victoria, East Africa, 1952–1953 illustrating the well mixed conditions during the southeast monsoon (June, July), the development of weak stratification beginning in August. Near surface stratification in August and September is due to heating and that at depth may be due to a gravity current. The rising isotherms in October and December 1952 were accompanied by upwelling of water with low oxygen concentrations. The seasonal thermocline develops in November–December and persists until May. The holomixis in December–January is likely due to downwelling of the seasonal thermocline. Figure redrawn from Fish (1957) and in MacIntyre, S. (2012) Stratification and Mixing in Tropical African Lakes. In: Bengtsson, L., Herschy, R.W., and Fairbridge, R.W. (Eds), Encyclopedia of Lakes and Reservoirs, pp. 737–743, with kind permission from Springer Science+Business Media B.V.

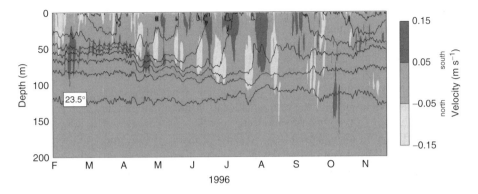

Figure 18.7 Simulated time-series temperatures in Lake Tanganyika from end of the rainy season, during the southeast monsoon, and in the warming period thereafter showing rising isotherms during the monsoon indicative of upwelling, and high frequency waves associated with lower frequency waves. Wave-induced transports associated with first vertical mode seiches are modeled and show a two-week periodicity. From Verburg *et al.* (2011: 920), copyright 2000 by the American Society of Limnology and Oceanography, Inc; used with permission.

Reductions in the transport of cool water to the monimolimnion by density driven flows in Lake Malawi have contributed to the warming of the deep waters of the lake (Vollmer *et al.* 2005). In Lake Tanganyika, a convective circulation in the upper mixed layer and metalimnion is induced by spatial gradients in temperature (Verburg, Antenucci, and Hecky 2011). Talling (1963) had hypothesized that gravity currents would contribute to the stratification in Lake Victoria. Recent field studies and analysis provide evidence for these flows in Lake Victoria (S. MacIntyre, J.R. Romero, and G. Silsbe, unpublished data). Cooler water is formed to the south due to north-to-south differences in evaporation, but, in contrast to the convective flows in Lake Tanganyika, the large-scale flows are likely wind driven. If aspects of climate change influence the overall heat budgets of these lakes, or enhance or depress the temperature contrasts, the volume of water transported by convective processes may change with consequences for the ecological functioning of the lakes.

18.3   Mixing Dynamics in a shallow Great Lake—Lake Victoria

18.3.1   Diel stratification and mixing

Diel stratification and mixing, first noted in Nyanza Gulf, Lake Victoria (Worthington 1930), is one of the predominant attributes of tropical African lakes (Figures 18.1, 18.2). Lake temperatures are nearly isothermal in the morning, increase rapidly in the upper water column during heating, reach a maximum by early afternoon, and then decrease beginning in late afternoon becoming isothermal at night. Variability is induced by changes in wind or cloud cover. As solar radiation decreases in the afternoon, mixed layer deepening is facilitated by increased evaporation and the resulting convective mixing. In the main basin of Lake Victoria, convective mixing at night during the southeast monsoon can reach to the lake bottom; at the end of the monsoon it generally penetrates between 5 to 20 m, and the depth of mixing increases as the lake warms to its maximum temperatures in March and April. By then a seasonal thermocline has formed, and nocturnal mixing penetrates 30 m to 50 m. The overturning, which begins at night persists for several hours each day below the diurnal thermocline.

The duration and vertical extent of the diurnal thermoclines relative to the depth of the euphotic zone influences species composition of autotrophs. The depressed N:P ratios in Lake Victoria, with the ratio smaller in inshore than offshore waters (Hecky 1993, North *et al.* 2008), favor cyanobacteria, which fix nitrogen,. The diffuse attenuation coefficient in inshore waters is $\sim 1.5\,m^{-1}$ (Silsbe, Hecky, and Guildford 2006; North *et al.* 2008). Euphotic zone depths (z_{eu}) are then of order 3 m, shallower than typical depths of the diurnal thermocline (Figure 18.2), and similarly favor cyanobacteria. Offshore, diffuse attenuation coefficients are three times lower, and z_{eu} is of order 10 m, deeper than typical diurnal thermoclines (Figure 18.1). While the development of a diurnal thermocline creates a region in the water column conducive to cyanobacteria the continued mixing below the diurnal thermocline can maintain diatoms in suspension and expose them to a adequate light. Due to the lake's warm surface waters, high evaporation rates, and variable cloud cover, diurnal thermoclines typically form after 09:00 hours and may break down by noon or 15:00 hours. Thus, the circulation induced by evaporation may allow continued success of diatoms if their growth rates, which are higher than those of cyanobacteria, exceed their settling rates. The depth of nocturnal mixing combined with horizontal advection influences abundance of phytoplankton (Silsbe, Hecky, and Guildford 2006). Thus, horizontal variations in the diel circulation regimes influence the phytoplankton ecology of Lake Victoria.

18.3.2 Seasonality

Fish's (1957) time-series measurements of temperature in Lake Victoria indicated that seasonal stratification could develop despite temperature differences of only 1 °C to 2 °C between the surface and bottom waters. His time series data (Figure 18.6), as well as that of Talling's (1966), established that Lake Victoria is holomictic during the southeast monsoon, becomes weakly stratified over its entirety in the calmer, sunnier, and more humid period that follows, and that a seasonal thermocline develops beginning at the end of the short rains or start of the northeast monsoon in December in association with decreasing relative humidity. Diurnal thermoclines remain a prevalent feature, with the depth of nocturnal mixing depending on the extent of night time cooling. Seasonal stratification is, in general, disrupted in late May, early June with the onset of the southeast monsoon with its cool, dry air masses and higher winds. Occasional holomixis to the north during stratification (e.g., during January in Figure 18.6), was attributed to wind-induced downwelling of the thermocline. A cross-basin transect in May 1995 validated this hypothesis (S. MacIntyre, J.R. Romero, and G.M. Silsbe, unpublished data). Importantly, observations in other lakes have shown that, on cessation of the wind, the subsequent up and downwelling of the thermocline can induce considerable mixing (Boegman, Ivey, and Imberger 2005; MacIntyre *et al.* 2009). Stratification to the north in Lake Victoria was considerably reduced after the event in May 1995. The loss of seasonal stratification in Lake Victoria may be initiated by increased winds which cause the thermocline to upwell with concomitant mixing, but deep mixing due to heat loss is ultimately responsible. By June, heat losses, particularly to the south and west, are three fold higher than in other seasons.

The cross basin difference in heat loss sets the stage for restratification towards the end of the southeast monsoon by differential cooling.

18.3.3   Differential heating and cooling

Diurnal heating and nocturnal cooling may lead to inshore-offshore exchange. Algal biomass is highest in the nearshore embayments, which causes increased attenuation of light and greater heating, and decomposition rates near the bottom are high such that hypoxia can develop in the day and persist at night if convective mixing is weak (Figure 18.2). MacIntyre, Romero, and Kling (2002) and Silsbe, Hecky, and Guildford (2006) showed that wind-induced currents in the upper water column, not differential heating, induce offshore flow, and gravity currents form due to cooling at night such that hypoxic, nutrient rich water can flow into the larger channels that connect to the main basin. Internal wave motions and land breeze conditions are then likely to cause flow offshore (Fish 1957).

Differential cooling may also lead to larger scale transport (Figure 18.8). Talling (1963) showed that differential cooling in Lake Albert, facilitated by a shelf region to the southwest that is nearly 20 km long and less than 10 m deep, was a likely cause of stratification deep in the water column during the short and long rains and hypothesized that such gravity currents would occur in Lake Victoria. A decrease in near bottom temperatures accompanies the restratification of Lake Victoria at the end of the southeast monsoon (Figure 18.6). While brief injections of cool water could occur due to upwelling of any water unmixed at the end of the southeast monsoon (e.g., early August 1953 in Figure 18.6), the cool water persists at depth for a month or more in Fish's (1957) and Talling's (1966) time series, suggesting another source. Cross-basin differences in net heat fluxes computed from meteorological data at six stations around the lake are largest during the southeast monsoon, synoptic surveys during the southeast monsoon show the expected north-to-south temperature differences, and assessment of the time scales for advection of gravity currents relative to rates of mixing provide evidence for these flows in Lake Victoria (S. MacIntyre, J.R. Romero, and G.M. Silsbe, unpublished data).

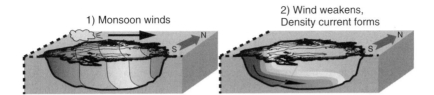

Figure 18.8   Schematic of north to south temperature differences created in Lake Victoria during the southeast monsoon. On relaxation of the winds, either gravitational readjustment leads to a gravity current that contributes to stratification to the north, or, following Song (2000), the gyre structure in the south is relaxed allowing northward flow of cool water. These currents appear to form when cooling is greater to the south during the southeast monsoon. Lower winds and higher relative humidity and a reduction in the cross-basin net heat losses, as occurred around 1980, would reduce the north–south temperature contrasts, suppress the formation of such density currents, and potentially decrease the stratification in the lower water column allowing a larger volume of anoxic water. (See insert for color representation.)

18.3.4 Temporal changes

Hecky (1993) and Hecky *et al.* (1994, 2010) demonstrated that thermal stratification was stronger and more persistent in Lake Victoria in the early 1990s relative to measurements from the 1960s, oxygen concentrations were higher in the day, and the volume of anoxic water increased in the lower water column. Eutrophication had been ongoing since the 1920s with increased land use in the basin, and the introduction of the Nile perch (*Lates nilotica*) in the 1950s set the stage for eventual changes in the food web. Hecky *et al.* (2010) indicated that within-lake changes became pronounced around 1980. Concentrations of $\delta^{13}C$ and biogenic silica (Si) increased abruptly in the sediments in inshore waters, and concentrations of total phosphorus increased offshore presumably due to offshore transport of organic matter produced inshore. N:P ratios plummeted in offshore waters, indicating phosphorus was being mobilized from the sediments. The Nile perch fishery exploded at that time, and abundance and diversity of haplochromines rapidly decreased. Hecky *et al.* (2010) inferred that the lake had undergone a regime shift due to the shallowing of the oxycline, related greater fluxes of phosphorus from the sediments but likely loss of inorganic nitrogen by denitrification, and increased turbidity caused by higher algal biomass, which made foraging for food difficult for the visual haplochromines. The diatom *Aulacoseira*, which had been a dominant in the lake during the southeast monsoon decreased in abundance and has since disappeared from the pelagic (R.E. Hecky, personal communication). The loss of this diatom was a response to the declining silica concentrations in the lake, but may also have been a response to a change in mixing dynamics. Hecky *et al.* (2010) identified a change in climate as the cause for the apparent abrupt changes in the lake, and Kolding *et al.* (2008) noted a decrease in wind stress beginning in 1976, an increase in precipitation and lake level, and that winds increased again by the mid-1990s.

Time-series temperature and oxygen diagrams indicate that the seasonal patterns of stratification dynamics did not change from the 1950s until the mid-1990s, but there were some between-year differences (S. MacIntyre, J.R. Romero, and G.M. Silsbe, unpublished data). The temperature differences between 50 m and 60 m were strongest in the time series records from 1951–1952 (Figure 18.6) and 1960–1961, with increased stratification intermittent and associated with upwelling (Talling 1966). Apparent loss of stratification occurred as the thermocline downwelled (Figure 18.6). Oxygen concentrations decreased in the lower water column during upwelling, indicating a pool of anoxic water existed farther offshore. Stratification was pronounced below 45 m in 1994 with hypoxic water below 50 m. In contrast, during 1991–1992, the thermocline during the northeast monsoon and long rains was between 30 m and 40 m with hypoxic waters below (Hecky *et al.* 1994). The time series, then, imply that conditions within the lake shifted between two regimes. In the first, temperatures decreased near the bottom towards the end of the southeast monsoon, and the seasonal thermocline was below 40 m and constrained the formation and areal extent of anoxic water. In contrast, in the alternate condition, the seasonal thermocline and oxycline were higher in the water column and poor oxygen conditions extended to previously oxygenated depths.

The between-year differences in seasonal stratification indicate that, despite general warming in the Victoria region (Lehman 1998), the controls on mixing dynamics are more complex. The regime shift described by Hecky *et al.* (2010) likely resulted from a regime shift in the Indian Ocean that began in 1976 (Clark *et al.* 2003). Temperatures shifted up by 0.2 °C. Air temperatures, which had been decreasing over African since the early 1940s, began to increase in 1976 (Hulme *et al.* 2001). The Pacific Decadal Oscillation changed from warm to cool in that year, so the changes may have been linked to its phase shift. Relative humidity increased, air temperatures and wind speeds decreased over Lake Victoria (Figure 18.9). Variability was much higher before and after that period, particularly in wind speeds during the long rains and southeast monsoon. Monthly average heat losses, the sum of latent and sensible heat fluxes and

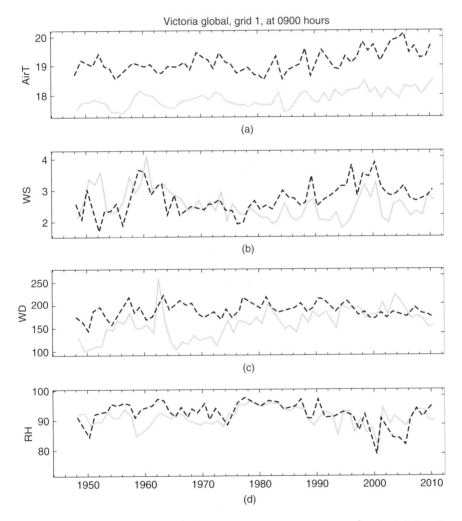

Figure 18.9 Time series of (a) air temperature, °C, (b) wind speed, m s^{-1}, (c) wind direction, degrees, and (d) percentage relative humidity offshore of Bukoba, Uganda (0 °N, 32.5 °E), from 1948 until 2010 obtained from NCEP global reanalysis data. Southeast monsoon, gray line, June–August, and long rains, black dashed line, February–May. Reanalysis data were provided by the NOAA/OAR/ESRL PSD, Boulder, Colorado, USA, from their Web site at http://www.esrl.noaa.gov/psd/ (accessed July 11, 2012) with the reanalysis project described in Kalnay *et al.* (1996).

net long wave radiation, assuming a constant lake surface temperature of 24 °C, were 150 W m^{-2} at night near Bukabo, Uganda (S. MacIntyre and J. Romero, unpublished data). In contrast, they were 200 Wm^{-2} to 250 Wm^{-2} during windier periods. The between year differences result from variations in the surface meteorology over Lake Victoria as it responds to decadal scale variations in Indian Ocean temperature coupled with interannual variations in phase of El Nino and the Indian Ocean Dipole (IOD) (S. MacIntyre and J. Romero, unpublished data). The differences in the north to south net heat fluxes at night from 1976–1982 are half of those computed for windier years. It is reasonable to infer a shallowing of the depth of nocturnal mixing and that the north to south temperature differences were smaller and did not favor formation of a gravity current. Conditions would have been similar during years with low winds in the early 1990s. As a result, the seasonal thermocline was shallower allowing a greater volume of anoxic water. Further, with lighter winds during the northeast monsoon, the heat introduced into the lake was not mixed downwards, further limiting the development of a seasonal thermocline. In response to higher winds in 1994, near bottom waters cooled at the end of the southeast monsoon and the seasonal thermocline was once again low in the water column constraining the volume of anoxic water (S. MacIntyre, J.R. Romero, and G.M. Silsbe, unpublished data). The shift in global climate and consequent change in the mixing dynamics within the lake, with the stage set by prior land use changes and introduced species, was the trigger for the regime shift in Lake Victoria.

A combination of processes affecting trade winds, the Walker circulation, sea-surface temperatures and large-scale sea-surface temperature differences moderate the timescales of variability in the Indian Ocean (Meehl and Arblaster 2011). This variability, and the resulting variability in surface meteorology over Lake Victoria (Figure 18.9), will likely cause Lake Victoria to shift between states in which the anoxic volume is extensive and years in which it is less extensive. Cross-lake surveys indicated reduction of the anoxic volume during the stratified period in some years (Kolding *et al.* 2008). Such conditions would reduce the loading of phosphorus and enable greater mixing, productivity would likely be lower, and conditions would be more favorable for haplochromines. Hecky *et al.* (2010) hypothesized that Lake Victoria could transition to hypertrophy, an even more deleterious state than the present one. With continued eutrophication, the amount of phosphorous released from the sediments would be even higher. Internal wave motions coupled with a shallow oxycline would lead to frequent fish kills and would be disastrous for all who depend on the lake for their sustenance. Such a state may result from a return to conditions with low winds and cool air temperatures. They may be a result of the phase of the PDO, which would imply infrequent shifts, but processes with shorter timescales are exerting control and causing considerable variability. That is, the increased temperatures in the Indian Ocean in 1976 occurred after a an El Nino-positive IOD event followed by a strongly negative IOD and La Nina followed by a positive IOD (Clarke *et al.* 2003). The frequency of IOD events has been increasing since the 1960s in concert with the more rapid increase in sea surface temperature of the Indian Ocean (Nakamura *et al.* 2009). Thus, wind forcing and the volume of anoxic water will likely vary with a periodicity of one to five years.

The Lake Victoria ecosystem is dependent on conditions within the Indian Ocean, and future regime shifts will depend upon its dynamics. Air temperatures have been warming in the Victoria region since 1980 (Figure 18.9), and these increases have been accompanied by increased solar radiation and net long wave radiation, both indicative of decreased cloud cover. Although warming of the lake could be predicted from these changes, wind speeds also increased in some seasons, and surface energy budgets indicate that as the lake warms, cooling by evaporation increases considerably. Annual estimates of surface temperatures in Lake Victoria show temperatures ranged from 24°C to 25.3°C from 1985 until 2009 and with a slight upward trend until 2010 (S. Hook and C. Wilson, personal communication, Jet Propulsion Laboratory, Pasadena, CA). Higher annual temperatures were more likely in El Nino years with low winds and often higher cloud cover, and lower temperatures were more prevalent in La Nina years with low values of the Indian Ocean Dipole. Thus, surface temperatures are responding to changes in the nearby Indian Ocean and to atmospheric cycles, but the temperature range has been limited so far. Air temperatures over Africa are expected to increase by 1.2°C to 5°C over the next 70 years (Hulme *et al.* 2001), so some warming will occur. The two stratification regimes in the lake result from near-decadal scale changes in ocean dynamics combined with inter-annual variability related to the Indian Ocean Dipole and ENSO cycles. Typically, temperatures in the Indian Ocean shift upward about every five to ten years after a pronounced El Nino event followed by a La Nina-negative IOD event. Indian Ocean temperatures are either stable or oscillate in response, and the wind fields over Lake Victoria have shown related variability. Thus, as the Indian Ocean warms, Lake Victoria will likely transition between states with a lesser and greater anoxic volume. With the increased eutrophication and the importance of the now intense fishery on structuring the food web, the implications of the physical variability on ecosystem function and species diversity remain to be determined.

18.4 Mixing Dynamics in a Meromictic Great Lake—Lake Tanganyika

The physical limnology of Lake Tanganyika is described in Coulter and Spigel (1991). Results of a UNOPS Project, "Pollution Control and Other Measures to Protect Biodiversity of Lake Tanganyika" and an FAO project "Research for the Management of the Fisheries on Lake Tanganyika" have further extended our understanding of the hydrodynamics and ecosystem ecology of Lake Tanganyika (e.g., Huttula 1997; Plisnier *et al.* 1999). Huttula (1997), Verburg and Hecky (2003), and Verburg, Antenucci, and Hecky (2011) review the meteorology around the lake and point out the windier, sunnier conditions to the south and, due to the lake effect, the reduced solar insolation and higher relative humidity to the north (Figure 18.3). The higher winds during the southeast monsoon create conditions favorable for upwelling of the thermocline to the south as well as mixed layer deepening due to cooling (Figures 18.4, 18.5). Coulter's (1963) measurements provided the first evidence of an internal wave field, the upwelling to the south during the southeast monsoon, and resulting increased primary production in near shore embayments to the south. On relaxation of the southeast trade winds, upwelling occurs to the north, which often leads to an increase in productivity to the north (Plisnier *et al.* 1999). The Chimbanfula waves that form to the south may be the surge predicted by Coulter and Spigel (1991). The time series temperature data obtained from 1993–1996 showed internal wave motions and the between year and climate-related variability in depths of upwelling (Huttula 1997). The deeper upwelling in 1993 than in 1995 was likely due to the cooler, somewhat windier conditions in 1993 (Figure 18.5). The depth of upwelling in 1993 was similar to that observed by Coulter (1963), but his sampling did not extend beyond 150 m. The vertical extent of the metalimnion also varied between years, being a sharper, more clearly defined feature during the rainy season in 1993–1994 than in 1995–1996. The thermocline downwells to the north during the southeast monsoon causing mixed layer deepening. The associated cooling eventually causes entrainment of metalimnetic waters. Internal wave displacements vary with depth and horizontal location and induce periodic flows (Figures 18.5, 18.7). Internal wave displacements were as large as 100 m near 300 m depth. As stratification decreases with depth, internal waves with even larger amplitudes are likely deeper in the water column. Kelvin waves were not expected within narrow, near equatorial lakes, but Naithani and Deleersnijder's (2004) modeling provides evidence for these waves in Lake Tanganyika and is supported by Antenucci's (2005) scaling analysis. While changes in nutrient limitation in the littoral region have occurred in July and are attributed to the basin-scale upwelling (McIntyre *et al.* 2006; Corman *et al.* 2010), the nearshore up and downwelling of Kelvin waves, as well as that from diel variations in wind magnitude, should contribute to controls on productivity in the littoral zone. The nonlinear waves predicted by Coulter and Spigel (1991) were present to the south in 1995 but not in 1993 (P. Verburg and S. MacIntyre, unpublished data). Thus, the increased stratification in 1995 may have contributed to a more complex internal wave field and resulting vertical and horizontal transport. Horizontal convective circulation also causes lake-wide transport (Verburg, Antenucci, and Hecky 2011). This circulation is driven by the greater heating to the north than to the south (Figures 18.3, 18.10). As a result, warm water travels north to south near the surface in the latter months of the southeast monsoon and is in a direction opposite to the prevailing winds. Cooler water flows north in the metalimnion and modulates the upwelling there. This motion supplies nutrients from the south to northern waters. Increases in primary production will depend on subsequent mixing events which may be more frequent when the metalimnion upwells to the north in Oct-Nov (Figure 18.4). The horizontal convective flows also increase the stratification to the south and may reduce the duration of mixing during the southeast monsoon.

18.4.1   Ecosystem response to climate change

Both paleolimnological approaches and estimates of primary production are being used to determine the ecosystem response to climate change in Lake Tanganyika. Verburg, Hecky, and Kling's (2003) and O'Reilly *et al.*'s (2003) analysis indicates primary productivity decreased within the lake as a consequence of recent climate change. They assumed productivity depends on depth of vertical mixing during the southeast monsoon, and that increases in stratification and decreases in the depth of mixing would lead to reduced nutrient fluxes and reduced productivity. Evidence for change within the lake includes a shallower depth of anoxia, an increase in transparency of the lake waters, increased concentrations of silica in the upper water column, and a decrease in the number of diatoms (Verburg, Hecky, and Kling 2003). On the basis of $\delta^{13}C$, pelagic, pristine waters indicate decreased primary productivity, whereas there is a trend towards increasing productivity in near-shore waters where land use changes have occurred (R.E. Hecky, personal communication). Tierney *et al.* (2010) report a

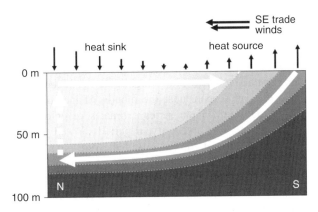

Figure 18.10 Schematic of horizontal convective circulation in Lake Tanganyika. The thermocline upwells to the south in response to the southeast trade winds and as illustrated in Figures 18.4 and 18.5. However, rather than the wind blowing surface water to the north, as would be expected, the greater heating to the north than the south generates a flow that moves southward against the prevailing winds. To balance the surficial southward flow, water moves northward in the metalimnion. Modeling indicates these flows occur in July and August and cause considerable north–south exchange. From Verburg *et al.* (2011), p. 923, Copyright 2000 by the American Society of Limnology and Oceanography, Inc); used with permission.

reduced flux of diatoms to the sediments. Although Tierney *et al.*'s (2010) regressions over the last 1300 years show a statistically significant negative relation between air temperature and biogenic silica (Si) (r = −0.57), they also find a statistically significant negative response to changes in humidity (r = −0.39) as inferred from charcoal. The more recent period is a wetter one with higher humidity. As discussed for Lake Victoria, higher relative humidity is associated with cooler air temperatures, even as temperatures rise, and lower wind speeds. Thus, the initial driver for changes in productivity in the lake may be a reduction in evaporation due to the more humid, wetter conditions, and possibly less windy conditions, in the twentieth century relative to conditions prior to 1900.

Direct measurements of primary production have been too intermittent to determine whether the changes in the lake have caused a decrease in productivity. Measurements of primary productivity in Lake Tanganyika, as in the other African Great Lakes, are rare (Melack 1980; Sarvala *et al.* 1999), and most estimates are based on photosynthesis versus irradiance measurements rather than *in situ* measurements (Hecky and Fee 1981; Stenuite *et al.* 2007). Stenuite *et al.* (2007) reported increased algal biomass during the dry season when winds are higher, and higher productivity in the south basin. They reported annual production 20% lower than Hecky and Fee (1981), but similar to Melack (1980). Sarvala *et al.* (1999) reported values 47% to 127% higher than those of Hecky and Fee (1981). Hecky and Fee's (1981) measurements in the north basin in October–November, 1975, included an event that caused chlorophyll *a* concentrations to increase fourfold and the assimilation rate of carbon to decrease by a factor of 3.5. The changes may have been the result of the upwelling of dark adapted phytoplankton, and the high diel and between station variability they observed indicates the difficulty in discriminating climate change related changes in ecosystem function from sparse within lake measurements.

Use of remote sensing combined with bio-optical models allows basin-wide estimates of primary productivity and an appreciation of climatic controls (Bergamino *et al.* 2007, 2010). A whole-lake estimate of primary production for 2003 using a MODIS based bio-optical time series, $646 \pm 142 \, \mathrm{mg \, C \, m^{-2} \, d^{-1}}$, was within the range of other measurements. Their study highlights the higher productivity during the dry season than the wet season, near the mouths of incoming rivers, in shallow embayments, and to the south. Time series estimates of chlorophyll at the north, central, and southern basins over a seven year period showed the enhanced biomass to the south during the southeast monsoon and the one month delay in maximum biomass to the north. Dual peaks occurred in the northern and central portions of the lake, with the first likely due to upwelling from cooling, and the second at the northern station likely due to the upwelling of the thermocline (Figure 18.4), from a second vertical mode wave (P. Verburg and S. MacIntyre unpublished data), or the arrival in the metalimnion of nutrients transported northward in the horizontal convective circulation. The second peak at the lake center was attributed to rainfall. Shifts in the timing of these peaks in algal biomass occurred in response to the intensity of winds over the lake.

Studies assessing controls on rainfall over east and southern Africa in the recent period may prove useful in understanding the changes that may have caused the decrease in diatom productivity in Lake Tanganyika in the past century. During wet years, less moisture is exported from east Africa and more moisture arrives from the southern Congo, with the changes attributed to modified atmospheric circulation (Mapande and Reason 2005). Most importantly, during wet years, the duration of the rainy season is longer. If the increased moisture over Lake Tanganyika (Tierney *et al.* 2010) occurred with a lengthening of the rainy season, the decreased diatom productivity would be the result of a shorter period of upwelling. This interpretation is supported by recent paleolimnological studies in Lake Chala, Tanzania, which show the thickness of varves enriched in diatoms versus organic matter depends on rainfall and associated wind patterns (Wolff *et al.* 2011). Their results indicated winds were reduced when rainfall was higher. Thus, in Lake Tanganyika, an overall decrease in winds during the current, more humid period would have contributed to the reported decrease in depth of upwelling. The increased stratification reported within Lake Tanganyika for the period 1913 to 1975 (Verburg, Hecky, and Kling 2003, O'Reilly *et al.* 2003), and inferred to the present from temperature proxies in the sediments (Tierney *et al.* 2010), may then result from greater heat captured during the rainy season combined with a shorter, less intense period of upwelling.

Examination of the time series of NCEP reanalysis data (Kalnay *et al.* 1996) indicates that surface energy budgets must be computed by season and by time of day in order to understand ongoing changes. Heat content peaks in May prior to the southeast monsoon (Huttula 1997). The NCEP data shows increased heating, increased solar radiation, and variations in net long wave radiation depending on time of day with no clear change in wind speeds during the three months prior to the southeast monsoon. Thus, over time, more heat will be captured in this period. Variability in winds and other meteorological variables during the southeast monsoon is linked with sea surface temperatures in the Indian Ocean and the El Nino Southern Oscillation. Thus, interannual variability in depth of wind induced and cooling induced upwelling during the southeast monsoon will occur, but will work against increasing stratification within the lake.

Changes in the length of the wet versus the dry season, and any changes in stratification during the wet season which influence settling rates of organic matter, may influence the contribution of remineralization to the lake's function. Lewis (1973) discusses the importance of atelomixis for recycling and formation of layered communities of plankton within tropical lakes; such layering occurs in Lake Tanganyika (Salonen *et al.* 1999). Kilham and Kilham (1990) illustrated the importance of recycling within warm tropical lakes; and Hecky *et al.*'s (1981) observation that remineralization exerts a strong control on productivity in Lake Tanganyika is supported by Plisnier *et al.*'s (1999) finding that nutrient concentrations increased in all basins at depths just below the mixed layer. Mean concentrations of soluble reactive phosphorus on the order of $1\,\mu M$ and of nitrate of $10\,\mu M$ were found at $60\,m$, and maximum concentrations were at least three times higher. The lake's carbon budget seasonally depends on heterotrophs (Stenuite *et al.* 2007), and this dependence may be greater now than in the prior, more arid and presumably windier period described by Tierney *et al.* (2010). With warming temperatures, heterotrophy may increase. Increasing dependence on the microbial loop may decrease the efficiency of carbon transfers to higher trophic levels, which would be a concern for maintenance of the fishery (Sarvala *et al.* 1999). Thus, remineralization within and below the mixed layer needs to be quantified to understand controls on nutrient availability under a warming climate. Further, if the current warm period is one with an overall increase in rainfall (Hulme *et al.* 2001), land-use practices will exert an increasing impact on the lake. Changes in species composition of chironomids near the mouths of rivers flowing into Lake Tanganyika provide evidence of localized eutrophication due to land use changes (Cohen *et al.* 2005). Thus, understanding controls on both winds and rainfall is of major importance for making predictions about the future ecosystem dynamics within Lake Tanganyika.

Evaluation of the NCEP reanalysis data for Lake Tanganyika shows variability similar to that for Lake Victoria since 1950 during the southeast monsoon (Figure 18.9) but overall heating. Analysis of paleolimnological records indicates that variability in winds and rainfall has increased with previous and current warming of the southwestern Indian Ocean (Wolff *et al.* 2011). Indian Ocean sea surface temperatures began warming in 1960, thus the current period differs from that which caused the initial decrease in diatom productivity. Significant interannual and decadal variability will occur in rainfall and winds such that Lake Tanganyika will cycle between years with deep upwelling and high diatom productivity and years in which increases in rainfall and remineralization will control productivity. Overall, though, the reanalysis data indicates the lake will continue to warm, and with increasing stratification, decreases in productivity will occur in pristine waters as inferred with the paleo-record. Quantifying ecosystem-level change within the lake requires systematic multiyear studies and models. The importance of using remotely sensed data is highlighted by the two recent papers by Bergamino *et al.* (2007, 2010). It may be difficult to assess the variability in primary productivity from space due to mixing associated with internal waves and, similarly, from the layering of phytoplankton as occurs with stratification. Hence, ground-based measurements must complement remotely sensed data. Dominant controls will be between year variations in depth of upwelling (see, for example, Naithani *et al.* 2011), with variability also induced by between-year variations in the amplitudes and nonlinearity of the internal wave field (Figure 18.5) and in the intensity of the horizontal convective circulation and associated south–north nutrient fluxes.

Three dimensional hydrodynamic modeling, as in Verburg, Antenucci, and Hecky (2011), would capture some of the important dynamics, but quantifying internal wave effects on mixing requires use of nonhydrostatic models (Zhang, Fringer, and Ramp 2011). Similar nonhydrostatic modeling would be of value for Lakes Malawi and Kivu. Modeling should include the variability introduced by ENSO, the Indian Ocean Dipole, and the effects of variability in temperature in the South Atlantic Ocean as it modifies airflow from the west and southwest and variability in temperature in the Indian Ocean as it affects attributes of the air masses coming from the east (Camberlin, Janicot, and Poccard 2001; Mapande and Reason 2005).

18.5   Lake Kivu—a chemically stratified lake

Lake Kivu, located in a region with active volcanoes, and Lakes Monoun and Nyos, two maar lakes in the Cameroon, share the distinction of having large concentrations of carbon dioxide in their lower water columns (Kling *et al.* 2005; Tassi *et al.* 2009). Due to the relatively recent explosions of Lakes Monoun and Nyos, they have been called "killer lakes." Lake Kivu has deep water supersaturated with CO_2 relative to STP and additionally has elevated methane concentrations (see also Chapter 24). Chemical stratification imposed by warm, saline CO_2-rich deep springs as well as its 500 m depth ensure that Lake Kivu is meromictic. Temperature and conductivity both increase below ~ 100 m resulting in double diffusive convection (Newman 1976; Schmid, Busbridge, and Wuest 2010). If the concentrations of dissolved gases exceed the threshold set by hydrostatic pressure, they will come out of solution, and an explosive plume will form as occurred in the lakes in the Cameroon. The nearly two million people living around Lake Kivu are currently at risk, and recent measurements indicated the gas concentrations are increasing (Schmid, Busbridge, and Wuest 2010). The flux of organic matter from surface waters resulting from increased productivity has contributed to the increased gas concentrations (Pasche *et al.* 2009; 2010). A pilot plant is in operation to determine the feasibility of methane extraction for commercial use and, by venting the gases, of reducing the danger for the people living around the lake. Funding is in place for a commercial-scale development of several hundred megawatts of electricity production.

Rainfall, through its influence on the fluxes of water into the lake from the underground springs, which are meteoric in origin, and into surface waters, and on the depth of the lake relative to the outlet, may play a predominant role in the future functioning of the Lake Kivu ecosystem. Like the other African Great Lakes, climate controls on the intensity of wind, cooling, and rainfall will determine interannual variability in mixing regime in the upper water column, phytoplankton species composition, and primary productivity, with upwelling to the south inducing blooms (Haberyan and Hecky 1987; Sarmento, Isumbisho, and Descy 2006). The lake is heating under the influence of global warming (Lorke *et al.* 2004), but it is uncertain how that affects depths of upwelling. The contribution of wind-induced upwelling to the lake's productivity is likely much less than for the other deep lakes due to its shorter fetch. In addition, air temperatures are sometimes warm in some years instead of cool during the southeast monsoon, with the warming mitigating deepening by evaporative cooling. Consequently, increases in productivity from nutrient loading from rainfall may make a larger contribution to annual productivity in Lake Kivu than in the deeper lakes.

Land-use changes in the basin have included a loss of forests, but the nutrient supply from incoming streams is reported to be modest compared with vertical advection of nutrients attributed to inflows from the subterranean springs (Pasche *et al.* 2010). Subterranean inflows are presumed to have increased since the early 1960s due to increased rainfall in the basin. Verification of the extent of nutrient loading to surface waters from incoming rivers requires additional studies during periods with high discharge. Rainfall is predicted to increase in the Kivu region, although there could be some diminution by the increased rainfall over the Indian Ocean (Hulme *et al.* 2001; Gu *et al.* 2007). It is rainfall, with its influence on nutrient loading to surface waters and on vertical advection from below, which will exert the largest overall control on the lake's productivity and maintenance of the fishery. Methane concentrations will increase in the lower water column due to increased productivity and the climate controlled inflows from the underground springs. Thus, predicted increases in rainfall and anticipated increased loading of dissolved gases to the lower water column further mandate the venting of the greenhouse gases at depth to reduce the risk of catastrophic gas release.

18.6 Changes in mixing dynamics and ecosystem dynamics due to changes in climate and land use

Interannual variations in the mixing dynamics in the African Great Lakes are linked with the El Nino Southern Oscillation (ENSO) with increased rainfall during El Nino years and higher winds during La Nina years (Nicholson and Yin 2002; Clark, Webster, and Cole 2003; Wolff *et al.* 2011). As the Indian Ocean warms, ENSO and the related Indian Ocean Dipole are changing phase more rapidly than in the past (Nakamura *et al.* 2009). This cycling will cause interannual variability in the depth of upwelling in the deep lakes and mixing to the bottom in the shallow lakes with concomitant effects on primary productivity. The warming in the deep lakes, however, will contribute to reduced productivity where waters are pristine and shifts to picoplankton. In years in which the Indian Ocean Dipole is strongly positive and the western equatorial Indian Ocean is anomalously warm, large increases in rainfall occur with flooding and considerable increases in lake level (Birkett, Murtugudde, and Allan 1999; Clark, Webster, and Cole 2003). Such events are intermittent, and, depending on latitude, prolonged drought may occur in the intervening years. These shifts in wind and rainfall determine the natural variability in primary productivity in the East African Great Lakes (Wolff *et al.* 2011).

   Climate warming, and the variability induced by the atmospheric cycles, is co-occurring with explosive population growth in East Africa and ever increasing demands upon the lakes and the surrounding watersheds (see also Chapter 11). The paleo record indicates that land use changes began to alter nutrient loading into Lake Victoria by the 1920s and into Lake Malawi by the 1940s (Hecky *et al.* 2010; Otu *et al.* 2011). Lake Victoria rapidly transitioned from oligo-mesotrophic to eutrophic during the large-scale climate transition in the late 1970s with devastating consequences for the endemic haplochromines. Consequences were positive for the commercial fishery (Kolding *et al.* 2008), but negative for water quality with nearly 70% of the mean phytoplankton biomass now consisting of cyanophytes with dominance by *Anabaena*, *Cylindrospermopsis*, *Planktolyngbya* and *Microcystis* species (Kling, Mugidde, and Hecky 2001). Improvements in land-use management are required to avoid the lake's shifting to a hypereutrophic state. Sediment cores indicate eutrophication is occurring in disturbed nearshore regions of Lake Tanganyika and in much of the southern basin of Lake Malawi with accelerated change in Lake Malawi since 1980 (Otu *et al.* 2011; R.E. Hecky, personal communication). The changes in diatom community structure in Lake Malawi in one decade exceeded those due to 100 years of climate variability and mimic those seen as a result of cultural eutrophication in Lake Victoria. The changes are largely attributed to run off via the Linthipe River, which flows through the increasingly deforested terrestrial environment to the southwest (Otu *et al.* 2011). Perturbations will be increased during future high rainfall events similar to those documented previously when the IOD and/or ENSO were in a positive phase and the western equatorial Indian Ocean was anomalously warm (Birkett, Murtugudde, and Allan 1999; Clark, Webster, and Cole 2003). Similar to Lake Victoria, increases in cyanophytes are expected with increased eutrophication. Similar shifts are a likely response to disturbance and warming in the other African Great Lakes. Some of the larger cyanobacteria are toxic, and the smaller forms cause shifts towards the microbial food web with less energy transferred to higher trophic levels which support fisheries (see also Chapter 11). Hence, improved land use management is required to combat the negative consequences of warming, intermittent high rainfall, and ongoing land use change. The combined use of remotely sensed data, improvements in the grid of meteorological stations in East Africa, reanalysis data from the U.S. National Center for Environmental Prediction, new products to identify rainfall anomalies (FEWS Net, Famine Early Warning System Network), and within-lake sampling will enable modeling, predictions, and development of strategies to manage and protect these Great Lakes.

18.7   Acknowledgements

This chapter results from the excellent research being conducted by a number of research groups working on the African Great Lakes. The new data and analysis for Lake Victoria results from a collaboration with Jose R. Romero, with financial support provided by National Science Foundation grants Division of Environmental Biology (DEB) 95–53064, 93–18085, 93–17986, 97–26932, 01–08572, 06–40953, Ocean Sciences (OCE) 99–06924, support by the Centre for Water Research, University of Western Australia for J. Romero, and a Blaustein Visiting Professorship from the Department of Environmental Earth System Science, Stanford University, to S. MacIntyre. Thanks are given to J.M. Melack, B. Emery, R.E. Hecky, J.F. Talling, G.W. Kling, P. Ramlal, F. Bugenyi, R. Mugidde, M. Magumba, R. Ogutu, P. Njuru, G. Silsbe, L. Domingo, J. Vidal, the captain and crew of the R/V *Ibis*, and the Fisheries Research Institute in Uganda.

References

Antenucci, J.P. (2005) Comment on "Are there internal Kelvin waves in Lake Tanganyika?" by Jaya Naithani and Eric Deleersnijder. *Geophys. Res. Letters*, **32**, L22601, doi:10.1029/2005GL024403.

Beadle, L.C. (1981) *Inland Waters of Tropical Africa*, Longman Group Limited, London, pp. 231–44.

Bergamino, N., Horion, S., Stenuite, S., *et al.* (2010) Spatio-temporal dynamics of phytoplankton and primary production in Lake Tanganyika using a MODIS based bio-optical time series. *Remote Sense. Environ.*, **114**, 772–80.

Bergamino, N., Loiselle, S.A., Cozar, A., *et al.* (2007) Examining the dynamics of phytoplankton biomass in Lake Tanganyika using empirical orthogonal functions. *Ecol. Model.*, **204**,156–62.

Birkett, C., R. Murtugudde, and T. Allan. 1999. Indian Ocean climate event brings floods to East Africa's lakes and the Sudd Marsh. *Geophys. Res. Lett.*, **26**,1031–4.

Boegman, L., Ivey, G.N., and Imberger, J. (2005) The degeneration of internal waves in lakes with sloping topography. *Limnol. Oceanogr.*, **50**, 1620–37.

Camberlin, P., Janicot, S., and Poccard, I. (2001) Seasonality and atmospheric dynamics of the teleconnection between African rainfall and tropical sea-surface temperature: Atlantic vs. ENSO. *Int. J. Climatology*, **21**, 973–1005. doi: 10.1002/joc.673.

Clark, C.O., Webster, P.J., and Cole, J.E. (2003) Interdecadal variability of the relationship between the Indian Ocean zonal mode and East African coastal rainfall anomalies. *J. Climate*, **16**, 548–54.

Cohen, A.S., Palacios-Fest, M.R., McGill, J., *et al.* (2005) Paleolimnological investigations of anthropogenic environmental change in Lake Tanganyika: I. *An introduction to the project*. *J. Paleolimn.*, **34**, 1–18. doi: 10.1007/s10933-005-2392-6.

Corman, J.R., McIntyre, P.B., Kuboja, B., *et al.* (2010) Upwelling couples chemical and biological dynamics across the littoral and pelagic zones of Lake Tanganyika, East Africa. *Limnol. Oceanogr.*, **55**, 214–24.

Coulter, G.W. (1963) Hydrological changes in relation to biological production in southern Lake Tanganyika. *Limnol. Oceanogr.*, **8**, 463–77.

Coulter, G.W., and Spigel, R.H (1991) Hydrodynamics, in *Lake Tanganyika and Its Life* (ed. G.W. Coulter), Oxford University Press, Oxford, pp. 49–75.

Eccles, D.H. (1974) An outline of the physical limnology of Lake Malawi (Lake Nyasa). *Limnol. Oceanogr.*, **19**, 730–42.

Fish, G.R. (1957) *A Seiche Movement and its Effect on the Hydrology of Lake Victoria*, Colonial Office Fishery Publications, London.

Gu, G., Adler, R.F., Huffman, G.J., and Curtis, S. (2007) Tropical rainfall variability on interannual-to-interdecadal and longer time scales derived from the GPCP monthly product. *J. Clim.*, **20**, 4033–46, doi: 10.1175/JCLI4277.1.

Haberyan, K.A. and Hecky, R.E. (1987) The late Pleistocene and Holocene stratigraphy and paleolimnology of Lakes Kivu and Tanganyika. *Paleogeography, Paleoclimatology, and Paleoecology*, **61**, 169–97.

Hamblin, P.F., Bootsma, H.A., and Hecky, R.E. (2003) Surface meteorological observations over Lake Malawi/Nyasa. *J. Great Lakes Res.*, **29**, 19–33.

Hecky, R.E. (1993) The eutrophication of Lake Victoria. *Verh. Internat. Verein. Limnol.*, **25**, 39–48.

Hecky R.E., Bugenyi, F.W.B., Ochumba, P., *et al.* (1994) Deoxygenation of the deep water of Lake Victoria, East Africa. *Limnol. Oceanogr.*, **39**, 1476–81.

Hecky, R.E. and Fee, E.J. (1981) Primary production and rates of algal growth in Lake Tanganyika. *Limnol. Oceanogr.*, **26**, 532–47.

Hecky, R.E., Fee, E.J., Kling, H.J. and Rudd, J.W.M. (1981) Relationship between primary production and fish production in Lake Tanganyika. *Trans. Am. Fish. Soc.*, **100**, 336–45.

Hecky, R.E., Muggide, R., Ramlal, P.S., *et al.* (2010) Multiple stressors cause rapid ecosystem change in Lake Victoria. *Freshwater Biology*, **55** (Suppl. 1), 19–42.

Hulme, M., Doherty, R., Ngara, T., *et al.* (2001) African climate change: 1900–2100. *Climate Res.*, **17**, 145–68.

Huttula, T. (1997) Flow, Thermal Regime and Sediment Transport Studies in Lake Tanganyika, Kuopio University Publications, Kuopio.

Imberger, J. and Patterson, J.C. (1989) Physical Limnology. *Adv. App. Mech.* **27**, 303–475.

Kalnay, E., Kanamitsu, M., Kistler, R., *et al.* (1996) The NCEP/NCAR 40-year reanalysis project. *Bull. Amer. Meteor. Soc.*, **77**, 437–70.

Kilham, P. and Kilham, S.S. (1990) Endless summer: internal loading processes dominate nutrient cycling in tropical lakes. *Freshwat. Biol.*, **23**, 379–89.

Kling, G.W., Evans, W.C., Tanyileke, G., *et al.* (2005) Lakes Nyos and Monoun: Defusing certain disaster. *Proc. Nat. Acad. Sci.*, **102**, 14185–90.

Kling, H.J., Mugidde, R.M., and Hecky, R.E. (2001) Recent changes in the phytoplankton community of Lake Victoria in response to eutrophication, in *The Great Lakes of the World (GLOW): Food-Web, Health, and Integrity* (eds. M. Munawar and R.E. Hecky), Backhuys Publishers, Leiden, pp. 47–65.

Kolding, J., Van Zwieten, P., Mkumbo, O., *et al.* (2008) Are the Lake Victoria fisheries threatened by exploitation or eutrophication? Towards an ecosystem-based approach to management, in *The Ecosystem Approach to Fisheries* (eds. G. Bianchi and H.R. Skjoldal), CAB International, Wallingford and FAO, Rome, pp. 309–54.

Lehman, J.T. (1998) Role of climate in the modern condition of Lake Victoria. *Theor. Appl. Climatol.*, **61**, 29–37.

Lewis, W.M. Jr., (1973) The thermal regime of Lake Lanao (Philippines) and its theoretical implications for tropical lakes. *Limnol. Oceanogr.*, **18**, 200–17.

Lorke, A., Tietze, K., Halbwachs, M., and Wuest, A. (2004) Response of Lake Kivu stratification to lava inflow and climate warming. *Limnol. Oceanogr.*, **49**, 778–83.

MacIntyre, S. (2012) Stratification and mixing in tropical African lakes, in *Encyclopedia of Lakes and Reservoirs* (eds. L. Bengtsson, R. Herschy, R. Fairbridge), Springer, Dordrecht.

MacIntyre, S., Flynn, K.M., Jellison, R., and Romero, J.R. (1999) Boundary mixing and nutrient flux in Mono Lake, *CA. Limnol. Oceanogr.*, **44**, 512–29.

MacIntyre, S., Romero, J.R., and Kling, G.W. (2002) Spatial-temporal variability in mixed layer deepening and lateral advection in an embayment of Lake Victoria, East Africa. *Limnol. Oceanogr.*, **47**, 656–71.

MacIntyre, S., Clark, J.F., Jellison, R.S., and Fram, J.P. (2009) Turbulent mixing induced by non-linear internal waves in Mono Lake, *CA. Limnol. Oceanogr.*, **54**, 2255–72.

Maderich, V.S., Van'Heijst, G.J.F., and Brandt, A. (2001) Laboratory experiments on intrusive flows and internal waves in a pycnocline. *J. Fluid Mech.*, **432**, 285–311.

Mapande, A.T., and Reason, C.J.C. (2005) Internannual rainfall variability over western Tanzania. *Int. J. Climatol.*, **25**, 1355–68.

McIntyre, P.B., Michel, E., Olsgard, M. (2006) Top-down and bottom-up controls on periphyton biomass and productivity in Lake Tanganyika. *Limnol. Oceanogr.*, **51**, 1514–23.

Meehl, G.A. and Arblaster, J.M. (2011) Decadal variability of Asian-Australian monsoon—ENSO–TBO relationships. *J. Climate*, **24**, 4925–40, doi: 10.1175/2011JCLI4015.1

Melack, J.M. (1980) An initial measurement of photosynthetic productivity in Lake Tanganyika. *Hydrobiologia*, **72**, 243–7.

Mercier, F., Cazenave, A., and Maheu, C. (2002) Interannual lake level fluctuations (1993–1999) in African from Topex/Poseidon: connections with ocean–atmosphere interactions over the Indian Ocean. *Global Plan. Changes*, **32**, 141–63.

Monismith, S.G., Imberger, J., and Morison, M.L. (1990) Convective motions in the sidearm of a small reservoir. *Limnol. Oceanogr.*, **35**,1676-1702.

Naithani, J. and Deleersnijder, E. (2004) Are there internal Kelvin waves in Lake Tanganyika? *Geophys. Res. Letters*, **31**, L06303, doi:10.1029/2003GL019156.

Naithani, J., Plisnier, P.D., and Deleersnijder, E. (2011) Possible effects of global climate change on the ecosystem of Lake Tanganyika. *Hydrobiologia*, **671**, 147–63, doi: 10.1007/s10750-011-0713-5.

Nakamura, N., Kayanne, H., Iijima, H., *et al.* (2009) Mode shift in the Indian Ocean climate under global warming stress. *Geophys. Res. Letters*, **36**, L23708, doi:10.1029/2009GL040590.

Newell, B.S. (1960) The hydrology of Lake Victoria. *Hydrobiologia*, **15**, 363–83.

Newman, F.C. (1976) Temperature steps in Lake Kivu: A bottom heated saline lake. *J. Phys. Oceanogr.*, **6**, 157–63.

Nicholson, S.E. and Yin, X. (2002) Mesoscale patterns of rainfall, cloudiness, and evaporation over the Great Lakes of East Africa, in *The East African Great Lakes: Limnology, Paleolimnology, and Biodiversity* (eds. E.O. Odada and D.O. Olago), Kluwer Academic Publishers, Dordrecht, pp. 93–120.

North, R.L., Guildford, S.J., Smith, R.E.H., *et al.* (2008) Nitrogen, phosphorus, and iron colimitation of phytoplankton communities in the nearshore and offshore regions of the African Great Lakes. *Verh. Internat. Verein. Limnol.*, **30**, 259–64.

O'Reilly, C.M., Slin, S.R., Plisner, P.-D., *et al.* (2003) Climate change decreases aquatic ecosystem productivity of Lake Tanganyika, Africa. *Nature*, **424**, 766–8.

Otu, M.K., Ramlal, P., Wilkinson, P., *et al.* (2011) Paleolimnological evidence of the effects of recent cultural eutrophication during the last 200 years in Lake Malawi, *East Africa. J. Gr. Lakes Res.*, **37**, 61–74.

Pasche, N., Alunga, G., Mills, K., *et al.* (2010) Abrupt onset of carbonate deposition in Lake Kivu during the 1960: response to recent environmental changes. *J. Paleolimnol.*, **44**, 931–46.

Pasche, C. Dinkel, B. Muller, M. *et al.* (2009) Physical and biogeochemical limits to internal nutrient loading of meromictic Lake Kivu. *Limnol. Oceanogr.*, **54**, 1863–73.

Patterson, G., Wooster, M.J., and Sear, C.B. (1998) Satellite-derived surface temperatures and the interpretation of the 3-dimensional structure of Lake Malawi, Africa: the presence of a profile-bound density current and the persistence of thermal stratification. *Verh. Int.Verein. Limnol.*, **26**, 252–5.

Plisnier, P.-D., Chitamwebwa, D., Mwape, L., *et al.* (1999) Limnological annual cycle inferred from physical-chemical fluctuations at three stations of Lake Tanganyika. *Hydrobiologia*, **407**, 45–58.

Salonen, J. Sarvala, M. Jarvinen, V., *et al.* (1999) Phytoplankton in Lake Tanganyika—vertical and horizontal distribution of *in vivo* fluorescence. *Hydrobiol.*, **407**, 89–103.

Sarmento, H., Isumbisho, M., and Descy, J.P. (2006) Phytoplankton ecology of Lake Kivu (eastern Africa). *J. Plankton Res.*, **28**, 815–29.

Sarvala, J.K., Salonen, M., Jarvinen, E., *et al.* (1999) Trophic structure of Lake Tanganyika: carbon fluxes in the pelagic food web. *Hydrobiologia*, **407**, 149–73.

Schmid, M., Busbridge, M., and Wuest, A. (2010) Double-diffusive convection in Lake Kivu. *Limnol. Oceanogr.*, **55**, 225–38.

Scholz, C.A., Cohen, A.S., Johnson, T.C., *et al.* (2011) Scientific drilling in the Great Rift Valley: The 2005 Lake Malawi Scientific Drilling Project—An overview of the past 145 000 years of climate variability in Southern Hemisphere East Africa. *Paleogeography, Paleoclimatology, Paleoecology*, **303**, 3–19.

Silsbe, G.M., Hecky, R.E., and Guildford, S.J. (2006) Variability of chlorophyll *a* and photosynthetic parameters in a nutrient-saturated tropical great lake. *Limnol. Oceanogr.*, **51**, 2052–63.

Song, Y. (2000) A Numerical Modeling Study of the Coupled Variability of Lake Victoria in Eastern Africa and the Regional Climate. Ph.D. thesis, North Carolina State University, Raleigh NC.

Spigel, R.H. and Coulter, G.W. (1996) Comparison of hydrology and physical limnology of the East African Great Lakes: Tanganyika, Malawi, Victoria, Kivu and Turkana (with reference to some North American Great Lakes), in *The Limnology, Climatology and Paleoclimatology of the East African Lakes* (eds. T.C. Johnson, and E.O. Odada), Gordon and Breach Publishers, Amsterdam, pp. 103–40.

Stenuite, S., Pirlot, S., Hardy, M.-A., *et al.* (2007) Phytoplankton production and growth rate in Lake Tanganyika: evidence of a decline in primary productivity in recent decades. *Freshwat. Biol.*, **52**, 2226–39, doi:10.1111/j.1365-2427.2007.01829.x.

Talling, J.F. (1963) Origin of stratification in an African rift lake. *Limnol. Oceanogr.*, **8**, 68–78.

Talling, J.F. (1966) The annual cycle of stratification and phytoplankton growth in Lake Victoria (East Africa). *Int. Revue Ges. Hydrobiol.*, **51**, 545–621.

Talling, J.F. and Lemoalle, J. (1998) *Ecological Dynamics of Tropical Inland Waters*, Cambridge University Press, Cambridge.

Tassi, F., Vaselli, O., Tadesco, D., *et al.* (2009) Water and gas chemistry at Lake Kivu (DRC): Geochemical evidence of vertical and horizontal heterogeneities in a multibasin structure. *Geochemistry, Geophysics, Geosystems*, **10**, Q02005, doi: 10.1029/2008GC002191.

Tierney, J.E., Mayes, M.T., Meyer, N., *et al.* (2010) Late-twentieth-century warming in Lake Tanganyika unprecedented since AD 500. *Nat. Geosci.*, doi:10.1038/NGEO865.

Verburg, P. (2007) The need to correct for the Suess effect in the application of δ^{13}C in sediment of autotrophic Lake Tanganyika, as a productivity proxy in the Anthropocene. *J. Paleolimnol.*, **37**, 591–602, doi: 10.1007/s10933-006-9056-z.

Verburg, P., Antenucci, J.P., and Hecky, R.E. (2011) Differential cooling drives large-scale convective circulation in Lake Tanganyika. *Limnol. Oceanogr.*, **56**, 910–26.

Verburg, P. and Hecky, R.E. (2003) Wind patterns, evaporation, and related physical variables in Lake Tanganyika, *East Africa. J. Great Lakes Res.*, **29**, 48–61, doi: 10.1016/S0380-1330(03)70538-3.

Verburg, P., and Hecky, R.E. (2009) The physics of the warming of Lake Tanganyika by climate change. *Limnol. Oceanogr.*, **54**, 2418–30.

Verburg, P., Hecky, R.E., and Kling, H.J. (2003) Ecological consequences of a century of warming in Lake Tanganyika. *Science*, **301**, 505–7.

Vollmer, M.K., Bootsma, H.A., Hecky., R.E., *et al.* (2005) Deep-water warming trend in Lake Malawi, East Africa. *Limnol. Oceanogr.*, **50**, 727–32.

Wells, M.G. and Sherman, B. (2001) Stratification produced by surface cooling in lakes with significant shallow regions. *Limnol. Oceanog.*, **46**, 1747–59.

Williams, A.P. and Funk, C. 2010. A westward extension of the warm pool leads to a westward extension of the Walker circulation, drying eastern Africa. *Clim. Dyn*, doi: 10.1007/s00382-010-0984-y.

Wolff, C., Haug, G.H., Timmermann, A., *et al.* (2011) Reduced interannual rainfall variability in East Africa during the last ice age. *Science*, **333**, 743–7.

Worthington, E.B. (1930) Observations on the temperature, hydrogen-ion concentration and other physical conditions of the Victoria and Albert Nyanzas. *Int. Rev. Ges. Hydrobiol. U. Hydrogr.*, **24**, 328–57.

Zhang, Z., Fringer, O.B., and Ramp, S.R. (2011) Three-dimensional non-hydrostatic numerical simulation of nonlinear internal waves generation and propagation in the South China Sea. *J. Geophys. Res.*, **116**, C05022, doi:10.1029/2010JC006424.

19

Effects of Climate Change on New Zealand Lakes

David P. Hamilton[1], Chris McBride[1], Deniz Özkundakci[1],
Marc Schallenberg[2], Piet Verburg[3], Mary de Winton[3], David Kelly[4],
Chris Hendy[5], and Wei Ye[1]

[1]Department of Biological Sciences, University of Waikato, New Zealand
[2]Department of Zoology, University of Otago, New Zealand
[3]National Institute of Water and Atmosphere Ltd (NIWA), New Zealand
[4]Cawthron Institute, New Zealand
[5]Chemistry Department,University of Waikato, New Zealand

19.1   Introduction

This chapter outlines the potential impacts of climate change on New Zealand (NZ) lakes and examines how these effects may be mitigated. The chapter briefly outlines the origin and nature of NZ lakes, considers the historical climate to which the lakes have been exposed, and examines the potential impact of a future climate, up to 2100. The isolated nature of the land masses of NZ (including North, South and Stewart Islands, as well as several offshore islands) has led to specific distinguishing features of its freshwater flora and fauna, including limited representation of some lifeform types found elsewhere, local endemism, and sensitivity to the direct and indirect impacts of exotic species. In providing a context with which to evaluate the impacts of climate change in NZ lakes we repeatedly draw comparisons with the massive landscape changes that have occurred in the past 150 years as a result of human settlement and widespread conversion of forests, wetlands and lakes into agricultural land, or use of freshwater for other economic benefits such as hydro power.

19.2   Geographical and climate perspective

New Zealand is located over a wide, mid-latitudinal band (34° to 47 °S) in the southern hemisphere. The climate is temperate maritime and is characterized by modest seasonal

Climatic Change and Global Warming of Inland Waters: Impacts and Mitigation for Ecosystems and Societies,
First Edition. Edited by Charles R. Goldman, Michio Kumagai and Richard D. Robarts.
© 2013 John Wiley & Sons, Ltd. Published 2013 by John Wiley & Sons, Ltd.

temperature oscillations and strong westerly winds, which alternate with subtropical high-pressure systems (Maunder 1971; Sturman and Tapper 2006). Annual mean air temperature varies from about 9 °C in the far south to 16 °C in the far north. The interaction of westerly winds with the perpendicularly aligned land mass results in marked variations in rainfall across the land mass. Rainfall across the mid part of the South Island may vary from 3–4 m year^{-1} on the west coast to >12 m year^{-1} in the southern Alps and <0.4 m year^{-1} in the arid interior (Wratt *et al.* 1996).

Interannual variations in climate are predominantly associated with the Interdecadal Pacific Oscillation (IPO) (Salinger, Renwick, and Mullan 2001) and the El Niño-Southern Oscillation (ENSO) cycle (Mullan 1995). El Niño periods are characterized by increased intensity of westerly winds, often reinforcing the high annual mean rainfall in westerly regions but coincident with drought east of the Southern Alps (South Island) as well as in eastern regions of the middle of North Island. Periods of La Niña tend to be more variable but are typically warmer, with higher rainfall in the north-east and lower rainfall in the south-west. Overlaying the ENSO cycle is the Interdecadal Pacific Oscillation (IPO). A positive IPO phase leads to more intense El Niño events with predominance of westerly winds and frontal systems compared with a negative IPO phase corresponding to a La Niña phase.

The effect of ENSO events on lakes has not been studied in great detail but Lake Ellesmere (in Māori, Te Waihora), a large (surface area, $A = 186\,\text{km}^2$), shallow (mean depth, $\bar{z} = 1.9\,\text{m}$) lake on the east coast of the Canterbury region of South Island, experiences lower total rainfall in about two-thirds of summers when there is an El Niño phase but no specific pattern of higher or lower summer rainfall during La Niña. The largest NZ lake, Taupo ($A = 616\,\text{km}^2$, $\bar{z} = 90\,\text{m}$) did not undergo its regular annual winter mixing cycle in 1998 when there was an intense El Niño phase (see Section 19.11).

19.3   Historical climate

Along with the rest of the world, the NZ climate has changed continuously for as long as climatic proxies are available. Over the past 600 000 years the periodicity of major climatic changes has been about 100 000 years, but within this period there have been significant excursions ranging in length from multimillennia to multi-decadal (Hays, Imbrie, and Shackleton 1976; Imbrie and Imbrie 1979). Multi-millennial climatic changes have been identified using excursions of the oxygen isotope ratios of foraminifera in marine sediments, which were first described by Emiliani (1966). During the last glacial/interglacial cycle the prominent interglacial intervals were designated marine oxygen-isotope (MOI) stage 5 (90 000 to 125 000 years ago) and MOI stage 1 (the Holocene, ~10 000 years ago to present), whilst MOI stages 2 and 4 represented prominent periods of glaciation.

Within NZ (at least from latitudes extending from mid South Island to mid North Island), there appears to have been little difference in the magnitude of climate change, with the long-term average temperature about 5 °C colder in MOI stages 2 and 4 than at present, even within 50 km of the piedmont glaciers, which developed at the base of the Southern Alps. The effect of such a decrease in mean temperature was a lowering of the snow line resulting in large-scale advance of glaciers from high mountains, particularly in the Southern Alps, to form coalescing piedmont glaciers in Westland

(South Island), some of which extended well beyond the present-day coastline, and the occupation of all of the finger lakes to the south and east of the Southern Alps. At the same time the accumulation of ice in large sheets in North America and Europe resulted in a drop of sea level by 120 to 140 m, combining the three main islands of NZ (South, North and Stewart Islands) into one contiguous land mass. Glacial erosion from the vastly increased volume of ice produced huge volumes of sediment ranging in size from boulders many meters in diameter to clay-sized silt, which was deposited as outwash extending from the glacier fronts to beyond the coastline, and wind-blown loess which smothered much of the non-glaciated landscape. With mean temperatures reduced by between 4 and 6 °C, vegetation cover altered drastically and the treeline was reduced to near current sea level over the southern half of NZ. Most of the South Island lakes would either have been occupied by ice or drained where coastal barriers were breached. Lakes of the Taupo Volcanic Zone (TVZ), North Island, would still have been in existence, although many of the current lakes were formed by Holocene eruptions. Most of the Waikato lowland lakes formed subsequent to the breakout flood following the Ouranui eruption ~22 000 years ago and restabilization and revegetation of the landscape ~15 000 years ago.

During MOI stage 2, three sets of glacial advances took place (30 000, 20 000 and 16 700 years ago), each of which completely occupied the finger lakes south and east of the Southern Alps. Each of the advances was matched by significant vegetation changes, with a decline in subalpine shrublands and increased dominance of grasses.

Warming from the last glacial maximum (LGM) appeared to have started in Antarctica ~18 000 years ago, 3000 years earlier than in the northern hemisphere, and was also marked by a temporary (1500 years ago) cooling known as the Antarctic Cold Reversal. By 14 700 years ago vegetation cover had largely recovered from its glacial depression. In the northern hemisphere a similar cooling (the Younger Dryas) occurred a little later. In NZ the late glacial readvance caused the Tasman Glacier to advance substantially 11 700 years ago, to the current location of Lake Pukaki, and for the Franz Josef Glacier to also advance substantially and form the Waiho Loop. No similar re-advances of the glaciers have taken place since that time.

Within the Holocene, climate continued to change continuously although deviations were of lower magnitude and shorter duration. Within the last 500 years the most significant climatic changes have been a period of cooling referred to as the "Little Ice Age" with maximum cooling attained about 1730 AD. This period lasted until the late nineteenth century when it was followed by rapid warming beginning in the middle of the twentieth century and continuing until the present day. Over the past 200 years, mean temperatures in NZ have risen by about 1.5 °C. Accurate air temperature data are available since 1910 (Mullan *et al.* 2008) and indicate an increase in temperature of about 0.9 °C from that time to 2000. This rate of increase is higher than that reported in many other locations around the globe and could be linked to low concentrations of sulfate aerosol pollution resulting in reduced global dimming in the South Pacific. In 1998 a strong El Niño phase was associated with this year being the warmest on record in NZ. Mean air temperature in NZ has, however, been relatively stable since 1970 (Renwick *et al.* 2010a), unlike the increases observed in the global mean (Hansen *et al.* 2010). A negative excursion of the IPO commencing around 1999–2000 may have moderated the climate change signal over NZ in the past decade or two. Compared with the 1971–2000 mean, annual mean temperatures in NZ were +0.4 °C

Figure 19.1 Tasman Lake at the foot of the Tasman Glacier, South Island, NZ, in 2010. Source: Wikipedia (http://en.wikipedia.org/wiki/Tasman_Lake).

and $-0.2\,°C$ in 2008 and 2009, respectively (Renwick *et al.* 2010a), and $0\,°C$ and $-0.5\,°C$ for Taupo in 2008 and 2009, respectively, compared with the local mean.

The average rate of retreat of the largest glacier in NZ, the Tasman, has been about $180\,m\ year^{-1}$ since the 1990s, leading to the formation of terminal Lake Tasman in 1973. This lake was 7 km long, 2 km wide and 245 m deep by 2008 and is expected to continue to expand until the disappearance of Tasman Glacier as climate warms (Figure 19.1).

The NZ coastline is 15 100 km in length and NZ ranks ninth in the world for ratio of coastline length to land area. There is an average of 7.4 lakes, wetlands and lagoons per 100 km of coastline which are vulnerable to a 1 m rise in sea level (Schallenberg, Hall, and Burns 2003). As the sea pushes inland with climate warming, the main effect on these ecosystems will be salinization. The last time that this occurred was during the mid-Holocene sea level highstand, ~4000–6000 years ago (Gibb 1986), during which time a coastal freshwater lake in South Island experienced sufficient marine influence to shift its waters to brackish. The lake, with resident freshwater mussel, *Hyridella menziesii*, was at the time extensively colonized by the estuarine clam, *Austrovenus stutchburyi*, which requires a salinity of ~14 ppt to survive (Marsden 2004).

19.4   Future climate

The ensemble results of global climate models yield an increase in global mean temperature of around $3\,°C$ by the end of this century, under an A1B emissions scenario (see definition below), which is in the mid-range for all emissions scenarios (1.1 to $6.4\,°C$). The $3\,°C$ increase under the A1B scenario equates to a mean increase over NZ of around $2.1\,°C$ (Figure 19.2a, Mullan *et al.* 2008), with the reduced rate of warming for NZ attributed largely to the moderating effect of the surrounding Pacific Ocean; of the major oceans across the globe southern oceans are likely to be the most resilient to a warming climate. Of particular note, however, is an increase in the intensity of the ENSO cycle and a higher frequency of westerly winds. El Niño events are likely to lead to more rainfall in the wetter western land mass and lower rainfall in the east (Figure 19.2b). The frequency of intense rainfall events is likely to increase (Pall, Allen, and Stone 2007) whilst warmer temperatures and more rainfall (as opposed to snowfall) are predicted to reduce the snow pack in the Southern Alps (Mullan *et al.* 2008; Hendrikx *et al.* 2009), possibly affecting the seasonal hydrological cycles and the duration of ice-free season in high-altitude lakes.

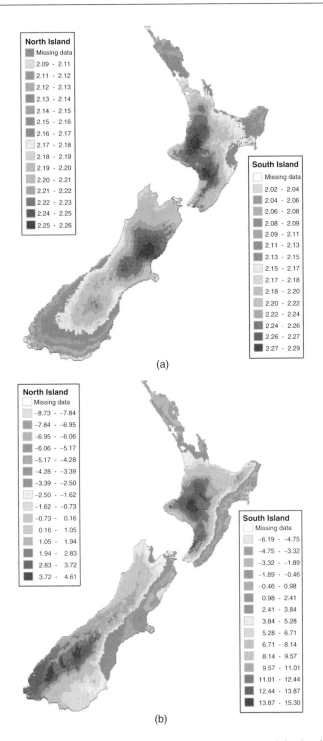

Figure 19.2 Change in (a) temperature (°C) and (b) percentage precipitation in 2091–2099 compared with baseline of 1965–1995, based on ensemble output and statistical downscaling from 12 GCM models. (See insert for color representation.)

Other statistical measures associated with a warming climate are likely to follow, such as reduced number of frost days, higher frequency of heat waves and reduced soil moisture, but one outcome of changes to weather circulation patterns and intensification of the ENSO cycle is likely to be an increase in the frequency of extreme winds (Mullan *et al.* 2008). This change may have implications for surface mixed layer depths of deep lakes, which are markedly deeper in NZ than for lakes at comparable latitudes in the northern hemisphere (Davies-Colley 1998), and also for shallow coastal and lowland lakes whose submerged vegetation can be influenced by extreme wind events. For example, a massive storm ("the Wāhine") destroyed extensive submerged fringing weed beds (mostly *Ruppia* spp. and *Potamogeton* spp.) in shallow ($\bar{z} = 1.9$ m) Lake Ellesmere in 1968. Weed beds have not re-established to any substantial extent since the storm and black swan (*Cygnus atratus*) numbers have decreased from up to 80 000 prior to the storm to <10 000 nowadays.

Based on records from four long-term tidal gauges, sea levels in NZ have increased approximately 1.6 mm year^{-1} during the twentieth century; similar to the global average of 1.7 ± 0.3 (standard error) mm year^{-1} (Church and White 2006). The International Panel for Climate Change (IPCC) projections for sea-level rise, which are generally considered conservative, range from 18 to 59 cm across six different emission scenarios, with more extreme sea level rise (56–200 cm) projected based on recently-observed rapid declines in ice mass in Greenland and Antarctica. These more extreme sea-level rise scenarios would obliterate many of the coastal lakes and lagoons which currently open either naturally or artificially (by breaching coastal barrier substrates) to the sea, and would lead to saline intrusion in other currently freshwater systems.

19.5   Overview of lake types and formation processes

There are 3820 lakes in NZ that have a surface area larger than one hectare, collectively representing about 1.3% of the land area (Leathwick *et al.* 2010). These lakes are of widely varying types and origins. Many of the lakes in the central North Island area are volcanic crater lakes, while the majority of the lakes near the Southern Alps are associated with glacial scour and retreat. Coastal lakes, lagoons and wetlands are widespread across both North and South Islands with approximately 74 per 1000 km of coastline which have their mean water level within 1 m of mean sea level (Schallenberg, Hall, and Burns 2003). Artificial lakes are also extremely important for storage-water for hydropower generation (∼65% of the electricity supply of NZ), and major river systems including the Waitaki (South Canterbury), Clutha (Otago) and Waikato (Waikato region) have a series of dams at intervals along their lengths (approximately 25 hydro lakes nationally). Each of these rivers has a large storage source, some of which have artificially regulated water levels, including Lake Taupo for the Waikato River, lakes Pukaki, Tekapo and Ohau for the Waitaki River, and lakes Hawea, Wanaka and Wakatipu for the Clutha River. The largest power station in NZ is associated with Lake Manapouri and together with a hydro power station at Lake Te Anau, where lake levels are also artificially regulated, these stations capitalize on the very high rainfall in this area of the south-west of South Island known as Fiordland. A large number of smaller shallow man-made lakes, totaling 544, have been created for both domestic water use in urban areas, and for water storage for agricultural irrigation.

Shallow lakes in NZ are polymictic (frequently alternating periods of mixing and stratification) (Hamilton, Hawes, and Davies-Colley 2004). Deep lakes are almost uniformly monomictic (one vertical circulation period per year) with the exception of some high-altitude lakes in the central South Island which freeze over in winter (Hamilton, Hawes, and Davies-Colley 2004; Hamilton *et al.* 2010), though there is no current record of a dimictic circulation pattern (two periods of complete vertical circulation per year) in these lakes. According to the classification scheme of Lewis (2000), lakes at >40° latitude (i.e., encompassing the South Island and lower part of the North Island) should be dimictic. The anomaly of mixing patterns in NZ lakes compared with Lewis' (2000) classification reflects the temperate windy climate in NZ in which lakes mix more deeply than those of large continental land masses and are not subject to calm cold winters or especially hot summers (Davies-Colley 1988). Interestingly, Lake Taupo ($\bar{z} = 90$ m) has deviated from monomictic behavior and remained vertically stratified (amictic) in exceptionally warm winters such as 1998, when there was a strong El Niño phase (see Section 19.11).

Most lakes in NZ were formed in the past 20 000 years but they have been shaped by a wide variety of geological and climatic processes (Lowe and Green 1987). Glacial lakes of the South Island represent 38% of all NZ lakes and were formed during Pleistocene glaciations. Volcanic lakes occur in the Taupo Volcanic Zone where massive volcanic eruptions have led to a localized concentration of deep lakes, some of which are directly influenced by geothermal inputs. Shallow coastal lakes formed either through shoreline erosional and depositional processes as riverine lagoons (291 lakes), or by the blockage of small river valleys by windblown sand as coastal dune lakes (319 lakes), are numerous along the coastline (Leathwick *et al.* 2010). Many coastal lakes are eutrophic from impacts of agricultural development in their catchments and reduced water levels from draining, but may also be increasingly affected by inundation of saline waters with projected increases in sea level with climate change. There are approximately 106 lakes under 3 m elevation above sea level, and just over 300 under 10 m elevation above sea level.

19.6 Climate change and impacts on endemic and exotic flora and fauna

Regulations around importation of organisms have effectively prevented potential freshwater pests entering NZ officially, but illegal introduction routes remain (Champion and Clayton 2000). These routes may be enhanced by changing international trade, altered human migration, demographics and additional pest pathways and vectors associated with climate change. New Zealand's isolation and maritime borders have largely protected against chance freshwater introductions. The freshwater fauna of NZ are characterized by a high level of endemism and, in the case of Galaxiid fish in particular, also a high level of local endemism (Allibone *et al.* 2012). However, the NZ aquatic flora has many species in common with Australia, but with a narrower representation of lifeforms (e.g., absence of water lily types). Continued establishment of "colonizer" plants (de Lange *et al.* 2009) from Australia such as bladderwort (*Utricularia gibba*), *Grateola pedunculata* and *Juncus polyanthemus* point to seed transport by waterfowl migration as an occasional pathway between the two land masses. In the case of naturalized black swan (*Cygnus atratus*) populations, however, there is debate as to the extent to which intentional introductions in the 1860s were supplemented with chance migrations from south-east Australia, possibly in association with extreme westerly winds. The frequency of chance migrations appears low, however, even with alterations to weather patterns, and humans remain the major vector associated with introductions of exotic pests. Climate change may, therefore, indirectly increase exposure of NZ lake ecosystems to exotic species through chance migrations but more importantly, it is likely to alter the probability of naturalization or affect the resilience of resident communities to invaders (Walther *et al.* 2009).

The most immediate threat to lake ecosystems comes from the wide range of alien freshwater organisms already present in the country, in the aquarium trade (Champion and Clayton 2001), as well as naturalized populations of invasive weeds, fish and invertebrates (Champion *et al.* 2004; Duggan, Green, and Burger 2006). However, colonization opportunity for potential pests in NZ has been limited by management, such as by bans on sale and distribution, including a National Pest Plant Accord.

Unlike contiguous terrestrial systems, each lake and catchment presents introduction barriers to aquatic species, which commonly require anthropogenic-mediated transfers (e.g., illegal fish releases, weed contaminated equipment). A lakescape of increased waterbody proximity and connections could enhance spread by invasive freshwater pests via "stepping stone" invasion (MacIsaac *et al.* 2004). Therefore, signaled increases in water storage structures and distributional infrastructure in NZ driven by climate change and limited water resources (Ministry for the Environment 2011) are likely to increase the opportunities for spread and establishment of alien aquatic species (Rahel and Olden 2008). In addition, a greater frequency of extreme rainfall events under climate change scenarios may lead to increased flood-mediated transfer (Walther *et al.* 2009) of aquatic species between water bodies. Positive associations between impoundments and alien species have also been noted (Johnson, Olden, and Vander Zanden 2008), and invasion of non-indigenous calanoid copepods in NZ has been facilitated by new waterbody construction and low inherent biotic resistance in artificial water bodies (Banks and Duggan 2009).

19.7 Climate change impacts on fish

The influence on lake ecosystems of invasive species is more likely through indirect rather than direct climate-change effects, mediated through enhanced eutrophication and its symptoms, changes in available habitat (e.g., reduced oxygenated zone), and altered competitive abilities and community structure (see also Chapters 6, 10, 15, and 16). For example, suggested outcomes for fisheries involving greater herbivory and omnivory, lower piscivory, higher growth rate, earlier maturity (although possibly shorter lifespan) and greater fecundity (Jeppesen *et al.* 2010) have ramifications for NZ lakes, where a range of introduced cyprinids and percids are implicated in water quality deterioration (Rowe 2007).

Of the approximately 35 documented fish species native to NZ more than 85% are unique to this country (McDowall 2000). A further 21 fish species have been deliberately introduced, leading to Leprieur *et al.* (2008) considering NZ to be a global hotspot in terms of impacts on endemic species from exotic invasive fishes (Leprieur *et al.* 2008). Most fish introductions have been intended to provide recreational fishing opportunities similar to those of northern hemisphere sports fisheries. Whilst some species such as brown trout (*Salmo trutta*) and rainbow trout (*Oncorhynchus mykiss*) are highly valued and even have statutory protection under NZ's Resource Management Act of 1991, other species such as koi carp (*Cyprinus carpio*) and catfish (*Ameiurus nebulosus*) have become pests whose impacts extend to water quality as well as native flora and fauna. Nowadays 76% of non-diadromous freshwater fish taxa are considered threatened or at risk (Allibone *et al.* 2012) and the 'highly valued' introduced predatory salmonids have had major predatory effects on native fish (McIntosh, 2000; McDowall, 2003, 2006).

Habitat changes have also taken place in NZ at extremely high rates since European colonization began to increase rapidly in the mid 1800s. Around 90% of wetlands in NZ have been lost, mostly drained for agriculture, and each of the three largest NZ rivers (Clutha, Waitaki and Waikato) has a series of hydro dams along its length. In these cases obligate diadromous fish species have been severely impacted by hydro dams blocking access to the sea or to upstream freshwater habitats. Recent apportionment of funding by the NZ government for increases in water storage for irrigation and hydropower (Ministry for the Environment 2011), as well as a massive expansion of the dairy industry from the early 1990s and its associated impacts on lowland river water quality (Larned *et al.* 2004), have intensified pressures on native fish for the foreseeable future, and on water quality of lowland lakes and rivers more generally. The arid eastern and interior regions of the South Island and the mid-latitudes of eastern North Island are likely to be increasingly drought-prone in a future climate and these are also the regions where water resources are being most intensively used for irrigation. Reductions in discharge are likely to occur in braided eastern rivers of the South Island due to the combined effects of increasing water abstraction and lower rainfall, with the possibility that persistent river-mouth closures or reduced frequency of floods could prevent diadromous species from completing their life cycle.

McDowall (1992) examined how a warming climate might affect freshwater fish in NZ. He considered that some diadromous species may move further south, depending on the availability of habitat, some species with restricted ranges could become extinct, and others may have sufficient behavioral or genetic plasticity to adapt sufficiently rapidly to survive the changing conditions. Ultimately climate change is likely to act synergistically with substantial existing and proposed freshwater habitat and flow modifications, and in some cases (e.g., long-finned eel; *Anguilla dieffenbachii*) commercial fishing pressure, to intensify pressures on native fish.

As mentioned above, some fish introduced to NZ are now major pests and several of these are highly tolerant to elevated water temperatures. Common carp (*Cyprinus carpio*) and brown bullhead catfish (*Ameiurus nebulosus*) are reasonably widespread in lakes and rivers in the upper North Island (Hicks, Ling and Wilson 2010) and, whilst illegal introductions via human vectors remain the major mechanism for further spread of these species, high water temperatures could facilitate the spread of them, particularly into central and southern parts of North Island and South Island. On the other hand, Western mosquitofish (*Gambusia affinis*) and goldfish (*Carassius auratus*) are already widespread throughout NZ and will likely benefit from any increase in water temperature. Populations of temperature-restricted alien species have also established in NZ thermal areas from aquarium "escapees" including fish (e.g., guppy: *Poecilia reticlata*) and snails (e.g., Malaysian trumpet snail: *Melanoides tuberculata*) (Duggan 2002; Wilding and Rowe 2008). Although it is unlikely that future elevations in temperature will be sufficient to see expansions in these species, additional tropical/subtropical organisms may be subsequently released through the aquarium trade.

19.8   Climate change impacts on aquatic plants and macroinvertebrates

It is hypothesized that climate change may increase disturbance intensity in fresh waters, making them easier to invade (Rahel and Olden 2008). Examples of invasions by major weeds in relatively pristine NZ lakes (e.g., *Lagarosiphon major* in Lake Wanaka), however, suggest their competitive abilities already put them at a significant competitive advantage and disturbance is unlikely to be a factor in introduction or establishment. Nevertheless, climate change scenarios of increased lake nutrient status and turbidity could further favor growth performance of tall-growing weed species (e.g., *Egeria densa, Ceratophyllum demersum, L. major*) over native plants because these weeds can "escape" light limitation by forming a dense canopy close to the water surface (Tanner, Clayton, and Wells 1993). Indeed, *E. densa* appears most invasive in eutrophic conditions but it has also been associated with weed bed collapses and transitions to a resilient devegetated turbid state (Schallenberg and Sorrell 2009). Phenotypic plasticity in invasive species has been suggested to increase the range of environments under which they can establish or invade. Comparisons of photosynthesis and "fitness" measures for an invasive strain of *C. demersum* from NZ and a noninvasive strain showed the invasive strain acclimated better to elevated temperatures (Hyldgaard and Brix 2012), suggesting weed performance could be enhanced under climate change but distributions of a number of the submerged lake weeds (e.g., *E. densa, C. demersum*) that reproduce vegetatively have been best explained by variables characterizing human access and use (Compton *et al.* 2012). The advantage of an extended growing season for invasion by lake weeds under climate change (Rooney and Kalff 2000) would not be pronounced in NZ compared with more continental climates due to the current perennial nature of lake vegetation.

Although the potential NZ range of the tropical floating weeds water hyacinth (*Eichornia crassipes*) and salvinia (*Salvinia molesta*) would theoretically be extended by climate change, an ongoing management program effectively prevents expansion (Clayton 1990). Climate change would have limited impact on the threat that major freshwater pests such as zebra or quagga mussels (*Dreissena polymorpha, D. bugensis*) pose for NZ lake environments, as the climate of this country already falls within their known tolerance range.

Resilience of native lake species to competitive exclusion under climate change scenarios is unknown, but, interestingly, several NZ freshwater biota have demonstrated an extended range and are regarded as invasive in other geographic areas, including the NZ mudsnail *Potamopyrgus antipodarum* (Loo, MacNally, and Lake. 2007), the amphibious macrophyte *Glossostigma cleistanthum* (Les, Capers, and Tippery 2006) and pygmyweed *Crassula helmsii* (Dawson 1996), suggesting tolerance of a wider set of environmental conditions than is currently experienced in NZ.

19.9 Effects of climate change on shallow NZ lakes

Schallenberg and Sorrell (2009) identified 37 shallow lakes in NZ that had undergone regime shifts from a clearwater state to a turbid state. These lakes were located throughout much of NZ, from Northland to Otago in the south. The authors also identified 58 other lakes that had maximum depths and mean annual air temperatures within the ranges of the lakes which had undergone regime shifts, but the second group of lakes had not been reported to have undergone such regime shifts. This dataset allows the examination of the effect of temperature on the tendency for lakes to undergo such regime shifts. Using the lakes from the Schallenberg and Sorrell (2009) dataset, Figure 19.3 shows the number of regime shifting and non-regime shifting lakes in relation to mean annual air temperature. Regime shifting lakes were reported with a range of mean annual air temperatures of 8.7 to 15.8 °C. There was a clear unimodal pattern in the proportion of lakes which underwent regime shifts in relation to temperature (Figure 19.3). This appears to indicate that there is an optimum temperature range of 10–13 °C within which regime shifts tend to occur in NZ lakes. Schallenberg and Sorrell (2009) found strong correlations between regime shifts and land use, the presence of an invasive macrophyte and the presence of benthivorous, herbivorous and planktivorous fish. Therefore, the unimodal relationship in Figure 19.3 may be influenced by other, indirect factors. However, Figure 19.4 suggests that climate warming may shift the prevalence of regime shifting southward in NZ.

In contrast, supported by the historical analysis of trends in NZ sea levels by Hannah (2004), the predictions of sea-level rise would appear to be more robust than those for air temperature alone. Studies from Lake Waihola, Otago, South Island, have demonstrated recently that a climate-induced saline intrusion, which raised the salinity of the lake to 5 ppt, was sufficient to cause major changes to the zooplankton abundance and community structure of the lake (Schallenberg, Hall, and Burns 2003). The salinity increase reduced zooplankton numerical abundance 5-fold, and taxonomic richness three-fold while temporarily extirpating *Daphnia carinata*, a keystone species in the lake food web and a major grazer of phytoplankton and seston. This commonly observed sensitivity of *Daphnia* and other cladocerans to salinity (see references in

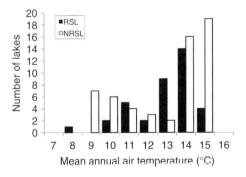

Figure 19.3 Number of regime shifting lakes (RSL) and non-regime shifting lakes (NRSL) in relation to mean annual air temperature.

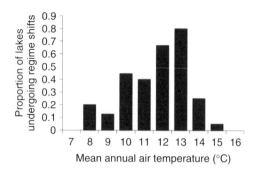

Figure 19.4   Proportion of lakes sampled in each temperature class which underwent regime shifts.

Schallenberg, Hall, and Burns 2003) could play a pivotal role in instigating regime shifts in salinizing shallow lakes (Jeppesen *et al.* 1994).

The predominant role of salinity variations in structuring zooplankton communities was also reported from an intermittently opened barrier-bar lagoon in southern NZ (Duggan and White 2010). In this system, zooplankton were sampled across multiple marine incursions and the oligohaline taxa appeared to recover readily when freshwater conditions were restored, indicating that re-population was possible after temporary salinization, either as a result of the production of resting eggs or due to the availability and utilization of freshwater refugia within the lagoon system. Such resilience would not be expected to persist with rising sea levels due to global climate change.

Although not studied formally, similarly strong effects of salinization occur on the communities of benthos and hyperbenthos in NZ's coastal lakes and lagoons. In contrast to the typically insect-dominated freshwater invertebrate communities found in NZ lakes and wetlands, invertebrates in estuarine systems tend to be dominated by hyperbenthic mysids and amphipods and by benthic polychaetes and snails. Chironomid larvae appear to be the main salinity-tolerant benthic insect larvae and a few odonate taxa seem to tolerate moderate salinities.

The fish taxa found in NZ's coastal lakes and wetlands are generally either diadromous (or facultatively diadromous) and, therefore, show high tolerances for salinity. Even the few taxa which are anadromous show moderately high salinity tolerances. Therefore, the strongest effects of salinization on fish communities in coastal lakes and wetlands are likely to be indirect, most likely due to the progressive colonization of these systems by estuarine/marine fish species and/or to changes in their available prey and habitats.

The moderately diverse aquatic macrophyte flora of shallow NZ lakes and wetlands is poorly represented in coastal lakes and lagoons which experience saline intrusions. However, invasive exotic macrophytes such as *L. major* and *E. canadensis* are also poorly represented. As a result, where submerged aquatic macrophyte communities persist in brackish coastal lakes and lagoons, they generally comprise few species of native aquatic macrophytes, including *Ruppia* spp., *Stuckenia pectinata*, and *Myriophyllum triphyllum*. These species all rely on periods of low salinity for survival. Therefore, accelerating sea level rise will cause these communities to disappear from currently brackish environments and will allow them to colonize the current freshwater habitats which will become brackish as sea level rises.

Remane and Schlieper (1971) proposed that the taxonomic diversity of estuarine environments is generally lower than that of freshwater and marine environments. Their model appears to hold for most biological communities in NZ coastal lakes and lagoons. Although robust data so far only exist for zooplankton communities, observations indicate that invertebrate and macrophyte communities in coastal lakes and wetlands will lose biodiversity as marine intrusions begin to affect them. Some keystone species (e.g., *Daphnia* sp.) will be affected and consequently, salinity increases will cause major ecological regime shifts in many of these ecosystems.

19.10 Effects of climate change on high-altitude NZ lakes

It is recognized that ecosystems dependent upon seasonal patterns of snowpack and ice are likely to be highly sensitive to future climate change (see reviews by Hauer *et al.* 1997; Melack *et al.* 1997; Battarbee 2002; see also Chapter 13). Small changes in temperature can result in significant effects on hydrological processes and seasonality of ice cover, which in turn can affect processes controlling communities and their interactions within the food web.

A significant number of lakes occur in alpine regions of NZ associated with the Southern Alps as well as high-altitude volcanic crater lakes on the North Island. There are more than 600 lakes larger than one hectare above 1000 m elevation, with the highest being crater lakes of the Central Volcanic Plateau region above 2500 m. The duration of the annual cycle in which precipitation would fall as snow under a warmer climate would likely influence the amount of spring snow-melt which occurs in these catchments. It is also expected that patterns of wind direction will alter (Mullan *et al.* 2008), meaning that the amount of precipitation on the eastern and western portions of mountain ranges will change with climate change.

Relatively little work has been conducted in NZ alpine lakes to potentially understand these effects on aquatic communities. Although alpine systems are expected to have relatively simple food webs, many of the species are endemic and do not occur in lower altitude freshwater environments (Kilroy *et al.* 2006), thus there are significant unique biodiversity values in these environments. Some recent experimental work conducted by Greig *et al.* (2012) demonstrated that warming can have complex effects on whole-system metabolism in alpine tarns as well as changing the dynamics between insects and their predators through changes in emergence timing. Low temperature is known to limit the occupation of these environments by fish species, thus warming could also change community interactions by enabling invasion by fish into systems not presently inhabited. The climbing abilities of some of the NZ diadromous galaxiid species such as koaro (*Galaxias brevipinnis*) would facilitate this potential.

In conclusion, alpine lakes are a significant feature of NZ lakes, and it is expected they are likely to be highly sensitive to future changes in climate. These will most likely be mediated through changes in precipitation patterns, changes in snowfall, and possibly by changes in the duration of ice cover, as in other alpine areas. The limited research conducted to date suggests that changes will mainly be manifested through alterations to community interactions and predator–prey dynamics, and possibly by invasion by species presently occurring in lower altitude freshwater environments.

19.11   Case study: Lake Taupo

Lake Taupo (maximum depth 160 m) is in the central North Island at latitude 38.81°S. In the period 1994 to 2009, the warmest years, based on local air temperature records, were in 1998 and 1999 (Figure 19.5a). In 1998 and 1999 none of the monthly temperature profiles demonstrated that surface water temperatures were identical to those of bottom waters (e.g., surface versus 130 m; Figure 19.5b). Thus the tendency for vertical mixing of the water column, which characterizes winter in other deep monomictic NZ lakes, may have been incomplete. Measurements of dissolved oxygen at 130 m also suggest incomplete mixing, a least in 1998, when there was a much-reduced winter peak in oxygen (Figure 19.5c). This winter peak is normally associated with circulation of cold water throughout the water column. This water is generally close to saturation and may even be supersaturated due to coincident annual peaks in phytoplankton biomass and productivity in winter. It was on this basis that Vincent (1983) described the lake as a hybrid temperate-tropical system because the widespread pattern of a spring-summer peak of phytoplankton productivity in temperate lakes was reversed. Besides 1998 there have been other years (for example 2001) when dissolved oxygen at 130 m has decreased to <7 mg L^{-1} but it would likely take several years of continuous stratification to make hypolimnetic waters anoxic in this oligotrophic lake which has low rates of organic matter deposition.

Should 1998 be more typical of the temperature regime of Lake Taupo under a warming climate, then it is possible that the lake could remain stratified for several

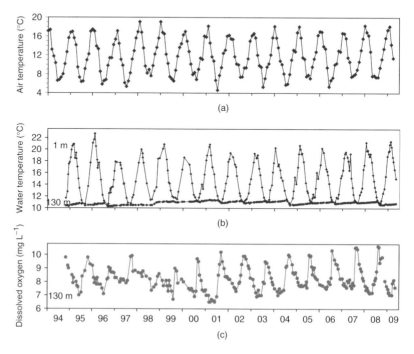

Figure 19.5   (a) Air temperature (Taupo Airport), (b) water temperature at depths of 1 m and 130 m, and (c) dissolved oxygen concentration at 130 m at a central station in Lake Taupo, 1994–2009.

years, interrupted intermittently in years of complete winter mixing. A similar scenario has been hypothesized for Lake Tahoe (California, USA; see Chapter 14) where incomplete vertical mixing in winter has also been noted on two or three occasions since the 1970s, and infrequent vertical mixing with climate warming could be of insufficient frequency to prevent development of anoxia during extended periods of stratification. With anoxia there is potential for rapid increase in trophic status of the lake due to bottom-sediment releases of phosphate and ammonium (e.g., see Chapter 6). Like several lakes of the TVZ, however, Taupo has geothermal heat inputs, which are evident by small increases in water temperature at depth (130 m) during stratification (Figure 19.5b). Thus hypolimnetic waters gradually warm during stratification, leading to reduced vertical density stability. A warming climate is therefore more likely to result in Lake Taupo not becoming fully amictic but almost certainly deviating from the idealized mixing regime implied by monomixis; perhaps a thermal regime more closely approximating a hybrid temperate-tropical system (cf. Vincent 1983).

19.12 Case study: Lake Pupuke

Lake Pupuke is a deep (maximum depth 55 m) crater lake of area 0.11 km^2 in the city of Auckland at latitude $36.780\,°\text{S}$. Whilst measurements are less regular in Pupuke than in Lake Taupo, it appears that mixing may have been incomplete in some years, notably in the warm 1998 and 1999 El Niño years (Figure 19.6a). Rates of oxygen consumption in bottom waters (38 and 50 m) during the stratified season are far higher than in Lake Taupo, corresponding to the eutrophic status of this lake, and there is a period of at least seasonal anoxia. In 1998 and 2000 mixing appeared to be incomplete, with dissolved oxygen levels at 50 m remaining close to zero but at 38 m reaching levels close to saturation and similar to those at 1 m depth (Figure 19.6b). Between these years it appeared that mixing was even further reduced and readings at 38 m remained close to zero throughout winter. It was not until winter 2001 that oxygen in deeper waters (38 and 50 m) was replenished and that a more regular pattern of monomixis was re-established, at least until 2005.

19.13 Case study: surface temperature in monomictic and polymictic Rotorua lakes

Lakes in the Rotorua region were formed between 200 000 years ago (Lake Rotorua) and as recently as AD 1886 (Lake Rotomahana) from inundation of volcanic calderas and from lava flows cutting off river drainage basins. Because of the wide variety of morphologies of these lakes their mixing regimes are either monomictic or polymictic, but some (such as Lake Rotowhero—Forsyth et al. 1974) are fully vertically mixed during each day and have high water temperature ($> 35\,°\text{C}$) due to geothermal heating from bottom sediments. Water temperature records were examined in four of these lakes for the period January 1990 to June 2011: polymictic Rotorua (eutrophic, $\bar{z} = 10.8 \text{ m}$, $A = 80.5 \text{ km}^2$), polymictic Rerewhakaaitu (mesotrophic, $\bar{z} = 7.0 \text{ m}$, $A = 5.17 \text{ km}^2$), monomictic Rotoiti (mesotrophic, $\bar{z} = 31.5 \text{ m}$, $A = 33.7 \text{ km}^2$) and monomictic Tarawera (oligotrophic, $\bar{z} = 50.0 \text{ m}$, $A = 41.2 \text{ km}^2$). All measurements

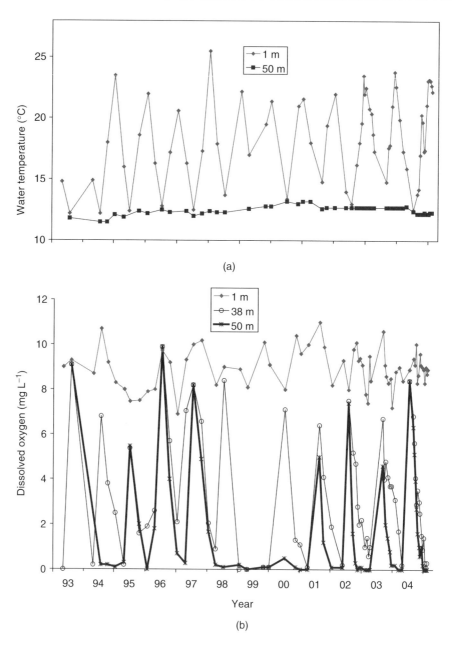

Figure 19.6   (a) Water temperature at 1 and 50 m depth, and (b) dissolved oxygen concentration at 1, 38 and 50 m depth at a central station in Lake Pupuke, Auckland, 1993–2005.

were assigned to the day of the year the measurement was taken, and a polynomial model (Microsoft® Excel) was fitted to minimize error between time of year and surface water temperature (Figure 19.7). The residual (error) of modeled and observed water temperature was then calculated for each measurement, and the resulting time-series was analyzed for any trend by linear regression (Figure 19.8). Different

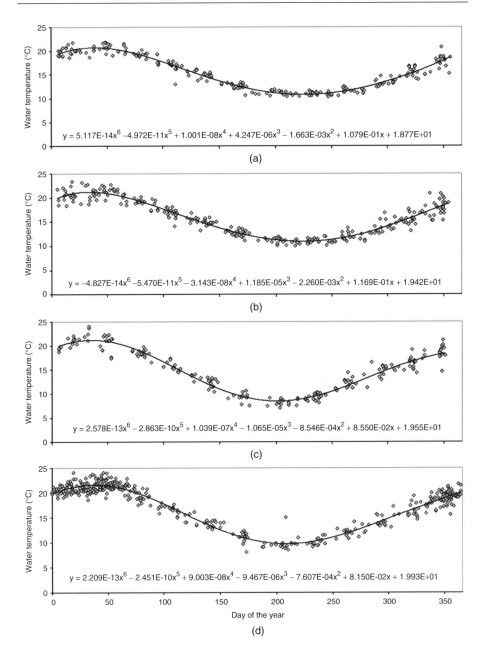

Figure 19.7 Surface water temperature by day of year based on records from January 1990 to June 2011 and polynomial fit for (a) Lake Tarawera ($R^2 = 0.95$, $p < 0.01$), (b) Lake Rotoiti ($R^2 = 0.93$, $p < 0.01$), (c) Lake Rerewhakaaitu ($R^2 = 0.94$, $p < 0.01$) and (d) Lake Rotorua ($R^2 = 0.93$, $p < 0.01$), Rotorua region, North Island.

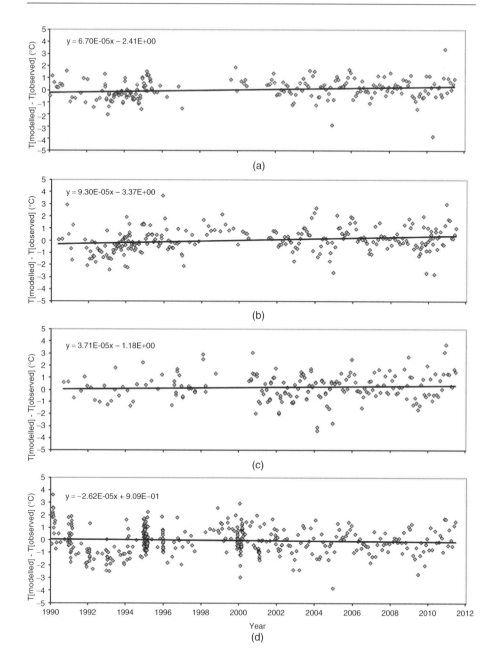

Figure 19.8 Time-series of the residual of observed and modeled water temperature (see Figure 19.7) based on records from January 1990 to June 2011 and linear fit for (a) Lake Tarawera ($R^2 = 0.04$, n.s.), (b) Lake Rotoiti ($R^2 = 0.05$, n.s.), (c) Lake Rerewhakaaitu ($R^2 = 0.005$, n.s.) and (d) Lake Rotorua ($R^2 = 0.003$, n.s.), Rotorua region, North Island.

frequencies of measurement along the temperature time series, evident in all lakes except Rotorua, could potentially bias the residual trend line and inferences drawn below should be interpreted with caution, particularly given the limited time frame for which measurements are available, especially compared with well-monitored lakes in the northern hemisphere.

Although no statistically significant ($p<0.05$) trends were observed, visual observation of the time series revealed some interesting features. Slopes of regressions for monomictic lakes Tarawera and Rotoiti (0.00009 and 0.00007 $°C\ day^{-1}$, respectively) were higher than those of polymictic lakes Rerewhakaaitu and Rotorua (0.00004 and $-0.00003\ °C\ day^{-1}$, respectively). The percentage of variation explained by the polynomial fit for day of year was also higher in the monomictic lakes ($R^2 = 0.04$ (Tarawera) and 0.05 (Rotoiti)) compared with the polymictic ones ($R^2 = 0.005$ (Rerewhakaaitu) and 0.003 (Rotorua)). The upper limit for the rate of temperature increase (equivalent to 0.33 $°C$ per decade in Lake Tarawera) is still within the error limits observed in the annual maximum or minimum surface water temperatures observed in the Rotorua lakes but is substantially less than the rate of increase observed in many northern hemisphere lakes for which surface water temperatures have warmed considerably faster than air temperatures in the past 1–2 decades (see Chapters 2, 3, 7 and 12).

19.14   Case study: bottom-water dissolved oxygen in Lake Rotoiti

Dissolved oxygen at 60 m depth in Lake Rotoiti provides a good indication of concentrations throughout the hypolimnion of this lake as geothermal heating from its bottom sediments creates vertical mixing that largely homogenizes this water layer. A similar analysis technique was used for 60 m oxygen measurements in Lake Rotoiti to that used above for surface water temperature in the four Rotorua lakes. Measurements were assigned to day of the year and a polynomial model was fitted to relate the measurements to time of year (Figure 19.9a). The time series between the residual (error) of modeled and observed oxygen was then examined for any trend by linear regression (Figure 19.9b), again with the caveats noted above for the non-randomized nature of the time-series record. In the case of 60-m dissolved oxygen in Lake Rotoiti a significant ($p<0.05$) negative slope was noted for the linear regression, with levels of dissolved oxygen declining by about 0.16 mg L^{-1} per decade. This rate of decline appears to have slowed considerably (cf. Hamilton, Hawes, and Gibbs 2006) compared with what might be interpreted from intermittent records between the 1950s and 1980s when Vincent, Gibbs, Dryden (1984) noted that hypolimnetic waters had changed from c. 50% saturation (1950s; Jolly 1968) to being anoxic at the end of the seasonally stratified period. Lake Rotoiti has changed from oligotrophic to mesotrophic/eutrophic in the period from the 1950s to 2000s, at least partly due to eutrophication of its major inflow source—Lake Rotorua—as wastewater from the city of Rotorua was discharged to this lake until 1990 (Von Westernhagen, Hamilton, and Pilditch 2010). Eutrophication of Rotoiti has probably obscured any other major changes in the lake ecosystem in the past few decades—with the possible exception of introductions of exotic weeds (see Section 19.6)—and in 2008 a wall was constructed at the western end of Lake Rotoiti to divert the Lake Rotorua inflow directly towards the outlet of Rotoiti to attempt to remedy the eutrophication problem in Rotoiti.

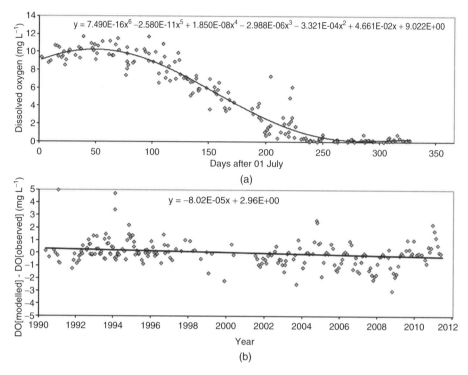

Figure 19.9　(a) Bottom-water (60 m) dissolved oxygen concentration by day of year based on records from to 1976 to 2007 (polynomial fit: $R^2 = 0.94$, $p<0.01$) and (b) time-series of the residual of observed and modeled bottom-water dissolved oxygen (linear fit: $R^2 = 0.03$, $p<0.05$) for Lake Rotoiti, Rotorua region, North Island.

19.15　Case study: modeling effects of land use and climate change for Lake Rotorua

A one-dimensional coupled hydrodynamic-ecological model, DYRESM-CAEDYM, developed at Centre for Water Research, University of Western Australia, has previously been applied to three NZ lakes to examine the relative impacts of climate change and external nutrient loading on lake water quality (Trolle *et al.* 2011). The simulated lakes included two in the Rotorua region of North Island; polymictic Rotoehu (eutrophic, $A = 7.9\,\text{km}^2$, $\bar{z} = 8.2\,\text{m}$) and monomictic Okareka (oligotrophic to mesotrophic, $A = 3.34\,\text{km}^2$, $\bar{z} = 20\,\text{m}$), and eutrophic Lake Ellesmere in Canterbury, South Island (polymictic, $A = 186\,\text{km}^2$, $\bar{z} = 1.9\,\text{m}$). The model had several simplifying assumptions including a truncated food web and limited number of phytoplankton taxa as state variables. The results were broadly consistent amongst lakes, with a 2100 climate (A2 scenario; air temperature warming 2.5–2.7 °C) being similar to an increase in nutrient loading ranging from 25 to 50%, with lakes with lower trophic status (Okareka in particular) being somewhat more sensitive to both increases in nutrient loading and climate change.

Here we demonstrate application of the model to eutrophic, polymictic Lake Rotorua ($A = 80\,\text{km}^2$, $\bar{z} = 10.8\,\text{m}$) to evaluate the relative impacts of climate change and land use change. Model input data (daily inflow, outflow and meteorology) were derived from measurements from 2001 to 2009, with independent periods of calibration and validation within the whole duration of the simulation. Climate-change scenarios included simulations of future water quality for the period 2091–2099 without any climate change (i.e., with the measured 2001–2009 climate) and with the measured climate (i.e., air temperature, humidity and rainfall) perturbed based on an A1B scenario, equivalent to a mean annual increase in air temperature of 2.27 °C for this region, slightly lower than the increases in temperature of the A2 scenario used by Trolle *et al.* (2011). The A1B scenario is generally acknowledged to provide a conservative or mid-range projection of climate change by the end of the twenty-first century. A further series of scenarios included nutrient reductions, both with and without climate change. These scenarios were generated by altering the inflow data input in the simulations. Only reductions in external nutrient load are considered here because management actions are underway to address eutrophication of the lake from the past three or four decades. Details of the nutrient load reductions are given in Rutherford, Palliser, and Wadhwa (2011) but, briefly, they include land-use change (conversions of intensive agricultural land to forestry) in order to reduce nitrogen loads by 250, 300 and 350 t year^{-1} in three separate simulations. Reductions in phosphorus load are made according to the land-use change, and changes in internal loading are inferred from knowledge of the composition of the bottom sediments under historical nutrient loading estimates (see Trolle, Hamilton, and Pilditch 2010).

The 2091–2090 climate-change scenario, with no alteration to the external nutrient load, produced a mean increase in surface water concentrations of total phosphorus by 35.5%, total nitrogen by 9.0% and chlorophyll *a* by 3.6% compared with unaltered climate (2001–2009) and nutrient loading. Table 19.1 shows two metrics of critical importance to ecological processes in Lake Rotorua; concentrations of cyanobacteria in surface waters and dissolved oxygen in bottom waters. The most dramatic change due to climate change alone is in the relative number of days with dissolved oxygen <2 mg L^{-1}; these occurrences are 56% less frequent in 2001–2009 than under the simulated climate of 2091–2099. Under the three nutrient reduction scenarios (land use to reduce external nitrogen loads by 250, 300 and 350 t year^{-1}, and corresponding phosphorus load reductions), however, it is possible to not only negate the negative impact of climate change on bottom-water dissolved oxygen concentrations, but to reverse it relative to identical simulations (2091–2099 climate) which have no reduction in nutrient loading (Table 19.1).

Days when chlorophyll *a* associated with cyanobacteria is greater than 20 mg m^{-3} have been used as a proxy for when there may be water quality problems in Lake Rotorua associated with blooms of cyanobacteria. Similar to dissolved oxygen, the simulations indicate that nutrient load reductions may more than offset the effects of climate change; for a 2091–2099 climate and land-use change equivalent to 350 t N year^{-1} load reduction there is a 56% reduction in days with cyanobacteria >20 mg m^{-3} relative to no nutrient load reduction, while a 2001–2009 climate and no nutrient load reduction yields only a 3% reduction (Table 19.1).

Table 19.1 Comparisons of cyanobacterial concentration (as equivalent chlorophyll a concentration) and dissolved oxygen for five different simulations of Lake Rotorua with DYRESM-CAEDYM. Simulations include: 2091–2099 climate with no change in present (2001–2009) nutrient loading (CC-0), present climate (2001–2009) and no change in nutrient loading (0–0), and climate change and reduced nutrient loading (CC-250, CC-300 and CC-350; see text). The output statistics are for the total number of days and proportion of days with cyanobacteria chlorophyll >20 mg m^{-3} (Cyan d>20 and Cyan p>20, respectively), the relative change in days with cyanobacteria chlorophyll >20 mg m^{-3} referenced against the CC-0 case (Cyan relative), the total number of days and proportion of days with bottom-water dissolved oxygen concentrations <2 mg L^{-1} (DO d<2 and DO p<2) and the relative change in days with dissolved oxygen <2 mg L^{-1} (DO relative) referenced against the CC-0 case.

	CC-0	0-0	CC-250	CC-300	CC-350
Cyan d>20	1158	1091	812	614	461
Cyan p>0 (%)	40	37	28	21	16
Cyan relative (%)	0	-3	-14	-37	-56
DO d<2	373	163	18	0	0
DO p<2 (%)	13	6	1	0	0
DO relative (%)	0	-56	-95	-100	-100

19.16   Management challenges and mitigation measures

Examples from North American, European and Asian lakes (see Chapters 2, 3, 6, 8, 10, 12, 14 and 24) and modeling studies of three NZ lakes of varying morphology and trophic status (Trolle *et al.* 2011) as well as Lake Rotorua (this chapter) suggest that climate warming can stimulate eutrophication. However, Visconti, Manca, and De Bernardi (2008) report the warming effect as an increase in *Daphnia* biomass, rather than an increase in phytoplankton biomass. Similarly, the density of *Daphnia* in spring increased strongly with increasing lake temperature, but winter phytoplankton biomass only correlated weakly with temperature in a shallow English lake (Carvalho and Kirika 2003). *Daphnia* is a keystone species in lake planktonic food webs and its ability to control phytoplankton blooms underpins water quality in many lakes. Thus, the temporal coupling of phytoplankton production with *Daphnia* grazing pressure is a key trophic linkage that may be affected by climate warming and requires further investigation (see also Chapters 7 and 11).

Similarly, better understanding and models of changes to catchment hydrology and biogeochemistry with climate change are required as most lake models that simulate climate change have rudimentary or no representation of effects on the catchment. Return periods for major flood events will approximately halve under A1B and A2 scenarios, which could have major implications for the delivery of sediments and nutrients from lake catchments. Nevertheless, the weight of evidence suggests that lakes will respond to increased temperatures in a way that may mimic increases in trophic status. Trolle *et al.* (2011) indicate that a 25% to 50% reduction in nutrient loads to three NZ lakes may be required to offset the increased phytoplankton biomass induced by warming by 2100, whilst Lake Rotorua simulations (not shown) suggest a 7–12% reduction in nutrient load (50–85 t N year^{-1}) would be required.

Polymictic NZ lakes such as Rotorua have yet to show any clear directional change in surface water temperature (nonsignificant decrease in Rotorua and increase in Rerewhakaaitu over the past 20 years) but could be highly sensitive to a changing climate due to increases in frequency and duration of stratification events and, especially in eutrophic systems, greater potential for anoxia of bottom waters and associated nutrient release events (Burger *et al.* 2008). Strong westerly winds normally associated with the El Niño phase of ENSO evidently did not offset the strong warming that took place in 1998–1999 in NZ when winter mixing of the water column in deep lakes such as Taupo and Pupuke was of reduced duration or incomplete. This future balance of altered wind speed and direction (Mullan *et al.* 2008) with higher air temperature may be critical from several perspectives and is not well understood or readily predictable at present, and there has not been an analysis of the temporal coherence of strong ENSO phases across NZ lakes or subsets of the lakes. The balance may, for example, impact upon regime shifts in shallow lakes (specifically by changing exposure of submerged plants to wind-wave-induced shear stresses), frequency and duration of stratification and mixing in polymictic lakes in particular, for which there could be concomitant changes in bottom-water anoxia and sediment nutrient releases, and potentially radically alter mixing behavior of deep monomictic lakes to intermittent with increased probability of development of anoxic hypolimnia even in oligotrophic systems.

In all of the above considerations it is evident that some degree of protection from impacts of climate change may be afforded by appropriate land use and management in lake catchments to arrest and ultimately decrease external nutrient load increases, particularly given the potential for increasing storm events to increase catchment sediment and nutrient loads (also see Chapter 24). The effects of reduced nutrient and sediment loads, as well as reduced organic matter deposition, could offset the increased probability of anoxia in bottom waters with climate change. It could also reduce heat trapping of solar radiation in surface waters through reduced concentrations of suspended minerals, phytoplankton and other suspended organic particulates. Finally, in shallow regime-shifting lakes whose vegetation may be susceptible to the turbulent forces of increased wind speeds with climate change (Mullan *et al.* 2008), it could increase the resilience of the vegetated clear-water phase over the turbid, phytoplankton-dominated phase.

For the moment climate change remains but one (minor) aspect in wider global changes that affect alien invasions in NZ lakes, and interplays with increased human population pressure, international trade, and habitat alteration. This escalating global change, as a whole, is seen as driving the emergence of a "homogocene era" (sensu Strayer 2010), coined to represent increasing similitude and loss of local biotic distinctiveness worldwide. However, increased pressure on natural water resources in NZ to provide ecosystem services, such as irrigation and hydro power, is likely to intensify on two fronts with climate change reducing rainfall and soil moisture in intensively farmed eastern areas and with the NZ government signaling that it will make substantial future investment in irrigation and hydro power development. Thus, we can expect a synergistic effect of climate change on currently existing pressures on water resources at catchment and waterbody scales.

Table 19.2 Predicted effects of anthropogenic climate change in New Zealand and key limnological impacts.

Effect	Predicted magnitude of effect	Reference	Limnological impact	Reference
Warming	0.1– 1.4 °C by 2030s	Wratt et al. (2004)	Eutrophication	Visconti et al. (2008), Trolle et al. (2011)
	0.2– 4.0 °C by 2080s	Wratt et al. (2004)	Phenology	Winder and Schindler (2004a, b)
Precipitation	Increase except eastern North Island and northern South Island	Hennessey et al. (2007)	Decreased water residence times Increased external nutrient loading	
Wind	Midrange projections by 2080: 60% increase in mean westerly component of wind speed	Wratt et al. (2004)	Increased turbulence /resuspension Deeper mixing Increased aeolian dust (phosphorus)	Hamilton and Mitchell (1997) Davies-Colley (1984) McGowan et al. (1996)
Sea level	By 2100: 0.18–0.59 m increase relative to year 2000	Hennessey et al. (2007)	Salinization of coastal lakes and lagoons and disappearance of some	Schallenberg et al. (2003) Duggan and White (2010)

While there is only a modest level of information on the limnological impacts of changing climate on NZ lakes and wetlands, particularly given the lack of obvious temperature signals in the past 30–40 years, a substantial amount of work has been done to predict the changes of key climatic drivers of ecosystem structure and function. Table 19.2 lists some of the predicted changes in key climatic drivers. Warming, precipitation and wind effects are estimated from statistically downscaled global circulation models. Thus, the magnitudes of these effects may be subject to substantial error and should be considered provisional (see Koutsoyiannis et al. 2008). The predictions in Table 19.2 have been used as a basis for discussing in detail (see above) three of the likely limnological impacts of global climate change on NZ lakes: (i) increases in the duration of thermal stratification, possibly changing some polymictic lakes to monomictic and some monomictic lakes to amictic or rarely mixed, (ii) potential for regime shifts in shallow lakes, and (iii) effects of sea-level rise on the ecological structure and functioning of coastal lakes, wetlands and lagoons.

Hydro power constitutes a large fraction of the total electricity demand in NZ (Renwick et al. 2010b). and climate change therefore presents a risk to the security of this energy source. The balance of increased rainfall and reduced snowpack with climate change complicates assessments of how hydro storages will be affected, but there is likely to be altered seasonality of inputs to hydro storage reservoirs, which will affect the alignment of hydro power availability and demand (McKerchar, Pearson, and Fitzharris 1998). Evaporation will also be altered with changing climate due to changes in water temperature as well as water vapor content. Renwick *et al.* (2010c) projected increases in evaporation losses from 6 to 129% of present under different future climate scenarios for Lake Ellesmere, located in the relatively dry coastal region of Canterbury, South Island. This could have important implications for the artificial opening regime currently used to manage water levels in the lake.

19.17 Conclusions

The longest continuous regional record of lake water temperature in NZ is for the Rotorua lakes in central North Island. Over 2–3 decades, there has been no significant change in seasonally corrected surface-water temperature in the deep, monomictic or shallow polymictic Rotorua lakes. Similarly for the largest NZ lake, Taupo, also in the geothermally active area of the Taupo Volcanic Zone and Lake Pupuke (Auckland), there have not been significant changes in water temperature since the early 1990s. Important climate signals have, however, been reflected in changes in mixing and hydrological regimes in some lakes. For example, an intense El Niño phase in 1998–1999 coincided with reduced vertical mixing in winter, leading to greater depletion of dissolved oxygen in bottom waters of lakes Taupo and Pupuke. A warming climate may not necessarily produce multi-year periods without seasonal mixing in several of the deep monomictic TVZ lakes, however, because of progressive geothermal warming of bottom waters, which gradually erodes the water column thermal gradient. Similar geothermal warming is absent from the deep monomictic lakes of the subalpine region of South Island for which a warming climate may slightly extend the period of stratification in these oligotrophic lakes. Climate events acting over a more prolonged period, such as the Interdecadal Pacific Oscillation, also play an important role in the hydro-ecology of NZ lakes, with changes in rainfall and runoff to lakes of the TVZ acting at decadal time scales. By contrast to effects of climate, the TVZ lakes have shown quite profound changes due to modifications of land use associated with increased land area and intensity of pastoral farming within their catchments, leading to symptoms of lake eutrophication. For the many shallow coastal and lowland lakes in NZ that are polymictic, the effect of a warming climate may be more profound than for deep lakes, as periods of temporary stratification will become more regular and prolonged, with potential for increased deoxygenation of bottom waters in eutrophic systems. The resulting increases in nutrient inputs from bottom sediments may increase their trophic status; many of these lakes are already highly impacted by anthropogenic activities. An important consideration for the large number of NZ lakes that are tidally influenced is changes in sea level with climate change. Coastal lakes, wetlands and lagoons with mean water level within 1 m of mean seawater levels will be particularly vulnerable to marine ingress, leading to loss of some systems and major ecological and food web changes from saline intrusion in others. Our modeling across lakes of different mixing regime and trophic status indicates that the effects of an indicative conservative warming scenario may produce water quality effects similar to moderate to substantial increases in nutrient load. The modeling results suggest that both long-term climate change as well as short-term extreme climate events should be built into the future management plans for NZ lakes to reduce involving reduction of catchment nutrient and sediment loads.

The effect of climate change on NZ lakes is likely to be a synergistic one, with major existing pressures from alien species, water extraction and eutrophication, acting in tandem with additional pressures from forecasted increases in irrigation and water impoundment as well as new alien species introductions, to negatively impact upon lake ecological integrity and biodiversity. Actions to improve land management and in some cases make land use change, will be necessary to reduce nutrient and sediment loads to lakes and provide greater resilience to storm events, whilst also increasing surveillance, control and eradication efforts for noxious alien freshwater invaders.

References

Allibone, R., David, B., Hitchmough, R., *et al.* (2012) Conservation status of New Zealand freshwater fish, 2009. *New Zeal. J. Marine and Freshwater Res.*, **45**, 301–2, doi: 10.1080/00288330.2010.514346.

Banks, C.M. and Duggan, I.C. (2009) Lake construction has facilitated calanoid copepod invasions in New Zealand. *Divers. Distrib.*, **15**, 80–7, doi:10.1111/j.1472-4642.2008.00524.x.

Battarbee, R.W., Grytnes, J.A., Thompson, R., *et al.* (2002) Comparing palaeolimnological and instrumental evidence of climate change for remote mountain lakes over the last 200 years. *J. Paleolimnol.*, **28**, 161–79, doi:10.1023/A:1020384204940.

Burger, D.F., Hamilton, D.P., and Pilditch, C.A. (2008) Modelling the relative importance of internal and external nutrient loads on water column nutrient concentrations and phytoplankton biomass in a shallow polymictic lake. *Ecol. Model.*, **211**, 411–23, doi:10.1016/j.ecolmodel.2007.09.028.

Carvalho, L., and Kirika, L. (2003) Changes in shallow lake functioning: response to climate change and nutrient reduction. *Hydrobiologia*, **506–509**, 789–96, doi: 10.1023/B:HYDR.0000008600.84544.0a.

Champion, P.D., and Clayton, J.S. (2000) *Border Control for Potential Aquatic Weeds. Stage 1—Weed Risk Model. Science for Conservation 141*. Department of Conservation, Wellington.

Champion, P.D. and Clayton, J.S. (2001) *Border Control for Potential Aquatic Weeds. Stage 2 Weed Risk Assessment. Science for Conservation 185*. Department of Conservation, Wellington.

Champion, P., Rowe, D., Smith, B., *et al.* (2004) Identification guide: freshwater pests of New Zealand. NIWA Information Series No. 55, National Institute of Water and Atmosphere (N.Z.).

Church, J.A. and White, N.J. (2006) A twentieth century acceleration in global sea-level rise. *Geophys. Res. Lett.*, **33**, L01602, doi:10.1029/2005GL024826

Clayton, J. (1990) Impact of climate change on aquatic plants, in *The Impact of Climate Change on Pests, Diseases, Weeds and Beneficial Organisms Present in New Zealand Agricultural and Horticultural Systems* (eds. R.A. Prestidge, and R.P. Pottinger), New Zealand Ministry for the Environment, Wellington, pp. 113–16.

Compton, T.J., De Winton, M., Leathwick, J.R., and Wadhwa, S. (2012) Predicting spread of invasive macrophytes in New Zealand lakes using indirect measures of human accessibility. *Freshwat. Biol.*, **57**, 938–48, doi:10.1111/j.1365-2427.2012.02754.x

Davies-Colley, R.J. (1988) Mixing depths in New Zealand lakes. *New Zeal. J. Mar. Fresh.*, **22**, 517–27, doi:10.1080/00288330.1988.9516322.

Dawson, F.H. (1996) *Crassula helmsii*: attempts at elimination using herbicides. *Hydrobiologia*, **340**, 241–5, doi: 10.1007/BF00012762.

De Lange P.J., Heenan, P.B., Norton, D.A., *et al.* (2009) *Threatened Plants of New Zealand*, Canterbury University Press, Christchurch.

Duggan, I.C. (2002) First record of a wild population of the tropical snail *Melanoides tuberculata* in New Zealand natural waters. *New Zeal. J. Mar. Fresh.*, **36**, 825–30, doi: 10.1080/00288330.2002.9517135.

Duggan, I.C. Green, J.D. and Burger, D.F. (2006) First New Zealand records of three non-indigenous zooplankton species: *Skistodiaptomus pallidus, Sinodiaptomus valkanovi, and Daphnia dentifera*. *New Zeal. J. Mar. Fresh.*, **40**, 561–9, doi: 10.1080/00288330.2006.9517445.

Duggan, I.C., and White, W.A. (2010) Consequences of human-mediated marine intrusions on the zooplankton community of a temperate coastal lake. *New Zeal. J. Mar. Fresh.*, **44**, 17–28, doi: 10.1080/00288331003641661.

Emiliani, C. (1966) Isotopic palaeotemperatures. *Science*, **154**, 851–7, doi: 10.1126/science.154.3751.851.

Forsyth, D.J., and McColl, R.H.S. (1974) The limnology of a thermal lake: Lake Rotowhero, New Zealand: I. General description and water chemistry. *Hydrobiologia*, **43**, 313–32, doi: 10.1007/BF00015354.

Gibb, J.G. (1986) A New Zealand regional Holocene eustatic sea-level curve and its application to determination of vertical tectonic movements. A contribution to IGCP-Project 200. *Royal Soc. N. Z. Bull.*, **24**, 377–95.

Greig, H.S., Kratina, P., Thompson, P.L., *et al.* (2012) Warming, eutrophication, and predator loss amplify subsidies between aquatic and terrestrial ecosystems. *Glob. Change Biol.*, **18**, 504–514, doi:10.1111/j.1365-2486.2011.02540.x

Hamilton, D.P., and Mitchell, S.F. (1997) Wave-induced shear stresses, plant nutrients and chlorophyll in seven shallow lakes. *Freshwat. Biol.*, **38**, 159–68, doi: 10.1046/j.1365-2427.1997.00202.x.

Hamilton, D.P., Hawes, I., and Davies-Colley, R. (2004) Physical and chemical characteristics of lake water, in Freshwaters of New Zealand (eds. J. Harding, P. Mosley, C. Pearson, and B. Sorrell), New Zealand Hydrological Society, Christchurch, pp. 21.1–21.20.

Hamilton, D.P., Hawes, I., and Gibbs, M.M. (2006) Climatic shifts and water quality response in North Island lakes, New Zealand. *Verh. Internat. Verein. Limnol.*, **29**, 1821–4, doi: 10.1029/2003GL019166.

Hamilton, D.P., O'Brien, K.R., Burford, M.A., *et al.* (2010) Vertical distributions of chlorophyll in deep, warm monomictic lakes. *Aquat. Sci.*, **72**, 295–307, doi: 10.1007/s00027-010-0131-1.

Hannah, J. 2004. An updated analysis of long term sea level change in New Zealand. *Geophys. Res. Lett.*, **31**, L03307, doi:10.1029/2003GL019166

Hansen, J., Ruedy R., Sato, M. and Lo, K. (2010) Global surface temperature change. *Rev. Geophys.*, **48**, doi: 8755-1209/10/2010RG000345

Hauer, F.R., Baron, J.S., Campbell, D.H., *et al.* (1997) Assessment of climate change and freshwater ecosystems of the Rocky Mountains, *USA and Canada. Hydrol. Process*, **11**, 903–24, doi:10.1002/(SICI)1099-1085(19970630)11:8<903::AID-HYP511>3.0.CO;2–7.

Hays, J.D., Imbrie, J., and Shackleton, N.J. (1976). Variations in the earth's orbit: Pacemaker of the ice ages. *Science*, **194**, 1121–32, doi: 10.1126/science.194.4270.1121

Hellmann, J.J., Byers, J.E., Bierwagen, B.G., and Dukes, J.S. (2008) Five potential consequences of climate change for invasive species. *Conserv. Biol.*, **22**, 534–43, doi: 10.1111/j.1523-1739.2008.00951.x.

Hendrikx, J., Clark, M., Hreinsson, E.O., *et al.* (2009) Simulations of seasonal snow in New Zealand: past and future. Proceedings of the 9th International Conference on Southern Hemisphere Meteorology and Oceanography, Melbourne, February 2009. American Meteorological Society.

Hennessy, K., Fitzharris, B., Bates, B.C., *et al.* (2007) Australia and New Zealand. climate change 2007: Impacts, adaptation and vulnerability, in *Fourth Assessment Report of the Intergovernmental Panel on Climate Change Contribution of Working Group II* (eds. M.L. Parry, O.F. Canziani, J.P. Palutikof, *et al.*), Cambridge University Press, Cambridge.

Hicks, B.J., Ling, N., and Wilson, B (2010) Introduced fish, in *Waters of the Waikato: Ecology of New Zealand's Longest River* (eds. K.J. Collier, D.P. Hamilton, W. Vant, and C. Howard-Williams), Waikato/University of Waikato, Hamilton, pp. 209–28.

Hyldgaard, B. and H. Brix (2012) Intraspecies differences in phenotypic plasticity: Invasive versus non-invasive populations of *Ceratophyllum demersum*. *Aquat. Bot.* **97**, 49–56, doi: 10.1016/j.aquabot.2011.11.004.

Imbrie, J. and Imbrie, K.P. (1979) *Ice Ages: Solving the Mystery*, Macmillan, London.

Jeppesen, E., Meerhoff, M., Holmgren, K., *et al.* (2010) Impacts of climate warming on lake fish community structure and potential effects on ecosystem function. *Hydrobiologia*, **646**, 73–90, doi: 10.1007/s10750-010-0171-5.

Jeppesen, E., Søndergaard, M., Kanstrup, E., *et al.* (1994) Does the impact of nutrients on the biological structure and function of brackish and freshwater lakes differ? *Hydrobiologia*, **275/276**, 15–30, doi: 10.1007/BF00026696.

Jolly, V.H. (1968) The comparative limnology of some New Zealand lakes. I. Physical and chemical. New Zeal. J. Mar. Fresh, **2**, 214–59, doi: 10.1080/00288330.1968.9515236.

Johnson, P.T.J., Olden, J.D., and Vander Zanden, M.J. (2008) Dam invaders: impoundments facilitate biological invasions into freshwaters. *Front. Ecol. Environ.*, **6**, 357–63, doi: 10.1890/070156.

Kilroy, C., Biggs, B.J.F., Vyverman, W., and Broady, P.A. (2006) Benthic diatom communities in subalpine pools in New Zealand: Relationships to environmental variables. *Hydrobiologia*, **561**, 95–110, doi: 10.1007/s10750-005-1607-1.

Koutsoyiannis, D., Efstratiadis, A., Mamassis, N., and Christofides, A. (2008) On the credibility of climate predictions. *Hydrolog. Sci. J.*, **53**, 671–84, doi: 10.1623/hysj.53.4.671.

Leathwick, J.R., West, D., Gerbeaux, P., *et al.* (2010) Freshwater Ecosystems of New Zealand (FENZ) Geodatabase, Department of Conservation, Hamilton, http://www.doc.govt.nz/conservation/land-and-freshwater/freshwater/freshwater-ecosystems-of-new-zealand/ (accessed July 23, 2012).

Larned, S.T., Scarsbrook, M.R., Snelder, T.H., *et al.* (2004) Water quality in low-elevation streams and rivers of New Zealand: Recent state and trends in contrasting land-cover classes. *New Zeal. J. Mar. Fresh.*, **38**, 347–66, doi: 10.1080/00288330.2004.9517243.

Leprieur, F., Beauchard, O., Blanchet, S., *et al.* (2008) Fish invasions in the world's river systems: when natural processes are blurred by human activities. *PLoS Biology*, **6**, 404–10, doi: 10.1371/journal.pbio.0060028.

Les, D.H., Capers, R.S., and Tippery, N.P. (2006) Introduction of *Glossostigma* (Phrymaceae) to North America: a taxonomic and ecological overview. *Am. J. Bot.*, **93**, 927–39, doi: 10.3732/ajb.93.6.927.

Lewis, W.M. 2000. Basis for the protection and management of tropical lakes. Lakes & Reservoirs: Research and Management **5**, 35–48, doi: 10.1046/j.1440-1770.2000.00091.x.

Loo, S.E., MacNally, R., and Lake, P.S. (2007) Forecasting New Zealand mudsnail invasion range: model comparisons using native and invaded ranges. Ecol. Appl. **17**, 181–9, doi: 10.1890/1051-0761(2007)017[0181:FNZMIR]2.0.CO;2.

Lowe, D.J., and Green, J.D. (1987) Origins and development of the lakes. *NZ DSIR Bull.*, **241**, 1–64.

Marsden, I.D. (2004) Effects of reduced salinity and seston availability on growth of the New Zealand little-neck clam *Austrovenus stutchburyi*. *Mar. Ecol. Prog. Ser.*, **266**, 157–71, doi: 10.3354/meps266157.

MacIsaac, H.J., Borbely, J.V.M., Muirhead, J.R., and Graniero, P.A. (2004) Backcasting and forecasting biological invasions of inland lakes. *Ecol. Appl.*, **14**, 773–83, doi:10.1890/02-5377.

McDowall, R.M. (1992) Global climate change and fish and fisheries: What might happen in a temperate oceanic archipelago like New Zealand. *Geojournal*, **28**, 29–37, doi:10.1007/BF00216404.

McDowall, R.M. (2000) The Reed Field Guide to New Zealand Freshwater Fishes, Reed Publishing, Auckland.

McDowall, R.M. (2003) Impacts of introduced salmonids on native galaxiids in New Zealand upland streams: a new look at an old problem. *Trans. Am. Fish. Soc.*, **132**, 229–38, doi: 10.1577/1548-8659(2003)132<0229:IOISON>2.0.CO;2.

McDowall, R.M. 2006. On size and growth in freshwater fish. *Ecol. Freshwater Fish*, **3**, 67–79, doi: 10.1111/j.1600-0633.1994.tb00108.x.

McGowan, H.A., Sturman, A.P., and Owens, I.F. (1996) Aeolian dust transport and deposition by foehn winds in an alpine environment, Lake Tekapo, New Zealand. *Geomorphology*, **15**, 135–46, doi: 10.1016/0169-555X(95)00123-M.

McIntosh, A.R. (2000) Aquatic predator–prey interactions, in *New Zealand Stream Invertebrates: Ecology and Implications for Management* (eds. K.J. Collier, and M.J. Winterbourn), New Zealand Limnology Society, Christchurch, NZ, pp. 125–56.

McKerchar, A.I., Pearson, C.P., and Fitzharris, B.B. (1998) Dependency of summer lake inflows and precipitation on spring *SOI*. *J. Hydrol.*, **205**, 66–80, doi: 10.1016/S0022-1694.

Maunder, W.J. (1971) The climate of New Zealand—physical and dynamic features, in *World Survey of Climatology* (ed. J. Gentilli), Elsevier, Amsterdam, p. 213–27.

Melack, J.M., Dozier, J., Goldman, C.R., *et al.* (1997) Effects of climate change on inland waters of the Pacific Coastal Mountains and Western Great Basin of North America. *Hydrol. Process.*, **11**, 971–92, doi: 10.1002/(SICI)1099-1085(19970630)11:8<971::AID-HYP514>3.0.CO;2-Y.

Ministry for the Environment (2011) Cabinet paper: *Fresh Start for Fresh Water—Forward Work Programme*, http://www.mfe.govt.nz/cabinet-papers/fresh-start-fresh-water-forward-work.pdf (accessed 22 May 2012). Ministry for the Environment, NZ.

Mullan, A.B. (1995) On the linearity and stability of Southern Oscillation-climate relationships for New Zealand. *Int. J. Climatol.*, **15**, 1365–86, doi: 10.1002/joc.3370151205.

Mullan, B., Wratt, D., Dean, S., *et al.* (2008) *Climate Change Effects and Impacts Assessment: A Guidance Manual for Local Government in New Zealand*, 2nd edn, Ministry for the Environment, Wellington.

Pall, P., Allen, M.R., and Stone, D.A. (2007) Testing the Clausius-Clapeyron constraint on changes in extreme precipitation under CO_2 warming. *Clim. Dynam.*, **28**, 351–63, doi: 10.1007/s00382-006-0180-2.

Parker, B.R., Vinebrooke, R.D., and Schindler, D.W. (2008) Recent climate extremes alter alpine lake ecosystems. *PNAS*, **105**, 12927–31, doi: 10.1073/pnas.0806481105.

Rahel, F.J. and Olden, J.D. (2008) Assessing the effects of climate change on aquatic invasive species. *Conserv. Biol.*, **22**, 521–33, doi: 10.1111/j.1523-1739.2008.00950.x.

Remane, A., and Schlieper, C. (1971) The Biology of Brackish Waters, 2nd edn., Wiley Interscience, New York.

Renwick, J., Clark, A., Griffiths, G., *et al.* (2010a) *State of the Climate 2010, A Snapshot of Recent Climate in New Zealand*, NIWA Science and Technology Series No. 56, National Institute of Water and Atmosphere, Wellington.

Renwick, J., Mladenov, P., Purdie, J., *et al.* (2010b) The effects of climate variability and change upon renewable electricity in New Zealand, in Climate Change Adaptation in New Zealand: Future Scenarios and some Sectoral Perspectives (eds. R.A.C. Nottage, D. S. Wratt, J. F. Bornman, and K. Jones) New Zealand Climate Change Centre, Wellington, NZ, pp. 70–81.

Renwick, J., Horrell, G., McKerchar, A., *et al.* (2010c) Climate change impacts on Lake Ellesmere (Te Waihora), NIWA Client Report WLG2010-49, Project: ENC10301, National Institute of Water and Atmosphere, Wellington.

Rooney, N. and Kalff, J. (2000) Inter-annual variation in submerged macrophyte community biomass and distribution: the influence of temperature and lake morphometry. *Aquat. Bot.*, **68**, 321–35, doi: 10.1016/S0304-3770(00)00126-1.

Rowe, D.K. (2007) Exotic fish introductions and the decline of water clarity in small North Island, New Zealand lakes: a multi-species problem. *Hydrobiologia*, **583**, 345–58, doi: 10.1007/s10750-007-0646-1.

Rutherford, K., Palliser, C., and Wadhwa, S. (2011) Prediction of nitrogen loads to Lake Rotorua using the ROTAN model. NIWA Client Report: HAM2010-134. National Institute of Water and Atmosphere, Hamilton.

Salinger, M.J., Renwick, J.A., and Mullan, A.B. (2001) Interdecadal Pacific Oscillation and South Pacific climate. *Int. J. Climatol.*, **21**, 1705–21, doi: 10.1002/joc.691

Schallenberg, M., Hall, C.J., and Burns, C.W. (2003) Consequences of climate-induced salinity increases on zooplankton abundance and diversity in coastal lakes. *Mar. Ecol. Prog. Ser.*, **251**, 181–9, doi: 10.3354/meps251181.

Schallenberg, M. and Sorrell, B. (2009) Regime shifts between clear and turbid water in New Zealand lakes: Environmental correlates and implications for management and restoration. *New Zeal. J. Mar. Fresh.*, **43**, 701–12, doi: 10.1080/00288330909510035.

Strayer, D.L. (2010) Alien species in fresh waters: ecological effects, interactions with other stressors, and prospects for the future. *Freshwat. Biol.*, **55**, 152–74, doi: 10.1111/j.1365-2427.2009.02380.x.

Sturman, A.P. and Tapper, N.J. (2006) *The Weather and Climate of Australia and New Zealand*, 2nd edn, Oxford University Press, Melbourne.

Tanner, C.C., Clayton, J.S., and Wells, R.D.S. (1993) Effects of suspended solids on the establishment and growth of *Egeria densa*. *Aquat. Bot.*, **45**, 200–310, doi: 10.1016/0304-3770(93)90030-Z.

Trolle, D., Hamilton, D.P., and Pilditch, C.A. (2010) Evaluating the influence of lake morphology, trophic status and diagenesis on geochemical profiles in lake sediments. *Appl. Geochem.*, **25**, 621–32, doi: 10.1016/j.apgeochem.2010.01.003.

Trolle, D., Hamilton, D.P., Pilditch, C.A., *et al.* (2011) Predicting the effects of climate change on trophic status of three morphologically varying lakes: Implications for lake restoration and management. *Environ. Modell. Softw.*, **26**, 354–70, doi: 10.1016/j.envsoft.2010.08.009.

Vincent, W.F. (1983) Phytoplankton production and winter mixing: contrasting effects in two oligotrophic lakes. *J. Ecol.*, **71**, 1–20.

Vincent, W.F., Gibbs, M.M., and Dryden, S.J. (1984) Accelerated eutrophication in a New Zealand lake: Lake Rotoiti, Central North Island. *New Zeal. J. Mar. Fresh.*, **18**, 431–40, doi: 10.1080/00288330.1984.9516064.

Von Westernhagen, N., Hamilton, D.P., and Pilditch, C.A. (2010) Temporal and spatial variations in phytoplankton productivity in surface waters of a warm-temperate, monomictic lake in New Zealand. *Hydrobiologia*, **652**, 57–70, doi: 10.1007/s10750-010-0318-4.

Visconti, A., Manca, M., and De Bernardi, R. (2008) Eutrophication-like response to climate warming: an analysis of Lago Maggiore (N. Italy) zooplankton in contrasting years. *J. Limnol.*, **67**, 87–92, doi: 10.4081/jlimnol.2008.87.

Walther, G.-R., Roques, A., Hulme, P.E., *et al.* (2009) Alien species in a warmer world: risks and opportunities. *Trends Ecol. Evol.*, **24**, 686–93, doi: 10.1016/j.tree.2009.06.008.

Wilding, T. and Rowe, D. (2008) FRAM: fish risk assessment model for the importation and management of alien freshwater fish in New Zealand. NIWA client report: HAM 2008–029, National Institute of Water and Atmosphere, Hamilton, NZ.

Williams, J.W. and S.T. Jackson (2007) Novel climates, no-analog plant communities, and ecological surprises: Past and future. *Front. Ecol. Evol.*, **5**, 475–82, doi: 10.1890/070037.

Winder, M., and Schindler, D.E. (2004a) Climate effects on the phenology of lake processes. *Glob. Change Biol.*, **10**, 1844–56, doi: 10.1111/j.1365-2486.2004.00849.x.

Winder, M., and Schindler, D.E. (2004b) Climate change uncouples trophic interactions Ecology, **85**, 2100–6, doi: 10.1890/04-0151.

Wratt, D.S., Ridley, R.N., Sinclair, M.R., *et al.* (1996) The New Zealand Southern Alps Experiment. *B. Am. Meteorol. Soc.*, **77**, 683–92, doi: 10.1175/15200477(1996)077<0683:TNZSAE>2.0.CO;2

Wratt, D.S., Mullan, A.B., Salinger, M.J., *et al.* (2004) *Climate Change Effects and Impacts Assessment: A Guidance Manual for Local Government in New Zealand*, New Zealand Climate Change Office, Ministry for the Environment, Wellington.

20

Global Change Effects on Antarctic Freshwater Ecosystems: The Case of Maritime Antarctic Lakes

Antonio Quesada and David Velázquez

Department Biología, Universidad Autónoma de Madrid, Spain

20.1   Introduction

Current global circulation models (GCMs) predict an increase in air temperatures of several degrees by the end of the twenty-first century (Blunden, Arndt, and Baringer 2011), combined with large changes in the regional distribution and intensity of precipitation. These changes will also be accompanied by massive disruption of the cryosphere, the suite of ice-bearing environments on Earth. These shifts in climate forcing appear to have already begun, and the onset of changes in the physical, chemical and biological attributes of lakes is affecting their ability to maintain their present-day communities.

Global circulation models predict that the fastest and most pronounced warming will be in the high latitudes because of a variety of feedback processes that amplify warming in the north and south polar regions. These feedback processes include the capacity of warm air to store more water vapor, itself a powerful greenhouse gas, and the reduction of albedo as a result of melting snow and ice, leaving more solar energy available for heating (Vincent 2009). This warming is taking place at variable rates, both within the Arctic (see Chapters 1 and 2) and Antarctica. Maritime Antarctica (the Antarctic islands and Peninsula) has experienced some of the most rapid air temperature increases on Earth (2 °C since the 1960s; Quayle *et al*. 2002), while a slight cooling has been documented for the Antarctic continental interior (Steig *et al*. 2009 and references therein).

Lakes, as integrators and sometimes amplifiers of the processes taking place in the catchment, are considered indicators of the variations that may occur in the

Climatic Change and Global Warming of Inland Waters: Impacts and Mitigation for Ecosystems and Societies,
First Edition. Edited by Charles R. Goldman, Michio Kumagai and Richard D. Robarts.
© 2013 John Wiley & Sons, Ltd. Published 2013 by John Wiley & Sons, Ltd.

landscape unit (Quayle *et al.* 2003; Quesada *et al.* 2006) and sentinels of small variations (MacKay *et al.* 2009). Considering global warming, the differences in air temperature have a direct effect on the length of the ice-free period in the lakes of high latitude regions, which may be a few weeks longer during the short summers in maritime Antarctica and most of the Antarctic Peninsula but nonexistent in the permanent ice-covered lakes on the Antarctic continent.

The diversity of Antarctic freshwater bodies provides a suite of scenarios that may help to improve our understanding of the effects of global temperature change on terrestrial and limnetic ecosystems. Over the last few decades, Antarctic research has highlighted the continent's importance in the overall understanding of global processes. Limnetic polar ecosystems have several levels of relevance for ecology and biodiversity (Laybourn-Parry and Pearce 2007). Rogers (2007) demonstrated that past climate change has been critical in setting the patterns of diversity and population structure observed in Antarctic communities today. Most of the polar biota has never been exposed to temperatures as high as those forecasted by GCMs. Therefore, there are many homologous species that can take advantage of this new situation while others are likely to suffer. The potential loss of the autochthonous biota under warmer, less oxic conditions could trigger short-term (years to decades) disruptions in food webs until new biota are established. Chown and Convey (2007) also revealed that, despite their low diversity, Antarctic terrestrial and, perhaps, limnetic communities show a high level of spatial and temporal variation, and the consequences of climate change in such communities demand more attention from the scientific community with the aim of anticipating their impacts on Antarctic ecosystems. This variability suggests that future Antarctic conservation and management plans focused on preserving ecosystem integrity must adopt a continental perspective.

The effects of climate change on different ecosystems are widely researched, and their influence on lakes has been investigated and modeled in the northern hemisphere since the late 1990s (Rouse *et al.* 1997; Schindler and Curtis 1997; Winder and Schindler 2004; MacKay *et al.* 2009; see also Chapter 2 of this volume). However, lakes in the southern hemisphere have not been investigated so extensively (Kosten *et al.* 2012). Antarctic lakes, however, are simplified ecosystems with reduced trophic levels, in which the changes can produce clear effects with fewer confounding variables. The conclusions drawn from the study of these lakes may help us to identify causality and, thus, provide basic information for modeling efforts at other latitudes.

Here, a conceptual model is proposed that considers the impact of changing variables, primary effects (physical and chemical factors) and secondary effects on biological variables. The conceptual development suggested includes air temperature, ice-cover period, precipitation, the availability of liquid water near lakes and nutrient inputs and outputs. We describe the model in detail using a typical maritime Antarctic lake, comparing the present and forecasted springtime conditions as well as the mid-summer climax period.

20.2   Antarctic lakes

As noted above, climate change is affecting lakes differently in maritime Antarctica and the Antarctic Peninsula region than in continental Antarctica. In fact, physically and ecologically, the lakes from those regions are quite different and must be considered separately.

20.2.1 Continental Antarctic lakes

Continental Antarctic lakes are different from lakes in the rest of the world, both because of their biological diversity and because of the severity of the physical and chemical factors associated with them (Priscu *et al.* 1998). This severity leads to perennial ice cover, with remarkable ice thickness up to several meters. The very low air temperature also restricts the development of catchment biota and slows down biogeochemical processes, although weathering processes trigger ion fractionation (abiotic ion retention) under these extremely arid and cold conditions (Simmons *et al.* 1993). These are key factors in controlling the ecology and, particularly, the microbial processes of such lakes (Ellis-Evans 1996). Additionally, the stability provided by the perennial ice cover exacerbates the chemical stratification observed in these lakes, with many examples of meromixis (for example, Ace Lake and other meromictic lakes in the Vestfold Hills area; Gibson 1999; Lauro *et al.* 2011).

Subglacial lakes are a particular type of lake, characterized by being permanently covered by extremely thick ice layers (from hundreds to thousands of meters). More than 380 subglacial lakes have been identified in continental Antarctica (Siegert *et al.* 2011). A warming climate will likely increase the liquid water availability and modify the ice flows and the hydrology of these under-ice aquatic ecosystems (Siegert *et al.* 2001; Siegert 2005). However, abrupt variations are not expected in continental Antarctic aquatic ecosystems, including both supra- and subglacial lakes, according to the current GCM scenarios.

20.2.2 Maritime Antarctic lakes

Most lakes in maritime Antarctica are relatively shallow (less than 10 m deep) and small (some tens of thousands of square meters in surface area) and, except for those littoral lakes exposed to marine animals, are typically oligo- or ultra-oligotrophic. The ice cover lasts between 260 and 320 days a year (Rochera *et al.* 2010) and is between 100 and 140 cm thick at its maximum (Figure 20.1). Most of the primary producers in the lakes at this latitude are benthic and can be aquatic mosses, microbial mats and epiphytic phytobenthos or a mixture of these organisms. This dominance of the benthic compartment is a paradigm for lakes in polar regions (for example, Moorhead, Schmeling, and Hawes 2005; Vincent and Laybourn-Parry 2008) due to the low concentration and fluxes of nutrients in the water column, but its proximity to the primary source of autochthonous nutrients: the sediments (Andersen *et al.* 2011). Phytoplankters are typically scarce and conformed by picocyanobacteria, chrysophytes, and other phytoflagellates (Toro *et al.* 2007). Antarctic lakes are fishless, so the higher trophic level is made up of crustaceans. The copepod *Boeckella poppei* is the most abundant zooplankton, although the anostracean *Branchinecta gainii* may also play an important role in maritime Antarctic lakes (Pociecha 2007). It has been suggested that the ciliate *Balanion planctonicum* is a key player in the matter and energy transfers in Antarctic lakes (Camacho 2006). Limnopolar Lake (Byers Peninsula, Livingston

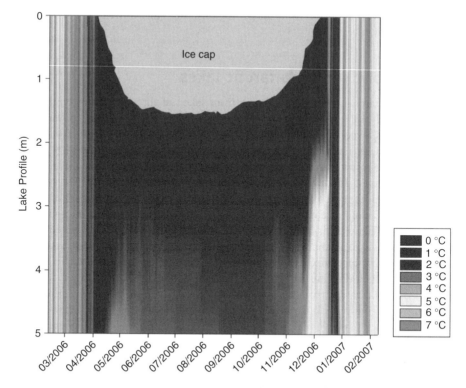

Figure 20.1   Thermal structure of Limnopolar Lake, Byers Peninsula, South Shetland Islands, from February 2006 to February 2007. Data were gathered at <1 m space resolution and at 30 min time resolution by means of a number of Hobo temperature loggers (Onset) hanging from a buoy. (See insert for color representation.)

Island, South Shetland Islands) has been the subject of several studies and will be used in this chapter as an example of the processes taking place in maritime Antarctic lakes (Toro *et al.* 2007; Villaescusa *et al.* 2010) and the potential effects of climate change.

20.3   Changes in physical parameters

20.3.1   Ice phenology

In polar and alpine regions, as well as in cold temperate areas, the ice cover drives a very important part of lake dynamics (see Chapters 2, 3 and 12). During the period when the lake is ice covered, the system can be considered almost a closed system, with very little influence from the watershed in terms of matter and energy. The low temperatures in polar regions ensure that the ice cover largely coexists with a frozen watershed with no water flow. Lake ice in maritime Antarctica is very often covered by a thick layer of fresh snow (up to a meter), physically isolating the lake, so that even strong winds do not affect the water column and very little sun energy can penetrate. However, in lakes deeper than 1.5 m, a water layer remains liquid year round. Under these conditions, the lake becomes heterotrophic and some mixotrophic organisms can survive and remain active during this long period (Unrein *et al.* 2005). In spring, when the sun starts melting snow, some irradiance can reach the water column, depending on the ice and snow characteristics (thickness, humidity, transparency). These first photons can act as triggers, preparing populations with phototrophic-driven metabolisms for a mixing situation. Typically in these regions, before the ice completely melts, a phytoplankton bloom takes place under the ice (Butler *et al.* 2000; Lopez-Bueno *et al.* 2009) and this primary production fuels the consumers (Winder and Schindler 2004).

Lakes in maritime Antarctica are responsive to small changes in air temperature because the air temperature in the region is close to the freezing point of water, with average summer temperatures slightly above the freezing point (Rochera *et al.* 2010). As a result, the freezing and melting season can be shifted considerably in time with a small air temperature variation (Rochera *et al.* 2010), producing remarkable changes in the ice cover. This can be important in polar regions where the ice-free period is short and a few days may be crucial for the survival of limnetic organisms. For instance, in Byers Peninsula, a particularly cold summer led to a 47% reduction in the duration of the ice-free period in Limnopolar Lake (Rochera *et al.* 2010), illustrating the extreme sensitivity of these lakes to subtle changes in air temperature. Other variables, such as waterbody size and other characteristics in the watershed, can account for a significant fraction of the ice thickness and duration (30–40%) (Palecki and Barry 1986; Livingstone 1997). Under a scenario of climate change, we hypothesize that ice phenology, and thus lake dynamics, will change, amplifying the warm temperature effects experienced throughout the watershed.

Lake ice phenology has crucial effects on lake ecology due to its influence on water temperature, water column stability, nutrient availability, gas exchange and light availability (MacKay *et al.* 2009). The physical structure of ice (thickness and other characteristics) is of great relevance in explaining the light regime and the availability of light for organisms living in the ice-covered water column (Belzile *et al.* 2001). In Limnopolar Lake, as well as in some other maritime Antarctic lakes, snow covers the ice for a significant period of time, almost completely extinguishing the incident irradiance (López-Bueno *et al.* 2009) (Figure 20.2). Light availability exerts a strong control on the phytoplankton bloom that takes place under ice cover before it melts (Tanabe *et al.* 2008), and variations in ice characteristics may modify the bloom timing and intensity, in turn affecting the consumers coupled to this primary production (Winder and Schindler 2004). Water-column stability is also significant because the lakes under ice cover are quite stable and often exhibit inverse stratification causing the warmer bottom layer to be nutrient rich while the cold surface is extremely dilute. The mixing process is responsible for the availability of bottom nutrients in the complete water column; in this way, water stability variations will also affect shifts in nutrient availability in the water column. Additionally, changes in the dark period will lead to variation in the anaerobic/microaerobic mineralization of organic matter, which produces inorganic nutrients and further increases their availability in the water column.

Snow and lake ice are effective attenuators of UV radiation (Belzile *et al.* 2001) and variations in the length of the ice-free period also cause changes in biological UV exposure down the water column. Thus, earlier melt out would mean a larger exposure to UV radiation (UVA+UVB) for Antarctic organisms, not only because of the UV reflection and attenuation characteristics of the snow and ice but also because of the ozone hole, which is at its maximum at the end of the Antarctic winter period or during springtime (Blunden, Arndt, and Baringer 2011).

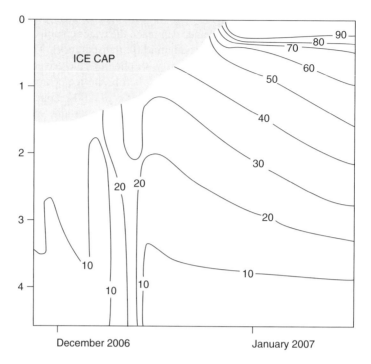

Figure 20.2 Light regimen during austral summer. Transmission percentage of photosynthetic active radiation (PAR) was determined using a LI-COR air and underwater PAR sensor every week from late November 2006 to late January 2007 with an average vertical resolution of 0.1 m at Limnopolar Lake (Byers Peninsula, South Shetland Islands).

In this way, slight variations in ice phenology can represent cascading effects on lake dynamics, from chemical, physical and biological perspectives. Climate change is impacting the ice phenology in the northern hemisphere, where the ice-cover period has been reduced by over 12 days (on average) during the last century (Magnuson *et al.* 2000). Interannual variability of ice-cover duration is also influenced by climate change, and in fact the variability has increased since 1950 in the northern hemisphere by 12% (Magnuson *et al.* 2000). In maritime Antarctica this decrease in the duration of the ice-cover period has been found to be more intense on Signy Island (Quayle *et al.* 2002), with a reduction of 63 days between 1980 and 1993, although interannual variability is also quite important. Forecasts for the Canadian Boreal Shield predict substantial increases in the length of the ice-free period, up to 29 days by 2080 (Keller 2007). However, these values could be higher in maritime Antarctica, where models forecast a greater temperature increase (Doran *et al.* 2002).

Global warming will increase water temperature because the ice-free period will be longer and thus water will absorb more solar radiation and also because of the higher radiative input from a higher air temperature. In any case, only a few of the lakes in maritime Antarctica will develop summer stratification due to the typical shallowness of these water bodies and the strong winds that are constantly present in the region. Therefore, climate change will represent a whole waterbody warming, accelerating biological processes both in the planktonic and benthic compartments.

20.4 Watershed changes

Because lakes are the downstream integrators of their catchments, they are responsive to variations in landscape properties triggered by climate change, as has been demonstrated in the Arctic (Serreze *et al.* 2000). In the northern hemisphere, climate change is promoting important changes in the landscape due to changes in the permafrost distribution and thickness of the active layer (Schuur *et al.* 2009). In maritime Antarctica, permafrost distribution and characteristics have been recently investigated, although most available data refer to the northern hemisphere (de Pablo *et al.* 2013). The data show that maritime Antarctic permafrost might be discontinuous, with an active layer that is quite deep, probably up to 1 m. Additionally, important hydrological features have been described, such as under-snow relevant water flow during the thaw that would impact landscape features (de Pablo *et al.* 2013) establishing a direct connection between climate change and variations in lake hydrology (Quesada *et al.* 2006 and references therein), including water balance and solute inputs as well as particle movements (for details see Quesada *et al.* 2006 and Rouse *et al.* 1997). Other aspects of geomorphology, such as massive permafrost melting, ground slumping and thermokarst formation, will also influence catchment hydrology and thus lake dynamics.

20.5 Changes in chemical parameters

Antarctic lakes, which are subject to perennial or short ice-free periods, low precipitation and inflows from their catchment areas, may prompt anoxic bottom water conditions and nutrient release from the sediments. These conditions may result in increased concentrations of available phosphorus with a consequent rise in fertility and/or trophic status. When precipitation is higher, as in maritime Antarctica, the runoff mobilizes more limiting elements, which in turn become more available. In a climate-change scenario, with an ice-free catchment for a longer period, runoff will have higher mobilizing intensity, resulting in greater nutrient availability for the primary producers within the lake. In some maritime Antarctic lakes, such as Limnopolar Lake, bacterial production might be considered the basis of the planktonic food web. In these lakes, photoautotrophic production is mainly benthic, with planktonic production being quite marginal. Higher nutrient availability in the water column, derived from a longer runoff, will most likely represent a variation in the balance between photoautotrophic and heterotrophic metabolism, although temperature could also exert an important influence under this scenario (Robarts, Sephton, and Wicks 1991). In a maritime Antarctic lake, Quayle *et al.* (2002) showed that from 1980 to 1995 there was an important enrichment of dissolved reactive phosphorus (DRP), ammonium and alkalinity (5.0, 2.5 and 2.2 fold, respectively), which is attributable to the climate change that is taking place in this region. The nitrogen cycle will be affected in a more complex way because, aside from the atmospheric N deposition, in these regions N_2 fixation by microbial mats could represent an important contribution (Fernández-Valiente *et al.* 2001; Fernández-Valiente *et al.* 2007). However, microbial mats are distributed in semi-aquatic environments (Velázquez *et al.* 2013) and subject to hydrological variations.

Dissolved organic carbon (DOC) concentration is considered a key variable affecting lake function because of its optical characteristics, absorbing PAR and UV, and as well as its role as the fuel for microbial food webs. Dissolved organic carbon in lakes comprises a combination of C from autochthonous and allochthonous origins (predominantly allochthonous origins) originating from watershed entities (Roberts *et al.* 2009). However, in Antarctic lakes, the catchments are almost barren soils with no or very slightly developed vegetation depending on the studied site, and thus the allochthonous C contribution can be quite low. In maritime Antarctic oligotrophic lakes, the catchments may have more vegetation, with mosses, lichens, microbial mats and, in some cases, even small numbers of vascular plants, and thus, the allochthonous C contribution could be even higher than the autochthonous and be heavily impacted by climate change. The factors affecting DOC input into lakes and the effects of climate change on DOC concentrations are also controversial (see, e.g., Schindler and Curtis 1997; Keller 2007), but, given the crucial role of DOC in lakes, its dynamics should be investigated in detail in Antarctic lakes (Quayle and Convey 2006).

Other elements that are not considered nutrients, such as monovalent or divalent major ions, are expected to increase in the lakes as a result of the longer ice-free periods, which would lead to longer runoff periods and also higher evaporation rates (Schindler and Curtis 1997), and thus an increase in conductivity is expected. However, the increase in ion concentrations during ice formation due to the salt-exclusion process may be reduced because of ice thinning by global warming, counteracting the potential ion increase. In any case, the typical low conductivity of maritime Antarctic lakes leads to the idea that the balance between both processes could be of minor relevance, with minor or negligible biological effects. Therefore, the effects of climate change on the chemistry of Antarctic lakes will depend on the hydrological changes, and, aside from runoff, the retention time will control lake chemistry.

20.6   Changes in the biota

Water availability is relevant because it will modify the hydraulic residence time of lakes. In the case of an increase in water flow, additional water will dilute nutrient loads as well as the organisms. In the case of a reduction in water availability, retention time will increase.

Higher primary production will increase the water column light absorption characteristics because of an increase in phytoplankton density, thus reducing the amount of light reaching a lake bottom. As a result, there will be a critical depth below which light may be limiting for photosynthesis; although, it may not be critical for most lakes, which are typically <10 m deep.

The forecasted scenario of a longer ice-free period with more light and nutrient availability and a higher temperature may induce phytoplankton community structure succession (Hawes 1985). A potential decrease in the dominance of Chrysophytes and small phytoflagellates and an increase in larger phytoplankters, such as green algae, which are more nutrient demanding, faster growing and less edible for zooplankton, may occur. Early opening of lakes allows water and sediments to absorb more solar energy, creating a forcing effect. During winter heat is transferred from sediments to the water column. Heat loss is reduced by ice cover insulation, and water warming is amplified compared with air temperature increases. Climate warming may change the pace of the biological cycles of egg-laying animals and those with larval stages (Schindler and Scheuerell 2002). Moreover, in ice-dominated ecosystems, some important processes, such as egg hatching or maturation, might be associated with annual cycles. It has been suggested that an earlier ice-off period may uncouple the biological cycles of phytoplankters and the zooplankters consuming them (Winder and Schindler 2004). Keller (2007) suggested that warmer temperatures may change interactions between aquatic biota because of the speed of changes in some crucial processes, such as reproduction or egg hatching. These interaction effects can interrupt energy and matter flow between different trophic levels, producing critical situations in lake dynamics.

At the beginning of the summer season, in maritime Antarctica such as the South Shetland areas, when the ice-cap starts melting, light reaches the water column (Figure 20.3) and the phytoplankton community may bloom (Butler *et al.* 2000; Lopez-Bueno *et al.* 2009), with maximum chlorophyll *a* concentrations typically below $1 \mu g L^{-1}$. This fraction is dominated by potential mixotrophic organisms: Chrysophytes (*Ochromonas* sp.), Cryptophytes (*Plagioselmis nanoplanctica*) and Prymnesiophytes (*Chrysochromulina parva*), although non-mixotrophic organisms, such as the Chlorophyte *Ankyra ancora*, may also be present (Unrein *et al.* 2005; Rochera *et al.* 2010). Later in the summer season, chlorophyll *a* values decrease and remain below $0.1 \mu g L^{-1}$.

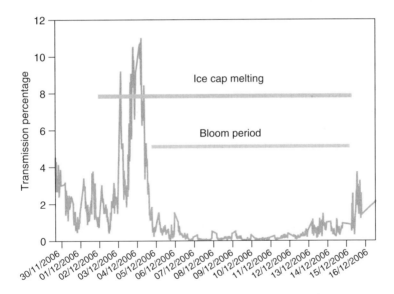

Figure 20.3 Light transmission in Lake Limnopolar at the end of spring, before the ice thaw. At the beginning of the sampling season, the ice was approximately 100 cm thick but was quite transparent with a few centimeters of snow cover. During the first week of December, an important snow melt increased the % of transmission reaching the bottom of the lake. High temperatures and rain melted the ice surface, making white ice with relatively high absorptive characteristics, until the ice thickness was reduced at the end of the second week of December and melted completely a few days later. The data were collected by a 2 Π LI-COR underwater PAR sensor (UW190-S) integrating the irradiance received every 30 min and placed on the lake bottom (4 m) connected to a LI-COR 1000 data logger. The % transmission was calculated using data from a sky PAR sensor placed onshore 2 m above the ground and integrating PAR in the same periods.

These variations may have relevant consequences for the biological dynamics of a lake. From the viral community perspective, which follows bloom development, a clear ecological succession from a small-sized bacterial infecting community (ssDNA-viruses) to a large-sized algal infecting community (dsDNA-viruses, mainly Phycodnavirus) (Lopez-Bueno *et al.* 2009) occurs. The latter can potentially infect the plankton bloom as well as the benthic diatoms coming from the deepest layers. Recent work on viruses in polar lakes (Lauro *et al.* 2011) demonstrated a high abundance of viruses and high rates of infection, implying that they may play an important role in genetic exchange in these extreme environments and in limiting bloom development.

The longer ice-free season promoted by global change will make the growth season longer and also allow more irradiance (both PAR and UV) to enter the water column, affecting phytoplankters and the very abundant phytobenthic communities. It is expected that both terrestrial and limnetic primary producers will be active for a longer period after the shifts in the ice phenology, although perhaps at lower rates because of the enhanced UVB impact during the ozone hole period. Higher temperatures may also accelerate bacterial remineralization of the primary producers' biomass, which, together with higher permafrost melting rates, can drive more nutrients to become available for primary producers in the system. An enhanced primary production will represent a higher production of organic matter, which will be partially converted into dissolved organic matter (DOC), which in some maritime Antarctic lakes is apparently the base for higher trophic levels, from bacteria to ciliates and crustaceans (Camacho 2006). Under this scenario, it is expected that the trophic level of maritime Antarctic lakes will increase to a certain extent, and, in fact, Quayle *et al.* (2002) already found a considerable increase (over twofold in 15 years) in the chlorophyll *a* concentration in a lake in this region, which they attribute to climate change. In parallel, an increase in primary production will take place, contributing more to the increased input of nutrients into the lake and making it difficult to predict shifts based on current knowledge. Therefore, some biological responses and shifts can be identified from the climate change prognoses for the next century (IPCC 2007).

20.7   Proposed model for the response of Antarctic maritime lakes to global climate change

Global climate change in the maritime Antarctic region is clearly affecting air temperature, which has noticeably increased over the last 50 years, affecting ice phenology. Most models also predict changes in precipitation, both in intensity and in seasonality; however, these models are based on precipitation records that unfortunately are very fragmentary in the area and, apparently, very geographically limited in this climatically complex region (Bañón *et al.* 2013). Thus, the proposed model will be based only on the well-documented temperature increase, while the responses based on precipitation changes will remain uncertain.

Our model is based on the conceptual framework adapted from Rouse *et al.* (1997), in which the primary effects are based on a temperature increase that will reduce ice cover duration and thus allow more runoff because the catchment will be ice free for longer periods (Figures 20.5d, f). The direct effects of these changes will impart a decrease in the duration of the inverse stratification period (Figures 20.4 and 20.5.b), and subsequent earlier mixing. This earlier mixing, together with higher weathering and biological decomposition, will lead to higher nutrient inputs. The biological consequences of these variations will be based on a longer growing season with more nutrients and more light available for longer periods (in shallow systems)

RESPONSE OF MARITIME ANTARCTIC LAKES

PRIMARY EFFECTS:

    (1) Temperature - increase

    (2) Run-off - increase

SECONDARY EFFECTS (PHYSICAL-CHEMICAL):

    (1) Ice-free season - Increase

        Decrease stratification in lakes

        Inverse stratification

        Water column mixing

    (2) Decomposition & weathering - Increase

        Nutrient supply - Increase

    (3) Water residence times – Decrease

        Dilution of nutrients – Decrease

HIGHER-ORDER EFFECTS (BIOTA):

    (1) Primary production   **-Autotrophic growth**

                                        -Imported POM increase

        Longer growing season

        Enhanced nutrient supply

        Underwater light- Increase

        Terrestrial primary production – Increased

    (2) Secondary production

        Temperature favors increased production

        Mismatched events

        Organisms' dispersal abilities – Variable (possible invasions)

Figure 20.4 Potential response of maritime Antarctic lakes to temperature and run-off increases.

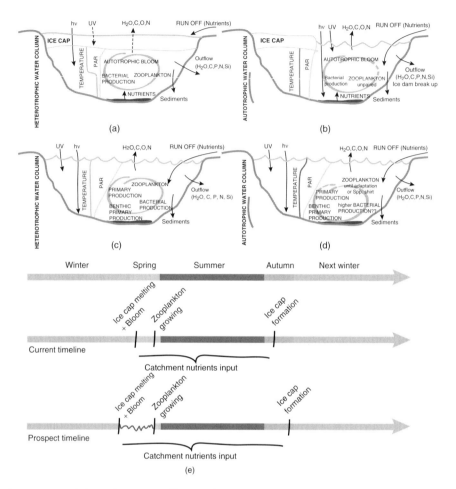

Figure 20.5 Schematic process diagram for a proposed maritime Antarctic lake's model. Mass flows: Grey arrows indicate terrestrial flux exchange of water and nutrients, the green loop represents the relationship between functional groups, the black arrows represent the flux exchange of chemical constituents of the lake with the atmosphere and benthos. (a) Descriptions of current relationships during early spring. (b) Scenario proposed for early spring under a global change scenario. (c) Current growing season scenario. (d) Scenario proposed for a growing season using current global change prognoses. (e) Timeline of seasonal shifts, where the main points of lake dynamics are addressed and the proposal for maritime Antarctic lakes in Antarctica. The time lapse that will cause a mismatch between a primary producer's bloom and zooplankton growth is marked here. (See insert for color representation.)

but also a higher contribution of terrestrial production. Secondary production will be strongly affected by these changes (Figure 20.5).

Spring time is a crucial period for maritime Antarctic lakes, and it will be one of the most affected periods. The ice cover will be thinner and will disappear earlier, increasing PAR availability and harmful UV effects (Figure 20.5b) and causing the water column to change from heterotrophic to autotrophic. The trophic web might become uncoupled because of the different timing between crustaceans' life cycles and under ice blooms of primary producers. This scenario will lead to changes in the summer climax, with higher nutrient input, higher primary production and perhaps a reduced grazing period until the existing zooplankters adapt or species change. In the summers, heterotrophically dominated lakes will shift to autotrophically dominated systems (Figure 20.5e).

20.8 Conclusions

Lakes are sentinels of watershed changes and the maritime Antarctic lakes in particular are highly responsive to environmental changes, especially temperature. The average summer temperatures in the region are near the freezing point, making the ice condition extremely sensitive to slight changes in air temperatures. Moreover, the relatively simple trophic food webs cause slight changes in the activity or timing of one of the functional groups to affect a system's dynamics as a whole. Aside from changes in precipitation, global warming will likely increase the trophic level of maritime Antarctic lakes and will lead to species changes, both in primary producers and consumers. Polar lakes, particularly maritime Antarctic lakes, provide exceptional scenarios for forecasting changes not only in the region but also at temperate latitudes, yielding data that facilitate the generation of models for more complex ecosystems. In lakes in which ice cover is persistent enough, future prognoses should be made carefully, as the hysteresis shown by those systems may result in different scenarios in each water catchment without any regional pattern. Species invasions from temperate latitudes could lead the system to eventually have greater species biodiversity that will mask further species losses.

As Wall *et al.* (2011) suggested, "The scientific community needs to move to Antarctic observational networks that address how Antarctic ecosystems are responding to a changing world." This is an invaluable opportunity to acquire the knowledge needed to generate adequate policies that may be adapted to global change scenarios at moderate latitudes.

20.9 Acknowledgements

The authors want to express their thanks to all the members of LIMNOPOLAR Project, with special thanks to Carlos Rochera and Antonio Camacho for their valuable work on the studied lakes. UTM and the Spanish Navy BIO Las Palmas provided the logistics for the field work, and without them all the work would have been impossible. The Spanish Ministerio de Ciencia e Innovación funded this work with the grants REN2000-0435, GCL 2005-06549-CO2-01 and POL2006-06635.

References

Andersen, D.T., Sumner, D.Y., Hawes, I., *et al.* (2011) Discovery of large conical stromatolites in Lake Untersee, Antarctica. *Geobiology*, **9**, 280–93, doi:10.1111/j.1472-4669.2011.00279.x

Bañón, M., Justel, A., Velázquez, D., and Quesada, A. (2013) Regional Weather Survey as Tool for Landscape Studies of Maritime Antarctica. *Antarctic Science*. In press.

Belzile, C., Vincent, W.F., Gibson, J.A.E., and Van Hove, P. (2001) Bio-optical characteristics of the snow, ice and water column of a perennially ice-cover lake in the High Arctic. *Can. J. Fish. Aquat. Sci.*, **58**, 2405–18.

Blunden, J., Arndt, S., and Baringer, M.O, (2011) State of the climate in 2010. *Bull. Amer. Meteorol. So.*, **92**, S1–S266.

Butler, H.G., Edworthy, M.G., and Ellis-Evans, J.C. (2000) Temporal plankton dynamics in an oligotrophic maritime Antarctic lake. *Freshwat. Biol.*, **43**, 215–30, doi: 10.1046/j.1365-2427.2000.00542.x.

Camacho, A. (2006) Planktonic microbial assemblages and the potential effects of metazoo-plankton predation on the food web of lakes from the maritime Antarctica and Sub-Antarctic Islands. *Rev. Environ. Sci. Biotech.*, **5**, 167–85.

Chown, S.L. and Convey, P. (2007) Spatial and temporal variability across life's hierarchies in the terrestrial Antarctic. *Phil. Trans. R. Soc. B*, **362** (1488), 2307–31.

De Pablo, M.A., Blanco, J.J., Molina, A., *et al.* (2013) Active layer interannual variability at the Limnopolar lake CALM site on Byers Peninsula, Livingston Island, Antarctica. Antarctic Science. In press.

Doran, P.T., Priscu, J.C., Lyons, W.B., *et al.* (2002) Antarctic climate cooling and terrestrial ecosystem response. *Nature*, **415**, 517–20.

Ellis-Evans, J.C. (1996) Microbial diversity and function in Antarctic freshwater ecosystems. *Biodiversity and Conservation*, **5**, 1395–431.

Fernández-Valiente, E., Camacho, A., Rochera, C., *et al.* (2007) Community structure and phys-iological characterization of microbial mats in Byers Peninsula, Livingston Island (South Shetland Islands, Antarctica). *FEMS Microbiol. Ecol.*, **59**, 377–85.

Fernández-Valiente, E., Quesada, A., Howard-Williams, C., and Hawes, I. (2001) N_2-fixation in cyanobacterial mats from ponds on the McMurdo Ice Shelf, Antarctica. *Microb. Ecol.*, **42**, 338–49.

Gibson, J.A.E. (1999) The meromictic lakes and stratified marine basins of the Vestfold Hills, East Antarctica. *Antarctic Sci.*, **11**, 175–92.

Hawes, I. (1985) Light climate and phytoplankton photosynthesis in maritime Antarctic lakes. *Hydrobiologia*, **123**, 69–79. doi: 10.1007/bf00006616.

IPCC (2007) *Intergovernmental Panel on Climate Change. The Physical Science Basis*, Cam-bridge University Press: Cambridge.

Keller, W. (2007) Implications of climate warming for boreal shield lakes: A review and synthesis. *Environ. Rev.*, **15**, 99–112, doi: 10.1139/a07-002.

Kosten, S., Huszar, V.L.M., Bécares, E., *et al.* (2012) Warmer climates boost cyanobacte-rial dominance in shallow lakes. *Glob. Change Biol.*, **18**, 118–26. doi: 10.1111/j.1365-2486.2011.02488.x.

Lauro, F.M., DeMaere, M.Z., Yau, S., *et al.* (2011) An integrative study of a meromictic lake ecosystem in Antarctica. *ISME Journal*, **5**, 879–95. doi: 10.1038/ismej.2010.185.

Laybourn-Parry, J. and Pearce, D.A. (2007) The biodiversity and ecology of Antarctic lakes: models for evolution. *Philos. Trans. Royal Soc. B.: Biological Sci.*, **362**: 2273–89.

Livingstone, D.M. (1997) Break-up dates of Alpine lakes as proxy data for local and regional mean surface air temperatures. *Climate Change*, **37**, 407–39.

Lopez-Bueno, A., Tamames, J., Velázquez, D., *et al.* (2009) High diversity of the viral com-munity from an Antarctic lake. *Science*, **326**, 858–61, doi: 10.1126/science.1179287.

MacKay, M.D., Neale, P.J., Arp, C.D., *et al.* (2009) Modeling lakes and reservoirs in the climate system. *Limnol. Oceanogr.*, **54**, 2315–29.

Magnuson, J.J., Robertson, D.M., Benson, B.J., *et al.* (2000) Historical trends in lake and river ice cover in the northern hemisphere. *Science*, **289** (5485), 1743–6.

Moorhead, D., Schmeling, J., and Hawes, I. (2005) Modelling the contribution of benthic microbial mats to net primary production in Lake Hoare, *McMurdo Dry Valleys. Antarctic Sci.*, **17**, 33–45. doi: 10.1017/s0954102005002403.

Palecki, M.A. and Barry, R.G. (1986) Freeze-up and break-up of lakes as an index of temperature changes during the transition seasons. A case study for Finland. *J. Clim. Appl. Meteorol.*, **25**, 893–902.

Pociecha, A. (2007) Effect of temperature on the respiration of an Antarctic freshwater anostracan, *Branchinecta gainii* Daday 1910, in field experiments. *Polar Biol.*, **30**, 731–4, doi: 10.1007/s00300-006-0230-6.

Priscu, J.C., Fritsen, C.H., Adams, E.E., *et al.* (1998) Perennial Antarctic lake ice: An oasis for life in a polar desert. *Science*, **280**, 2095–8.

Quayle, W.C., and Convey, P. (2006) Concentration, molecular weight distribution and neutral sugar composition of doc in maritime Antarctic lakes of differing trophic status. *Aquat. Geochem.*, **12**, 161–78, doi: 10.1007/s10498-005-3882-x.

Quayle, W.C., Convey, P., Peck, L.S., *et al.* (2003) Ecological responses of maritime Antarctic lakes to regional climate change, in *Antarctic Peninsula Climate Variability: Historical and Paleoenvironmental Perspectives* (eds. E. Domack, A. Leventer, A. Burnet, R. *et al.*), American Geophysical Union, Washington, D.C., pp. 159–70.

Quayle, W.C., Peck, L.S., Peat, H., *et al.* (2002) Extreme responses to climate change in Antarctic lakes. *Science*, **295**, 645.

Quesada, A., Vincent, W.F., Kaup, E., *et al.* (2006) *Landscape control of high latitude lakes in a changing climate*, in *Trends in Antarctic Terrestrial and Limnetic Ecosystems*, (eds. D. M. Bergstrom, P. Convey, A. Huiskes), Springer, Dordrecht, pp. 221–52.

Robarts, R.D., Sephton, L.M., and Wicks, R.J. (1991) Labile dissolved organic-carbon and water temperature as regulators of heterotrophic bacterial activity and production in the lakes of sub-Antarctic Marion Island. *Polar Biol.*, **11**, 403–13.

Roberts, P., Newsham, K.K., Bardgett, R.D., *et al.* (2009) Vegetation cover regulates the quantity, quality and temporal dynamics of dissolved organic carbon and nitrogen in Antarctic soils. *Polar Biol.*, **32**, 999–1008, doi: 10.1007/s00300-009-0599-0.

Rochera, C., Justel, A., Fernández-Valiente, E., *et al.* (2010) Interannual meteorological variability and its effects on a lake from maritime Antarctica. *Polar Biol.*, **33**, 1615–28. doi: 10.1007/s00300-010-0879-8.

Rogers, A.D. (2007) Evolution and biodiversity of Antarctic organisms: A molecular perspective. *Phil. Trans. Roy. Soc. B-Biol. Sci.*, **362**, 2191–214. doi: 10.1098/rstb.2006.1948.

Rouse, W.R., Douglas, M.S.V., Hecky, R.E., *et al.* (1997) Effects of climate change on the freshwaters of Arctic and subArctic North America. *Hydrol. Proc.*, **11**, 873–902. doi: 10.1002/(sici)1099-1085(19970630)11:8<873::aid-hyp510>3.0.co;2-6.

Schindler, D.E. and Scheuerell, M.D. (2002) Habitat coupling in lake ecosystems. *Oikos*, **98**, 177–89.

Schindler, D.W. and Curtis, P.J. (1997) The role of DOC in protecting freshwaters subjected to climatic warming and acidification from UV exposure. *Biogeochemistry*, **36**, 1–8. doi: 10.1023/a:1005768527751.

Schuur, E.A.G., Vogel, J.G., Crummer, K.G., *et al.* (2009) The effect of permafrost thaw on old carbon release and net carbon exchange from tundra. *Nature*, **459**, 556–9. doi: 10.1038/nature08031.

Serreze, M.C., Walsh, J.E., Chapin, F.S., *et al.* (2000) Observational evidence of recent change in the northern high-latitude environment. *Climate Change*, **46**, 159–207, doi: 10.1023/a:1005504031923.

Siegert, M.J. (2005) Lakes beneath the ice sheet: The occurrence, analysis, and future exploration of Lake Vostok and other Antarctic subglacial lakes. *Ann. Rev. Earth and Planetary Sci.*, **33**, 215–45. doi: 10.1146/annurev.earth.33.092203.122725.

Siegert, M.J., Ellis-Evans, J.C., Tranter, M., *et al.* (2001) Physical, chemical and biological processes in Lake Vostok and other Antarctic subglacial lakes. *Nature*, **414**, 603-609. doi: 10.1038/414603a.

Siegert, M.J., Kennicutt II,, M.C., and Bindschadler, R.A. (eds.) (2011) *Antarctic Subglacial Aquatic Environments*, American Geophysical Union, Washington DC.

Simmons, G.M., Vestal, J.R., and Wharton, R.A. (1993) Environmental regulators of microbial activity in continental Antarctic lakes, in *Physical and Biogeochemical Processes in Antarctic Lakes* (eds. Green, W.J. and Friedmann), American Geophysical Union, Washington, D.C.

Steig, E.J., Schneider, D.P., Rutherford, S.D., *et al.* (2009) Warming of the Antarctic Ice-sheet surface since the 1957 International Geophysical Year. *Nature*, **457**, 459–62.

Tanabe, Y., Kudoh, S., Imura, S., and Fukuchi, M. (2008) Phytoplankton blooms under dim and cold conditions in freshwater lakes of East Antarctica. *Polar Biol.*, **31**, 199–208.

Toro, M., Camacho, A., Rochera, C., *et al.* (2007) Limnological characteristics of the freshwater ecosystems of Byers peninsula, Livingston Island, in maritime Antarctica. *Polar Biol.*, **30**, 635–49, doi: 10.1007/s00300-006-0223-5.

Unrein, F., Izaguirre, I., Massana, R., *et al.* (2005) Nanoplankton assemblages in maritime Antarctic lakes: Characterization and molecular fingerprinting comparison. *Aquat. Microb. Ecol.*, **40**, 269–82. doi: 10.3354/ame040269.

Velázquez, D., Lezcano, M.A., Frias, A., and Quesada, A. (2013) Ecological relationships and stoichiometry within a maritime Antarctic watershed. Antarctic Science. In press.

Villaescusa, J.A., Casamayor, E.O., Rochera, C., *et al.* (2010) A close link between bacterial community composition and environmental heterogeneity in maritime Antarctic lakes. *Internat. Microbiol.*, **13**, 67–77. doi: 10.2436/20.1501.01.112.

Vincent, W.F. (2009) Effects of climate change on lakes, *in Encyclopedia of Inland Waters* (ed. G. E. Likens), Elsevier, Oxford UK, vol. **3.**, pp. 55–60.

Vincent, W.F., and Laybourn-Parry, J. (eds.) (2008) *Polar Lakes and Rivers: Limnology of Arctic and Antarctic Aquatic Ecosystems*, Oxford University Press, Oxford.

Wall, D.H., Lyons, W.B., Chown, S.L., *et al.* (2011) Long-term ecosystem networks to record change: An international imperative. *Antarctic Sci.*. **23**, 209, doi: 10.1017/s0954102011000319.

Winder, M. and Schindler, D.E. (2004) Climatic effects on the phenology of lake processes. *Glob. Change Biol.*, **10**, 1844–56, doi: 10.1111/j.1365-2486.2004.00849.x.

Part II

Impacts on Societies

21

Adaptation to a Changing Climate in Northern Mongolia

Clyde E. Goulden and Munhtuya N. Goulden

Institute for Mongolian Biodiversity and Ecological Studies, Academy of Natural Sciences, Philadelphia, PA, USA

The annual mean temperature of Mongolia has increased by 2.14 °C since the 1930s and 1940s. This warming and associated climate changes are affecting ecosystems of Mongolia in ways that are apparent to the nomadic herders of Mongolia who depend on pastures and water resources for their livelihoods. We have interviewed more than 100 herder families to learn of their perception of climate change and its impacts on their environment, and have discussed means of adaptation to these recent changes.

21.1 Background

The present scientific consensus about the long-term impacts of "global warming" is that impacts will continue to increase and change our environment and ecosystems well into the middle of this century before mitigation can bring our climate back to a pattern similar to what we have experienced in the past. This is because of the long time lag in reducing atmospheric levels of green house gases following a reduction of emissions. Until then, warming will have major impacts on our environment. To avoid the most negative consequences of the impacts, which now appear inevitable, local peoples must learn different behavior patterns that will minimize the effects of climate change's negative impacts. To understand what these impacts are and how best to adapt to them we need to study actual changes in parts of the world where the climate has already begun to warm significantly and to evaluate their impact on the environment. Northern boreal regions of the world have already warmed significantly, more rapidly than temperate areas, producing thermal differences among regions that are already disrupting expected climate patterns and modifying our ecosystems. In the continental United States the average warming has been less than 1 °C over the last century, but in contrast, Alaska, parts of Canada and northern Asia (Siberia and Mongolia), the

Climatic Change and Global Warming of Inland Waters: Impacts and Mitigation for Ecosystems and Societies,
First Edition. Edited by Charles R. Goldman, Michio Kumagai and Richard D. Robarts.
© 2013 John Wiley & Sons, Ltd. Published 2013 by John Wiley & Sons, Ltd.

rate of the warming trend is greater. The Hydrometeorology Institute of Mongolia has estimated that temperature warming in Mongolia has already increased the mean annual temperature of the country by 2.14 °C over the last 65 years (Dagvadorj *et al*. 2009; see also Nandintsetseg, Greene, and Goulden 2007 for temperature changes in the Lake Hovsgol Region).

We must understand what these changes mean to our environment and the ecosystems we are dependent upon, and how to adapt to the changes and their negative consequences:

- What changes in climate patterns are occurring in these northern regions that have already experienced substantial recent climate warming?
- What are the environmental consequences or impacts of the changing climate to ecosystems?
- How can local peoples adapt to the changing climate patterns to minimize impacts on themselves and their environment?

To address these questions we have been interviewing nomadic herders from two different areas of northern Mongolia, where the taiga forest of Siberia and the vast central Asian steppe converge, to learn of herder perceptions of climate changes and how it affects their future livelihoods. Water resources are a critical part of the changes that are occurring in northern Mongolia and will be discussed here in the context of what changes in water resources herders have perceived and how the changes will affect local herders.

We interviewed 39 herder families living in tributary valleys of Lake Hovsgol, an ancient lake in northern Mongolia and a "sister lake" to Siberia's Lake Baikal. We interviewed 65 herder families in Hentii, northeast of Ulaanbaatar and approximately 800 km east of Hovsgol. The interviews consisted of a series of open-ended questions that allowed herders to freely respond. Answers were recorded, translated from Mongolian, and formatted for qualitative analysis with NVivo software (version 9). Long-term meteorological data sets, collected by Mongolia's Hydrometeorology Institute, were obtained for towns near the herder families and analyzed for trends in different climate parameters such as trends in number of summer days (number of days >25 °C), annual and daily precipitation trends (number of days with rains greater than 10 mm or 20 mm), number of days with winds > 4.5 m s^{-1} (wind speeds that can carry dust or sand grains and cause dust storms) and other parameters related to seasonal temperature extremes and precipitation.

21.1.1 The study area

Mongolia is located between China and Russia on a high plateau (the Mongolian Plateau) with a mean altitude of 1500 m. The country ranges from latitudes 41°35'N to 52°09'N and is largely surrounded by mountains. Mongolia has a continental climate with limited rainfall and with very cold dry winters and moderately warm summers when 90 to 95% of the precipitation occurs. Precipitation is highest in the northern part of the country, seldom-exceeding 500 to 600 mm per year, and least in the south where it is <25 mm per year. Vegetation zones in the moister north are a mix of taiga forest on north-facing slopes and steppe on south-facing slopes with permafrost on north facing slopes and under stream channels. South of this zone the vegetation is primarily steppe and this gradually gives way to semi-desert and in the southern part of Mongolia to the desert, the Gobi. The annual average air temperature is 0.7 °C compared with the average temperature of the earth of 14.4 °C. Air temperature can change as much as 30 °C in one day with the arrival of brisk northwesterly winds (Batjargal 2007). Temperature extremes range from −50 °C during the winter in the north, to +50 °C in the Gobi during the summer. The low rainfall makes it difficult to cultivate most grain crops; agricultural crops do not grow well in Mongolia without irrigation.

Ninety percent of the total annual precipitation falls between April and September, most during July and August. Although annual precipitation is low, its intensity can be high; sometimes 40–65 mm may fall in a one- or two-hour period. The amounts and timing of summer rains are quite variable (Goulden, Nandintsetseg, and Ariuntsetseg 2011). For example, at Hatgal in northern Mongolia, 2002 was a very dry year with total rainfall of 160 mm while 2003 was a wet year with total rainfall over 450 mm causing major flooding in headwater valleys. Moderate rains of ca. 250 mm with modest stream flow occurred in 2004 and 2005 but in 2006 a heavy one to two day rain in early July when the active layer above permafrost had just begun to thaw resulted in widespread flooding in the tributary valleys entering Lake Hovsgol. The annual rain total in 2006 was about 350 mm—not a significant deviation from average rainfall but the timing and intensity of the rain resulted in major valley flooding that altered many of the stream channels.

Drought occurs one year in five in the southern part of the country, and one year in 10 in the northern part, on average (Batjargal 1992, 2007). The worst recent drought occurred along with consecutive *Dzud* (see below) years from 1999 to 2002 and affected 50 to 70% of Mongolia (Batima *et al.* 2005). The drought had a large impact on reducing the growth of plant biomass in the steppe areas while the *Dzud* deprived livestock of winter forage and resulted in the death of more than 12 million domestic animals. Drought caused many herders who had lost their water resources in pasture areas to move to more permanent water sources, resulting in heavy overgrazing and pollution of nearby rivers and lakes. Some winters can be particularly difficult for animals because heavy snows or ice cover resulting from daytime melting snow and night-time freezing can restrict the animals' ability to get to forage. When these conditions persist for several days or weeks it is called a *Dzud*.

Most of the country is steppe and as a result livestock grazing is the main activity and source of support for countryside people. Nomadic herders tend mixed herds of sheep, goats, horses, cows, yaks or camels; the people of the Mongolian steppe have been dependent on nomadic livestock grazing for over 3000 years. Nomadism has been successful as long as herders make regular movements from grazed to fresh ungrazed pastures and can find sufficient water for their animals. Livestock grazing also occurs on south-facing mountain slopes in the northern part of the country. Historically, many of the movements appear to have been made on specific dates, seasonal in the north but more frequent in the south. In the southern semi-desert and some areas of the desert grazing herds are widespread because of limited water resources and poor grass growth, so that herds must move more frequently than on a seasonal basis.

Herders are very aware of climate variability and its impact on pasture and water resources; survival of their animals is very dependent on a herder's ability to anticipate changes. Herders often say "any change in the weather is bad," particularly if the change is unpredictable. They must make careful decisions about where to move animals in the winter and how much winter forage (*hadlan*) must be collected to feed animals during heavy snows. Prejvalskii, the great Russian explorer, described this ability; "Mongols have an extraordinary ability to forecast strong winds, storms, rains and they can find their lost horses and camels with a help of tiny signs and symptoms." (p. 32, Dagvadorj *et al*. 2009). How will climate warming affect their ability to make such predictions when future weather patterns and water resources are very likely to be altered?

Landscape and climate conditions play key roles in water resource availability, land use and sustainable development throughout Asia but particularly in Mongolia. Nomadic pastoralism is the most suitable form of existence in this semi-arid to arid landscape because over-grazing can quickly cause the loss of good plant food species and reduce plant cover (Humphrey and Sneath 1999). In the northern part of Mongolia where herders move their animals seasonally, they often move to the mountains during the colder months because thermal inversions make higher altitudes warmer than in the lowlands. Water for the animals is supplied by snow. They return to their lowland pastures near rivers, lakes and ponds in late spring. They do not graze their animals in the mountains during the summer because there is little water present and they can preserve the mountain grasses for winter pasture use. Individual species in the mixed herds of Mongolian nomadic herders in the north graze in different habitats; sheep and cashmere goats graze on hill slopes; cows, yaks, and horses graze near the streams. The herders live in portable felt covered dome tents called a "ger" (Russian "Yurt") that can be set up or dismantled in 1–2 hours, and placing all of their possessions on a two-wheeled cart or the back of yaks or camels, they move their animals to new pastures. They prefer to be near a small city where they can sell milk products, beef, or mutton, or near a road where a truck will stop to buy their cashmere and sheepskins. Above all, they must be near water.

They may not follow the animals on a daily basis for as they say, "the animals know where to go" and return in the evening for milking or for sheep, goats and young cows and yaks, the safety and warmth of the corral. As the animals go farther for better pasture, the herders follow, sometimes moving the ger locally, or will make longer movements when pastures become heavily grazed. They must make these longer movement's two to four or more times each year to avoid serious damage to their pastures. These longer movements coincide with seasonal weather changes so they have distinct grazing pastures and often will move as much as 20 km to autumn, winter, or spring pastures. As the summer rains end they move near to winter pastures to collect and store fodder (*hadlan*) to feed young animals. Hadlan is essential as a food supplement for their animals during harsh winters when snow and ice cover pasture grasses. Occasional severe cold winters (*Dzud*) with long periods of snow and ice-cover over pasture plants requires large amounts of *hadlan*; without it many animals will die. The winter of 2009–10 was a *Dzud* year throughout most of Mongolia and many animals (at least eight to ten million) died, already weakened by the cold weather and poor or inadequate grass.

Herders of the valleys of Lake Hovsgol have ample water because they are near the lake but the timing of rains and intensity can result in flooding of pastures or loss of hay areas near streams. They are more concerned about how many herders are moving into the area but climate is important to them as a chronic problem because it affects soil moisture, plant growth or flooding of pastures and how they prepare for winter. In Hentii, to the east, herders say the region has experienced a ten-year drought and view weather and water availability as an acute problem. In contrast, for people in the U.S., climate change is a more distant problem because we do not recognize the impacts of the modest warming ($+1\,^\circ$F). Mongolian nomadic herders are more dependent upon their environment and are well aware of the climate changes they are experiencing and the damage the changes are doing to their grazing pastures and water availability in the steppe and forest/steppe areas where they live.

Of the more than 100 nomadic herder families we have interviewed regarding their perceptions of local changes, when asked what is having the greatest impact on their lives, livestock overgrazing or weather, 90% of the families responded, the "weather" (Tsag-Agaar). They use the term "weather" to describe day-to-day changes in temperature, rain, snow, and wind but they are very well aware of long-term climate trends, the average of years of weather changes. They view change in climate often as occurring since they were children. This can be expressed as differences in the clothing they or their parents wore during the winter or summer when they were young, or dates used when they formerly moved to seasonal grazing pastures.

21.2 Impacts of climate change

Mongolian herders have seen major recent changes in the climate that impact pastures and water resources that are of great concern to them. Climate change in Mongolia is more than just warming; it is associated with a number of other changes that are affected by regional temperature differences. Many of the changes that concern the herders are consistent with the changes that are predicted by climate-change models, such as increased frequency of wind gusts, greater intensity of rains (Allan and Soden 2008), late spring snows, impacts on plant growth and increased variation in the availability of water resources.

Now they make seasonal movements based on weather changes independent of the calendar; for example, snow and ice melt in the early spring or during late summer when rains are less frequent and they must begin to collect *hadlan* or hay as a winter food supplement. Increasingly, the drying of springs, streams and ponds, which are the main water supply for the family and their livestock, requires herders to move herds closer to rivers and large lakes to insure that the animals have water. What is their perception of climate changes and what do they believe the future climate will be?

Mongolian nomads must be sensitive to the weather; they quickly learn where the rains have been sufficient to support good plant growth. They try to anticipate bad winters and gather sufficient *hadlan* to ensure the survival of their animals and prepare a corral (*buuts*) with a good ground layer of manure, and caulk the walls of a shelter with manure to protect the animals from winter storms and very cold nights. The collection of *hadlan* is dependent on the timing and amount of the summer rains that provide the soil moisture critical to the growth of the grasses and plants that make the animals "fat and strong."

Their awareness of the environment is evident in their detailed description of rains and impact on pastures; they have specific words for what they perceive to be different kinds of rain (*boroo*) and know how each kind of rain affects pasture soils and stream flows. *Namiraa boroo* is a light rain often lasting from a few minutes to a few hours but does little to moisten the soil. *Zuser boroo* is a light to moderate long rain lasting two to three days to a week, which is good in the summer for increasing soil moisture but is not good during the late summer when it is time to cut grass for *hadlan* because they do not want to cut and store wet grass. *Shivree boroo* is a light rain that lasts several days and can moisten deep soil if the rain is long enough. In contrast to these lighter rains, *aadar boroo* is a very intense heavy rain that lasts only five to 20 minutes. Often there are strong winds before an *aadar* rain begins; the very strong gusting winds can blow *gers* over and increase soil erosion. This type of rain does not moisten the soil because it is so intense that most of the rain water runs off into stream channels causing floods before very much of it can soak into the ground. *Aadar boroo* can be very cold, causing animals and humans to die from hypothermia if caught outside in an *aadar* rain without proper clothing, and Mongolians fear this.

21.3   Herder responses and meteorological data

A detailed analysis of herder responses and meteorological data is now being prepared in a separate manuscript and will be submitted for publication soon. We are presenting only a summary of the results here.

21.3.1   Seasonal changes

The herders we interviewed told us they no longer use calendar dates for when to move to summer pastures and instead use the timing of snow melt or the beginning of summer rains to move to summer pastures because the timing of the onset of summer is so variable. Winters have been milder with less snow but when it snows it can be a very heavy snow. "We no longer wear the heavy deels (the traditional Mongolian dress and outer garment extending from the neck to below the knees) and heavy winter boots or fur hats my parents wore, the winter is too mild for such heavy clothing now." But recently there have been colder winters (Dagvadorj *et al.* 2009). "Summer is shorter with a few extremely hot days that burn the pasture turning the grass brown and poor as a food for animals. These hot days are generally followed by days with strong gusty winds and cold rains. Spring is short and autumn starts early but snows can begin in September or October, earlier than November when snows formerly began."

21.3.2   Rains

Have the rains changed? "We used to have light, warm rains that lasted 3 to 4 days." "We called them *shivree boroo* or *namiraa boroo*. As kids, we loved to go out in the rain wearing little clothing to run and play. Now the rains are different, they are patchy and more like *aadar boroo*, beginning suddenly with strong gusty winds." "The rains are cold, heavy and intense. It is too cold to go out in the rain without a heavy deel or coat and the animals can get hypothermia (*Osgokh*) from being out in the rain." "Fortunately, the rains are short only lasting five to ten minutes, or we could have a natural disaster."

Few herders we queried believed an *aadar* rain adds very much moisture to the soil: "The soils are dry and hard. Very little rain water gets into the soil; most of it runs off the land causing erosion of soils." Or, "The rain runs away." We have not been able to identify changes in rains in analysis of the long-term climate data information. Most rainfall records are reported on a daily basis whereas the herders all say that *aadar* rains are short, lasting only 5 to 15 or 20 minutes. Therefore, short rains cannot be distinguished in the available daily rainfall data sets. This is not a refutation of the occurrence of *aadar* rains; it simply means that the data as reported cannot be used to determine the occurrence or absence of short intense rains.

An increase in intense, heavy rain is one of the important predictions of the model studies that predict climate change (Allan and Sodan 2008). If the rainwater is not increasing soil moisture by soaking into the ground, plant growth will be seriously affected in this semi-arid climate. Flooding of local streams is also a serious problem for herders because pastures and hay growing areas can be lost to flooding; frequent flooding can turn a good pasture into a wetland. In addition, most floods can be followed by very low water periods when some streams can be virtually dry due to poor groundwater recharge from rains. As a result, according to the herders, stream flows are changing.

Permafrost thaw has impacts on hill slopes and water availability; periglacial phenomena such as pingos, icings and solifluction, are widespread throughout northern Mongolia. Solifluction is particularly bad on hill slopes and could be dangerous for the animals as the surface soil layers slump downward. Another factor is that permafrost is not a flat table-like frozen layer of soils but is highly uneven with depressions and swales that collect large amounts of water in the depressions (Etzelmuller *et al.* 2006). This water in the active layer above permafrost is available to plant roots, and if near the soil surface can be an important source of water for growing plants, particularly in drought years. As permafrost thaws and the active layer descends, the pockets of water become less available to plants with short roots, such as the grasses. This could lead to an increase in the number of bushes and forbs that are undesirable as foods for grazing animals.

The ground water that remains in the soil as the active layer thaws provides a summer flow to streams but once permafrost thaws substantially, there will be less ground water to sustain stream flows.

21.3.3 Pastures

What is the biggest impact on your lives? Thirty percent of herders in the Hovsgol valleys believe that overgrazing is a major problem; whereas, 80% of herders at Hovsgol, and 90% in Hentii believe that weather and long-term climate change are responsible for poor plant growth and are the major threat to their livelihoods. The impacts on vegetation of a long-term warming trend are similar to those of overgrazing, i.e., encouraging the growth of low-nutrition weeds and bushes and poor plant growth so animals are unable to gain strength quickly and accumulate fat reserves for the cold winter. This can be a problem during cold winters when animals must draw on their reserves and herders depend upon the cutting and storage of late summer grasses as supplemental winter feed (*hadlan*). The amount of *hadlan* collected and stored for the winter is affected by poor grass growth and supplemental feeding of grazing animals is essential to animal survival. Hadlan is needed for winter-feed, particularly during cold, heavy snow *Dzud* winters when animals cannot feed on last summer's plants because they are covered by snow and ice.

21.3.4   Unpredictable climate and an unpredictable future

Most of the herders we interviewed think the climate will continue to worsen. In the past Mongolians have been known to be very effective in predicting future climate patterns, which are quite critical for deciding what pastures they should move to for winter grazing of their animals. Now, however, herders say that the climate has become more variable and unpredictable. They say that it is difficult to know what the weather will be tomorrow or next week. During the rainy season they claim that cold, intense rains occur suddenly, without warning, and they worry about being caught outside during a cold rain without proper clothing. In the past they could predict such changes, but now storms appear suddenly without warning and they are unable to anticipate such rapid changes.

21.4   Adaptations to climate change

During our interviews with the herders we asked them how they would adapt to the changes in their environment.

21.4.1   More frequent movements

Herders tend to be concentrated near towns and roads and more frequent movement may not be possible unless they move away from the towns and roads. Near towns, more frequent movement are possible only if there are fewer herders present in the areas that we visited because most of the available pastures are already occupied by herders who have several seasonal pastures that they can choose from depending on the quality of the plant growth. Moving animals before they have a serious impact on vegetation is extremely important to maintain plant cover in pastures and reduce erosion of top soils. Other adaptations to climate change were as follows.

21.4.2   Collect more *hadlan*

Collecting more *hadlan* is impossible for herders with herd sizes as much as 300 to 1000 animals. Two points were clear from the interviews: the number of livestock a herder lost during the recent *Dzud* was inversely correlated with the amount of hay or *hadlan* collected, and second, drought or the change in rains affected the growth of the grasses so that *hadlan* was more difficult to collect and herders had to find new areas to cut *hadlan*, requiring greater time and effort. Herders that have many sheep and goats have the greatest difficulty in collecting sufficient *hadlan* and appeared to lose the most animals during the *Dzud* of 2010. The amount of hay available is critical if herders want to use this as a means to protect their animals in the future. But virtually all herders told us they can only collect sufficient hay to feed young or sick animals during the winter. Reducing herd sizes will be necessary if the collection of winter forage becomes a primary alternative to feeding animals during winter. Large herds of 1000 to 2000 animals should be discouraged because of their impact on pastures and pollution of water bodies. This is essential for reducing impacts on streams and ponds and sustaining clean-water resources.

21.4.3 Select best animals

Herders suggested that they should cull their herds and eliminate weak and sick animals. This would result in a temporary reduction of herd sizes that could be beneficial. In Dadal, in northern Hentii region, herders have been experimenting with raising a hybrid Kazakh cow in smaller herd sizes of 30 to 40 animals because it produces more milk. Dadal herders have very few sheep and goats. They are able to collect sufficient *hadlan* for their few animals and, during the 2010 *Dzud*, lost very few animals. This is one viable alternative to the large herds of most nomad families but there is a question as to whether these animals can survive in the semi-arid steppe areas of Mongolia.

21.4.4 Eliminate cashmere goats

Goats are hard on steppe plants because they pull up the whole plant and eat it, leaving no roots or plant base for regrowth. As a result, they can be very destructive to a pasture. However, cashmere is the major cash crop for most herders so this will be a difficult decision for most herder families.

21.4.5 *Otor*

Herders without permanent or high quality pastures find that frequent movements are best for their animals. However, these herders must leave their families and gers behind and live in small tents remaining for only a few days at one site. The herders refer to this as *Otor* and because it is more common during the winter months, it can be very difficult given the extremely cold winters of Mongolia. *Otor* is very difficult on individuals and only younger herders without young children will go to *Otor*. It would not seem to be a universal adaptation for the problems caused by climate change.

21.4.6 Give up herding

As most of the herders we interviewed thought the weather will continue to worsen led us to ask younger herders, "Do you want your children to grow up to be herders?" All replied "No! I do not want my children to be herders, it is too difficult and the future is uncertain." We recently learned that many herders in Hentii are "hedging their bets" for the future by selling animals (rather than seeing them die during bad storms) and are using the money to buy apartments in town. If many herders decide to sell their animals and give up herding, it is difficult to predict what will happen in Mongolia in the future if the weather does continue to worsen.

21.4.7 How can the government help the herders?

The government needs to be involved in improved weather forecasting and greater use of radio and television weather reports to warn herders of incoming storms that could harm herd animals. The government should also be prepared to augment feed during periods of *Dzud* to reduce animal losses.

Tractors are very helpful in collecting *hadlan*, and donations of tractors should be encouraged from donor countries and agencies.

21.4.8 Water resources and impacts on streams

Herders cannot alter rainfall patterns but they can build small check dams in runoff depressions that will hold up runoff before it enters streams and causes flooding. Herders can also build small ponds to retain water for animals to drink. Larger reservoirs may minimize flooding and maintain low water flows between heavy rains.

21.5 Acknowledgement

This research was funded by NSF-PIRE award (OISE 0729786).

References

Allan, R.P. and Soden, B.J. (2008) Atmospheric warming and the amplification of precipitation extreme. *Science*, **321**, 1481–4.

Batima, P., Natsagdorj, L., Gombluudev, P., and Erdenetseg, B. (2005) *Observed Climate Change in Mongolia*, Assessments of Impacts and Adaptations of Climate Change Working Paper No.12, Columbia University, New York, NY.

Batjargal, Z. (1992) *Desertification in Mongolia*, RALA Report No. 200, RALA, Reykjavik.

Batjargal, Z. (2007) *Fragile Environment, Vulnerable People and Sensitive Society*, Kaihatsu-sha Col, Ltd., Tokyo.

Dagvadorj, D., Natsagdorj, L., Dorjpurev, J., and Namkhainyam, B. (2009) *Mongolia Assessment Report on Climate Change 2009*, Ministry of Nature, Environment and Tourism, Ulaanbaatar.

Etzelmuller, B., Flo Heggem, E.S., Sharkhuu, N., *et al.* (2006) Mountain permafrost distribution modeling using a multi-criteria approach in the Hovsgol Area, *Northern Mongolia. Permafrost and Periglac. Process.*, **17**, 91–104.

Goulden, C. E., Nandintsetseg, B., and Ariuntsetseg, L. (2011) The geology, climate and ecology of Mongolia, in *Mapping Mongolia: Situating Mongolia in the World from Geologic Time to the Present* (ed. P. Sabloff), University of Pennsylvania Press, Philadelphia PA, pp. 87–103.

Humphrey, C. and Sneath, D. (1999) *The End of Nomadism?* Duke University Press, Durham NC.

Nandintsetseg, B., Greene, J.S., and Goulden, C.E. (2007) Trends in extreme daily precipitation and temperature in the Lake Hövsgöl basin area, *Mongolia. Int. J. Climatology*, **27**, 341–7.

22

Managing the Effects of Climate Change on Urban Water Resources

Gabriela da Costa Silva
School of Environment and Sustainability, University of Saskatchewan, Canada

22.1 Introduction

Economic development and population growth have affected the resilience of social and environmental systems, especially in cities. Rapid urbanization and rural–urban migration has demanded and will continue to demand effective preparedness to face environmental changes. Climate change affects all water-related sectors and requires effective water governance and management. Integrated Water Resource Management (IWRM) strategies, ideally at a watershed scale, are thereby required to respond to the consequences of climate change. Integrated Water Resource Management focuses on water resources conservation while considering its social, economic and ecological values, and accounting for the equilibrium between interests and the needs of all stakeholders (Conca 2005). One of the greatest challenges for water governance nowadays is building adaptive capacity at the urban scale. Megacities undergoing water stress from global climate change are of rising concern. This chapter discusses whether and how current climate change action plans developed by some megacities' governments have incorporated adaptation and mitigation goals for water stresses that will develop in cities as explored in the chapter's first section as the consequences of climate change on water resources tend to be magnified in cities. The second section argues that the combination of urban demographic growth, poor infrastructure and low administrative and technical capacity increases the vulnerability of city populations to environmental hazards. This section also describes the criteria used for data selection and analysis, and details the ten worlds' most populated megacities and their climate change action plans. Finally, the last section summarizes the current challenges of megacities to adapt to water-related climate change issues.

Climatic Change and Global Warming of Inland Waters: Impacts and Mitigation for Ecosystems and Societies,
First Edition. Edited by Charles R. Goldman, Michio Kumagai and Richard D. Robarts.
© 2013 John Wiley & Sons, Ltd. Published 2013 by John Wiley & Sons, Ltd.

22.2   Building adaptive capacity to face urban water stress

There is growing scientific evidence recognizing that human activities affect the resilience of human and ecological systems (for example, greenhouse-gas emissions, carbon-based energy systems) (IPCC 2007). Water resources have been and will continue to be subjected to innumerable stresses. According to the Third Assessment report of the Intergovernmental Panel on Climate Change (IPCC 2001), the impacts of climate change on water resources tend to: (i) change precipitation regimes affecting stream flow and ground water recharge, (ii) increase the magnitude and frequency of floods, (iii) increase irrigation demand due to higher temperatures, and (iv) reduce water quality by changes in biochemical patterns from higher temperatures. For example, monitoring of lake depth has proved to be an effective indicator of climate change; warming temperatures cause eutrophication of lakes and reduces water quality (Ambrosetti and Barbanti 1999). Changes in climate can manifest slowly (for example, increased interannual and seasonal variability) or rapidly (for example, catastrophic shifts in ecosystems), accelerating the need to develop effective adaptive strategies (Tompkins and Adger 2004).

Adaptive capacity is the ability of individuals or groups to transform an undesirable regime into a desired one to manage and promote ecological resilience (Gunderson *et al.* 2006). Adaptation is thereby the "adjustment in natural or human systems, in response to actual or expected climatic stimuli or their effects, which moderates harm or exploits beneficial opportunities" (IPCC 2007). Because adaptation strategies are designed for specific social and ecological systems (for example, at the city scale, in a developing country, and so forth), communities and institutions are likely to respond differently to change. For instance, in developing areas, where poor communities are more vulnerable to environmental changes and public administrations typically have low financial capacity to face these issues, adaptation strategies look more appealing to reduce vulnerability even if they do not address structural environmental problems (for example, lack of an integrated sewerage system).

The impacts of climate change and the challenges of adaptation in cities are undeniable. Cities are, on the one hand the locus of many environmental stressors (for example, growing urban population, proximity to the coast, urbanization and industry practices, and so forth), most of which can be seriously impacted by climate change (e.g., heat island, floods, and so forth). On the other hand, cities are the place for innovative technologies and governance practices to face the consequences of climate change. Adaptive governance at the urban scale can therefore be an advantage. At the local level, stakeholders and decision makers, whether with private or public interests, can interact closely to incorporate adaptive strategies into new or reformed local policy and practices (Corfee-Merlot *et al.* 2011; Hunt and Watkiss 2011). Nevertheless, it is still a challenge for administrations to co-manage differences between national, regional and urban governance guidelines towards building sustainable cities (Hodson and Marvin 2010).

Adaptation strategies are urgently needed considering the current rates of population growth and economic development that have been increasing (and will continue to increase) the need to provide water, food and energy. Recent environmental interventions at the urban scale have applied the "eco-city" approach, which defines the necessity to balance the economic, environmental and social dimensions of urban development through integrated policies and sectoral activities to increase communities' life standards (Surjan and Shaw 2008). While building eco-cities to increase urban resilience, measures to avoid illegal construction on flood-prone areas (for example, illegal housing built on river banks), infrastructure floods (for example, old zones served with obsolete sanitation infrastructure), and inadequate land use zoning (for example, paving of known flood plains) are necessary and are still being implemented, especially in the developing world (White 2002). In order to better manage the pressures on the water resources, the global agenda for water policy has adopted the Integrated Water Resources Management (IWRM) framework at the watershed scale. Watershed management is frequently a struggle to combine scientific and policy interests to improve urban livelihoods. Watersheds are key indicators of environmental degradation that tend to be more impacted by climate change in cities (Palmer *et al.* 2009). In a growing urbanizing and climate changing world, urban watersheds seem to be the most appropriate units for water management because cities face specific problems and needs that may transcend the physical and administrative boundary of watersheds.

22.3 Adapting to climate change: the action plans of ten megacities

The most critical events of climate change will probably take place at the urban scale, where growing demographic densification joins poor urban resilience. Cities have been recognized as the ideal geopolitical scale for adaptation and mitigation strategies to tackle climate change impacts (UN WWAP 2009b; Angel *et al.* 2011). Urban areas worldwide are the centre of population growth, economic development, and thereby increasing energy and water demand. In 2010 over 50% of the world's population lived in cities (UNDESA 2010), which accounted for 80% of global greenhouse gas emissions (Hoornweg, Sugar, and Gomez 2011). By 2025, 37% of the global population will be living in cities with over 500 000 people; within this group, 47% in megacities (over 10 million) and 43% in large cities (5 to 10 million) (UNDESA 2010). Urban densification would not threaten natural resources if cities were prepared to change (e.g., climate change, economic development, waves of industrialization, etc.). When compared to rural or suburban areas, dense city centers can yet reduce energy demand and thereby GHG emissions (Hoornweg *et al.* 2011). Indeed, cities tend to be more equipped to face environmental adversities due to the concentration of knowledge, expertise, technology and financing opportunities. Within this setting, mayors, and local and regional leaders have been encouraged to develop water-related actions to respond to the consequences of climate change on the sustainability of human settlements (UN WWAP 2009b).

The degree of urban vulnerability to extreme weather events is heavily influenced by the quality of urban infrastructure and housing, including the extent to which they can reduce environmental risk, and the level of preparedness among the community and important emergency services (Mosha 2011). High levels of risk are especially found in rapidly urbanizing megacities of poor countries where increasing numbers of families live in slums with unhealthy infrastructure (Gruebner *et al.* 2011). The most direct effects of climate change on these minorities are the material and human losses from flooding and landslides (Kavindranath and Jayant 2002). In order to reduce water-related risks in cities, co-responsible public administrators (for example, municipal, provincial and even national) should consider the "urban hydrological cycle" by integrating a land-use planning perspective (White 2002). Densely populated cities have been experiencing improvements in housing, transportation, and other urban services especially in emerging economies. The vulnerability of populations in relation to urban water systems has been decreasing. At the global scale, from 1990 to 2008, among the people who gained access to improved sanitation and drinking water, 64% and 59%, respectively, live in urban areas (UN WWAP 2009a). Despite these urban improvements, climate change has been pressuring governments to take action into reducing (or mitigating) socio-environmental vulnerability to cope with emerging environmental hazards.

Many horizontal networks and coalitions have been acting at the national and international levels to build more participatory decision making for adaptation efforts, for example, the C40 Large Cities Climate Change Leadership Group, ICLEI— Local Governments for Sustainability, etc. The C40 is, however, one of the most active global–local initiatives. Since 2005, its representatives have been working together towards reducing greenhouse gas (GHG) emissions. According to the Seoul Declaration (2009), the C40 cities have set

a common goal of transforming themselves into low-carbon cities, by cutting greenhouse gas emissions to the largest extent possible, by adapting themselves to the unavoidable climate change consequences, by making cities less vulnerable to climate change, and by enhancing cities' capacity for remediation.

Amongst the policies and measures cited in this declaration, the improvement of water-resource management is referred to as a way to adapt cities to inevitable climate change effects and to increase human and environmental security especially by conducting forecasting analysis. Most of the C40 cities have established their Climate Change Action Plan (hereafter CCAP), with the main goal typically being to build low-carbon cities.

In this section, I selected the ten world's most populated megacities (over 10 million) according to the United Nations' Department of Economic and Social Affairs at *World Urbanization Prospects: The 2009 Revision* (UNDESA 2010). They were selected from the UN data on the 30 largest urban agglomerations (Table 22.1). The UN demographic projections show that by 2025 the increase of 50.77 million residents in these cities will represent, for example, more than the population of Canada projected to be 39.64 million by 2021 considering a high growth scenario (CANSIM 2010). From 1990 to 2000, the world's large cities have registered significant high (25%; 2 to 4% annual growth rate) and moderate (50%; 1 to 2% annual growth rate) growth compared to small, intermediate and big cities. Within this period, populations living in large cities increased heavily in Asia especially in China and India (UN Habitat 2009). A similar trend is expected to 2025, when seven of the ten most populated megacities will be located in Asia (Figure 22.1). Indeed, projections show that global urban areas will be concentrated in South and Central Asia, Northern and Sub-Saharan Africa (Angel *et al.* 2011). The selected megacities are mainly located in Asia (70%) followed by the Americas (30%). Eighty percent of the selected megacities are participating within the C40.

Table 22.1 Selected world's megacities and population forecasting, 2010–2025.

Megacities	2010 Rank and population (millions)		2015 Rank and population (millions)		2020 Rank and population (millions)		2025 Rank and population (millions)		Population growth (%)
Tokyo	1	35.67	1	37.05	1	37.09	1	37.09	3.98
Delhi	2	19.04	2	24.16	2	26.27	2	28.57	50.04
São Paulo	3	19.02	3	21.80	3	23.72	3	25.81	35.70
Mumbai	4	18.97	4	21.30	4	21.63	4	21.65	14.13
Mexico City	5	18.84	5	20.08	5	20.48	6	20.94	11.12
New York	6	15.92	6	19.97	6	20.37	7	20.71	30.11
Shanghai	7	14.98	7	17.84	7	19.09	9	20.64	37.76
Kolkata (Calcutta)	8	14.78	8	16.92	8	18.72	8	20.11	36.08
Dhaka	9	13.48	9	16.62	9	18.45	5	20.02	48.49
Karachi	10	12.79	10	14.82	10	16.69	10	18.73	46.41
Total		183.49		210.56		222.51		234.26	27.67

Source: Modified from UNDESA (2010). The numbers presented under "megacities" refer to "urban agglomerations," which refer "to the de facto population contained within the contours of a contiguous territory inhabited at urban density levels without regard to administrative boundaries. It usually incorporates the population in a city or town plus that in the sub-urban areas lying outside of but being adjacent to the city boundaries" (UNDESA 2010).

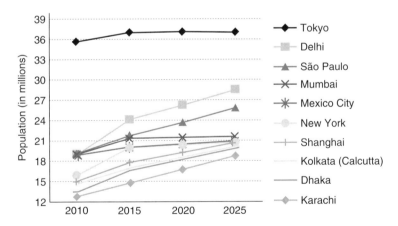

Figure 22.1 Projected population growth in selected world's mega cities, 2010–2025. Modified from UNDESA (2010).

The state of the mega cities' strategies to face climate changes (Table 22.2) shows that all these cities have developed a CCAP at the national level except for Karachi; Pakistan does not have a government initiative to adapt to climate change. Half of the selected cities have created a CCAP at the city level. Within this group 40% are located near to large water bodies such as a sea, river or delta, which represents 30% of the total.

Table 22.2   State of selected world's megacities and their CCAP (Climate Change Action Plan).

Megacity	Country	Located near a large water body	CCAP at the city level	Period
Tokyo	Japan	x	x	2007–2017
Delhi	India	–	x	2010–2012
São Paulo	Brazil	–	x	2009–2017
Mumbai	India	x	–	–
Mexico City	Mexico	–	x	2008–2012
New York	USA	x	x	2008–2017
Shanghai	China	x	–	–
Kolkata (Calcutta)	India	x	–	–
Dhaka	Bangladesh	x	–	–
Karachi	Pakistan	x	–	–

Urban water resources management and planning are of major concern for climate-change scenarios. One valuable review on the main studies for urban adaptation in large cities (mostly in OECD countries) revealed that the majority focus on sea-level rise (including coastal flooding) followed by extremes in temperature rise (Hunt and Watkiss 2011). Improvements in sanitation, urban infrastructure and disaster preparedness are central to reducing socio-environmental risk and building urban resilience to change. For these reasons, local approaches need to be developed for each world city, especially the ones facing a decrease in water quantity and quality, insufficient drainage capacity leading to floods, unsustainable sanitation, unhealthy human settlements, etc. (UN WWAP 2009b). For the purpose of this review, my aim is to answer questions about the selected climate change action plans:

- Do the goals outline water resources management? If they do, to what extent do they discuss the watershed scale?
- Does the background information contribute to derive adaptation and/or mitigation strategies at the watershed scale?
- Do the adaptation and/or mitigation strategies apply an integrative (i.e., cross-sectoral) approach for water resources?

Among the five megacities with a CCAP, four have established an integrated goal to prepare for climate change. Their CCPA's main goal refers to (i) participation of civil society and government, (ii) improvement of communities' living conditions, (iii) development with less risks to the populations, and of course (iv) building climate-change preparedness. One exception is Tokyo's plan; it is centered uniquely on the reduction of GHG emissions (Table 22.3). The following sections detail these plans.

22.4   Integrating agendas in Delhi, Mexico City and São Paulo

Climate-change action plans should ideally incorporate an integrative approach. In Delhi, Mexico City, and São Paulo the CCAP applied cross-sectoral management

Table 22.3   Goals of selected CCAP (Climate Change Action Plan).

Megacity	CCAP's title	Overall goal
Delhi	Climate Change Agenda for Delhi	To "encourage all civil society groups and government departments to forge a set of programs which can make Delhi a megacity which balances the needs of development, aspirations of the people and addresses the concerns of our planet, mother earth."
Mexico City	Mexico City Climate Action Program	To "integrate, coordinate, and encourage public actions in the capital city to diminish environmental, social, and economic risks stemming from climate change and to promote the welfare of the population through the reduction and capture of greenhouse gas emissions."
New York	New York City Climate Change Program Assessment and Action Plan	To "1) take into account the potential risks of climate change on the city's water supply, drainage, and wastewater management systems, and 2) integrated GHG emissions management to the greatest extent possible."
São Paulo	Guidelines of the City of São Paulo Action Plan for the Mitigation and Adaptation to Climate Changes	To "respond to climate change and improve city infrastructure to raise the quality of life of the Paulistano people."
Tokyo	Tokyo Climate Change Strategy. A Basic Policy for the 10-Year Project for a Carbon-Minus Tokyo	To "reduce Tokyo's greenhouse gas emissions by 25% from the 2000 level by 2020."

and planning to design adaptation and mitigation strategies for all urban sectors (e.g., water and energy supply, sanitation, transportation, and so forth). Mexico City and São Paulo also conducted inventory studies based on the IPCC framework. Delhi's CCAP or the *Climate Change Agenda for Delhi 2009–2012* (Government of National Capital Territory of Delhi 2009) was launched in January 2009 and is the local proposal of India's National Action Plan on Climate Change (NAPCC) (Government of India 2008). It covers six of the eight areas in the NAPCC except for agriculture and the Himalayan ecosystem. This "green agenda" recommends 65 actions divided in five chapters on air pollution, water, greening, noise and municipal waste management. The water agenda brings 11 strategies among which five focus on the improvement of sewerage systems (house connections to the official sewerage lines, waste water treatment for industry, and so forth). The plan focuses on the construction of a reliable sanitation system, which confirms the poor water sanitation situation in Delhi as in most Indian cities (Revi 2008). One of the top priorities for local and national authorities is the restoration of Yamuna River, the most polluted river in Delhi, and also its main water supplier. Conflicts between its basin states' authorities on its water-resources management are mentioned in the Delhi CCAP, which reveals a contentious situation in terms of building more effective water governance.

In Mexico, the *Mexico City Climate Action Program 2008–2012* (Government of Mexico City 2008) was designed at the city level with the support of the World Bank in order to give continuity to existing municipal government programs aimed at mitigating greenhouse gas (GHG) emissions, in particular the *Mexico City's General Development Program 2007–2012* and *Mexico City's Green Plan 2008–2012*. The program was conceived in three parts: (i) greenhouse-gas mitigation (with 26 actions), (ii) climate-change adaptations (three lines of actions) and (iii) climate-change communications and educational actions (six actions). The mitigation actions rely heavily on the reduction of GHG emissions and CO_2 capture, and are grouped as follows: five for energy (energy efficiency, sustainable buildings, and so forth), seven for water (improvement of sewerage systems, water saving programs, and so forth), ten for transportation (implementation of subway, streetcars, and so forth), and four for waste (modernization and automation of waste transfer stations, and so forth). The adaptation actions were designed, firstly, to identify primary threats and conduct vulnerability analysis; secondly, to integrate an adaptation perspective into existing Mexico City government plans; and, finally, to implement the adaptation actions to diminish climate change risks to local communities. These actions are organized into two groups of early and medium-term responses, respectively.

With regards to water-sustainable management, Mexico City's CCAP interacts with other municipal initiatives such as the Environmental Agenda, the Green Plan, the Local Strategy of Climate Action, the Sustainable Water Management Program and the Program of Energy Efficiency in Water Systems. Amongst the actions proposed for the water sector, those that have the most significant impact on GHG emission mitigation refer to: (i) the Home Water Savings Program (for example, to increase the use of sanitary equipment of low water consumption through an informational campaign); (ii) the improvement of water supply and sanitation infrastructure (for instance, leak suppression, pipe rehabilitation and sectioning, energy efficiency of pumping systems); (iii) the improvement of hydroelectric power generation; (iv) the reduction of mud emissions from city biological treatment plants; and (v) the reduction of methane emissions from city septic tanks. At the micro-basin scale, specific adaptation actions are of early term response (for example, urban ravines) and medium-term response (such as soil and water conservation projects in urban and rural areas).

Unlike all the other megacities, in São Paulo the city government defined its CCPA first through legal means and then through a plan. The Municipal Law 14.933 sanctioned in June 2009 "enforces the municipal strategy for climate change in São Paulo" (São Paulo City Hall 2009). The main goal is "to reduce 30% of aggregate anthropic (human) emissions in the city until 2012." The macro strategies proposed by the law were based on the Inventory of Greenhouse Gas (GHG) Emissions of the City of São Paulo established in 2005. The *Guidelines of the City of São Paulo Action Plan for the Mitigation and Adaptation to Climate Changes* was only conceived in May 2011 and basically followed the structure of the municipal law (São Paulo City Hall 2011). By incorporating the concept of "compact city" the plan suggests the development of a "compact São Paulo," or "one with public policies that induce the implementation of sustainable regional centers, that is, the creation of areas of greater density with road and transportation support, and that have environmental and urban quality and offer suitable support capacity where it was lacking" (São Paulo City Hall 2011). To rebuild São Paulo as a sustainable city, the CCPA addresses both adaptation and mitigation strategies with regards to transportation, energy, construction, land use, solid waste, and health.

Four out of nine priorities of São Paulo's CCPA refer to water management: (i) preservation of water sources and biodiversity; (ii) revitalization of the system of rivers and streams; (iii) structural and non-structural actions related to macro and micro drainage; and (iv) capturing and reuse of rainwater. At the architectural scale, in the plan's construction section, I found reference to the rational use of energy and water consumption standards in the construction, use, and operation of private and especially public administration buildings. It should include the capturing and reuse of rain water as well. This action should fully implement the *Municipal Program for Rational Water Use in Buildings* according to Municipal Law 14.018 and decrees 47.731 of 2006, and 14.403 of 2007. At the watershed scale, water conservation actions are encouraged to increase permeable areas and water absorption zones therefore reducing run off and the risk of flooding. The CCAP suggests the completion of specific programs already in implementation, such as the preservation of mangrove swamps and streams, and the removal of vulnerable communities living in areas of environmental risk. In many Brazilian cities, the poor are usually more exposed to risk (for instance, flooding, and water-borne diseases) because they tend to build their homes along the edges of water bodies, especially when sanitation is absent or poorly provided. In this regard, I found mention of improvements in the quality of the potable water at the Guarapiranga and Billings reservoirs. To avoid flooding, structural and nonstructural actions related to macro and micro drainage are suggested for the city's six main watersheds (Aricanduva, Cabuçu de Baixo, Ipiranga, Verde, Cordeiro and Morro do S) so as to reduce human vulnerability. The CCAP also defines a pilot project at the watershed scale: the implementation of non-structural land use interventions in the Aricanduva watershed.

22.5 The New York City plan: predicting climate change at the watershed scale

The *New York City DEP Climate Change Program Assessment and Action Plan* (hereafter NYC CCAP) is by far the most complete of the megacities' plans analyzed here (NYC DEP 2008). Launched in May 2008 by the Department of Environmental Protection, it brings historic information to predict climate change consequences on multiple urban sectors notably at the watershed scale. The scale of intervention is the City of New York and its Watershed Region (i.e., Delaware, Catskill and Croton Watersheds). The NYC CCAP is structured in six chapters with focus on the potential impacts of climate change, potential adaptation strategies, greenhouse gas emissions, and mitigation actions. It considers short-term and long-term strategies for infrastructure and policy planning. For the water sector, the plan identifies and evaluates adaptation strategies for the impact of climate change on water supply systems, drainage, and wastewater management facilities, harbor water quality, and flood preparedness. Projections of climate change scenarios were developed at the watershed scale for the 2020s, 2050s, and 2080s and considered parameters of, for example, population growth, GDP growth, energy use, land-use changes, and so forth. In addition, watershed management models were developed to assess potential climate-related variables such as water quality, water quantity, maximum and minimum temperature, precipitation, sea level, streamflow, runoff, and sediment and nutrient loadings.

For the water supply system, which relies heavily on the Delaware and Catskill Watersheds, the proposed strategies considered the threats to water quantity (such as flood, drought and ecological change), water quality (e.g., increased water temperature and changes in precipitation patterns) and demand (e.g., changes in water availability for water supply systems). To maintain water quantity, the potential adaptation strategies include implementing demand-reducing conservation actions, enhancing alternative water sources such as groundwater, building regional pipe lines between New York, New Jersey, Connecticut, or Long Island, and using reservoir models to balance flows and releases. To maintain water quality standards, adaptive propositions refer, for example, to the filtered Croton system during turbidity events. Finally, to reduce water demand, small-scale approaches should be considered, such as installation of better hydrant-locking mechanisms, use of more spray caps, and imposition of in-City conservation measures, and so forth. In summary, all these strategies are directly or indirectly related to the PlaNYC's actions about water supply (City of New York, 2011). This plan to make a greener New York City establishes water-related strategies such as the continuation of the Watershed Protection Program to protect the water supply from hydrofracking for natural gas, the completion of the Catskill/Delaware Ultraviolet (UV) Disinfection Facility and the Croton Water Filtration Plant, and the enhancement of the Delaware and Catskill Aqueducts.

For the drainage and wastewater systems, the NYC CCAP measures the potential impacts of flooding events (for example, storm or combined sewer flooding), which are usually manifested by means of inland flooding and coastal inundation from high tides both from severe rainfall. Current (as of 2008) and projected (2080s) sea-level estimates were conducted. The potential water quality impacts on receiving waters (e.g., temperature rise) were also assessed, especially for New York Harbor and Long Island Sound. A few alternatives to counter these negative situations refer to the augmentation of the capacity of the existing collection system (for example, increasing the size of existing sewer pipes, installing new pipes), the implementation of storm water controls (for instance, encouraging the use of green roofs to reuse storm water), and the improvement in managing increased flows (e.g., frequently cleaning sewers and maintaining catch basins in flood-prone areas, enhancing drainage features for runoff control), among others. In this setting, adaptation strategies "to quantify, better understand, and monitor the impacts of climate change on ... water supply, storm water, and wastewater management systems" (NYC DEP 2008: 45) were developed: (i) to build an integrated modeling project to determine the severity of changes on water quality (e.g., temperature rise), water demand and water quantity (e.g., probabilities of refill and drawdown); (ii) to create a long-term resilience plan for system management based on the identification of system-related vulnerabilities to severe weather events, precipitation pattern changes, future sea level rise and coastal flooding.

22.6   Tokyo—let's become a low-energy society!

Contrary to all climate-change adaptation strategies examined here, Tokyo's plan does not refer to water resources management either at the city or at the watershed scale.

The *Tokyo Climate Change Strategy. A Basic Policy for the 10-Year Project for a Carbon-Minus Tokyo (2007–2017)* was launched by the Tokyo Metropolitan Government (TMG) in June 2007. This plan is part of the *Tokyo's Big Change—The 10-year Plan* (December 2006), in which urban development goals were defined to prepare the candidate city to host the 2016 Olympics. The Tokyo CCAP is recognized by the TMG as an essential instrument to implement "concrete measures on climate change and renewable energy" besides other public initiatives aiming to build energy efficient cities in terms of carbon dioxide target (for example, *Tokyo's Big Change: The 10-year Plan 2006*), and renewable energy target (for example, *Tokyo Renewable Energy Strategy 2006, Tokyo Environmental Master Plan 2008*).

The Tokyo CCAP announces from its title the unique goal to reduce greenhouse gas emissions so as to enhance "Tokyoites' independence in terms of energy." It brings five initiatives for climate-change mitigation without mention of adaptation strategies, such as: (i) to apply energy efficient technologies, (ii) to increase the use of renewable energy in the city, (iii) to create a sustainable transport system, (iv) to develop environmental technologies and businesses, and (v) to encourage a carbon minus lifestyle. The aim is to build a low-energy society; therefore all initiatives refer to carbon-dioxide reduction in relation to, for example, private enterprises, households, urban development, etc. As far as water management is concerned, water is only mentioned in relation to energy consumption (for example, hot water supply, heating and cooling systems) especially at the architectural scale for residential use.

Although the Tokyo region is widely known for its remarkable performance in water management and planning, it is vulnerable to water-related hazards such as drought (e.g., the Arakawa River), flood and water quality degradation (for example, sub-basins of the Tone River system) (UN WWAP 2003). Besides this, according to the Bureau of Waterworks of the TMG, the average water consumption per day is forecasted to increase continuously (Tokyo Waterworks 2006). For this reason, the Tokyo Waterworks policy vision document analyzes the impacts of the disconnection of water service in Tokyo and defines goals towards supplying safe and tasty (high quality) tap water (for instance, by increasing water treatment technology in the Tone River system), shifting from a receiving tank system to a direct water supply system, preserving "water conservation forest," and so forth. With respect to climate-change strategies, like the Tokyo CCAP, Tokyo Waterworks confirms the need to reduce the emission of carbon dioxide in water treatment facilities.

22.7   Don't they need a climate change action plan? The cases of Mumbai, Kolkata (Calcutta), Karachi, Dhaka and Shanghai

The megacities of Mumbai, Kolkata (Calcutta), Karachi, Dhaka and Shanghai do not have a climate change action plan and yet they desperately need one. Together they represent a population of 75 million people (UNDESA 2010), which is, for example, twice the population of Canada (CANSIM 2010). Indian cities (e.g., Mumbai and Kolkata) are among the most vulnerable to climate change in the world. There, the populations grow in an unsustainable urban development pattern. Problems such as high urban poverty, low housing conditions, and an increase in rural–urban migration, have put local governance to the test (Verhagen 2005). Indian cities have poorly engaged the national initiatives to adaptation; the IPCC agenda and methods of analysis have been emphasized despite local complexities. Changes are expected to cause regional temperature rise, precipitation decline, and drought. River basins such as the Ganga, Narmada, Krishna, and Kaveri will be affected by severe changes in river hydrology. In addition, the melting of the Himalayan glaciers will heavily change river flow in the Brahmaputra valley and the Indo-Gangelic plain. The cities of Delhi, Dhaka, Kolkata, and Karachi (in Pakistan) depend on the Brahmaputra/Yarlung Tsang-po waters for irrigation and drinking water. Expected changes in these rivers will certainly compromise the socioeconomic development of these communities. In general, Indian cities are not prepared to tackle the potential consequences of climate change (Revi 2008). It is noteworthy, however, that Delhi has recently launched a climate-change agenda (Government of National Capital Territory of Delhi 2009).

In Mumbai, low-income groups are incapable of affording the high land prices and thus end up living in slums and informal settlements (Patel, d'Cruz, and Burra 2002). Around 74% of the total population in Greater Mumbai lives in poor housing conditions, which are more vulnerable to floods, health hazards, and so forth. Mumbai alone is constantly under the threat of floods as a result of the combination of low-lying areas and drainage system problems. In the city, a 1993 report identified 111 places as flood-prone areas. In 2005, over 1000 people died during severe flooding in Mumbai (Government of Maharashtra 2011). Even with the implementation of the Mumbai Disaster Management Plan 2003–2007, the city had difficulty in building adaptive capacity (De Sherbinin, Schiller, and Pulsipher 2007). Projections have shown that the city may face water shortages by 2050 (Revi 2008). In this setting, in 2009, the government of Maharashtra (Mumbai's state) approved the development of a climate change action plan following the directives of India's National Action Plan on Climate Change (Aghor 2009). The study *Assessing Climate Change Vulnerabilities and Adaptation Strategies for Maharashtra* was launched in March 2010 and has been developed by the Energy and Resources Institute (TERI) and the UK's Met Office (Government of Maharashtra 2010). The research outputs are expected to (i) provide regional climate projections for the time periods 2030s, 2050s and 2070s; (ii) conduct impact assessments for agriculture, human health, coastal urban systems, water and ecosystems; (iii) build a district level vulnerability index based on key indicators concerning the risk of communities to various climatic hazards, sensitivity and adaptive capacity; and (iv) propose adaptation strategies based on the outputs of the case studies applying the vulnerability index (Government of Maharashtra 2010). I believe that this plan will include some directives for Mumbai.

Like Mumbai, Kolkata has been facing severe problems related to water. Uncontrolled urban sprawl together with poor sanitation and drainage services seem to be the main cause for the disruption of urban watersheds (Verhagen 2005). In 2001, one-third of the population in the Calcutta Metropolitan Corporation (CMC) lived in slums, of which more than 5500 were located in Kolkata (UN Habitat 2003). Nevertheless, the city does not have any official initiative to reduce socio-environmental risk and build adaptive capacity to climate change despite the West Bengal State Disaster Management Policy and Framework 2005 from the Government of West Bengal, Kolkata's State (Government of West Bengal 2005).

The combination of unplanned urban sprawl and vulnerability to alterations in water resources is also present in Dhaka as well as elsewhere in Bangladesh (UN Habitat

2008). Dhaka has experienced severe flooding events since 1954. For example, eastern Dhaka was totally inundated in the floods of 1988, 1998 and 2004 (Mozaharul and Golam Rabbani 2007). There are main potential causes for these environmental hazards. First, Dhaka is entirely surrounded by rivers (Balu, Shitalakhya, Tongi, and Turag) and has its most urbanized area at only six to eight meters above mean sea level. Second, the city has serious problems of drainage congestion that occurs when the combined sewer system (domestic and storm water sewage) is not able to carry excessive rainfall. This situation usually causes flooding and water contamination, and may thus contribute to the occurrence of waterborne diseases among the most vulnerable communities. In addition to this, in a world of climate change, the melting of glaciers and snow in the Himalayas will probably increase flooding events in Dhaka and Bangladesh (Anisfeld 2010). Despite its history of flooding, Dhaka has no official plan to implement adaptation and mitigation actions to face climate change. The only government initiative to prevent against flooding was the Dhaka Integrated Flood Protection Project (1991–2000) with the support of the Asian Development Bank (ADB).

In Karachi and Shanghai the situation is similar. The city of Karachi as well as its Sindhi province and Pakistan as a whole do not have a climate-change action plan (S. Khan, personal communication, September 21, 2011). Water supply and management recommendations (for example, "to promote integrated watershed management") can only be found in Pakistan's National Environment Policy (Government of Pakistan 2005). In Shanghai, the population has been growing rapidly, as in many Chinese cities. China's rural–urban change has been challenging the capacity of local governments in providing basic urban services (Yeh 2011). The city is located on China's east coast at the mouth of the Yangtze River. It is very vulnerable to hurricane landfalls and flood. For instance, in 1998 the Yangtze River overflowed and caused serious damage including more than 3000 deaths. To rebuild the urban system, the Municipal Civil Defense Office implemented few punctual initiatives (Revi 2008). This disconnected way of structuring water-related services seems to follow the Chinese public policy practice of limited interaction between government departments and thereby public initiatives are developed in isolation (Cosiera and Shenb 2009). The Chinese Government Official Web Portal announced in 2007 the development of a climate change forecast (next 20 to 30 years) and action plan for Shanghai and east China to be finished by 2010. Nevertheless, I could not find any information about this plan (as of October 2011).

22.8 Challenges of adapting urban watersheds to climate change

Water resources and watershed management and planning are not always central to the selected megacities' climate-change adaptive strategies. Most CCAPs are structured to indeed build an "eco-city"—a city that should efficiently use water and energy, reduce carbon emissions and urban heat islands, and so forth. Ideally, climate-change strategies should guide local interventions to build urban resilience following an integrated framework that deals with the multiple natures of socio-environmental disruptions and threats.

Three main conclusions can be drawn from the analysis of these CCAPs. First, the plans' approaches vary from a one-sector (for example, energy supply) to a cross-sector approach (for example, transportation, sanitation, waste, and so forth). Second, strategies can be directly or indirectly related to water resources management. For instance, in Tokyo's plan, water is only mentioned in relation to energy consumption—there is no mention of water or watershed management. Initiatives that only concentrate on reducing carbon emissions are not sufficient to continue growing and developing in a sustainable way (Hodson and Marvin 2010). Last, strategies do not always address water-related concerns at the watershed scale. Both the CCAP of New York and São Paulo have clearly defined goals of urban water management at the watershed scale. However, only the NYC plan predicts the impacts of climate change at the watershed scale. Researchers have pointed out the effectiveness of NYC's integrative strategies to face the adversities of climate change (Birkmann *et al.* 2010; Corfee-Morlot *et al.* 2011; Hunt and Watkiss 2011). Besides New York, Delhi, Mexico City, and São Paulo have designed their CCAP within a more integrated approach with mention to designing adaptation and mitigation strategies for all urban sectors and to recognizing the watershed as the geographic unit for water resources management. Many of the CCPA (of New York, Mexico City and São Paulo) follow the IPCC structure—mapping the anticipated changes under severe and extreme events of climate change for air temperature, sea level, precipitation, etc. Table 22.4 summarizes the goals of selected CCAP's in relation to water resources management.

In conclusion, this chapter described whether and how some megacities' governments have been preparing to face the consequences of climate change on water resources. Questions of how effectively current climate change action plans have incorporated mitigation and adaptation goals for water stresses were discussed. Ideally, mitigation efforts should always be accompanied by adaptation strategies, because

Table 22.4   Goals of selected CCAP for urban water resources management.

Megacity	Goals of urban water management	Summary of main goal with reference to urban water management	Prediction of impacts on urban water resources
Delhi	Mentioned	To improve sewerage systems and to restore the Yamuna River	Not mentioned
Mexico City	Mentioned	To link policies and actions regarding efficient water use and supply to those dealing with risks	Not mentioned (inventory at the city scale for 1969–2002)
New York	Mentioned at watershed level	For the city's vital water systems to function under and be resilient to a range of potential future climate conditions	Mentioned at the watershed scale for 2020s, 2050s, 2080s
São Paulo	Mentioned at watershed level	To increase permeable areas and water absorption zones	Not mentioned (inventory at the city scale for 1933–2010)
Tokyo	Not mentioned	Not mentioned	Not mentioned

some climate-change impacts may occur in the long term (and are more suited for mitigation actions) while others may occur in the short and medium term (and are more suited for actions to adapt to or even to minimize these impacts) (Kavindranath and Jayant 2002). Nevertheless, the development of integrative approaches to urban water resources remains elusive for most of the selected megacities.

In a global urbanization scenario, questions of how cities invest in the efficient use of, for example, housing and infrastructure, are related to their level of development (Malpezzi 2011) and thereby to their capacity in building preparedness for climate change. This study showed that less-developed megacities (such as Indian cities) seem to be less prepared to address water-related climate change issues. Overall, most of the ten most populated world megacities seem to be unprepared to face the challenges of adapting water resources (and watershed) management and planning to climate change. Only half of the selected megacities have developed a climate change action plan—Delhi, Mexico City, New York, São Paulo, and Tokyo. Within the CCAP, background information about urban climate changes was assessed for Mexico City, New York and São Paulo. The New York City government was the only one to predict the impacts of climate variability at the watershed level in its CCAP.

This chapter showed that, despite these initiatives, there is still a lot of governance work to be done to prepare cities for climate change impacts. In this setting, one proactive government action to share best practices and city data at the global scale has been developed by the New York City Global Partners' Innovation Exchange Initiative. It has recognized four best practices related to building urban climate change preparedness in the selected megacities: (i) water leakage prevention controls and a green building program in Tokyo (City of New York 2010b, 2010d), landfill emissions control program in São Paulo (City of New York 2010c), and Mexico City's comprehensive climate change plan (City of New York 2010a). Finally, urban governments and communities have to join forces and efforts to address growing water stress and build climate resilient cities. Future research on community participation in decision making and capacity building within these climate-change action plans is in its early stages and still needs to be developed.

References

Aghor, A. (2009) State Approves Climate Change Action Plan, http://www.dnaindia.com /mumbai/report_state-approves-climate-change-action-plan_1285474 (accessed June 30, 2011).

Ambrosetti, W. and Barbanti, L. (1999) Deep water warming in lakes: an indicator of climate change. *J. Limon.*, **58**, 1–9.

Angel, S., Parent, J., Civco, D.L., and Blei, A.M. (2011) *Making Room for a Planet of Cities*, Lincoln Institute of Land Policy, Cambridge.

Anisfeld, S.C. (2010) *Water Resources*, Island Press, Washington.

Birkmann, J., Garschagen, M., Kraas, F., and Quang, N. (2010) Adaptive urban governance: new challenges for the second generation of urban adaptation strategies to climate change. *Sustainability Science*, **5**(2), 185–206.

CANSIM (Canadian Socio-Economic Information Management System from Statistics Canada) (2010) *Projected Population, by Projection Scenario, Sex and Age Group as of July 1, Canada, Provinces and Territories, Annual (Persons)*, http://www5.statcan.gc.ca/cansim /a05?lang=eng&id=520005&paSer=&pattern=052-0005&stByVal=2&csid= (accessed June 11, 2011).

City of New York (2010a) *Best Practice: Comprehensive Climate Change Plan*, The NYC Global Partners' Innovation Exchange, New York, http://www.nyc.gov/html/unccp/gprb /downloads/pdf/Mexicoper cent20City_Environment_ClimateChange.pdf (accessed October 15, 2011).

City of New York (2010b) *Best Practice: Green Building Program*, The NYC Global Partners' Innovation Exchange, New York, http://www.nyc.gov/html/unccp/gprb/downloads/pdf /Tokyo_GreenBuildings.pdf (accessed October 15, 2011).

City of New York (2010c) *Best Practice: Landfill Emissions Control*, The NYC Global Partners' Innovation Exchange, New York, http://www.nyc.gov/html/unccp/gprb /downloads/pdf/SaoPaulo_landfills.pdf (accessed October 15, 2011).

City of New York (2010d) *Best Practice: Water Leakage Prevention Controls*, The NYC Global Partners' Innovation Exchange, New York, http://www.nyc.gov/html/unccp/gprb /downloads/pdf/Tokyo_Energy_Waterper cent20Leakageper cent20Controls.pdf (accessed October 15, 2011).

City of New York (2011) *PlaNYC Update April 2011: A Greener, Greater New York*, The City of New York, New York NY.

Conca, K. (2005) *Governing Water: Contentious Transnational Politics and Global Institution Building*, MIT Press, Cambridge MA.

Corfee-Morlot, J., Cochran, I., Hallegatte, S., and Teasdale, P.J. (2011) Multilevel risk governance and urban adaptation policy. *Climatic Change*, **104**, 169–97.

Cosiera, M. and Shenb, D. (2009) Urban Water Management in China. *Int. J. Water Res. Dev.*, **25**, 249–68.

De Sherbinin, A., Schiller, A., and Pulsipher, A. (2007) The vulnerability of global cities to climate hazards. *Environment and Urbanization*, **19**, 39–64.

Government of India (2008) *National Action Plan for Climate Change of India (NAPCC)*, Government of India, Prime Minister's Council on Climate Change, New Delhi.

Government of Maharashtra (2010) *Assessing Climate Change Vulnerabilities and Adaptation Strategies for Maharashtra*, Government of Maharashtra, Department of Environment, Maharashtra, http://ccmaharashtra.org/index.php (accessed June 15, 2011).

Government of Maharashtra (2011) *Mumbai Disaster Management Plan*, Government of Maharashtra, Department of Relief and Rehabilitation, Maharashtra, http://mdmu .maharashtra.gov.in/pages/Mumbai/mumbaiplanShow.php (accessed June 15, 2011).

Government of Mexico City (2008) *Mexico City Climate Action Program 2008–2012*, Secretaría del Medio Ambiente del Distrito Federal, Mexico City.

Government of National Capital Territory of Delhi (2009) *Climate Change Agenda for Delhi 2009–2012*, Government of National Capital Territory of Delhi, Delhi.

Government of Pakistan (2005) *National Environment Policy*, Ministry of Environment Pakistan, Islamabad.

Government of West Bengal (2005) West Bengal State Disaster Management Policy and Framework, Department of Disaster Management West Bengal, Kolkata.

Gruebner, O., Staffeld, R., Khan, M.M.H., *et al.* (2011) Urban health in megacities: extending the framework for developing countries. *IHDP Update*, **1**, 42–9.

Gunderson, L.H., Carpenter, S.R., Folke, C., *et al.* 2006. Water RATs (resilience, adaptability, and transformability) in lake and wetland social-ecological systems. *Ecology and Society*, **11**, 16, http://www.ecologyandsociety.org/vol11/iss1/art16, accessed April 20, 2010.

Hodson, M. and Marvin, S. (2010) *World Cities and Climate Change: Producing Urban Ecological Security*, Open University Press, Milton Keynes.

Hoornweg, D., Sugar, L., and Gomez, C.L.T. (2011) Cities and greenhouse gas emissions: moving forward. *Environment and Urbanization*, **23**, 207–27.

Hunt, A. and Watkiss, P. (2011) Climate change impacts and adaptation in cities: a review of the literature. *Clim. Change*, **104**, 13–49.

IPCC (Intergovernmental Panel on Climate Change) (2001) *Climate Change 2001: Impacts, Adaptation and Vulnerability: Contribution of Working Group II to the Third Assessment Report of the IPCC*, Cambridge University Press, Cambridge.

IPCC (Intergovernmental Panel on Climate Change) (2007) *Climate Change 2007: Impacts, Adaptation, and Vulnerability. Contribution of Working Group II to the Fourth Assessment Report of the Intergovernmental Panel on Climate Change*, Cambridge University Press, Cambridge.

Kavindranath, N.H. and Jayant, A. (eds.) (2002) *Climate Change: Vulnerability, Impacts and Adaptation. Climate Change and Developing Countries*, Kluwer Academic Publishers, Dordrecht.

Malpezzi, S. (2011) Urban growth and development at six scales: An economist's view, in *Global Urbanization* (eds. E.L. Birch and S.M. Watcher), University of Pennsylvania Press, Philadelphia PA, pp. 48–66.

Medellín-Azuara, J., Harou, J.J., Olivares, M.A., *et al.* (2008) Adaptability and adaptations of California's water supply system to dry climate warming. *Clim. Change*, **87** (Suppl. 1), S75–S90.

Mosha, A.C. (2011) The effects of climate change on urban human settlements in Africa, in *Climate Change and Sustainable Urban Development in Africa and Asia* (eds. B. Yuen and A. Kumssa), Springer, New York, pp. 69–99.

Mozaharul, A. and Golam Rabbani, M.D. (2007) Vulnerabilities and responses to climate change for Dhaka. *Environ. Urbanization*, **19**, 81–97.

NYC DEP (New York City Department of Environmental Protection) (2008) *NYC DEP Climate Change Program Assessment and Action Plan*, NYC Department of Environmental Protection, New York NY.

Palmer, M.A., Lettenmaier, D.P., Poff, N.L., *et al.* (2009) Climate change and river ecosystems: protection and adaptation options. *Environ. Management*, **44**,1053–68.

Patel, S., d'Cruz, C., and Burra, S. (2002) Beyond evictions in a global city; people managed resettlement in Mumbai. *Environ. Urbanization*, **14**, 159–72.

Revi, A. (2008) Climate change risk: an adaptation and mitigation agenda for Indian cities. *Environ. Urbanization*, **20**, 207–29.

São Paulo City Hall (2009) *Municipal Law 14.933, Municipal Policy for Climate Change in Sao Paulo*, City Council São Paulo, São Paulo.

São Paulo City Hall (2011) *Guidelines of the City of São Paulo Action Plan for the Mitigation and Adaptation to Climate Changes*, The Municipal Committee on Climate Change and Ecoeconomy, and the Working Groups for Transportation, Energy, Construction, Land Use, Solid Waste and Health, São Paulo.

Surjan, A.K. and Shaw, R. (2008) "Eco-city" to "disaster-resilient eco-community": a concerted approach in the coastal city of Puri, India. *Sustain. Sci.*, **3**, 249–65.

Tokyo Waterworks (2006) *Tokyo Waterworks Principal Vision. STEP II. Safe Waterworks, Our Pride to the World*, Tokyo Waterworks, Tokyo, http://www.waterprofessionals.metro.tokyo.jp/pdf/tokyo_waterworks_principal_vision_step2.pdf (accessed October 11, 2011).

Tompkins, E.L. and Adger, W.N. (2004) Does adaptive management of natural resources enhance resilience to climate change? *Ecology and Society*, **9** (2), 10, http://www.ecologyandsociety.org/vol9/iss2/art10 (accessed 20 April, 2010).

UNDESA (United Nations Department of Economic and Social Affairs) (2010) *World Urbanization Prospects: The 2009 Revision*, UNDESA, Population Division, New York NY.

UN Habitat, United Nations Human Settlements Programme (2009) *Planning Sustainable Cities. Global Report on Human Settlements 2009*, Earthscan, London.

UN Habitat, United Nations Human Settlements Programme (2008) *State of the World's Cities 2008/2009—Harmonious Cities*, Earthscan, London.

UN Habitat, United Nations Human Settlements Programme (2003) *The Challenge of Slums: Global Report on Human Settlements 2003*, Earthscan, London.

UN WWAP (United Nations World Water Assessment Programme) (2003) Greater Tokyo, Japan, in *UN WWAP. UN World Water Development Report 1: Water for People, Water for Life*, UNESCO, Paris and Berghahn, New York NY, pp. 481–98.

UN WWAP (United Nations World Water Assessment Programme) (2009a) *United Nations World Water Development Report 3: Water in a Changing World (WWDR3)*, UNESCO-WWAP, Paris.

UN WWAP (United Nations World Water Assessment Programme) (2009b) *Messages to Urban Mayors and Local Governments*, UNESCO-WWAP, Paris.

Verhagen, J. (2005) Cities, lakes, and floods. The case of the Green Hyderabad Project, India, in *Urban Flood Management* (eds. A. Szöllösi-Nagi and C. Zevenbergen), A.A. Balkema, London, pp. 27–40.

Yeh, A. (2011) *Urban Growth and Spatial Development: The China Case*, in *Global Urbanization* (eds. E.L. Birch, and S.M. Watcher), University of Pennsylvania Press, Philadelphia PA, pp. 67–85.

White, R.R. (2002) *Building the Ecological City*, Woodhead Publishing Limited, Cambridge.

Part III

Mitigation Approaches

23

Water Management Preparation Strategies for Adaptation to Changing Climate

Balázs M. Fekete[1] and Eugene Stakhiv[2]

[1]*CUNY Environmental CrossRoads Initiative, The City College of New York, City University of New York, New York, USA*
[2]*IJC Upper Lakes Study and UNESCO-ICIWaRM, Institute for Water Resources, Alexandria, VA, USA*

23.1 Introduction

Water resources are severely impacted by climate variations and changing climate, therefore water-management infrastructure is always designed to cope with extreme events (both excess waters or prolonged water shortages). The traditional "bottom up" (Dessai and van der Sluijs 2007) approach to handle climate variation uncertainties is the continuous recording of the various states of the hydrological system and establishing the probability distribution of those states assuming that the varying observations belong to a stationary probability distribution. Water managers always recognized that this "stationarity" assumption might be invalid beyond certain time scales, but found it satisfactory in their typical multi-decadal design time frames of 30–50 years (Stakhiv 2011). The "stationarity" assumption was challenged recently suggesting that water managers will need to endorse new "top-down" approaches while preparing water-management infrastructures (both water supply or flood control) for long-term operation (Milly *et al.* 2008). The climate modeling and hydro-climate community's strong advocacy to incorporate Global Circulation Models (GCM) in water management decision making has met with equally strong reluctance from practitioners despite significant efforts to develop smart downscaling techniques to bridge the gap between GCM capabilities and the needs of field hydrologists.

Criticism from hydrologists (Koutsoyiannis *et al.* 2008; Anagnostopoulos *et al.* 2010) concerned with the apparent lack of GCMs ability to reproduce contemporary

Climatic Change and Global Warming of Inland Waters: Impacts and Mitigation for Ecosystems and Societies, First Edition. Edited by Charles R. Goldman, Michio Kumagai and Richard D. Robarts.
© 2013 John Wiley & Sons, Ltd. Published 2013 by John Wiley & Sons, Ltd.

climate conditions within error bounds suitable for hydrological applications, has met resistance (Anagnostopoulos *et al.* 2010; Kundzewicz and Stakhiv 2010) from the climate community. Instead hydrologists and water resources planners were recommended to reconsider their expectations of GCM outputs according to the following principles (Wilby 2010):

- quantify the uncertainty in the observed data used for model evaluation;
- compare like with like;
- select indicators of performance relevant to the hydrological applications;
- evaluate climate models relative to other components of hydrological uncertainty;
- test combined climate, downscaling and hydrological model skill using near-term applications.

Following these principles is much easier said than implemented. We tested state-of-the-art GCMs by applying their climate forcing estimates under contemporary and future climate conditions in a global hydrological modeling context.

The presented application is intended to assess the anticipated hydrological responses to the GCM simulated climate forcing contrasted with twentieth-century hydrography based on gridded climate data records from the Climate Research Unit (CRU) of East Anglia (New, Hulme, and Jones 1999, 2000), which was specifically designed and developed for GCM evaluation. We used the historical climate scenarios for the 1901–2000 period that overlap with the 1901–2002 time period of the CRU data. We used a global water balance/transport model (Vörösmarty *et al.* 1989; Vörösmarty, Federer, and Schloss 1998; Wisser *et al.* 2008, 2010) operating at 30′ spatial scale (which matches the CRU data and slightly higher than typical GCMs at 1–2.5° spatial resolution) and daily temporal resolution (where we used the daily model outputs from the GCM simulations and applied statistical downscaling of the monthly CRU data). Although, the water balance model simulations were carried out at daily time steps, we only used aggregated monthly data for GCM evaluation. This analysis can be viewed as a comparison of the GCM-simulated climate data to the CRU monthly gridded climate products, in the perspective of hydrological modeling, rather than as a real hydrological application. Therefore, we make no attempt to evaluate the performance and sensitivity of applied hydrological models by considering other hydrological observations (such as streamflow and peak flow).

In this regard, we satisfy both the first and second principles of Wilby (2010) by using observational data that was specifically designed as a validation target for GCMs and using it in a context that are comparable both in spatial and temporal resolutions. We selected very minimalistic indicators (in order to meet the third principle) to evaluate GCM performance and contrasted the predicted changes in those indicators by the GCMs under future climate scenarios for the 2006–2100 period. We selected the most pessimistic RCP 8.5 (Representative Concentration Pathway scenario for the upcoming Fifth Assessment Report of the Intergovernmental Panel on Climate Change) with largest changes in climate forcings. This scenario anticipates over 8.5 Wm^2 additional radiative forcing due to continuously rising accumulation of greenhouse gases in the atmosphere to an equivalent of 1370 ppm CO_2 concentration by 2100.

We applied a parsimonious water-balance model configuration to minimize the numerous uncertainties that are introduced via model sensitivity to input variables and parameterization. We tested the hydrological model output at the grid cell level providing evaluation by discharge categories. Discharge, which is closely linked to catchment area upstream, is an integrated measure over a larger area. While we didn't expect good performance for small rivers represented (which could be as small as a single grid cell along the underlying gridded network), we did anticipate improvement in GCM performance for larger rivers, which have catchment areas comparable in size to the typical GCM grid cell size that should eliminate the need for downscaling.

23.2 Hydrological modeling experiment

We carried out hydrological modeling experiments with contemporary and future climate forcings using the Water Balance/Transport Model (WBM$_{plus}$) developed by Vörösmarty et al. (1989), Vörösmarty, Federer, and Schloss (1998), and Wisser et al. (2008, 2010). WBM$_{plus}$ is a highly configurable modeling platform, operating on simulated topological gridded networks (Vörösmarty et al. 2000) at various resolutions (ranging between a few hundred meters to tens of kilometers grid cell size). WBM$_{plus}$ allows hydrological simulations with different degrees of complexity offering multiple implementations of core components (for example, nine methods for calculating potential evapotranspiration (Federer, Vörösmarty, and Fekete 1996; Vörösmarty, Federer, and Schloss 1998). For the present work, we chose a parsimonious configuration that only requires near surface air temperature and precipitation as climate input and subsequently has limited needs for land surface parameterization.

We carried out water-balance simulations using state-of-the-art GCM outputs for contemporary and future climate scenarios and contrasted a few key statistical characteristics of the simulated river flows (long-term mean, minimum and maximum monthly flows and their standard deviations) using observed twentieth-century forcings. Our goal was to evaluate the predicted changes in these key flow characteristics with respect to the GCM's contemporary performance.

23.2.1 Climate forcings from global circulation models

Our original plan was to download outputs for 5–6 global circulation models from the Earth System Grid (www.earthsystemgrid.org), which are part of the Coupled Model Intercomparison Project phase 5 (http://cmip-pcmdi.llnl.gov/cmip5, CMIP5). The assembling of CMIP5 data archive is still in progress as large modeling groups are continuously submitting new modeling results in support of the IPCC Fifth Assessment Report (AR5). We originally planned to assemble a baseline ensemble of multiple model outputs by a weighted averaging of the ensemble members according to their performance in depicting contemporary climate.

The Earth System Grid (ESG) is a unified data portal operated by large modeling centers. The ESG offers common web interface (which is uniform across the participating institutions) to search the available data and provide download capabilities. It lists 35 GCMs that shrinks rapidly when common model configurations are specified. We found 14 GCMs that had both "historical" and RCP8.5 simulations. Unfortunately, ESG centers don't mirror the data from one another; instead they redirect the users to other data centers according to the requested data. As a consequence, downloading multiple model outputs requires transferring large amounts of data across the globe rather than accessing the nearest data server. Furthermore, we encountered enormous difficulties in downloading data even from data centers in the US. For instance, our attempt to download GCM outputs from Community Climate System Model (CCMS), hosted by the National Centers for Atmospheric Research (NCAR), failed repeatedly. Neither the Goddard Institute of Space Sciences (NASA GISS) nor the Geophysical Fluid Dynamic Laboratory (NOAA-GFDL) had readily available GCM results for historical and RCP8.5 future trajectories. While we attempted to download ten GCMs out of the available 14, we only succeeded to complete five full data sets both for contemporary and future scenarios. Out of the five successfully downloaded simulations, we dropped two because of their inconsistencies with the other three models. The HadGEM2-ES model from the Hadley Center in the UK had a simplified handling of the annual cycle that used a 360 day calendar. The INMCM4 model from the Institute from Numerical Mathematics provided a 17-layer precipitation (unlike the single layer outputs from other models). Ultimately, we were able to utilize the CanESM2 model from the Canadian Centre for Climate Modeling and Analysis (CCCma), Victoria, BC, Canada and the MIROC5 model from the Atmosphere and Ocean Research Institute (AORI) at the University of Tokyo, Chiba, Japan and MRI-CGCM3 model from the Meteorological Research Institute, Tsukuba, Japan. Figure 23.1 shows the annual average temperature and precipitation under contemporary and future scenarios over the continental landmass according to these three models.

The three models showed clear shortcoming in depicting the contemporary climate and have significant bias compared to observed climate variations depicted by the gridded data products from the Climate Research Unit at the University of East Anglia (New, Hulme, and Jones 2000; Mitchell *et al.* 2004). While none of the models showed any clear trends during the twentieth century and they seemed to miss the bimodal temperature rise throughout the twentieth century (a rise in the first four decades followed by a modest decrease starting in the early 1940s and a renewed rise starting in the 1950s), all models showed almost a steady increase in both air temperature and, to a lesser degree, in precipitation for the twenty-first century.

23.2.2   Discharge responses to changing climate

We applied the GCM-modeled and CRU-observed forcing in a global water balance and transport model to assess the combined effect of changing air temperature and precipitation on the corresponding discharge regime. We carried out a cell-by-cell comparison of simulated river discharge responses over a $30'$ gridded network on a geographic longitude-latitude grid (Vörösmarty *et al.* 2000) by calculating the symmetric relative difference between GCM vs. CRU driven estimates (Q_{GCM} and Q_{CRU}, respectively) expressed as: $\frac{Q_{GCM}-Q_{CRU}}{|Q_{GCM}|+|Q_{CRU}|}$. The symmetric relative difference that can range between $\pm100\%$ regardless of overestimation and underestimation (in contrast with normal relative differences that can be in the $-100-\infty\%$ range) has identical behavior for overestimates and underestimates. We performed the same comparisons for mean annual, monthly minimum and maximum discharge calculated from the contemporary (1901–2000) and future (2006–2100) periods.

The first row in Figure 23.2 shows the results of the mean annual discharge comparisons for the three GCMs summarized by discharge categories and indicated that discharge estimates based on GCM forcing had large deviations from those based on

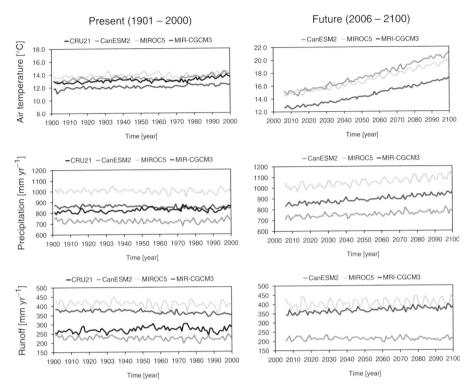

Figure 23.1 Continental, annual average air temperature, precipitation and corresponding simulated runoff over the twentieth century and predicted for the twenty-first century. The three climates model significantly departs from the observed air temperature and precipitation records depicted by the gridded observations from the Climate Research Unit of East Anglia (colored as black lines in the contemporary time series). The rapid rise of the air temperature and the somewhat more modest increase in predicted precipitation appears to cancel out resulting in a largely similar annual runoff regime compared to contemporary climate.

observed climate forcings from CRU. The second row in Figure 23.2 shows a similar relative difference between long-term mean annual discharge under present and future climate. The predicted changes for all discharge categories were well below the discrepancies between CRU observed versus GCM climate for contemporary estimates.

The same comparison carried out for monthly maximum and minimum flows (Figure 23.3) showed a similar pattern. We only present the maximum flow comparison since discrepancies were very similar. Clearly, the predicted changes in river discharge driven by global circulation models were far less than the differences between hydrological model simulations using GCM forcings versus observed climate data.

We performed a similar comparison considering the normalized standard deviations of mean annual and minimum and maximum flows under contemporary versus future climate. The discharge estimates from GCM climate forcings showed a similar lack of skill to reproduce the normalized discharge variations from observed climate forcings (Figure 23.4). The GCM performance was slightly better, depicting the normalized interannual variation of annual discharge over the twentieth century (Figure 23.4 first row), but the discrepancy was still higher than the predicted change for the twenty-first

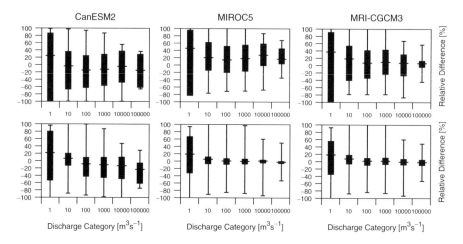

Figure 23.2 Relative difference between discharge estimates based on simulated versus observed climate forcing (first row) contrasted with the difference in discharge under present and future climate. The box plot shows the minimum, maximum, 15 and 85 percentiles and the mean differences on a cell-by-cell basis grouped by discharge categories.

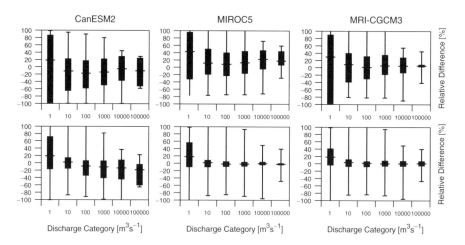

Figure 23.3 Relative difference between maximum monthly discharge estimates based on simulated vs. observed climate forcing (first row) contrasted with the difference in discharge under present and future climate.

century. The lack of GCM skills to capture the normalized discharge variations should be a concern for any downscaling and bias correction given the GCM's deficiencies in reproducing the twentieth-century climate in relative terms.

The poor correspondence between climate forcings from GCM and observations is well known to the scientific community (Ines and Hansen 2006; Sperna Weiland *et al.* 2010; NRC 2011). Capturing precipitation dynamics, which is the most critical input for hydrological computation, is one of the most important weaknesses of both GCMs and weather forecast models (Fekete *et al.* 2004), especially for water

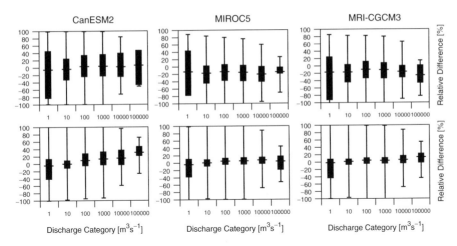

Figure 23.4 Relative difference between the contemporary normalized standard deviation of the annual discharge (1901–2000) estimates based on GCM versus CRU observed climate forcing (first row) contrasted with the future normalized standard deviation of the annual discharge (2006–2100) in the second row.

management purposes. The sensitivity of the hydrological system to change in precipitation, called climate elasticity of runoff (Fu, Charles, and Chiew 2007; Chiew *et al.* 2009), is found to be around one-to-three (1% change in precipitation yields 3% change in runoff (Chiew *et al.* 2009)). The experimental result was not surprising considering that climate elasticity expressed as $\varepsilon\,(P,R) = \frac{dR/R}{dP/P} = \frac{dR}{dP}\frac{P}{R}$ would yield $\varepsilon\,(P,R) = \frac{1}{\rho}$, assuming that the absolute change in precipitation (ΔP) translates to the same amount of absolute change in runoff (ΔR), which is the reciprocal of the runoff ratio $(\rho = R/P)$. Considering annual runoff $(\sim 300\,\text{mm yr}^{-1})$ and continental precipitation $(\sim 820\,\text{mm yr}^{-1})$ leads to a similar 1:2.7 ratio. The reasonable performance of this crude error propagation assessment in the hydrological systems indicates that, despite the numerous and often complex nonlinear processes in runoff generation, ultimately the excess water from the hydrological systems is still dominated by the amount of precipitation in a wide range of climate conditions, and the runoff response to climate change (primarily to precipitation) is reasonably linear.

Numerous "bias-correction" procedures have been proposed (Wilby, Hay, and Leavesley 1999; Piani *et al.* 2010; Hagemann 2011) to compensate for the GCM deficiencies ranging from simple "delta" or ratio method when contemporary climate data are adjusted according to the delta change or ratio between the present and future climate according to the GCM simulations (Lenderink, Buishand, and Van Deursen 2007) to more sophisticated statistical approaches (Piani *et al.* 2010; Hagemann 2011) that all depend on accurate knowledge of the contemporary conditions. Global Circulation Models and Regional Circulation Models (RCMs) have questionable value in reflecting anthropogenic forcings and future climate change, and, as a consequence, in their current state are not ready for operational use in relatively short-term (decadal) predictions needed by the water management community (Pielke and Wilby 2012). Global Circulation Models, on their own, often give unrealistic climate forcings, even for longer term climate change assessment that can only be made to better correspondence to real climate by incorporating contemporary observations.

The climate-modeling community at some point will have to face serious questions of how much the highly complex and chaotic climate system is predictable and how much improvement can be expected from refined models with detailed model physics. The World Modeling Summit for Climate predictions in 2008 (http://wcrp.ipsl.jussieu.fr/Workshops/ModellingSummit, accessed July 11, 2012) articulated the need for new supercomputing facilities that had a thousand times the capacity of the fastest high-performance computing facilities available to climate scientists (Dessai *et al.* 2009). The estimated $1 billion price tag is comparable to the 60 billion yen ($700 million in 2002) investment by Japan to build their Earth Simulator, which was the fastest computer from 2002–2004, but needed a complete replacement by 2009. Most modeling centers typically operate less expensive infrastructure (for example, NCAR's new supercomputing center in Wyoming, will have a total cost of $100 million, where $70 million will be spent on a new building, and the remaining $30 million will be needed for the computing equipment that will be 15–20 times faster than NCAR's existing facilities in Boulder, CO (http://www.wyomingbusiness.org/news/article/ncar-s-construction-on-schedule/5303, accessed July 11, 2012). The need for increasingly faster computing resources is undeniable, but the last 20 years may serve as a guide to what improvements can be expected from climate models. Since "Moore's law" dictates that CPU power doubles every 18 months (a rate that the computer industry was able to maintain steadily in the past) the computational power increased $2^{13} = 8192$ times in 20 years, which is a magnitude more than the World Modeling Summit participants were hoping for in upcoming years. Despite this increase in computing power, the uncertainties in climate models increased at the same time.

To put these investment costs into perspective, average satellite missions recommended for NASA (NRC 2007), planned for a three year operation, carry a $300–$400 million price tag. Alternatively, a $100 million annual budget is sufficient to operate 5000 discharge gauges (which would be the equivalent of operating one station per the typical 2.5° GCM grid box) providing the most accurate measurement of the hydrological cycle (Grabs, De Couet, and Pauler 1996; Gutowski, Chen, and Otles 1997) at the average cost of $20K/station (according to USGS and Environment Canada).

Considering the monotonic and rather gradual changes in climate according to the latest generation of GCMs (Figure 23.1) that are expected to remain within climate variation for precipitation and runoff responses, the stationarity assumption in dominant climate forcings seems to be justified (Stakhiv 2011), especially for water-management design purposes. Given the short (30–40 year) time span of water management designs, one could argue that continued monitoring and regular updates in stochastic probability distributions will provide sufficient adaptation pathways for water managers in the upcoming decades. Although stochastic (or black box) methods suggested by Anagnostopoulos *et al.* (2010) obviously lack the capability to foresee sudden changes in the probabilistic behavior of the represented system, regular updates would clearly allow their gradual adaptation. Deterministic modelers often criticize stochastic methods for the failure to represent the governing physics of the underlying observed data. While stochastic methods indeed lack the formal representation of the governing physical laws, implicitly they are just as much constrained by physical principles as deterministic models via the observations they are drawn from.

Stochastic simulation approaches are clearly inadequate, especially when sudden changes in the hydrological system results in nonlinear, sometimes step function responses, but these nonlinear behaviors are highly regional, and often in response to direct human activities (reservoir operation, water diversion, consumptive water uses, land use changes) that are not depicted by GCMs, and require deterministic hydrological modeling, potentially combined with regional climate modeling.

Water managers can likely accept stationarity in large-scale climate forcings (updated regularly with adequate observations) in the upcoming decades, but will need to prepare for intensive monitoring and assessment of changes in processes affecting the hydrological cycle. Better water management and better resource allocation of scarce water resources will require significant investment in monitoring infrastructure.

23.3 Contemporary observations

The period since the early 1990s has seen a steady decline in many aspects of our observational capabilities to monitor planet Earth (Shiklomanov, Lammers, and Vörösmarty 2002; Vörösmarty *et al.* 2001). This development is particularly troubling as it coincided with elevated interest in climate change and the desire to initiate actions to curb carbon emissions, which is regarded as the dominant anthropogenic force in changing our planet. The First Assessment Report from Intergovernmental Panel on Climate Change (IPCC 1991) provided detailed justification for more comprehensive monitoring and particularly emphasized the need for long-term observational records. In 2007 both the *in situ* monitoring network and the number of operational satellites were declining (NRC 2007).

The decline of *in situ* networks is partly attributed to the collapse of the Soviet Union and aligned countries, and civil wars in the developing world (mostly in Africa), but not limited to troubled countries. Developed nations like the United States or Canada also cut back on investment in various monitoring networks based on a series of budget cuts. The US Geological Survey justified these reductions in the stream gauge monitoring network with long observational records on the basis that the statistical characteristics are well established for those rivers and there is no need for continuous updates.

Global data centers (for example, UNEP GEMS/Water and WMO GRDC) continuously struggle to collect data and they are tasked with responsibilities well beyond their resources. The scientific community, which often expressed strong dissatisfaction with the data availability and the restrictive data policies (imposed on the data centers by donor partner agencies) by calling the data centers "data cemeteries," was not nearly as vocal advocating more international collaboration in collecting essential Earth observations. Instead, the lack of data sharing and investments in comprehensive *in situ* monitoring were regarded as insurmountable obstacles, and ironically were used as a justification for inferior remote-sensing alternatives. This surrender to perceived or real obstacles is particularly striking in the view of otherwise highly vocal advocacy to initiate much more costly mitigation strategies to combat climate change. One has to wonder, then, if international cooperation on operating adequate monitoring systems is deemed impossible, how can legally binding agreements on carbon emissions succeed?

Remote-sensing solutions, sometimes promoted in lieu of *in situ* monitoring, has only slightly better budgetary prospects. The National Research Council's (USA) review (NRC 2007), which provided recommendations for future research missions in the next decade and beyond, found similar declines in NASA's budget to fly new satellite missions. Successful missions like the Landsat satellites, which were in operation since 1972, were repeatedly threatened with elimination. The National Polar Orbiting Environmental Satellite Systems (NPOESS) to be operated by NOAA, which was meant to replace the aging NASA's Earth Observing System (EOS) satellites (Terra, Aqua), saw a series of budget cuts before it was entirely discontinued in February 2010 and will be replaced by separate, but yet-to-be-defined, civilian and military missions.

The Tropical Rainfall Measuring Mission (TRMM) satellite launched in 1997 (http://science.nasa.gov/missions/trmm, accessed July 11, 2012) and designed for three years operation is still in orbit after repeated efforts to extend its life time so the core satellite in the Global Precipitation Mission (its follow up) could have overlapping periods of operation. Despite the grossly extended life-time (TRMM flying for over nine years beyond its original mission), repeated delays in the GPM launch prevents simultaneous operation that would be essential to perform cross-calibration of the TRMM and GPM sensors and ensure observational continuity.

Instrumentation-based observational continuity is a severe problem for both *in situ* and remote sensing monitoring. Instrument changes and relocation plague the consistency of the retrieved information. Remote sensing appears more prone to repetitive calls for new innovation at the price of losing continuity with past observations, but *in situ* monitoring has also undergone numerous instrument upgrades (changing sensors, recording procedures, locations) without careful evaluation of how the equipment change would affect the observational records.

Numerous problems in current monitoring capabilities have been identified, ranging from instrument bias due to site location changes—for example, urban heat island effects (Ferguson and Woodbury 2007) and snow undercatch, particularly in windy conditions when precipitation gauges capture only a portion of the precipitation (Legates and DeLiberty 1993). The solution to most of these problems appears to be an engineering challenge rather than the lack of scientific knowledge. One would think that our current level of understanding of the climate and hydrological system would enable scientists and engineers to design and operate adequate monitoring networks. The recent scientific quest in finding the missing heat to explain the warming hiatus in the last ten years (Trenberth and Fasullo 2010) is not only a disturbing testimony of the apparent gaps in our monitoring, but ultimately it makes a travesty of the long-term credibility of climate science.

23.4   Conclusions

Water-resource management for a growing population will require significant improvement in resource allocation and planning for optimal operation of water infrastructure regardless of climate change. Water resources are particularly vulnerable to climate variability, which is expected to remain the dominating factor in the severity of the extreme events (drought or floods) exacerbated by climate change. While the large-scale climate change patterns are anticipated to remain gradual and monotonic, regional and local nonlinear responses to other drivers determining water-resource vulnerabilities will likely disturb the stationarity of the probabilistic characteristics of the hydrological systems. Water managers will need to deploy a range of deterministic regional hydrological and climate models that can depict those regional and local disturbances along with a new series of monitoring and surveying networks as part of an adaptive management strategy that are necessary to identify the alterations in the hydrological system.

The current generation Global Circulation Models are still in their infancy and inaccurate for operational decision making for water management and for designing future systems. Therefore, the best option will remain to operate adequate hydrometeorological monitoring networks to inform managers better about emerging trends, changes, and discontinuities. Stationarity in large-scale climate forcing will remain a reasonable approximation for decades to come that will allow adaptation to changing climate if properly updated regularly with new observations. Adequate monitoring records will inherently capture the changes in the hydrological system and will provide guidance for future generations to better understand the progression of climate change.

The intensified climate change debate in the period since the early 1990s has been coupled with an incongruent steady decline in monitoring capabilities (particularly *in situ*, but remote sensing as well), which is just the opposite of the recommendation of the First Assessment Report from IPCC. Our generation is on the verge of leaving mostly computer simulations to our descendants, who will need to continue chasing proxy data in sporadic satellite measurements and fragmented *in situ* observations to reconstruct climate change discontinuities. Operating improved monitoring networks is not only in the interest of our generation to support better water management and other climate-related policy decisions, but is our obligation to the next generation, even if taking the appropriate steps to mitigate climate change fails. Adaptation is necessary whether or not mitigation succeeds.

References

Anagnostopoulos, G.G., Koutsoyiannis, D., Christofides, A., *et al.* (2010) A comparison of local and aggregated climate model outputs with observed data. *Hydrol. Sci. J.*, **55**, 1094–110, doi:10.1080/02626667.2010.513518.

Chiew, F.H.S., Teng, J., Vaze, J., *et al.* (2009) Estimating climate change impact on runoff across southeast Australia: Method, results, and implications of the modeling method. *Water Resour. Res.*, **45**, 1–17, doi:10.1029/2008WR007338.

Dessai, S., Hulme, M., Lempert, R., and Pielke, R. Jr.,. (2009) Do we need better predictions to adapt to a changing climate? *Eos Trans. AGU*, **90**, 111, doi:10.1029/2009EO130003.

Dessai, S. and Van der Sluijs, J.P. (2007) *Uncertainty and Climate Change Adaptation—A Scoping Study*. Report NWS-E-2007-198, Copernicus Institute for Sustainable Development and Innovation, Utrecht University, Utrecht.

Döll, P., Kaspar, F., and Lehner, B. (2003) A global hydrological model for deriving water availability indicators: model tuning and validation. *J. Hydrol.*, **270**, 105–34. doi:10.1016/S0022-1694(02)00283-4.

Federer, C.A., Vörösmarty, C., and Fekete, B. (1996) Intercomparison of methods for calculating potential evaporation in regional and global water balance models. *Water Resour. Res.*, **32**, 2315–21.

Fekete, B.M., Vörösmarty, C.J., Roads, J.O., and Willmott, C.J. (2004) Uncertainties in precipitation and their impacts on runoff estimates. *J. Clim.*, **17**, 294–303.

Ferguson, G. and Woodbury, A.D. (2007) Urban heat island in the subsurface. *Geophys. Res. Let.*, **34**, 1–4, doi:10.1029/2007GL032324.

Fu, G., Charles, S.P., and Chiew, F.H.S. (2007) A two-parameter climate elasticity of streamflow index to assess climate change effects on annual streamflow. *Water Resour. Res.*, **43**, w11419.

Grabs, W., De Couet, T., and Pauler, J. (1996) *Freshwater Fluxes from the Continents into the World Oceans based on Data of the Global Runoff Data Base*, **GRDC Report 10**, Global Runoff Data Center, Koblenz, Germany.

Gutowski Jr.,, W.J., Chen, Y., and Otles, Z. (1997) Atmospheric water vapor transport in NCEP-NCAR reanalyses: comparison with river discharge in the central United States. *Bull. Am. Meteor. Soc.*, **78**, 1957–69.

Hagemann, S., Chen, C., Haerter, J.O., *et al.* (2011) Impact of a statistical bias correction on the projected hydrological changes obtained from three GCMs and two hydrology models. *J. Hydrometeor.*, **12**, 556–78.

Ines, A.V.M. and Hansen, J.W. (2006) Bias correction of daily GCM rainfall for crop simulation studies. *Agric. Forest Meteorol.*, **138**, 44–53.

IPCC (1991) *IPCC First Assessment Report*, Intergovernmental Panel on Climate Change, Cambridge University Press, Cambridge.

Koutsoyiannis, D., Efstratiadis, A., Mamassis, N., and Christofides, A. (2008) On the credibility of climate predictions. *Hydrol. Sci. J.*, **53**, 671–84, doi:10.1623/hysj.53.4.671.

Kundzewicz, Z.W. and Stakhiv, E.Z. (2010) Are climate models "ready for prime time" in water resources management applications, or is more research needed? *Hydrol. Sci. J.*, **55**, 1085–9, doi: 10.1080/02626667.2010.513211.

Legates, D.R. and DeLiberty, T.L. (1993) Precipitation measurement biases in the United States. *Water Resour. Bull.*, **29**, 855–61.

Lenderink, G., Buishand, A., and Van Deursen, W. (2007) Estimates of future discharges of the river Rhine using two scenario methodologies: direct versus delta approach. *Hydrol. Earth Syst. Sci.*, **11**, 1145–59, doi:10.5194/hess-11-1145-2007.

Milly, P.C.D., Betancourt, J., Falkenmark, M., *et al.* (2008) Stationarity is dead: Whither water management? *Science*, **319**, 573–4.

Mitchell, T.D., Carter, T.R., Jones, P.D., *et al.* (2004) A comprehensive set of high-resolution grids of monthly climate for Europe and the globe: the observed record (1901–2000) and 16 scenarios (2001–2100). Tyndall Centre for Climate Change Research, http://www.ipcc-data.org/docs/tyndall_working_papers_wp55.pdf (accessed June 30, 2012).

New, M., Hulme, M., and Jones, P.D. (1999) Representing twentieth century space-time climate variability: I. Development of a 1961–1990 mean monthly terrestrial climatology. *J. Climatol.*, **12**, 829–56.

New, M., Hulme, M., and Jones, P. (2000) Representing twentieth-century space-time climate variability. Part II: Development of 1901–1996 monthly grids of terrestrial surface climate. *J. Clim.*, **13**, 2217–38.

NRC (Committee on Earth Science and Applications from Space) (2007) *Earth Science and Applications from Space: National Imperatives for the Next Decade and Beyond*, National Research Council, National Academies Press, Washington, D.C.

NRC (Committee on Hydrological Sciences) (2011) Global change and extreme hydrology: Testing conventional wisdom committee on hydrologic science. National Research Council, US National Academies Press, Washington, D.C., p. 60.

Piani, C., Weedon, G.P., Best, M., *et al.* (2010) Statistical bias correction of global simulated daily precipitation and temperature for the application of hydrological models. *J. Hydrol.*, **395**, 199–215, doi:10.1016/j.jhydrol.2010.10.024.

Pielke Sr., R.A., and Wilby, R.L. (2012) Regional climate downscaling—what's the point? *EOS*, **93**, 52–3.

Shiklomanov, A.I., Lammers, R.B., and Vörösmarty, C.J. (2002) Widespread decline in hydrological monitoring threatens Pan-Arctic Research. *Eos Trans. AGU*, **83** 16–17.

Sperna Weiland, F.C., Van Beek, L.P.H., Kwadijk, J.C.J., and Bierkens, M.F.P. (2010) The ability of a GCM-forced hydrological model to reproduce global discharge variability. *Hydrol. Earth Syst. Sci.*, **14**, 1595–1621, doi:10.5194/hess-14-1595-2010.

Stakhiv, E.Z. (2011) Pragmatic approaches for water management under climate change uncertainty. *J. Amer. Wat. Resour. Assoc.*, **47**, 1183–96, doi:10.1111/j.1752-1688.2011.00589.x.

Trenberth, K.E., and Fasullo, J.T. (2010) Climate change. Tracking Earth's energy. *Science*, **328**, 316–17.

Vörösmarty, C., Askew, A., Grabs, W., *et al.* (2001) Global water data: A newly endangered species. *Eos. Trans. AGU*, **82**, 54 http://www.agu.org/pubs/crossref/2001/01EO00031.shtml.

Vörösmarty, C. J., Federer, C.A., and Schloss, A.L. (1998) Potential evaporation functions compared on US watersheds: Possible implications for global-scale water balance and terrestrial ecosystem modeling. *J. Hydrol.*, **207**, 147–69.

Vörösmarty, C.J., Fekete, B.M., Meybeck, M., and Lammers, R.B. (2000) Geomorphometric attributes of the global system of rivers at 30-minute spatial resolution. *J. Hydrol.*, **237**, 17–39.

Vörösmarty, C.J., Moore III,, B., Grace, A.L., *et al.* (1989) Continental scale models of water balance and fluvial transport: An application to South America. *Global Biochem. Cycles*, **3**, 241–65.

Wilby, R.L. (2010) Evaluating climate model outputs for hydrological applications. *Hydrol. Sci. J.*, **55**, 1090–3, doi:10.1080/02626667.2010.513212.

Wilby, R.L., Hay, L.E., and Leavesley, G.H. (1999) A comparison of downscaled and raw GCM output: implications for climate change scenarios in the San Juan River basin, *Colorado. J. Hydrol.*, **225**, 67–91, doi:10.1016/S0022-1694(99)00136-5.

Wisser, D., Fekete, B.M., Vörösmarty, C.J., and Schumann, A.H. (2010) Reconstructing 20th century global hydrography: a contribution to the Global Terrestrial Network- Hydrology (GTN-H). *Hydrol. Earth Syst. Sci.*, **14**, 1–24, doi:10.5194/hess-14-1-2010.

Wisser, D., Frolking, S.E., Douglas, E.M., *et al.* (2008) Global irrigation water demand: Variability and uncertainties arising from agricultural and climate data sets. *Geophys. Res. Let.*, **35**, doi:10.1029/2008GL035296.

24

In Search of Strategies to Mitigate the Impacts of Global Warming on Aquatic Ecosystems

Justin D. Brookes[1], Martin Schmid[2], Dominic Skinner[1], and Alfred Wüest[2]

[1]*School of Earth and Environmental Science, University of Adelaide, Australia*
[2]*Eawag, Surface Waters—Research and Management, Kastanienbaum, Switzerland*

24.1 Global warming and impacts on aquatic ecosystems

Elevated levels of atmospheric carbon dioxide (CO_2) are responsible for global warming changing the climate and hydrology. There is evidence that inland freshwaters have warmed as the atmosphere has warmed (Webb *et al*. 2008). In Central Europe, for example, rivers have experienced a temperature increase since the 1970s of $\sim0.04\,°C$ year^{-1}, almost identical to the air temperature increase during the same period. When added to the atmospheric warming of $\sim0.012\,°C$ year^{-1} prior to the 1970s, this equates to an overall temperature increase in the surface waters in Central Europe of $\sim2\,°C$ since the nineteenth century. Indeed, this warming trend has been observed globally (O'Reilly *et al*. 2003; Arhonditsis *et al*. 2004; Adrian *et al*. 2009), and global mean surface air temperatures have increased by $0.74\pm0.18\,°C$ over the past century (Trenberth *et al*. 2007). Recent climate change models have predicted that an increase in air temperature between 2 and 11 °C could be expected with a doubling of the atmospheric CO_2 concentration (Stainforth *et al*. 2005) but the likely temperature increase suggested by the IPCC is between 1.1 and 6.4 °C for 2090–2099 compared to 1980–1999 (Solomon *et al*. 2007). On a global scale, the water surface temperature is increasing in a similar trend (Schneider and Hook 2010;

Climatic Change and Global Warming of Inland Waters: Impacts and Mitigation for Ecosystems and Societies, First Edition. Edited by Charles R. Goldman, Michio Kumagai and Richard D. Robarts.
© 2013 John Wiley & Sons, Ltd. Published 2013 by John Wiley & Sons, Ltd.

van Vliet *et al*. 2011), and there are already effects on aquatic ecosystems on several fronts. Among the most fundamental effects are the changes to physical processes, as they are of a profound nature for all ecosystem processes.

Modification of lake heat budgets due to global warming has consequences for water quality, biogeochemistry and lake habitats (see also Chapters 2, 3, 6, 8, 12, 14, 16–20). A major concern is that an intensification of heating will increase thermal stratification within lakes, stabilizing the water column, depleting hypolimnetic oxygen, and modifying sediment biogeochemical cycling (Malmaeus *et al*. 2006; see also Chapters 6 and 14). One implication of this is a greater propensity for buoyant bloom-forming cyanobacteria (Paerl and Huisman 2008, 2009; Wagner and Adrian 2009), as was observed during the 2007 European summer heat waves (Jöhnk *et al*. 2008). Extreme events such as heat waves, drought, and floods are predicted to increase under climate change (Schär *et al*. 2004; IPCC 2011).

Changes to hydrology are likely to alter nutrient delivery to lakes as rainfall distribution as well as runoff and particle transport characteristics change. For example, in temperate southeastern Australia, a 10% reduction in rainfall is predicted over the next 50 years but the rainfall is likely to be distributed in fewer, more intense rain storms. This reduction is of concern since reduced rainfall is predicted to result in a 20% to 30% decrease in stream flow as the increased period between rain events will lead to subsaturated soils requiring greater rainfall before runoff results (Chiew *et al*. 1995). Furthermore, a 1 °C increase in temperature will result in a 2 to 3.5% reduction in mean annual stream flow (Chiew 2006). The southwest of Australia has seen a tremendous rainfall decline since the 1970s (Timbal *et al*. 2006), resulting in mean annual runoff into reservoirs decreasing from an average of $0.34 \, \text{km}^3 \, \text{year}^{-1}$ from 1911 to 1974 to an average of $0.06 \, \text{km}^3 \, \text{year}^{-1}$ for the period 2006 to 2010.

Other impacts of climate change on lakes are likely to influence water quality and habitat. Waterborne diseases (McMichael, Woodruff, and Hales 2006), invasive species (Rahel and Olden 2008), warm water pollution (Schindler 2001), and risk of wildfires (Williams, Karoly, and Tapper 2001) are all predicted to increase from climate change. Not surprisingly, the major focus of climate-change science relating to lakes has been on lakes as sentinels of change (Adrian *et al*. 2009; Williamson, Saros, and Schindler 2009) and exploration of how lake habitats will respond to a changing climate.

Climate change is not the only driver of recent temperature changes in natural waters. Damming accelerates the effect of warming of surface waters as longer residence times and subsequent longer exposure of water to the warmer atmosphere causes an additional increase in temperature. The retention of water in reservoirs means that flows are slower (Friedl and Wüest 2002), water remains longer in the landscape with extended surface area and, as a result, there is greater opportunity for water to absorb solar energy. This leads to warmer waters downstream of dams. However, lakes and rivers also present opportunities to mitigate some of the impact of climate change. In this paper we explore whether lakes, rivers, and other inland water bodies may be manipulated to address the warming trends that are expressed in aquatic habitats. The aim is to critically evaluate strategies and options that affect heat budgets of surface waters.

The opportunities for managers of lakes and rivers to mitigate climate change and the impacts of climate change are classified into three classes:

- those that lead to a reduction in temperature;
- those that influence the intensity of temperature stratification;
- those that affect greenhouse gas concentrations in the atmosphere.

24.2   First level—potential for reducing temperature in aquatic habitats

24.2.1   Heat pumps

A heat pump moves heat from an environmental compartment to an infrastructure compartment using mechanical work. In Europe this technology is increasingly employed for the heating of residential and commercial buildings. Of the newly built one-family houses in Switzerland today, $\sim$60% install a heat pump using groundwater (or less frequently surface water) as the source of heat. The government's goal for the next decade is to equip $\sim$400 000 buildings (out of a total of 1.4 million) with heat-pumps and in $\sim$3 decades, oil is expected to be phased out for heating altogether. The use of groundwater (including surface waters) and wastewater for thermal pumps has advantages over air-air heat pumps in cool climates such as Switzerland: The groundwater is of higher temperature, has smaller temperature differential with the exchanged air, and therefore has a greater efficiency. The additional benefits are that heat is exchanged from natural waters to houses with a net loss of heat from the ground- and river water.

Although, heat pumps still require energy and therefore still emit greenhouse gases, the CO_2 emissions per MJ are significantly reduced. If we were to take Switzerland as an example, where the energy-specific CO_2 emissions are low due to the mix of hydroelectric and nuclear energy, the cumulative CO_2 emissions for a heat pump are between $36-47$ g $(kWh)^{-1}$ ($10-13$ g MJ^{-1}). Comparatively, in other European countries supplied with a different electricity mix, cumulative CO_2 emissions for an operating heat pump range between $133-183$ g $(kWh)^{-1}$ ($37-51$ g MJ^{-1}; Dones *et al.* 2007).

What could potentially be the saving if heat pumps were applied systematically and what temperature decrease could be expected? In the European Union, $\sim$57% of total household energy consumption is used for room heating (Chwieduk 2003). If we assume an annual household consumption of 4200 kWh the annual energy consumption for heating is $\sim$2400 kWh. CO_2 released per kWh during electricity generation from natural gas is 450 g $(kWh)^{-1}$, oil 500 g $(kWh)^{-1}$, and black coal 800 g $(kWh)^{-1}$. Therefore, the adoption of heat pumps for residential heating equates to an annual reduction of CO_2 emissions of $1-2$ t CO_2 year^{-1} per household compared with heating by electricity generated with fossil fuels.

What is the cooling potential of heat pumps? Again by considering the example of Switzerland, where the total heating energy is $\sim$270 PJ year^{-1}, we can estimate for a steady-state heat balance how much the temperature of the local runoff water would be affected. If all of this heating was theoretically derived from heat pumps with water in conduction with the heat source then the amount of cooling that would occur in the 40 km^3 year^{-1} of annual water runoff from Switzerland would amount to 1.6 °C. Currently there are $\sim$200 000 installed heat pumps with an energy output of 9 PJ year^{-1}, which would theoretically cool the river water by only $\sim$0.05 °C under the assumption of a steady-state flow. Although we have ignored the complexity of the seasonal heat budget, the calculation demonstrates that we are far away from offsetting the climate-induced $\sim$2 °C of surface water warming of the past century (see above). Installing an order of magnitude more heat pumping will potentially offset some of the observed water temperature increases due to atmospheric warming. However, unfortunately the cooling effect on the waters would occur mainly during winter. Therefore, we cannot expect a substantial offsetting of climate warming of the natural waters, even for a perfect implementation of the heat pumping scheme.

An interesting example has been set on Lake San Murezzan (Schmid and Dorji 2008), where a single heat pump is using a substantial part of the entire warming potential of one small lake for supplying heat to a hotel and a school. In winter, the lake's hypolimnetic temperature is cooled by the heat pump by $\sim$0.2 °C indicating that only a few such installations of comparable dimension would be acceptable to remain consistent with the relevant federal bylaw. In this particular case, the winter temperature was not much affected by climate change because most of the hypolimnetic volume is still of a dimictic nature. This may be different for lakes which are turning from dimictic to monomictic or even oligomictic (Livingstone 2008).

24.2.2   Cooling by hydropower operations

Hydropower production causes water to be stored and released at different elevations and with different exposure to solar heating and atmospheric exchange. For example, deep or particle-rich water can reside in reservoirs without much influence from the atmosphere. Therefore, the vertical structuring of withdrawal towers in reservoirs allows thermal regimes of downstream rivers to be altered without much additional effort. Especially in clear-water reservoirs, withdrawal towers allow managers to choose water at its natural temperature and to reduce artificial seasonal temperature shifts.

What is the cooling potential of hydropower operations? The management of reservoirs affect the thermal regime of the downstream waters in various complex and sometimes counterintuitive ways. When analyzing river water temperature changes, it is useful to distinguish between (i) the reaches immediately below a reservoir, (ii) the abstraction sections, and (iii) the receiving rivers below the water release after power production.

The thermal effect of hydropower reservoirs is usually to warm the water. As the main reservoir inflows, consisting of cold water from melting snow and glacier fields, occur during the warmest season, the heat uptake by the reservoir can become very high (Meier and Wüest 2004). This is especially the case for turbid reservoirs, where particle-laden river water is denser than the receiving reservoir water and the inflows plunge deep in the reservoir, transferring surface water downwards and thereby maintaining the surface layer cold. Clear water reservoirs, in contrast, keep their temperature of maximum density near 4 °C. The intense sunlight in alpine areas can penetrate deep into clear water and thereby strongly warm and stratify a reservoir in the same manner as for natural lakes. If a reservoir outlet is at the surface, the downstream water becomes much warmer than in the original natural river. However, if water withdrawal is deep, the temperature change is small and the main effect is the release of the warmed water after summer.

In the abstraction section, heat exchange with the atmosphere and the underground hardly changes. However, as a consequence of the diversion, a river is shallower than natural and as the rate of temperature change varies with water depth, the abstraction section adjusts faster to ambient temperatures. As a result, temperatures in abstraction sections are higher than natural during summer and lower than natural during winter, whereas the annual average is usually not altered much.

Below water releases, in contrast, average temperatures are higher than natural in winter, and cooler than natural in summer (Figure 24.1) flattening the natural seasonal regime. Over the entire year there remains usually a heat deficit, as the water passing through turbines is cooler (electricity production) than in a natural river bed, where dissipation heats the water by 0.24 °C per 100 m elevation drop.

As a consequence of the alterations of temperature and particles, dammed alpine rivers show less deep plunging into natural lakes downstream of reservoirs. In the example of Lake Geneva, where the River Rhone (Figure 24.1) is substantially affected by hydropower operations, deep density currents support deep convective mixing in the following winter by warming the deepest layers. As result, upstream hydropower dams have an effect on heat storage and deep-convective mixing in downstream lakes.

24.2.3 Selection of suitable temperature water for fish migration

Warming of aquatic habitats is of particular concern for heat-sensitive species, such as salmonids. The continued success of salmon populations relies on the once-in-a-lifetime river migration of individuals to natal spawning grounds and successful return of offspring to sea. Consequently, they are extremely vulnerable to shifts in stream habitat that may impact spawning, food resources or migration.

Salmon are an important economic commodity with the four largest fisheries, Alaska, Japan, USSR and Canada, producing 800 000 tonnes of salmon in 1985 (Eagle, Naylor, and Smith 2004). Salmon farming is outcompeting the wild salmon fishery

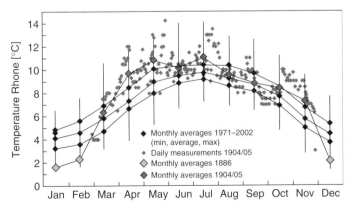

Figure 24.1 The current River Rhone temperature regime as a result of damming (and partly climate change) compared with data from one century ago. The black bars represent minima and maxima for the period 1982–2002. The hydropower operation has a "smoothing" effect on the temperature below the outlets in the Rhone River increasing the winter temperature to about 2 °C above natural and reducing summer temperatures by about 1 °C.

in some instances (Ruggerone *et al*. 2010) but the salmon fishing industry is still the biggest employer in the US state of Alaska (Eagle *et al*. 2004).

Energy and nutrient fluxes caused by spawning Atlantic salmon can be a significant contribution in rivers (Cederholm *et al*. 1999; Jonsson and Jonsson 2003). The marine inputs of nitrogen and phosphorus to the River Imsa, Norway were estimated to be 0.2% and 5% of the total river load, respectively, and are an important contributor to productivity of freshwater communities (Naiman *et al*. 2002) and adjacent terrestrial ecosystems (Hilderbrand *et al*. 1999a,b).

The decline of salmon numbers and associated nutrients has prompted concern for continued productivity in these rivers (Gresh *et al*. 2000; Schindler *et al*. 2003). Although declines of salmonid populations have been linked to overfishing, catchment modification, and pollution, climate change is increasingly implicated (Clews *et al*. 2010) and there is evidence of warming in the Pacific and Atlantic salmon habitat (Morrison *et al*. 2002; Friedland *et al*. 2003). Modified hydrology, varying discharge and lower summer flows may exacerbate the impact of warming (Clews *et al*. 2010).

Optimal growth temperature for most salmonid species varies little (Elliot 1981; Brett, Clarke, and Shelbourn 1982; Dickerson and Vinyard 1999; Edsall *et al*. 1999; Elliott and Hurley, 1999; Selong *et al*. 2001; Meeuwig *et al*. 2004; McMahon *et al*. 2007); however, thermal tolerance of species appears to differ and so some will be preferentially favored by warming temperatures displacing less heat-tolerant species such as cut-throat trout (Bear, McMahon, and Zale 2007). In general, temperatures exceeding 24 °C will be lethal (Johnstone and Rahel 2003) and optimal growth conditions are well below this (Selong *et al*. 2001).

Damming of watercourses for water supply and hydropower generation has often negatively affected migrating fish populations but these structures could now be used or modified to create the required in stream cooling to support these fisheries. Climate induced changes to a flow regime are also of concern. Gibson *et al*. (2005) identified that there could be dramatic changes to the hydrology with shifts in the timing of peak flows and reduced low flow velocities that persist for longer. They predicted this could lead to higher summer water temperatures, lower dissolved oxygen concentrations, and consequently reduced survival of larval fish.

Exploiting the vertical temperature stratification in reservoirs to extract water of the preferred temperature could mitigate some of the effects of global warming and increased water temperatures. Considerable effort has been spent assessing this mitigation strategy in the Fraser River, which drains a catchment of $230\,000\,\mathrm{km}^2$ (Foreman *et al*. 2001) and supports Canada's most economically valuable sockeye salmon fishery (McDaniels *et al*. 2010). This river has seen water temperatures increase by more than 1.5 °C over the last 40 years (Morrison, Quick, and Foreman 2002). Foreman *et al*. (2001) reported temperature at Hell's Gate on the lower Fraser River increased by 0.012 °C year^{-1} from 1941 to 1998 but there was strong interannual variability in flow and water temperature. Summers following El Niño events had $800\,\mathrm{m}^3\,\mathrm{s}^{-1}$ lower flows and temperatures of 0.9 °C below that in years with La Niña conditions.

Eight of the ten summers prior to 2009 were the warmest on record (McDaniels *et al*. 2010), which meant that the salmon encountered warmer water temperatures but there was also a shift in the timing of the salmon run (Cooke *et al*. 2004). Environmental flows need to consider water temperature but also food resources for growth and recruitment. Faster development of embryos may lead to early emergence of smaller fry. Early emergence may result in a mismatched timing with the spring bloom of zooplankton and limit feeding and growth opportunities for developing sockeye fry (McDaniels *et al*. 2010). However it remains that the returning adult individuals are considered most at risk from rising temperatures (McDaniels *et al*. 2010).

A proposal was developed in the 1990s to construct a two-level outlet adjacent to the Kenney Dam in British Columbia to achieve specific downstream water temperature objectives during sockeye salmon migration each summer. Modeling of Nechako River water temperatures showed that a maximum Kenney Dam release of $167 \, \mathrm{m^3 \, s^{-1}}$ at $10\,°C$ would be required to meet the downstream water temperature objectives (Mitchell, James, and Edinger 1995). The facility was never built, but it was the topic of a large and very emotional debate in British Columbia in the mid-1990s.

McDaniels *et al*. (2010) concluded that cold-water release to lower the water temperature of the Fraser main stem or spawning streams was not a viable option given the volumes of water that would be required. Deeper offtakes from Kenney Dam or construction of dams on other tributaries were viewed as impractical and likely to be ineffective. However, a panel of experts recommended temperature management investment in systems without lake influence (McDaniels *et al*. 2010), but it is unclear what strategy this would support other than the construction of spawning channels and off-channel spawning places. It could be argued that cold-water releases are one of the few options to maintain this fishery and strategic use of water could reduce further damage to the fishery until other management options are implemented.

Options remain for downstream releases but these need to balance economic benefit, electricity generation and ecological outcomes. Regions such as British Columbia with more abundant water resources ($312 \, \mathrm{km^3}$ in natural lakes and $209 \, \mathrm{km^3}$ in reservoirs; Schiefer and Klinkenberg 2004) may have more scope to make coldwater releases for temperature control achievable. Where environmental releases are required for downstream ecosystems the manipulation of discharge depth to optimize water temperature is a sound mitigation strategy to counteract warming. The strategy can also be used in reverse to maintain species that require warm water but are subjected to stress from cold hypolimnetic releases.

24.2.4 Shading river habitat

A simple strategy for cooling is to reduce the input of radiation. Energy from short- and long-wave radiation was generally the most important source of heat in the rivers studied by Webb and Zhang (1998), where net radiation accounted for more than 50% of the nonadvective daily heat gain. This was more typically the case with river sections with gradual slopes whereas in steep river sections the dissipation of kinetic energy can be a dominant heat source (Meier *et al*. 2003). Logically reducing input of radiation will decrease heat entering a river system.

Larson and Larson (1996) claimed, however, that shading does not control stream temperature, based on the notion that shade does not produce cooling and if the water is still in contact with air of greater temperature then energy will still be transferred to a water body. However, this view was not supported by the evidence of Webb and Zhang (1998) and Moosmann, Schmid, and Wüest (2005) Webb and Zhang (1998) in developing a heat budget for the river reaches in the Exe Basin (UK) showed that convective or sensible heat exchange between warm air and cooler streams comprised a relatively small portion of the total heat flux. They calculated that net radiation and friction contributed 56% and 22%, respectively, whereas sensible heat transfer, condensation and bed conduction contributed 13.2, 5.8 and 2.8%, respectively.

Strategies that consider shading of lakes with floating covers are impractical and may have water-quality implications because of reduced oxygen flux to the water from photosynthesis and reduced wind-induced mixing. However, shading strategies that use riparian vegetation along riverbanks are practical and have additional benefits of increasing bank stability (Simon and Collison 2002), light-covering for fishes near the banks, and reducing problematic attached algal growth (Mosisch *et al*. 2001).

Moosmann, Schmid, and Wüest (2005) estimated the potential of cooling a river by adding riparian vegetation. In this particular river, which meanders at low speed through open pastures, maximum summer temperatures were observed to exceed 25 °C, which was above the tolerance temperature for some of the fish species in the river, especially for the grayling (*Thymallus thymallus*). Model calculations showed that peak temperatures could be reduced by several degrees centigrade at the end of even relatively short shaded stretches of a few hundred meters length (Figure 24.2).

At smaller scales, the heat islands phenomenon associated with urbanization can increase heat effects as warm runoff from impervious surfaces travels rapidly to receiving streams (Herb *et al*. 2008). Attempted mitigation with unshaded storm water detention basins before release to streams has proved unsuccessful and so stream and wetland shading may be one of the few options for mitigation (Herb, Mohseni, Stefan 2009).

24.3 Second level—opportunities to modify heat distribution

The total heat budget of a lake will change with global warming but the distribution of heat energy within a lake is also important for many biogeochemical and biological processes. The preservation of stratification may be beneficial in some circumstances to maintain cool temperatures at the sediment surface and limit the rate of biogeochemical processes. On the other hand, a shallow mixed surface layer can also provide suitable conditions for bloom-forming cyanobacteria (Brookes *et al*. 1999). Stratified conditions leading to scums of bloom-forming cyanobacteria are promoted when energy is trapped within the surface layer and is greatest when particles or dissolved substances absorb radiation (Mazumder and Taylor 1994). The corollary to an increase in surface temperature of a lake is that radiative, sensible, and latent heat fluxes should also increase (Jones, George, and Reynolds 2005). The radiation absorbing properties of a lake can be manipulated to counteract climate impacts but the strategy may vary between lakes depending upon the requirement to minimize greatest environmental risk.

24.3.1 Nutrient reduction

There is a feedback between ecological processes and the thermal structure of a lake. Kumagai *et al*. (2000) performed a range of enclosure experiments in Lake Biwa, Japan, to see if nutrient addition and the resulting algal blooms could change the thermal stratification of a lake. They concluded that blooms of cyanobacteria could

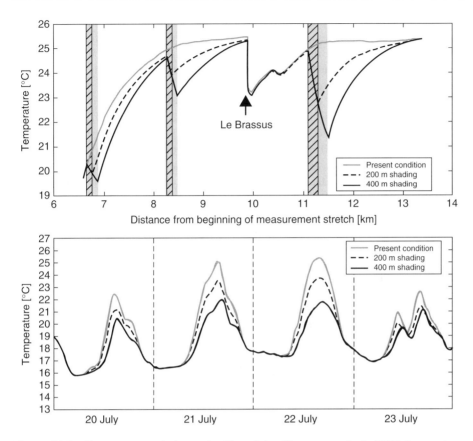

Figure 24.2   The upper panel shows the River Orbe (Moosmann *et al.* 2005) temperature development along the river at 3 p.m. on a hot day with grey: no shading, dashed (400 m): shading in the zones marked with black striped bars, and solid (800 m): shading in the zones marked with grey bars. The lower panel: temperature for four peak-summer days, as simulated at the end of the shaded stretch between 11.1 and 11.5 km in the upper panel. Gray: without shading; dashed: with a 200 m shaded stretch; solid: with a 400 m shaded stretch. The temperatures in the shaded sections drop by up to 5 °C during daytime (no changes at night time) relative to the current conditions of low riparian vegetation.

increase the available potential energy in the water column and create more stable stratification for their growth. This phytoplankton feedback on stratification can then be affected by nutrients (Kumagai *et al.* 2000) or grazing (Mazumder *et al.* 1990) as these are bottom-up and top-down controls on phytoplankton growth, respectively. The manipulation of phytoplankton community size is possible by controlling nutrients, however, while there have been some successes with biomanipulation for phytoplankton control it remains challenging and is not suitable in all situations. Boon *et al.* (1994) and DeMelo, France, and McQueen (1992) have advised that prudent lake management to reduce algal biomass should focus first on nutrient reduction and then consider biomanipulation.

Indeed, for Lake Constance there are both modeling studies (Rinke, Yeates, and Rothhaupt 2010) and historical data that implicate nutrient and phytoplankton in the intensification of stratification. Conversely, the reduction in nutrients that has occurred over the past three decades in Lake Constance has reduced surface temperature and stratification. Stitch and Brinker (2010) concluded that decreasing phosphorus concentrations will have a much stronger impact on phytoplankton productivity in Lake Constance than temperature change.

There are now notable examples of successful reoligotrophication, such as Lake Constance (Güde, Rossknecht, and Wagner 1998; Hase *et al*. 1998), Lake Veluwe (Ibelings *et al*. 2007), Lago Maggiore (Ruggiu *et al*. 1998), and Müggelsee (Köhler, Bernhardt, and Hoeg 2000). Investment in nutrient reduction will pay dividends but may take some time to be expressed in lake nutrient and chlorophyll concentrations. The ecological state of a lake during reoligotrophication may be a hysteresis as the restoration may also be affected by accumulated nutrients in sediment, the residence time of the lake and resilience of the alternate ecosystem state to change (Ibelings *et al*. 2007). There may also be food-web implications as the trophic structure is modified (Jeppesen, Jensen, and Søndergaard 2002).

Evidence suggests that the nutrient reduction task, to achieve greater water clarity, will become more difficult with further global warming. This is because increasing temperature encourages cyanobacterial growth (Paerl and Huisman 2008), greater phosphorus release from sediments (Jensen *et al*. 2006), longer phytoplankton growing seasons, and increasing hydrological variability, which can affect nutrient concentration through evaporative changes in lake volume (Özen *et al*. 2010).

Notwithstanding any hysteresis, reducing nutrient loads to lake ecosystems will reduce maximum chlorophyll biomass (Jeppesen, Jensen, and Søndergaard 2002; see also Chapter 10), reduce the incidence of problematic cyanobacterial blooms, and reduce heat capture within the surface layer of lakes. Returning lakes to oligotrophic status would make them less vulnerable to negative impacts of global warming as there would be less energy capture in surface water and reduction in issues associated with stratification such as cyanobacterial blooms (Brookes and Carey 2011) and anoxia (Matzinger *et al*. 2007).

24.3.2   Controlling particle inputs from glacier till

Where lakes are glacial fed, the till can impact heat capture and vertical distribution. Operating hydroelectric dams in alpine headwaters not only drastically cuts off high-flow occurrences, but also reduces downstream sediment transport as reservoirs trap particles and low competence streams carry smaller loads (Wüest 2010). In fact, on a global basis, greater than 50% of basin-scale sediment flux in regulated basins is trapped in artificial impoundments. Discharge-weighted sediment trapping due to large reservoirs ($>0.5\,\mathrm{km}^3$ volume) is 30%, with smaller reservoirs trapping about 23% of the total global sediment load (Vörösmarty *et al*. 2003).

24.3.3   Reforestation

A unique example, but one that highlights how catchment restoration can impact the heat budget of a lake, is the case of Clearwater Lake, Canada (Tanentzap *et al*. 2008). Clearwater Lake, in the 1970s, had a pH of 4 owing to high sulfur dioxide emissions from proximal metal smelters and an unusual thermal regime compared with similar lakes. In 1973 Clearwater Lake had bottom temperatures about $10\,^{\circ}C$ higher than would be expected in similar-sized lakes. However, with changes in the catchment the heat budget and temperature profile of the lake changed. With reforestation the mean wind speed has decreased from $5.5\,m\,s^{-1}$ to $3.6\,m\,s^{-1}$. Decreasing SO_2 emissions increased lake pH to 6.5 and light attenuation increased as the DOC concentrations increased. Since 1973 the temperature of the bottom water dropped from $15\,^{\circ}C$ to $8\,^{\circ}C$ and Tanentzap *et al.* (2008) concluded that both wind and increasing light attenuation were important in this cooling pattern.

Tanentzap (2008) concluded that Clearwater Lake demonstrated that any factors that influence local wind speeds and underwater light attenuation should be considered as modifiers of the effects of climate warming on lake thermal regimes. Reforesting degraded lake catchments would not only reduce local wind speeds, but would also increase shading of the lake shoreline as well as input streams, and may create a cooler microclimate. This can significantly affect the heat budget of a lake through differential heating and variable deepening of the surface mixed layer (Imberger and Parker 1985).

24.4   Third level—opportunities for lakes to reduce greenhouse gases

Aquatic systems are a major component of the global carbon cycle. Rivers and lakes receive terrestrial carbon mobilized from vegetation and soils in their catchments, then process this carbon as it is transported to the ocean (Cole *et al.* 2007). It is estimated that of the 29×10^8 tonnes of carbon that enters aquatic systems annually, 14×10^8 tonnes are evaded to the atmosphere, 6×10^8 tonnes are buried in sediments and the remaining 9×10^8 tonnes are transported into oceans (Tranvik *et al.* 2009). Of the carbon buried in the sediments of aquatic systems, the majority is in the organic form, whereas carbon sequestration by oceans is predominated by inorganic carbon. In fact, despite covering only 2% of the Earth's surface, lakes, reservoirs and peatlands bury approximately three times as much organic carbon as all of the world's oceans (Dean and Gorham 1998). This is due to a high accumulation of lacustrine sediments and a high preservation rate of material.

Global estimates suggest that carbon emissions from freshwater aquatic systems are of the same order of magnitude as emissions from global deforestation, the carbon uptake of oceans and emissions from fossil fuel combustion (Tranvik *et al.* 2009). Consequently, their importance in global carbon budgets, and society's mitigation response to global warming, cannot be overstated.

24.4.1   Carbon sequestration

Mitigation of climate change by soil carbon sequestration is a process that can be enhanced without compromising other aspects of farming or ecological systems. However, there is no comparatively simple opportunity to enhance this process in lakes. Increasing carbon production and sequestration through lake fertilization to increase phytoplankton productivity is a strategy that requires careful consideration and caution, despite eutrophic lakes showing net carbon retention rates (Hanson *et al.* 2004), because eutrophication is clearly linked to the increase of cyanobacteria and associated water-quality problems. Iron and nitrogen fertilization of oceans has been proposed (Martin, Gordon, and Fitzwater 1990) but has been dismissed as environmentally irresponsible (Gilbert *et al*. 2008).

The creation of artificial impoundments can increase carbon retention of a system compared with its preregulated state (Downing *et al*. 2008). However, these reservoirs release carbon from inundated soils and vegetation for decades after construction, so their net retention is questionable (Abril *et al*. 2005). Furthermore, methane emissions from reservoirs may be significant (St. Louis *et al*. 2000; DelSontro *et al*. 2010). Reservoirs that are used to generate hydropower often also emit methane downstream of the turbines (Giles 2006). However, these greenhouse gas emissions are generally an order of magnitude lower than emissions from a similar-sized fossil-fuel power station, with the possible exception of tropical reservoirs with low energy densities (Demarty and Bastien 2011).

There is likely to be a high degree of uncertainty when lake carbon cycles are extrapolated over a global scale and under a changing environmental regime such as increased CO_2. Coupling of phytoplankton photosynthesis with hydrodynamics and turbulence is the essential first step in making detailed inventories of the role of lakes in the global carbon cycle and elucidating mechanisms for increased carbon sequestration. Additionally, methods to alter the proportion of terrestrially derived carbon that is released to the atmosphere or buried in sediments can increase the sequestration of carbon in lakes without the associated adverse effects of eutrophication.

Globally, the majority of lakes, with the exception of highly productive eutrophic systems (Hanson *et al*. 2004), are supersaturated with respect to CO_2 and consequently act as a net source of carbon emissions (Cole *et al*. 1995). However, because most lacustrine carbon emission results from in-lake respiration (del Giorgio *et al*. 1999) by autotrophic producers and microbial decomposition of terrestrially derived allochthonous dissolved organic carbon (Jansson *et al*. 2000), decreasing catchment inputs of nutrients and dissolved organic carbon may reduce this respiration and concomitant release of carbon to the atmosphere. This process is independent of temperature (Sobek, Tranvik, and Cole 2005) so would remain a viable sequestration mechanism as atmospheric warming continues.

24.4.2   Methane capture or control

Methane concentrations in the deep water of permanently stratified lakes can, over long time scales, accumulate to high levels that can potentially be exploited as a renewable energy source. The most prominent example for this is Lake Kivu, where methane in the deep water has accumulated over several centuries to concentrations of up to $20\,mmol\,L^{-1}$, (Schmid *et al*. 2005; see also Chapter 18). Commercial exploitation of this methane has proved to be feasible and initial pilot projects are in operation. In the case of Lake Kivu, methane exploitation was also required for safety reasons, as a sudden catastrophic eruption of the gases dissolved in the lake would be disastrous. However, it is not necessarily a mitigation strategy for climate change as in the natural state only a very small fraction of the methane is released to the atmosphere (Borges *et al*. 2011), while most of it is oxidized to CO_2 in the surface mixed layer of the lake (Pasche *et al*. 2011). Lake Kivu is an exceptional case, however, and methane concentrations as well as amounts dissolved in other lakes are much smaller. It has been proposed that methane could economically be exploited even at micromolar concentrations (Ramos *et al.* 2009), but the feasibility of this approach still needs to be proved.

24.4.3 Burial of wood in aquatic systems

Greenhouse gas concentrations in the atmosphere will only be reduced if carbon can be locked away for generations. In lake sediments, allochthonous carbon that is high in recalcitrant compounds such as aliphatic polymers, lignin, and humic matter, is strongly resistant to anaerobic decomposition (Zehnder and Svensson 1986). Consequently, woody debris buried in the hypolimnia of lakes can sequester carbon over long periods of time. However, the volume of woody debris needed to bury 5% of annual anthropogenic carbon emissions is very large as suggested by the following estimates. Sequestering 5% of anthropogenic emissions is equivalent to burying approximately 40 million tonnes of carbon (Burgermeister 2007) in the anoxic hypolimnia of lakes every year. Assuming the carbon content of wood is 50% and the wood has an average density of $600 \, kg \, m^{-3}$ (the average between softwood and hardwood; Lamlom and Savidge 2003), 40 million tonnes of carbon would be locked up in $0.13 \, km^3$ of wood. If this wood were buried only in lakes and reservoirs with a surface area larger than $100 \, km^2$, which have a global area of approximately 1.6 million km^2 (Downing *et al.* 2006), then a depth of only 0.08 mm would be filled each year. After one century, 8 mm of the hypolimnia would be filled with recalcitrant woody debris. Clearly this offers some potential to sequester atmospheric carbon, but it is unclear whether burying wood in lakes is the best use of the resource.

24.5 Conclusions

Anthropogenic modifications to hydrology and land-use have fundamentally changed river and lake ecosystems and have challenged water security for human populations and aquatic biodiversity. The emerging threat of climate change may compound these challenges and expose aquatic ecosystems to even greater risk. There are opportunities to reduce this risk but these will require a change in management practices and considerable investment. Generally speaking, maximizing the adaptive capacity of aquatic systems to a changing climate requires that the resilience of river and lake ecosystems is enhanced by minimizing the adverse impacts from past anthropogenic modifications.

Developed, wealthy countries are able to overcome losses to water security with investment in water infrastructure such as desalination technology and reuse schemes. Investment should also target catchment and land-use practices, which can have multiple benefits. These benefits include the provision of quality water for potable supply, protection of biodiversity, and a decrease in the vulnerability of river and lake ecosystems to climate change impacts. Consequently, opportunities to mitigate climate change or climate-change impacts should form part of the investment strategy for water management. These are largely restricted to reducing nutrient, particle or DOC loadings, changing lake residence times, controlled water releases and shading with riparian vegetation. Technologies such as heat pumps may also play a role in managing warming in lakes and rivers.

Where there are clear economic drivers these technologies will be rapidly adopted. However, investment may also be required where the benefits accrue over decades. Environmental and societal values are difficult to equate economically but lakes and rivers may be manipulated to derive some of these values without compromising the ecosystems. It is demonstrated in this synthesis that a combination of natural and engineering processes can only partly mitigate some of the impacts of climate change. This contribution is far from an exhaustive analysis of what might be possible to counter rising water temperatures. However, by starting the discussion and proposing where some benefits may be gained, the science and management communities can work towards mitigating climate change impacts coincidentally with the implementation of adaptation strategies.

24.6 Acknowledgements

Justin Brookes was supported by the International Centre of Excellence in Water Resources Management (ICEWARM) and Eawag.

References

Abril, G., Guerin, F., Richard, S., *et al*. (2005) Carbon dioxide and methane emissions and the carbon budget of a 10-year old tropical reservoir (Petit Saut, French Guiana). *Glob. Biogeochem. Cycles*, **19**, GB4007, doi:10.1029/2005GB002457.

Adrian, R., O'Reilly, C.M., Zagarese, H., *et al*. (2009) Lakes as sentinels of climate change. *Limnol. Oceanogr.*, **54**, 2283–97, doi: 10.4319/lo.2009.54.6_part_2.2283.

Arhonditis, G.B., Brett, M.T., Degasperi, C.L., and Schindler, D.E. (2004) Effects of climatic variability on the thermal properties of Lake Washington. *Limnol. Oceanogr.*, **49**, 256–70, doi: 10.4319/lo.2004.49.1.0256.

Bear, E.A., McMahon, T.E., and Zale, A.V. (2007) Comparative thermal requirements of Westslope Cutthroat Trout and Rainbow Trout: Implications for species interactions and development of thermal protection standards. *Trans. Amer. Fish. Soc.*, **136**, 1113–21, doi: 10.1577/T06-072.1.

Boon, P.I., Bunn, S.E., Green, J.D., and Shiel, R.J. (1994) Consumption of cyanobacteria by freshwater zooplankton: Implications for the success of "top-down" control of cyanobacterial blooms in Australia. *Aust. J. Mar. Freshwat. Res.*, **45**, 875–87.

Borges A.V., Abril, G., Delille, B., *et al*. (2011) Diffusive methane emissions to the atmosphere from Lake Kivu (Eastern Africa). *J. Geophys. Res.—Biogeosci*., **116**, G03032, doi: 10.1029/2011JG001673.

Brett, J.R. (1952) Temperature tolerance in young Pacific salmon, genus *Oncorhynchus*. *J. Fish. Res. Bd. Can.*, **9**, 265–322.

Brett, J.R., Clarke, W.C., and Shelbourn, J.E. (1982) *Experiments on Thermal Requirements for Growth and Food Conversion Efficiency of Juvenile Chinook Salmon* Oncorhynchus tshawytscha. Canadian Technical Report of Fisheries and Aquatic Sciences, No. 1127, Government of Canada, Ottawa.

Brookes, J.D. and Carey, C.C. (2011) Resilience to blooms. *Science*, **334**, 46–7, doi: 10.1126/science.1207349.

Brookes, J.D., Ganf, G.G., Green, D., and Whittington, J. (1999) The influence of light and nutrients on buoyancy, filament aggregation and flotation of *Anabaena circinalis*. *J. Plank. Res.*, **21**, 337–41.

Burgermeister, J. (2007) Missing carbon mystery: Case solved? *Nature Reports*, **3**, 36–7, doi:10.1038/climate.2007.35.

Cederholm, C.J., Kunze, M.D., Murota, T., and Sibatani, A. (1999) Pacific salmon carcasses: essential contributions of nutrient and energy for aquatic and terrestrial ecosystems. *Fisheries Management/Habitat*, **24**: 6–15.

Chiew, F.H.S. (2006) Estimation of rainfall elasticity of streamflow in Australia. *Hydrol. Sci. J.*, **51**, 613–25, doi:10.1623/hysj.51.4.613.

Chiew, F.H.S., Whetton, P.H., McMahon, T.A., and Pittock, A.B. (1995) Simulation of the impacts of climate change on runoff and soil moisture in Australian catchments. *J. Hydrol.*, **167**, 121–47.

Chwieduk, D. (2003) Towards sustainable-energy buildings. Appl. Energy, **76**, 211–17, doi: 10.1016/S0306-2619(03)00059-X.

Clews, E., Durance, I., Vaughan, I.P., and Omerod, S.J. (2010) Juvenile salmonid populations in a temperate river system track synoptic trends in climate. *Glob. Change Biol*, **16**, 3271–83, doi: 10.1111/j.1365-2486.2010.02211.x.

Cole, J.J., Prairie, Y.T., Caraco, N.F., *et al.* (2007) Plumbing the global carbon cycle: integrating inland waters into the terrestrial carbon budget. *Ecosystems*, **10**, 171–84, doi: 10.1007/s10021-006-9013-8.

Cooke, S.J., Hinch, S.G., Farrell, A.P., *et al.* (2004) Abnormal migration timing and high en route mortality of sockeye salmon in the Fraser River, British Columbia. *Fisheries*, **29**, 22–33, doi:10.1577/1548-8446(2004)29[22:AMTAHE]2.0.CO;2.

Covich, A.P. (1993) Water and ecosystems, in *Water in Crisis* (eds. P.H. Gleick), Oxford University Press, New York, pp. 40–55.

Dean, W.E. and E. Gorham (1998) Magnitude and significance of carbon burial in lakes, reservoirs, and peatlands. *Geology*, **26**, 535–8.

Del Giorgio, P.A., Cole, J.J., Caraco, N.F., and Peters, R.H. (1999) Linking planktonic biomass and metabolism to net gas fluxes in Northern Temperate Lakes. *Ecology*, **80**, 1422–31.

DelSontro, T., McGinnis, D.F., Sobek, S., *et al.* (2010) Extreme methane emissions from a Swiss hydropower reservoir: contribution from bubbling sediments. *Environ. Sci. Techn.*, **44**, 2419–25, doi: 10.1021/es9031369.

Demarty, M. and Bastien, J. (2011) GHG emissions from hydroelectric reservoirs in tropical and equatorial regions: Review of 20 years of CH_4 emission measurements. *Energy Policy*, **39**, 4197–4206, doi:10.1016/j.enpol.2011.04.033.

DeMelo, R., France, R., and McQueen, D.J. (1992) Biomanipulation: Hit or myth? *Limnol. Oceanogr.*, **37**, 192–207.

Dickerson, B.R. and Vinyard, G.L. (1999) Effects of high chronic temperatures and diel temperature cycles on the survival and growth of Lahontan cutthroat trout. *Trans. Amer. Fish. Soc.*, **128**, 516–21.

Dones, R., Bauer, C., Bolliger, R., *et al.* (2007) *Life Cycle Inventories of Energy Systems: Results for Current Systems in Switzerland and UCTE Countries*. Ecoinvent Report No. 5, Paul Scherrer Institute Villigen, Swiss Centre for Life Cycle Inventories, Dübendorf CH.

Downing, J.A., Cole, J.J., Middelburg, J.J., *et al.* (2008) Sediment organic carbon burial in agriculturally eutrophic impoundments over the last century. *Global Biogeochem. Cycles*, **22**, GB1018, doi:10.1029/2006GB002854.

Downing, J.A., Prairie, Y.T., Cole, J.J., *et al.* (2006) The global abundance and size distribution of lakes, ponds, and impoundments. *Limnol. Oceanogr.*, **51**, 2388–97, doi: 10.4319/lo.2006.51.5.2388.

Eagle, J., Naylor, R., and Smith, W. (2004) Why farm salmon outcompete fishery salmon. *Marine Policy*, **28**, 259–70, doi: 10.1016/j.marpol.2003.08.001.

Edsall, T.A., Frank, A.M., Rottiers, D.V., and Adams, J.V. (1999) The effect of temperature and ration size on the growth, body composition, and energy content of juvenile Coho salmon. *J. Great Lakes Res.*, **25**, 355–63.

Einsele, G., Yan, J., and Hinderer, M. (2001) Atmospheric carbon burial in modern lake basins and its significance for the global carbon budget. *Global and Planetary Change*, **30**, 167–95, doi: 10.1016/S0921-8181(01)00105.

Elliott, J.M. (1981) Some aspects of thermal stress on freshwater teleosts, in *Stress and Fish* (ed. A.D. Pickering), Academic Press, New York, pp. 209–45.

Elliott, J.M. and Hurley, M.A. (1999) A new energetics model for brown trout, *Salmo trutta*. *Freshwat. Biol.*, **42**, 235–46, doi: 10.1046/j.1365-2427.1999.444483.x.

Friedl, G. and Wüest, A. (2002) Disrupting biogeochemical cycles—Consequences of damming. *Aquat. Sci.*, **64**, 55–65, doi: 1015-1621/02/010055-11.

Friedland, K.D., Reddin, D.G., McMenemy, J.R., and Drinkwater, K.F. (2003) Multidecadal trends in North American Atlantic salmon (*Salmo salar*) stocks and climate trends relevant to juvenile survival. *Can. J. Fish. Aquat. Sci.*, **60**, 563–83, doi:10.1139/f03-047.

Foreman, M.G.G., Lee, D.K., Morrison, J., *et al.* (2001) Simulations and retrospective analyses of Fraser watershed flows and temperatures. *Atmosph. Ocean*, **39**, 89–105, doi:10.1080/07055900.2001.9649668.

Gibson, C.A., Meyer, J.L., Poff, N.L., *et al.* (2005) Flow regime alterations under changing climate in two river basins: implications for freshwater systems. *River Res. and Appl.*, **21**, 849–64. doi: 10.1002/rra.855.

Gilbert, P., Azanza, R., Burford, M., *et al.* (2008) Ocean urea fertilization for carbon credits poses high ecological risks. *Mar. Poll. Bull.*, **56**, 1049–56, doi: 10.1016/j.marpolbul.2008.03.010.

Giles, J. (2006) Methane quashes green credentials of hydropower. *Nature*, **444**, 524–5, doi:10.1038/444524a.

Gresh, T., Lichatowich, J., and Schoonmaker, P. (2000) An estimation of historic and current levels of salmon production in the Northeast Pacific ecosystem: evidence of a nutrient deficit in the freshwater systems of the Pacific Northwest. *Fisheries*, **25**, 15–21, doi:10.1577/1548-8446(2000)025<0015:AEOHAC>2.0.CO;2.

Güde H., Rossknecht, H., and Wagner, G. (1998) Anthropogenic impacts on the trophic state of Lake Constance during the twentieth century. *Arch. Hydrobiol. Spec. Iss. Adv. Limnol.*, **53**, 85–108.

Hanson, P.C., A.I. Pollard, D.L. Bade, K. Predick, S.R. Carpenter, and J.A. Foley. 2004. A model of carbon evasion and sedimentation in temperate lakes. *Global Change Biol.*, **10**, 1285–98, doi: 10.1111/j.1529-8817.2003.00805.x.

Häse C., Gaedke, U., Seifried, A., *et al.* (1998) Phytoplankton response to re-oligotrophication in large and deep Lake Constance: photosynthetic rates and chlorophyll concentrations. *Arch. Hydrobiol. Adv. Limnol.*, **53**, 159–78.

Herb, W.R., Janke, B., Mohseni, O., and Stefan, H.G. (2008) Thermal pollution of streams by runoff from paved surfaces. *Hydrol. Proc.*, **22**, 987–99, doi: 10.1002/hyp.6986.

Herb, W.R., Mohseni, O., and Stefan, H.G. (2009) Simulation of temperature mitigation by a stormwater detention pond. *J. Amer. Wat. Res. Assoc.*, **45**, 1164–78, doi: 10.1111/j.1752-1688.2009.00354.x.

Hilderbrand, G.V, Hanley, T.A., Robbins, C.T., and Schwartz, C.C. (1999a) Role of brown bears (*Ursus arctos*) in the flow of marine nitrogen into a terrestrial system. *Oecologia*, **121**, 546–50, doi: 10.1007/s004420050961.

Hilderbrand, G.V., Schwartz, C.C., Robbins, C.T., *et al.* (1999b) The importance of meat, particularly salmon, to body size, population productivity, and conservation of North American brown bears. *Can. J. Zoo.*, **77**, 132–8, doi: 10.1139/z98-195.

Ibelings, B.W., Portielje, R., Lammens, E.H.R.R., *et al.* (2007) Resilience of alternative stable states during the recovery of shallow lakes from eutrophication: Lake Veluwe as a case study. *Ecosystems*, **10**, 4–16, doi: 10.1007/s10021-006-9009-4.

Imberger, J. and Parker, G. (1985) Mixed layer dynamics in a lake exposed to a spatially variable wind field. *Limnol. Oceanogr.*, **30**, 473–88.

IPCC (2011) *Climate Change 2001: The Scientific Basis. Contribution of Working Group I to the Third Assessment Report of the Intergovernmental Panel on Climate Change*, Cambridge University Press, Cambridge.

Jansson, M., Bergström, A., Blomqvist, P., and Drakare, S. (2000) Allochthonous organic carbon and phytoplankton/bacterioplankton production relationships in lakes. *Ecology*, **81**, 3250–5, doi: 10.1890/0012-9658(2000)081[3250:AOCAPB]2.0.CO;2.

Jaun, L., Finger, D., Zeh, M., *et al*. (2007) Effects of upstream hydropower operation and oligotrophication on the light regime of a turbid peri-alpine lake. *Aquat. Sci.*, **69**, 212–26, doi: 10.1007/s00027-007-0876-3.

Jensen, J.P., Pedersen, A.R., Jeppesen, E., and Søndergaard, M. (2006) An empirical model describing the seasonal dynamics of phosphorus in 16 shallow eutrophic lakes after external loading reduction. *Limnol. Oceanogr.*, **51**, 791–800, doi: 10.4319/lo.2006.51.1_part_2.0791.

Jeppesen, E., Jensen, J.P., and Søndergaard, M. (2002) Response of phytoplankton, zooplankton and fish to re-oligotrophication: An 11 year study of 23 Danish lakes. *Aquat. Ecosys. Health Manage.*, **5** (1), 31–43.

Jöhnk, K.D., Huisman, J., Sharples, J., *et al*. (2008) Summer heatwaves promote blooms of harmful cyanobacteria. *Global Change Biol.*, **14**, 495–512, doi: 10.1111/j.1365-2486.2007.01510.x

Johnstone, H. and Rahel, F.J. (2003) Assessing temperature tolerance on Bonneville Cutthroat Trout based on constant and cycling thermal regimes. *Trans. Amer. Fish. Soc.*, **132**, 92–9, doi:10.1577/1548-8659(2003)132<0092:ATTOBC>2.0.CO;2.

Jones I., George, G., and Reynolds, C. (2005) Quantifying effects of phytoplankton on the heat budgets of two large limnetic enclosures. *Freshwat. Biol.*, **50**, 1239–47. doi: 10.1111/j.1365-2427.2005.01397.x.

Jonsson, B. and Jonsson, N. (2003) Migratory Atlantic salmon as vectors for the transfer of energy and nutrients between freshwater and marine environments. *Freshwat. Biol.*, **48**, 21–7, doi: 10.1046/j.1365-2427.2003.00964.x.

Köhler, J., Bernhardt, H., and Hoeg, S. (2000) Long-term response of phytoplankton to reduced nutrient load in the flushed Lake Müggelsee (Spree system, Germany). *Arch. Hydrobiol.*, **148**, 209–29.

Kumagai, M., Nakano, S., Jiao, C., *et al*. (2000) Effect of cyanobacterial blooms on thermal stratification. *Limnol.*, **1**, 191–5.

Lamlom, S.H. and Savidge, R.A. (2003) A reassessment of carbon content in wood: variation within and between 41 North American species. *Biomass and Bioenergy*, **25**, 381–8, doi: 10.1016/S0961-9534(03)00033-3.

Larson, L.L. and Larson, S.L. (1996) Riparian shade and stream temperature: A perspective. *Rangelands*, **18**, 149–52.

Livingstone, D.M. (2008) A change of climate provokes a change of paradigm: taking leave of two tacit assumptions about physical lake forcing. *Int. Rev. Hydrobiol.*, **93** (4–5), 404–14.

Magnusson, J.J., Robertson, D.M., Benson, B.J., *et al*. (2000) Historical trends in lake and river ice cover in the Northern Hemisphere. *Science*, **289**, 1743–6, and Errata (2001) *Science*, **291**, 254. doi: 10.1126/science.289.5485.1743.

Malmaeus, J.M., Blenckner, T., Markensten, H., and Persson, I. (2006) Lake phosphorus dynamics and climate warming: A mechanistic model approach. *Ecol. Model.*, **190**, 1–14, doi: 10.1016/j.ecolmodel.2005.03.017.

Martin, J.H., Gordon, R.M., and Fitzwater, S.E. (1990) Iron in Antarctic waters. *Nature*, **345**, 156–8, doi:10.1038/345156a0.

Matzinger, A., Schmid, M., Veljanoska-Sarafiloska, E., *et al*. (2007) Eutrophication of ancient Lake Ohrid: Global warming amplifies detrimental effects of increased nutrient inputs. *Limnol. Oceanogr.*, **52**, 338–53, doi: 10.4319/lo.2007.52.1.0338.

Mazumder, A. and Taylor, W.D. (1994) Thermal structure of lakes varying in size and water clarity. *Limnol. Oceanogr.*, **39**, 968–76.

Mazumder, A., Taylor, W.D., McQueen, D.J., and Lean, D.R.S. (1990) Effects of fish and plankton on lake temperature and mixing depth. *Science*, **247**, 312–31, doi: 10.1126/science .247.4940.312.

McDaniels, T., Wilmot, S., Healey, M., and Hinch, S. (2010) Vulnerability of Fraser River sockeye salmon to climate change: A life cycle perspective using expert judgements. *J. Environ. Manag.*, **91**, 2771–80, doi: 10.1016/j.jenvman.2010.08.004.

McMahon, T.E., Zale, A.V., Barrows, F.T., *et al*. (2007) Temperature and competition between bull trout and brook trout: a test of the elevation refuge hypothesis. *Trans. Amer. Fish. Soc.*, **136**, 1313–26, doi:10.1577/T06-217.1.

McMichael, A.J., Woodruff, R.E., and Hales, S. (2006) Climate change and human health: present and future risk. *The Lancet*, **367**, 859–69, doi: 10.1016/S0140-6736(06)68079-3.

Meeuwig, M.H., Dunham, J.B., Hayes, J.P., and Vinyard, G.L. (2004) Effects of constant and cyclical thermal regimes on growth and feeding of juvenile cutthroat trout of variable sizes. *Ecol. Freshwater Fish*, **13**, 208–16, doi: 10.1111/j.1600-0633.2004.00052.x.

Meier, W., Bonjour, C., Wüest, A., and Reichert, P. (2003) Modelling the effect of water diversions on the temperature of mountain streams. *J. Environ. Engin.*, **129**, 755–64.

Meier, W. and Wüest, A. (2004) Wie verändert die hydroelektrische Nutzung die Wassertemperatur der Rhone? *Wasser Energie Luft*, **96**, 305–9.

Mitchell, C., James, C.B., and Edinger, J.E. (1995) Analysis of flow modification on water quality in Nechako River. *J. Energy Engin.*, **121**, 73–80.

Moosmann, L., Schmid, M., and Wüest, A. (2005) *Einfluss der Beschattung auf das Temperaturregime der Orbe*. Report, Eawag, Kastanienbaum.

Morrison, J., Quick, M.C., and Foreman, M.G.C. (2002) Climate Change in the Fraser River watershed: flow and temperature projections. *J. Hydrol.*, **263**, 230–44, doi: 10.1016/S0022-1694(02)00065-3.

Mosisch, T.D., Bunn, S.E., and Davies, P.M. (2001) The relative importance of shading and nutrients on algal production in subtropical streams. *Freshwat. Biol.*, **46**, 1269–1278. doi: 10.1046/j.1365-2427.2001.00747.x.

Naiman, R.J., Bilby, R.E., Schindler, D.E., and Helfield, J.M. (2002) Pacific salmon, nutrients, and dynamics of freshwater and riparian ecosystems. *Ecosystems*, **5**, 399–417, doi: 10.1007/s10021-001-0083-3.

O'Reilly, C.M., Alin, S.R., Plisnir, P.D., *et al*. (2003) Climate change decreases aquatic ecosystem productivity in Lake Tanganyika, Africa. *Nature*, **424**, 766–8.

Özen, A., Karapinar, B., Kucuk, I., *et al*. (2010) Drought-induced changes in nutrient concentrations and retention in two shallow Mediterranean lakes subjected to different degrees of management. *Hydrobiologia*, **646**, 61–72, doi: 10.1007/s10750-010-0179-x.

Paerl, H.W. and Huisman, J. (2008) Climate—blooms like it hot. *Science*, **320**, 57–8, doi: 10.1126/science.1155398.

Paerl, H.W. and Huisman, J. (2009) Climate change: a catalyst for global expansion of harmful cyanobacterial blooms. *Environ. Microb. Rep.*, **1**, 27–37, doi: 10.1111/j.1758-2229.2008.00004.x.

Pasche, N., Schmid, M., Vazquez, F., *et al*. (2011) Methane sources and sinks in Lake Kivu. *J. Geophys. Res.*, **116**, G03006, doi: 10.1029/2011JG001690.

Rahel, F.J. and Olden, J.D. (2008) Assessing the effects of climate change on aquatic invasive species. *Conserv. Biol.*, **22**, 521–33, doi: 10.1111/j.1523-1739.2008.00950.x.

Ramos F.M., Bambace, L.A.W., Lima, I.B.T., *et al*. (2009) Methane stocks in tropical hydropower reservoirs as a potential. energy source. *Clim. Change*, **93**, 1–13, doi: 10.1007/s10584-008-9542-6.

Rinke, K., Yeates, P., and Rothhaupt, K.O. (2010) A simulation of the feedback of phytoplankton on thermal structure via light extinction. *Freshwat. Biol.*, **55**, 1674–93. doi: 10.1111/j.1365-2427.2010.02401.x.

Ruggerone, G.T., Peterman, R.M., Dorner, B., and Myers, K.W. (2010) Magnitude and trends in abundance of hatchery and wild pink salmon, chum salmon, and sockeye salmon in the North Pacific Ocean. *Mar. Coast. Fish., Manag. Ecosys. Sci.*, **2**, 306–28, doi:10.1577/C09-054.1.

Ruggiu, D., Morabito, G., Panzani, P., and Pugnetti, A. (1998) Trends and relations among basic phytoplankton characteristics in the course of the long-term oligotrophication of Lake Maggiore (Italy). *Hydrobiologia*, **369/370**, 243–57, doi: 10.1023/A:1017058112298.

Schär, C., Vidale, P.L., Lüthi, D., *et al.* (2004) The role of increasing temperature variability in European summer heatwaves. *Nature*, **427**, 332–6, doi:10.1038/nature02300.

Schiefer, E. and Klinkenberg, B. (2004) The distribution and morphometry of lakes and reservoirs in British Columbia: a provincial inventory. *Can. Geogr.*, **48**, 345–55, doi: 10.1111/j.0008-3658.2004.00064.x.

Schindler, D.W. (2001) The cumulative effects of climate warming and other human stresses on Canadian freshwaters. *Can. J. Fish. Aquat. Sci.*, **58**, 18–29, doi: 10.1139/f00-179.

Schindler, D.E., Scheuerell, M.D., Moore, J., *et al.* (2003) Pacific salmon and the ecology of coastal ecosystems. *Frontiers Ecol. Environ.*, **1**, 31–7, doi: 10.1890/1540-9295 (2003)001[0031:PSATEO]2.0.CO;2.

Schmid, M., and Dorji, P. (2008) Permanent lake stratification caused by a small tributary—the unusual case of Lej da San Murezzan. *J. Limnol.*, **67**, 35–43.

Schmid, M., Halbwachs, M., Wehrli, B., and Wüest, A. (2005) Weak mixing in Lake Kivu: new insights indicate increasing risk of uncontrolled gas eruption. *Geochem. Geophys. Geosys.*, **6**, Q07009, doi:10.1029/2004GC000892.

Schneider, P., and Hook, S.J. (2010) Space observations of inland water bodies show rapid surface warming since 1985. *Geophys. Res. Let.*, **37**, L22405 5p., doi:10.1029/2010GL045059.

Selong, J.H., McMahon, T.E., Zale, A.V., and Barrows, F.T. (2001) Effect of temperature on growth and survival of bull trout, with application of an improved method for determining thermal tolerance in fishes. *Trans. Amer. Fish. Soc.*, **130**, 1026–1037. doi:10.1577/1548-8659(2001)130<1026:EOTOGA>2.0.CO;2.

Simon, A. and Collison, A.J.C. (2002) Quantifying the mechanical and hydrological effects of riparian vegetation on streambank stability. *Earth Surface Process. Landforms*, **27**, 527–46, doi: 10.1002/esp.325.

Sobek, S., Tranvik, L.J., and Cole, J.J. (2005) Temperature independence of carbon dioxide supersaturation in global lakes. *Glob. Biogeochem. Cycles*, **19** (2), article no. GB2003.

Solomon, S., Qin, D., Manning, M., *et al.* (2007) Technical summary, in *Climate Change 2007: The Physical Science Basis. Contribution of Working Group I to the Fourth Assessment Report of the Intergovernmental Panel on Climate Change* (eds. S. Solomon, D. Qin, M. Manning, *et al.*), Cambridge University Press, Cambridge.

Stainforth, D.A., Alna, T., Christensen, C., *et al.* (2005) Uncertainty in predictions of the climate response to rising levels of greenhouse gases. *Nature*, **433**, 403–6, doi:10.1038/nature0330.

Stitch, H.B. and Brinker, W. (2010) Oligotrophication outweighs effects of global warming in a large, deep, stratified lake ecosystem. *Global Change Biol.*, **16**, 877–88, doi: 10.1111/j.1365-2486.2009.02005.x.

St. Louis, V.L., Kelly, C.A., Duchemin, E., *et al.* (2000) Reservoir surfaces as sources of greenhouse gases to the atmosphere: a global estimate. *Biosci.*, **50**, 766–75, doi: 10.1641/0006-3568(2000)050[0766:RSASOG]2.0.CO;2.

Tanentzap, A.J., Yan, N.D., Keller, B., *et al.* (2008) Cooling lakes while the world warms: Effects of forest growth and increased dissolved organic matter on the thermal regime of a temperate, urban lake. *Limnol. Oceanogr.*, **53**, 404–10, doi: 10.4319/lo.2008.53.1.0404.

Timbal, B, Arblaster, J.M., and Power, S. (2006) Attribution of the late-twentieth-century rainfall decline in Southwest Australia. *J. Clim.*, **19**, 2046–62, doi: 10.1175/JCLI3817.1.

Tranvik, L.J., Downing, J.A., Cotner, J.B., *et al.* (2009) Lakes and reservoirs as regulators of carbon cycling and climate. *Limnol. Oceanogr.*, **54**, 2298–314, doi: 10.4319/lo.2009.54.6_part_2.2298.

Trenberth, K.E., Jones, P.D., Ambenje, P., *et al.* (2007) Observations: Surface and Atmospheric Climate Change, in *Climate Change 2007: The Physical Science Basis. Contribution of Working Group I to the Fourth Assessment Report of the Intergovernmental Panel on Climate Change* (eds. S. Solomon, D. Qin, M. Manning, *et al.*), Cambridge University Press, Cambridge.

Van Vliet, M.T.H., Ludwig, F., Zwolsman, J.J.G., *et al.* (2011) Global river temperatures and sensitivity to atmospheric warming and changes in river flow. *Wat. Resour. Res.*, **47**, W02544, doi: 10.1029/2010WR009198.

Vörösmarty, C.J., Meybeck, M., Fekete, B., *et al.* (2003) Anthropogenic sediment retention: major global impact from registered river impoundments. *Glob. Planet. Change*, **39**, 169–90, doi: 10.1016/S0921-8181(03)00023-7.

Vörösmarty, C.J., Sharma, K., Fekete, B., *et al.* (1997) The storage and aging of continental runoff in large reservoir systems of the world. *Ambio*, **26**, 210–19.

Wagner, C. and Adrian, R. (2009) Cyanobacteria dominance: quantifying the effects of climate change. *Limnol. Oceanogr.*, **54**, 2460–8, doi: 10.4319/lo.2009.54.6_part_2.246.

Webb, B.W., Hannah, D.M., Moore, R.D., *et al.* (2008) Recent advances in stream and river temperature research. *Hydrol. Proc.*, **22**, 902–18, doi: 10.1002/hyp.6994.

Webb, B.W. and Zhang, Y. (1998) Spatial and seasonal variability in the components of the river heat budget. *Hydrol. Proc.*, **11**, 79–101.

Williams, A., Karoly, D.J., and Tapper, N. (2001) The sensitivity of Australian fire danger to climate change. *Climatic Change*, **49**, 171–91, doi: 10.1023/A:1010706116176.

Williamson, C.E., Saros, J.E., and Schindler, D.W. (2009) Climate change. Sentinels of Change. *Science*, **323**, 887–8, doi: 10.1126/science.1169443.

Wüest, A. (2010) Downstream relevance of reservoir management, in *Alpine Waters* (ed. U. Bundi), Springer-Verlag. Berlin, doi: 10.1007/978-3-540-88275-6_12.

Zehnder, A.J.B. and Svensson, B.H. (1986) Life without oxygen—what can and what cannot. *Experientia*, **42**, 1197–205, doi: 10.1007/BF01946391.

25

Artificial Decomposition of Water into Hydrogen and Oxygen by Electrolysis to Restore Oxygen in Climate Change-Impacted Waters

Michio Kumagai[1] and Hiroyasu Takenaka[2]

[1]*Lake Biwa Sigma Research Center, Ritsumeikan University, Kusatsu, Japan*
[2]*Faculty of Engineering, Kinki University, Higashi-Hiroshima, Japan*

25.1 Introduction

Lake Biwa, the largest and oldest lake in Japan, contains almost 34% of Japan's surface fresh drinking water (Kumagai *et al*. 2003; see also Chapter 6). The area of the lake covers one-sixth of Shiga Prefecture. The annual amount of global solar radiation on Shiga prefecture is about 4.3×10^{15} kcal (here we use this unit for comparing with the human being energy scale) and about 2.2×10^{15} kcal is absorbed in forests and mountains, mainly for growing trees and grasses. The energy recovery from wood and charcoal is about 3.0×10^{11} kcal, which consumes only 0.014% of the incoming solar radiation. Forests, however, keep water inside and contribute to having great mass of fresh water as potential energy. On the other hand, about 6.4×10^{14} kcal is taken up by farming fields, of which the main crop is rice. Grains grown in Shiga Prefecture use about 7.6×10^{11} kcal, which corresponds to about 0.15% of the solar energy. It is remarkable to know that this energy covers about 73% of energy based on food consumption of this prefecture in a year. About 2.4×10^{14} kcal lands directly onto the roofs of houses. If we placed solar power panels on all roofs, we could obtain about 8.2×10^{12} kcal in a year and this would cover about 14% of necessary energy consumption in Shiga Prefecture. The solar energy flux over the surface of Lake Biwa is 6.8×10^{14} kcal or about the same as 7.9×10^{11} kWh of electricity. As the

Climatic Change and Global Warming of Inland Waters: Impacts and Mitigation for Ecosystems and Societies,
First Edition. Edited by Charles R. Goldman, Michio Kumagai and Richard D. Robarts.
© 2013 John Wiley & Sons, Ltd. Published 2013 by John Wiley & Sons, Ltd.

annual electricity energy consumed in Shiga Prefecture was about 1.25×10^{10} kWh in 2002, the solar energy reaching Lake Biwa is about sixtyfold the total energy consumption in the prefecture. Moreover, the incoming solar energy on the lake is equal to about 78% of electrical energy used in Japan. Of course, some energy is reflected back into the atmosphere, but almost 45% of the incoming energy is used to warm the lake water. Generally speaking, the input energy should balance with the output energy, but in the case of Lake Biwa the input energy since the late 1980s was about 3.05×10^{14} kcal and the output energy was 3.03×10^{14} kcal. The energy difference represents the heat accumulated in the lake. This energy storage is about 5.5×10^{13} kcal, which corresponds to a temperature increase of about $2.0\,^{\circ}$C. Some part of the solar radiation energy entering the lake is used by phytoplankton in photosynthesis for primary production. This is about 1.8×10^{12} kcal and represents 0.26% of the total incoming solar radiation, which is in good agreement with in situ measurements (Tsuda and Nakanishi 1992).

In addition, as mentioned by Sakamoto (Chapter 6), oxygen is depleted in the hypolimnion of Lake Biwa due to eutrophication and the reduction of vertical mixing caused by global warming. It became close to almost zero near the bottom in 2007. Combining excess energy input due to global warming and environmental deterioration in deep water areas, we introduce a test of a remediation plan for Lake Biwa as a trial for future lake resolutions.

25.2 New strategy based on energy and environmental conservation

In order to supply dissolved oxygen to the deep waters of Lake Biwa in the future, we conducted field experiments in 2007 to decompose water into hydrogen and oxygen by electrolysis, because we required an inexpensive oxygen supply for the lake and an efficient use of solar energy. Two different methods were tested: one was with a normal electrode and the other was the PEM method using a polymer electrolyte membrane as seen in Figure 25.1 (Kumagai 2008). This is also called a proton exchange membrane, which can generate hydrogen and oxygen from renewable energy sources by electrochemical reaction (Barbir 2005). We obtained

Polymer Electrolyte
Membrane: **PEM**

Electrodes

Figure 25.1 Two methods, PEM and electrode, for water electrolysis were tested in Lake Biwa.

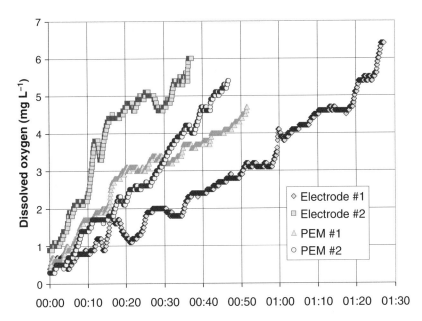

Figure 25.2 Promising results of water electrolysis experiments in the South Basin of Lake Biwa during July 18–20, 2007.

good recovery of dissolved oxygen near the bottom at 15 m depth at the dredged area in the South Basin of Lake Biwa (Figure 25.2).

We also evaluated the cost performance of these two methods and that of an ordinal aeration method when used in deeper water of 90 m in the North Basin of Lake Biwa. Aeration is a common method to increase the dissolved oxygen concentration of the water. We can use various systems such as either injecting air, mechanically mixing or agitating the water, or even injecting pure oxygen for various occasions. Management applications of hypolimnetic aeration include reduction of internal nutrient loading for eutrophication control, improvement of water quality for domestic use, and prevention of fish kill (Ashley 1983). In this case, we consider an aeration system using air injection to compare with the electrolysis methods.

The dead zone area found in 2007 (see Chapter 6) covered 20 km^2 with a 5 m thickness and a volume of nearly 10^8 m^3. We calculated the necessary costs to increase dissolved oxygen concentration by 1 ml L^{-1} for 10^7 m^3 of water. The PEM method to treat the water of 10^7 m^3 required 4.46 × 10^5 kWh in a year and its initial construction cost was 289 110 000 Yen (equivalent to about US$3 400 000). The electrode method needed 7.35 × 10^5 kWh of electricity at an initial cost of 463 480 000 Yen (equivalent to about US$5 450 000). On the other hand, ordinal aeration required annual electricity of 12.3 × 10^5 kWh at a cost of 833 000 000 Yen (equivalent to about US$9 800 000). This comparison shows that the PEM method was least expensive. The necessary area of solar panels is 50 × 50 m to supply electricity for PEM and 70 × 70 m for the electrode. The PEM method has another benefit in that it provides hydrogen for an additional energy source, of which the benefit is close to 5% of the initial cost and it may cover the running cost.

25.3   Application of an electrolysis system to save nature and society

In order to apply the technique described above, we designed a floating unit system equipped with 10 kW solar panels with a length of 10–15 m and a width of 5–6 m. These units were constructed to be easily connected with others to set up multiple unit, large scale systems and realize optimal efficiency.

The combination of 100 units not only provided 1000 kW of electrical power under normal situations but also may be useful for supplying energy and safe water under disaster occasions such as earthquakes and floods. This system can contribute to the realization of sustainable development as well as solving global warming.

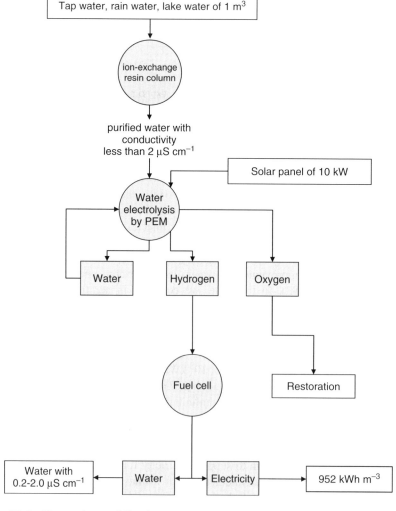

Figure 25.3   Flow pathway of floating systems to supply oxygen and electricity from natural waters.

For example, when we deploy 27 MW solar power systems on 100 different lakes and marine bays, we can reduce CO_2 emissions by 1.0×10^6 ton. If we can use oxygen produced from electrolysis for restoring "dead zones" (Diaz and Rosenberg 2008) and make the production cost of hydrogen 25% lower than the present cost, it might be possible to provide energy for almost the same price as fossil fuel. The use of fuel cells could be expanded in the 2020s and production of cheaper hydrogen could support the fundamental energy needs of our society. The spatial use of water surfaces has several merits such as less impact on surrounding environments because of the usage of unutilized solar energy, and easy transfer of units and technology to developing countries by marine transportation. Moreover, it is possible to supply energy and water directly to disaster areas from local water surfaces. Floating systems may be used for other functions such as keeping clean drinking water and food inside units to be used as part of disaster-relief infrastructure.

The unique benefits of this system are that oxygen and hydrogen are derived from natural water using solar energy and can help solve both anoxic water problems and energy shortages as seen in Figure 25.3. When we are able to use the reflection of sunlight from water surfaces as well as direct incident light on solar panels, the power efficiency can be higher than the equivalent system on land. We believe that taking account of the consistent flow of water and energy and their use is very important for ensuring the future development of global society.

25.4 Acknowledgement

This study was financially supported by Shiga Prefecture. We also show our sincere thanks to Prof. Yasuhiko Ito of Doshisha University for his suggestions and supports on the experiments.

References

Ashley, K.I. (1983) Hypolimnetic Aeration of A Naturally Eutrophic Lake: Physical and Chemical Effects. *Can. J. Fish. Aquat. Sci.*, **40**, 1343–59,

Barbir, F. (2005) http://www.sciencedirect.com/science/article/pii/S0038092X04002464-cor1 #cor1PEM electrolysis for production of hydrogen from renewable energy sources. *Solar Energy*, **78**, 661–9.

Diaz, R.J. and Rosenberg, R. (2008) Spreading dead zones and consequences for marine ecosystems. *Science*, **321**, 926.

Kumagai, M. (2008) Lake Biwa in the context of world lake problems. Baldi Lecture. *Verh. Internat. Verein. Limnol.*, **30**, 1–15.

Kumagai, M., Vincent, W.F., Ishikwa, K., and Aota, Y. (2003) *Lessons from Lake Biwa and other Asian lakes*, in *Freshwater Management—Global versus Local Perspectives* (eds. M. Kumagai, and W. F. Vincent), Springer-Verlag, Tokyo, pp. 4–33.

Tsuda, R. and Nakanishi, M. (1992) Light utilization by phytoplankton in the north basin of Lake Biwa. *Arch. Hydrobiol.*, **125**, 97–107.

26
Summary and Conclusions

Michio Kumagai

Lake Biwa Sigma Research Center, Ritsumeikan University, Kusatsu, Japan

26.1 Introduction

Our climate is changing. Scientists studying lake and river ecosystems of the world now have compelling evidence of change as they see the water temperature in many lakes and rivers rising, shifts in freshwater flora and fauna, and the expansion of "dead" zones with low dissolved oxygen concentrations. The Intergovernmental Panel on Climate Change predicted that air temperature will rise by $2-4\,°C$ by the end of this century (Intergovernmental Panel on Climate Change 2007), but the actual air temperature increase in the Far East for the last several decades is almost double this prediction (Kumagai 2008). In addition to addressing the causes of global warming, we have to face the many real problems occurring in inland water ecosystems.

In the case of Lake Biwa, the large, ancient lake in Japan that supplies water to more than 14 million citizens (see Chapter 6), many fish including the Isaza goby (*Gymnogobius isaza*, a deep-living species endemic to Lake Biwa) were killed in 2007. Since then, similar events have happened every year. We also found high concentrations of arsenic in the dead fish as well as in bottom sediments (Hirata *et al.* 2011). These were caused by anoxic conditions in the benthic boundary, which originated from the combination of eutrophication and global warming. Lake Biwa seems to have passed a tipping point, with a shift in the ecological regime due to climate change (Hsieh *et al.* 2010, 2011). In certain lakes, the dominant species of phytoplankton have changed from diatoms to cyanobacteria (see Chapter 11), and domestic fish have been replaced by exotic fish such as bluegill sunfish and large-mouth bass, which prefer warm water (Kamerath, Chandra, and Allen 2008; see also Chapters 14 and 15). What can we do to solve these steadily increasing environmental problems? Unfortunately, there is no perfect solution to overcome this crisis yet we must now work to reduce the potentially disastrous impacts for both ecosystems and human society.

As several authors in this book have suggested, we need ongoing monitoring systems based on good science to share information and data as we cannot generate sound policies and management plans to protect and save our inland waters without

Climatic Change and Global Warming of Inland Waters: Impacts and Mitigation for Ecosystems and Societies,
First Edition. Edited by Charles R. Goldman, Michio Kumagai and Richard D. Robarts.
© 2013 John Wiley & Sons, Ltd. Published 2013 by John Wiley & Sons, Ltd.

this fundamental support (see Chapters 5, 6, 13 and 23). Science remains the most important international language to exchange knowledge between water managers and the political forces that ultimately make decisions. If the data we obtain in science are wrong or are inadequate in scope, bad decisions are very likely to be made. To achieve this we need good analyses and models to evaluate the past and present situations in order to predict future changes (see Chapters 4, 8 and 9). Also scientists are often asked to provide future effective action plans to remediate aquatic environments that have undergone damage (see Chapters 22 and 25). We must recognize now that the very nature of our environment has been changing dynamically over the last few decades and assess both the most optimistic as well as the very pessimistic scenarios that are likely to play out during the next decades. In this chapter, the chapters composing this book on climate change impacts on inland waters are summarized and their key messages are highlighted to draw attention to possible adaptation and mitigation approaches that may reduce serious risks to these precious systems that are fundamental to the future of our planet.

26.2   Regime shifts in inland waters

Global air temperatures cooled until the beginning of the twentieth century, but then suddenly shifted to a warming trend after 1906 (Intergovernmental Panel on Climate Change 2007). In these years, meteorologists used the term "climate change" to describe this gradual air temperature rise and had up to now not paid much attention to the seriousness of this shift. The term "global warming" was first scientifically used by Broecker (1975), who described global water circulation as a belt of conveyers between surface and bottom layers in the oceans. His strong caution on global warming is finally getting the recognition it deserves. Only during the last several decades we have been facing the seriousness of this situation. Lovelock (1979, 2008) used the term "global heating" to give a severe warning of climate change to the public. He also explained that global air temperature can suddenly rise 6 °C if the CO_2 concentration in the air exceeds 500 ppm and becomes stable again despite further increases or decreases of atmospheric CO_2 (Lovelock and Kump 1994). This is of great concern because the present CO_2 concentration measured at Hawaii is already close to 370 ppm, and is continuing to rise at ever increasing rates. A warming to the extent predicted by Lovelock and Kump (1994) would represent a catastrophic regime shift in the atmosphere.

The sea ice covered area in the Arctic Sea has been decreasing and occupied only 50% of the area in September 2007 compared with the average between the 1950s and 1970s (Stroeve and Serreze 2008). It is still shrinking and it is said that regular use of the Northwest Passage for summer ship transportation could be realized within a few decades (Wang and Overland 2009). Following and parallel to these dramatic changes in the ocean, critical regime shifts in inland waters due to climate change are already well under way in the polar regions (Chapters 1, 2 and 20), with potential feedback effects on the rest of the planetary environment. Arctic and Antarctic ecosystems are extremely vulnerable to even small temperature changes, and it is a challenge to mitigate the impacts of anthropogenic climate change in these high-latitude regions (Chapter 1). The number of days that ice-covers form in lakes and rivers in the northern hemisphere are getting shorter, which can be directly related to the dramatic changes occurring in Arctic and temperate ecosystems alike (Magnuson *et al.* 2000; Quayle *et al.* 2002; see also Chapter 3). There is an urgent need to develop conservation plans for polar ecosystems to reduce the effects of multiple stressors in the face of climate change, and to provide refugia for vulnerable species and communities (Chapter 2).

It is important to recognize that inland waters form hydrological ecosystem networks that spread over the Earth, and changes occurring in the lakes, wetlands, and rivers of these provide us with key information to measure and understand the effects of climate change on land as well as aquatic ecosystems (Williamson *et al.* 2009). For example, climate change may accelerate extreme events such as floods or droughts in a catchment area and therefore have a great influence on river discharge and lake ecosystems. The water level of Lake Hovsgol, the largest fresh water lake in Mongolia, has risen by 1 cm every year since the 1970s due to permafrost melting as a direct result of global warming (Kumagai *et al.* 2006). The changes in water level may also affect fish reproduction, trophic relationships, littoral communities, the sedimentological regime and enhance methane ebullition (Chapter 16). Climate change can also increase surface water temperatures (Chapter 12) and suppress vertical mixing in lakes. Prolonged water- column stratification may induce internal nutrient loading such as phosphate and ammonium from bottom sediments (Chapters 6, 9 and 19). Strong stratification may reduce epilimnetic primary production (O'Reilly *et al.* 2003; see also Chapter 18), but sudden destratification may transfer nutrients from the hypolimnion to epilimnion and stimulate primary productivity (Chapters 14 and 15). However, there may be profound differences in the impact of climate change on deep versus shallow lakes, as noted for New Zealand lakes (Chapter 19).

The warming-induced changes in trophic structure can lead toward dominance of small-sized fish, lower zooplankton grazing, and stronger cyanobacterial dominance (Chapter 10). In the case of the Amazon basin, climate change is expected to decrease precipitation and increase temperature so that discharge and the extent of flooding are decreased (Chapter 17). The structure and function of inland water ecosystem are controlled by complex interactions among temporally spatially changing climate, human activities (see, for example, Chapters 6, 10, 14 and 19), biological forms and watershed characteristics (Schindler 2001). In order to extract the effects of climate change from such complicated processes, a paleo-limnological research approach can be used (Leavitt *et al.* 2009). This is based on the idea that lakes are integrators and sentinels of past climatic change. For example, Tsugeki, Ishida, and Urabe (2009) analyzed the ephippia and remains of *Daphnia galeata* in the bottom sediments of Lake Biwa and identified a regime shift of *D. galeata* due to winter warming from the 1970s to 2000s (Chapter 7).

26.3 Adaptation and mitigation to save inland waters

The definitions of adaptation and mitigation vary among authors, including among the contributors to this volume. We may say that adaptation can be related to passive management and mitigation by active management, as noted in Chapter 13. For example, adaptation is required as additional efforts to reduce external nutrient loading to levels lower than present-day conditions in order to maintain or improve the ecological level of steady state conditions (Chapter 10). Several management, or adaptation measures, can be proposed for north temperate freshwaters, for example, the re-establishment of riparian vegetation, re-establishment of lost wetlands, recreation of the meanders of channelized streams, and the planting of trees for shading and bank stabilization.

To mitigate the negative effects of climate change, three active directions for lakes and rivers are classified and some definite actions proposed (Chapter 24):

1. Potential reduction of temperature in aquatic habitats

   • heat pumps, cooling by hydropower operations, selection of suitable water temperature for fish migration, shading river habitat.

2. Modification of heat distribution

   • nutrient reduction, controlling particle inputs from glacial till, DOC control.

3. Reduction of greenhouse gases

   • carbon sequestration, methane capture or control, burial of wood in aquatic systems.

These approaches are not universally used in lakes and rivers. Some are effective and some should be used with care. These approaches can be found in several chapters of this book. Generally speaking, mitigation requires a very active management program, which often proves to be expensive. In many chapters of this book, authors have stressed the need for deterministic regional hydrological and climate models and for adequate hydro-meteorological monitoring networks for water management (Chapter 23).

As stated above, the clear differentiation between adaptation and mitigation is difficult. Two opposite approaches are discussed in this book. One is derived from interviews with Mongolian herders on how they would adapt their behavior to climate change (Chapter 21). These are not necessary governmental resolutions, but possible actions by individual herders such as:

• move more frequently;
• collect more *hadlan* (hay for winter);
• select the best animals;
• eliminate cashmere goats;
• go to *Otor* to find best pastures for their animals;
• give up herding and move to cities;
• receive improved weather forecasts; and
• improve the supply of water resources, with reduced impacts on streams for animals.

Another example is the Climate Change Action Plan (CCAP) for the many megacities that emerged in the last century, with the aim of building low-carbon cities (Chapter 22). Action plans of the world's ten most populated megacities (>10 million people) are compared. These are definitely large-scale management plans operated by governments, and each city has to develop its own management policy. The essence of this chapter is followed by the note that "Water resources and watershed management and planning are not always central to the selected mega cities' climate change adaptive strategies." Questions are how effectively current climate change action plans have incorporated adaptation and mitigation goals for reducing water stresses that will increasingly develop on a global scale. Ideally, mitigation efforts should always be accompanied by adaptation strategies because some climate-change impacts may only occur in the long term.

Four best practices related to building urban climate change preparedness in the selected megacities are listed as:

- water leakage prevention controls in Tokyo;
- Tokyo's green building program;
- Landfill emissions control program in São Paulo; and
- Mexico City's comprehensive climate change plan.

26.4 Conclusion

There are many attempts in the world to save inland waters from global warming on the basis of mitigation and adaptation strategies. However, we have yet to find the best solutions. The only solutions at present are to reduce the use of fossil fuels and improve our management strategies for inland water ecosystems. For this we will continue to need up-to-date and scientifically sound data urgently. We believe that the chapters presented in this book provide insights into the responses and changes of aquatic systems around the world to climatic change. In this way, they will serve in helping guide us to the most effective and environmentally sustainable solutions to promote mitigation and adaptation to save our precious inland waters.

26.5 Acknowledgements

This study was financially supported by Grants-in-Aid for Scientific Research (20244079) from the Japan Society for the Promotion of Science, and partially supported by the Global Environmental Research Fund (Fa-084) of the Ministry of the Environment, Japan and the Mitsubishi Foundation. We also strongly thank to the World Water and Climate Network (WWCN) which gave us the chance to edit this book and supported necessary expenses to meet and discuss for editing the chapters.

References

Broecker, W.S. (1975) Climatic Change: Are we on the brink of a pronounced global warming? *Science*, **189**, 460–3.

Hirata, S.H., Hayase, D., Eguchi, A., *et al.* (2011) Arsenic and Mn levels in Isaza (*Gymnogobius isaza*) during the mass mortality event in Lake Biwa. *Jpn. Environ. Poll.*, **159**, 2789–96.

Hsieh C.H., Ishikawa, K., Sakai, Y., *et al.* (2010) Phytoplankton community reorganization driven by eutrophication and warming in Lake Biwa. *Aquat. Sci.*, **72**, 467–83.

Hsieh C.H., Sakai, Y., Ban, S., *et al.* (2011) Eutrophication and warming effects on long-term variation of zooplankton in Lake Biwa. *Biogeosci.*, **8**, 1383–99.

Intergovernmental Panel on Climate Change (IPCC) (2007) Fourth Assessment Report (AR4) of the United Nations, Cambridge University Press, Cambridge.

Kamerath M., Chandra, S., and Allen, B.C. (2008) Distribution and impacts of warmwater fish in Lake Tahoe. *Aquat. Invasions*, **3**, 35–41.

Kumagai, M. (2008) Lake Biwa in the context of world lake problems. Baldi Lecture. *Verh. Internat. Verein. Limnol.*, **30**, 1–15.

Kumagai, M., Urabe, J., Goulden, C.E., *et al*. (2006) Recent rise in water level at Lake Hovsgol in Mongolia, in *The Geology, Biodiversity and Ecology of Lake Hovsgol (Mongolia)* (eds. C.E. Goulden, T. Sitnikova, J. Gelhaus and B. Boldgiv), Backhuys Publishers, Belgium, pp. 77–91.

Leavitt, P.R., Fritz, S.C., Anderson, N.J., *et al*. (2009) Paleolimnological evidence of the effects on lakes of energy and mass transfer from climate and humans. *Limnol. Oceanogr.*, **54**, 2330–48.

Lovelock, J. (1979) *Gaia, A New Look at Life on Earth*, Oxford University Press, Oxford.

Lovelock, J. (2008) A geophysiologist's thoughts on geoenginnering. *Phil. Trans. R. Soc. A*, 1–8.

Lovelock, J.E. and Kump, L.R. (1994) Failure of climate regulation in a geophysiological model. *Nature*, **369**, 732–4.

Magnuson, J. J., Robertson, D.M., Benson, B.J., *et al*. (2000) Historical trends in lake and river ice cover in the northern hemisphere. *Science*, **289**, 1743–6.

O'Reilly, C.M., Alin, S.R., and Plisnier, P. (2003) Climate change decreases aquatic ecosystem productivity of Lake Tanganyika, Africa. Nature, **424**, 766–8.

Quayle, W.C., Peck, L.S., Peat, H., *et al*. (2002) Extreme responses to climate change in antarctic lakes. *Science*, **295**, 645

Schindler, D.W. (2001) The cumulative effects of climate warming and other human stresses on Canadian freshwaters in the new millennium. *Can. J. Fish. Aquat. Sci.*, **65**, 18–29.

Stroeve, J. and Serreze, M. (2008) Arctic ice extent plummets in 2007. *EOS Trans. Am. Geophys. Un.*, **89**, 13–20.

Tsugeki, N.K., Ishida, S., and Urabe, J. (2009) Sedimentary records of reduction in resting egg production of *Daphnia galeata* in Lake Biwa during the twentieth century: A possible effect of winter warming. *J. Paleolimnol.*, **42**: 155–65.

Wang, M. and Overland, J.E. (2009) A sea ice free summer Arctic within 30 years? *Geophys. Res. Lett.*, **36**, 1–5.

Williamson, C.E., Saros, J.E., Vincent, W.F., and Smol, J.P. (2009) Lakes and reservoirs as sentinels, integrators, and regulators of climate change. *Limnol. Oceanogr.*, **54**, 2273–82.

Index